Ecology and Conservation of Lesser Prairie-Chickens

STUDIES IN AVIAN BIOLOGY

A Publication of The Cooper Ornithological Society

www.crcpress.com/browse/series/crcstdavibio

Studies in Avian Biology is a series of works published by the Cooper Ornithological Society since 1978. Volumes in the series address current topics in ornithology and can be organized as monographs or multi-authored collections of chapters. Authors are invited to contact the series editor to discuss project proposals and guidelines for preparation of manuscripts.

See complete series list on page [375].

Volume 48
Studies in Avian Biology
Cooper Ornithological Society

Ecology and Conservation of Lesser Prairie-Chickens

EDITED BY

David A. Haukos
U.S. Geological Survey
Kansas State University
Manhattan, KS

Clint W. Boal
U.S. Geological Survey
Texas Tech University
Lubbock, TX

CRC Press
Taylor & Francis Group
Boca Raton London New York

CRC Press is an imprint of the
Taylor & Francis Group, an **informa** business

Cover photo by Jonathan Lautenbach. Male Lesser Prairie-Chicken (*Tympanuchus pallidicinctus*) in Comanche County, Kansas, March 18, 2014.

CRC Press
Taylor & Francis Group
6000 Broken Sound Parkway NW, Suite 300
Boca Raton, FL 33487-2742

First issued in paperback 2020

© 2016 Cooper Ornithological Society
CRC Press is an imprint of Taylor & Francis Group, an Informa business

No claim to original U.S. Government works

ISBN-13: 978-1-4822-4022-1 (hbk)
ISBN-13: 978-0-367-65860-1 (pbk)

Visit the Taylor & Francis Web site at
http://www.taylorandfrancis.com

and the CRC Press Web site at
http://www.crcpress.com

To
Robert J. Robel, researcher, teacher, and sportsman
for
more than 50 years of resolute passion for Prairie Chickens

CONTENTS

EDITORS

David A. Haukos, PhD, is the unit leader of the U.S. Geological Survey, Kansas Cooperative Fish and Wildlife Research Unit, Kansas State University (Manhattan). Dr. Haukos earned his PhD from Texas Tech University (Lubbock). His research has centered on the ecology and conservation of the High Plains ecosystems and species. He has been involved with the conservation of and research on Lesser Prairie-Chickens since 1986, when he investigated the reproductive ecology of Lesser Prairie-Chickens in west Texas for his MS thesis. Since then, he has been associated with numerous research and conservation efforts related to the Lesser Prairie-Chicken, with publications on the ecology and management of the species throughout its range.

Clint W. Boal, PhD, is the assistant unit leader of the U.S. Geological Survey, Texas Cooperative Fish and Wildlife Research Unit, Texas Tech University. Dr. Boal earned his PhD from the University of Arizona (Tucson). His research background is in predatory bird ecology, the conservation of rare and decreasing species, and general avian ecology in the context of anthropogenic land changes. He has been involved with the conservation of and research on Lesser Prairie-Chickens since 2007, with numerous publications on the ecology and management of Lesser Prairie-Chickens in west Texas and New Mexico.

CONTRIBUTORS

MATTHEW R. BAIN
The Nature Conservancy
1114 Cty Road 370
Oakley, KS 67748
mbain@tnc.org

ANNE M. BARTUSZEVIGE
Playa Lakes Joint Venture
2575 Park Lane, Suite 110
Lafayette, CO 80026
anne.bartuszevige@pljv.org

GRANT M. BEAUPREZ
New Mexico Department of Game and Fish
Portales, NM 88130
grant.beauprez@state.nm.us

ADAM C. BEHNEY
College of Agricultural Sciences
Southern Illinois University
251 Life Science II, Mailcode 6504
Carbondale, IL 62901
abehney@siu.edu

PHILIP K. BORSDORF
Department of Natural Resources Management
Texas Tech University
15th and Detroit
Lubbock, TX 79409
borsy14@gmail.com

DAVID K. DAHLGREN
Wildland Resources
Utah State University
5230 Old Main Hill
Logan, UT 84322
dave.dahlgren@usu.edu

ALEX DANIELS
Playa Lakes Joint Venture
2575 Park Lane, Suite 110
Lafayette, CO 80026
alex.daniels@pljv.org

RANDY W. DEYOUNG
Caesar Kleberg Wildlife Research Institute
Texas A&M University–Kingsville, MSC 218
700 University Blvd
Kingsville, TX 78363
randall.deyoung@tamuk.edu

R. DWAYNE ELMORE
Natural Resource Ecology and Management
Oklahoma State University
008C Ag Hall
Stillwater, OK 74078
dwayne.elmore@okstate.edu

ARON A. FLANDERS
Partners for Fish and Wildlife Program
U.S. Fish and Wildlife Service
1434 NE 80th St
Stafford, KS 67578
aron_flanders@fws.gov

EDWARD O. GARTON
Department of Fish and Wildlife
University of Idaho
P.O. Box 44136
Moscow, ID 83844
ogarton@uidaho.edu

ALIXANDRA J. GODAR
Department of Natural Resources Management
Texas Tech University
15th and Detroit
Lubbock, TX 79409
alixandra.godar@ttu.edu

CODY P. GRIFFIN
Department of Natural Resources Management
Texas Tech University
15th and Detroit
Lubbock, TX 79409
cody.griffin@ttu.edu

BLAKE A. GRISHAM
Department of Natural Resources Management
Texas Tech University
15th and Detroit
Lubbock, TX 79409
blake.grisham@ttu.edu

CHRISTIAN A. HAGEN
Oregon State University
500 SW Bond Street, Suite 107
Bend, OR 97702
christian.hagen@oregonstate.edu

SEAN C. KYLE
Texas Parks and Wildlife Department
Texas Tech University
Box 42125
Lubbock, TX 79409
sean.kyle@tpwd.state.tx.us

LENA C. LARSSON
Sutton Avian Research Center
P.O. Box 2007
Bartlesville, OK 74005
llarsson@ou.edu

DUANE R. LUCIA
U.S. Fish and Wildlife Service
Texas Tech University
15th and Detroit
Lubbock, TX 79409
duane_lucia@fws.gov

PATRICIA MCDANIEL
109 Tanning Way
Clovis, NM 88101
phalaropecon@yahoo.com

MICHAEL A. PATTEN
Oklahoma Biological Survey and Department of
    Biology
University of Oklahoma
Norman, OK 73019
mpatten@ou.edu

MARKUS J. PETERSON
Department of Wildlife and Fisheries Sciences
Texas A&M University
College Station, TX 77843-2258
mpeterson@tamu.edu

JAMES C. PITMAN
Western Association of Fish and Wildlife Agencies
215 W 6th Ave, Suite 207
Emporia, KS 66801
jim.pitman@wafwa.org

RANDY D. RODGERS
Kansas Department of Wildlife, Parks, and
    Tourism
509 W 14th
Hays, KS 67601
randyr@ruraltel.net

DOUG D. SCHOELING
Oklahoma Department of Wildlife Conservation
1801 North Lincoln Boulevard
Oklahoma City, OK 73105
doug.schoeling@odwc.ok.gov

WILLIAM E. VAN PELT
Western Association of Fish and Wildlife Agencies
5000 West Carefree Highway
Phoenix, AZ 85086
bvanpelt@azgfd.gov

BETTY WILLIAMSON
P.O. Box 49
Pep, NM 88126
pepnm@hotmail.com

DAMON L. WILLIFORD
Caesar Kleberg Wildlife Research Institute
Texas A&M University–Kingsville,
    MSC 218
Kingsville, TX 78363
rook137@gmail.com

DONALD H. WOLFE
Sutton Avian Research Center,
P.O. Box 2007
Bartlesville, OK 74005
dwolfe@ou.edu

JENNIFER C. ZAVALETA
Department of Natural Resources Management
Texas Tech University
15th and Detroit
Lubbock, TX 79409
jenniferczavaleta@gmail.com

PREFACE

Grassland birds, as a group, are undergoing substantial population declines across North American prairies. The causes for the declines are numerous, but include loss of natural ecological drivers, conversion of native prairies for agricultural use, habitat fragmentation, energy exploration and development, natural fire suppression, unmanaged grazing, invasive vegetation, infrastructure associated with anthropogenic activities, and periodic droughts. Ultimately, these factors all contribute to the overarching threat—the loss of quality habitable space. Environmental changes may be especially detrimental to the prairie or "pinnated" grouse, the Sharp-tailed Grouse (*Tympanuchus phasianellus*), the Greater Prairie-Chicken (*T. cupido*), and, the subject of this volume, the Lesser Prairie-Chicken (*T. pallidicinctus*).

Lesser Prairie-Chickens were once widely distributed across the Southern Great Plains of Colorado, Kansas, New Mexico, Oklahoma, and Texas. The birds are mysterious, elusive, and worthy of considerable awe and respect considering the harsh environments in which they persist. Their spring mating rituals are spectacular, serving to define the subtle splendor and magnificence of the prairies that the birds inhabit. Lesser Prairie-Chickens have experienced substantial declines, both in terms of numbers of birds and the extent of area that the populations still occupy. Fully understanding the population trends is challenging because the species appears to have a boom-and-bust reproductive strategy; several years of poor reproductive success may be offset by one or two years of good productivity. Regardless, current evidence indicates that the species has experienced a persistent and distribution-wide population decline, especially since the Dust Bowl of the 1930s dramatically impacted the core range of Lesser Prairie-Chickens. Many grassland birds are migratory, but Lesser Prairie-Chickens are resident game birds that require large areas of prairie to complete their life histories and sustain their populations. Prairie chickens occupy prairie habitats year-round, which provides conservation opportunities and risks, in that management must take into account all aspects of the species life history in one area. Habitat management must address habitats needed for the display areas used for lekking behavior, high-quality nesting and brood-rearing areas, and sites that facilitate nonbreeding survival. Active management can be applied to address issues with grazing but may be difficult or impossible to implement for mitigating the effects of drought or climate change. Other factors focusing attention on Lesser Prairie-Chickens include that the species has been managed as an upland game bird, and wildlife agencies have sought to maintain populations that can sustain recreational harvesting. Indeed, much of the early research on prairie grouse was originally directed at game management issues. More recently, conservation concerns have also had a financial component from bird watchers with interests in natural history. Opportunities for the general public to view lekking behaviors of Lesser Prairie-Chickens from blinds or as part of spring festivals have provided economic benefits to rural communities and private landowners. Lesser Prairie-Chickens have gained a high profile as a "poster" species for recreational harvesting, nature tourism, and grassland conservation. Persistent population declines and resulting conservation concerns for the species led to a 1995 petition for the protection of Lesser Prairie-Chickens at the federal level under the Endangered Species Act. After a substantial review, the U.S. Fish and Wildlife Service issued a ruling providing Lesser Prairie-Chickens with federal protection as a threatened species in May 2014. However, the listing rule was vacated by judicial decision in September 2015, creating considerable regulatory uncertainty at the time of this volume.

The concept for this book stemmed from conversations regarding the research and management of Lesser Prairie-Chickens. We recognized that (1) a substantial amount of contemporary research had focused on Lesser Prairie-Chickens; (2) emerging issues, such as wind energy and climate change, were recognized as possible threats to the species; and (3) the species had been petitioned for federal protection. We concluded that there was a need to synthesize and compile the historic and contemporary state of knowledge on Lesser Prairie-Chickens into a single source. We developed an outline of contemporary and relevant topics to address in this volume and then approached potential authors for those chapters. Fortunately, we were not alone in our vision, and the concept of this volume was received with enthusiasm by a number of researchers and managers that we asked to serve as authors or referees. We thank the following individuals for taking time from their busy schedules to author the 18 chapters contributed to this volume of the series *Studies in Avian Biology*. The chapter authors are as follows: Matthew R. Bain, Anne M. Bartuszevige, Grant M. Beauprez, Adam C. Behney, Philip K. Borsdorf, David K. Dahlgren, Alex Daniels, Randy W. DeYoung, R. Dwayne Elmore, Aron A. Flanders, Edward O. Garton, Cody P. Griffin, Blake A. Grisham, Alixandra J. Godar, Christian A. Hagen, Sean C. Kyle, Lena L. Larsson, Duane R. Lucia, Patricia McDaniel, Michael A. Patten, Markus J. Peterson, James C. Pitman, Randy D. Rogers, Doug D. Schoeling, Betty Williamson, Damon L. Williford, Donald H. Wolfe, William E. Van Pelt, Jennifer C. Zavaleta, and ourselves.

In addition to authorship by leading researchers and managers working with Lesser Prairie-Chickens, the volume has also benefited from careful and diligent reviews by a host of external referees, whom we thank for their assistance in improving the manuscripts. We thank Roger Applegate, Anne Bartuszevige, Grant Beauprez, Leonard Brennen, Scott Carleton, Andy Chappell, David Dahlgren, Manual DeLeon, Dwayne Elmore, Pamela Ferro, Aron Flanders, Sam Fuhlendorf, Seth Gallager, Andrew Gregory, Blake Grisham, Christian Hagen, Chris Hise, Torre Hovick, Jeff Johnson, Chuck Kowaleski, Greg Kramos, Duane Lucia, Stephanie Manes, Patricia McDaniel, Gene Miller, Clay Nichols, Chris O'Meilia, Howard Parmen, Michael Patton, Reid Plumb, Jeffrey Prendergast, Christine Pruett, Nicole Quintana, Jim Ray, Jonathan Reitz, Mindy Rice, Terry Riley, Beth Ross, Brett Sandercock, Heather Whitlaw, Virginia Winder, and Don Wolfe.

Each chapter underwent external peer reviews by a minimum of two referees, and the final manuscript was edited by two volume editors, David Haukos and Clint Boal, with the assistance of Tracy Estabrook Boal serving as the initial volume copy editor. It was not feasible to obtain peer reviews solely from scientists who were not involved in this volume, but we made every effort to ensure that there were no conflicts of interest at any stage of the review process. The volume editors were authors on some chapters; in these cases, the manuscripts also underwent additional reviews by Kevin Whalen and Janine Powell of the U.S. Geological Survey. Once the final versions were submitted and found acceptable by us, we forwarded the revised manuscripts, along with original drafts, reviewer comments, and author responses, to the series editor Brett Sandercock. Sandercock reviewed all the manuscripts and made final editorial decisions regarding acceptance. We graciously thank him for his encouragement and guidance during the preparation of this book. We also acknowledge and thank Jonathan Lautenbach for providing the cover photo.

We hope this volume of *Studies in Avian Biology* serves as an overview and synthesis of the state of current knowledge on the ecology of Lesser Prairie-Chickens, and the challenges that natural resource managers continue to face in their conservation efforts. We also hope that the work serves as a stimulus for new research and management efforts toward this remarkable species of grassland bird that has survived for millennia but is now facing anthropogenic-driven challenges it was never equipped to deal with. Lesser Prairie-Chickens need our help, and the science in this volume will serve as a useful starting point for future research, management, and conservation.

DAVID A. HAUKOS
Manhattan, Kansas

CLINT W. BOAL
Lubbock, Texas

CHAPTER ONE

The Lesser Prairie-Chicken*

A BRIEF INTRODUCTION TO THE GROUSE
OF THE SOUTHERN GREAT PLAINS

Clint W. Boal and David A. Haukos

Abstract. North American prairies have experienced nearly two centuries of rapid change driven by anthropogenic actions. Change has progressed from loss of natural ecological drivers, extirpation of native grazers, institution of unmanaged or poorly managed domestic livestock grazing, suppression of natural fire, conversion of prairie to cropland, facilitation of invasive plant and animal species, increased settlement and associated infrastructure development, depletion of surface and groundwater stores, and, more recently, to the expansion of energy exploration and development. Ongoing actions have resulted in once contiguous prairie lands transitioning from fragmented, to degraded, and then outright loss. A consequence of the conversion and loss of prairie grasslands is the rapid decline in many grassland obligate bird populations. Prairie grouse (*Tympanuchus* spp.) in particular have experienced dramatic declines in abundance and range. Most recently, substantial conservation attention has centered on the Lesser Prairie-Chicken (*T. pallidicinctus*), which is the focus of this volume of *Studies in Avian Biology*. The recent listing of Lesser Prairie-Chickens for federal protection as a threatened species under the Endangered Species Act makes our summary volume on the ecology, conservation, and management of the species a timely contribution to the literature. Here, we provide a brief introduction to the Lesser Prairie-Chicken that will provide a foundation for the more detailed information provided in the following chapters of this volume.

Key Words: conservation, Endangered Species Act, Great Plains, prairie chickens, *Tympanuchus pallidicinctus*.

Prairies of North America have a history of poor regard. Throughout early exploration by Anglo-Europeans in the 1800s, the general sentiment was one of a dry, forlorn, and even worthless landscape where no settlement could persist except along large water courses (Savage 2004). Even the famous ornithologist John James Audubon was disparaging in his views of the prairie, referring to the landscape as *the most arid and dismal you can conceive of* (Savage 2004). It is ironic that now in the early part of the 21st century, North American prairies are one of the most endangered

* Boal, C. W. and D. A. Haukos. 2016. The Lesser Prairie-Chicken: a brief introduction to the grouse of the Southern Great Plains. Pp. 1–11 in D. A. Haukos and C. W. Boal (editors), Ecology and conservation of Lesser Prairie-Chickens. Studies in Avian Biology (no. 48), CRC Press, Boca Raton, FL.

natural resources on the continent (Samson et al. 2004). When entire ecoregions come under threat, the biotic communities that depend on them are likewise in peril (Samson and Knopf 1994, 1996). Thus, it is perhaps unsurprising that grassland obligate birds are experiencing rapid and wide-spread population declines across North America, which have been identified as an emerging conservation crisis (Brennan and Kuvlesky 2005).

Anthropogenic actions leading to fragmentation, degradation, and outright loss of natural habitats have been identified as the primary driver of population declines (Noss and Cooperrider 1994, George and Dobkin 2002, Brennan and Kuvlesky 2005). Yet, other drivers are at play as well, some for which management actions may be taken such as prescribed fire to replace natural fire periodicity, or managed grazing, and others where managers are largely helpless such as negative effects of drought or climate change. A contributing factor that presents both conservation challenges and opportunities is that ~85% of grasslands in the United States are in private ownership (North American Bird Conservation Initiative 2013).

Prairie grouse (*Tympanuchus* spp.) have experienced dramatic population declines across all three species. Once found in 20 states in the continental United States and four provinces in Canada, the Greater Prairie-Chicken (*T. cupido pinnatus*) is now present in only 11 states (Svedarsky et al. 2000); it was listed as endangered and, subsequently, found to be extirpated from Canada (Committee on the Status of Endangered Wildlife in Canada 2009). Still, it has fared better than its related subspecies, the Heath Hen (*T. c. cupido*) that went extinct in 1932 (Johnsgard 2002), and the Attwater's Prairie-Chicken (*T. c. attwateri*) that has languished as a federal-protected endangered species since 1967 (USFWS 1967). Sharp-tailed Grouse (*T. phasianellus*) are also declining across their range, with a petition put forward for the protection of a subspecies, the Columbian Sharp-tailed Grouse (*T. p. columbianus*), under the Endangered Species Act. Listing was not found to be warranted at the time of review but population declines continue (USFWS 2006).

Lesser Prairie-Chickens (*T. pallidicinctus*) are the most recent species of prairie grouse to receive widespread conservation attention and are the focus of this volume of *Studies in Avian Biology*. The recent listing of Lesser Prairie-Chickens as a federally protected threatened species makes this volume an especially timely contribution to the literature on this species (USFWS 2014). Although the listing rule was vacated by judicial decision in September 2015, creating considerable regulatory uncertainty at the time of this volume, the potential for the ruling to be overturned and restoration of the threatened species status exists. Here, we provide a brief review of the Lesser Prairie-Chicken and its basic ecology within the Great Plains ecosystems. Our goal is to provide an introduction to the species and current conservation challenges that will provide a foundation for the more detailed information provided in the following chapters by experts in the field.

LESSER PRAIRIE-CHICKEN

Description

The Lesser Prairie-Chicken is primarily a ground-dwelling terrestrial bird, generally taking wing only when flying to or from leks, when disturbed by predators or hunters, or during dispersal events (Hagen and Giesen 2005). Adults are about the size of a "football," measuring about 38–41 cm in length and weighing about 700–800 g (Hagen and Giesen 2005). Sexes are similar in plumage, with an overall grayish brown appearance, but closer inspection reveals an overall pattern of alternating bars of dark brown and buffy white, generally darker brown on the back. The chin and throat are generally buffy white and a horizontal white line extends from the chin, running below the eye and back to the auricular area of the head. Head markings are especially noticeable when males are displaying for females by inflating the red esophageal air sack on each side of the neck, erecting the yellow eye combs above their eyes and erecting their pinnae feathers (Figure 1.1). The pinnae are long feathers on the side of the neck, and males erect them fully overhead during display behaviors. Females also have pinnae, but they are substantively shorter (Hagen and Giesen 2005). Immature birds are similar to adults.

Taxonomy

Lesser Prairie-Chickens are gallinaceous birds classified in the order Galliformes and placed in the family Phasianidae and subfamily Tetraoninae, which include the prairie, forest, and tundra grouse. Lesser Prairie-Chickens are most closely related to the other prairie grouse species within the genus

Figure 1.1. A male Lesser Prairie-Chicken in full display at a lek in northwestern Kansas. (Photo by Jonathan Lautenbach.)

Tympanuchus. Related species include three subspecies of *T. cupido*, the Greater Prairie-Chicken (*T. c. pinnatus*), the extinct Heath Hen (*T. c. cupido*), and the endangered Attwater's Prairie-Chicken (*T. c. attwateri*). Also closely related is the Sharp-tailed Grouse (*T. phasianellus*), which consists of six living subspecies (*T. p. caurus, T. p. kennicotti, T. p. phasianellus, T. p. campestris, T. p. columbianus,* and *T. p. jamesi*) and one extinct subspecies (*T. p. hueyi*).

The Lesser Prairie-Chicken was first described as a subspecies of the Greater Prairie-Chicken (Ridgway 1873) but later named a full species (Ridgway 1885). However, as recently as the early 1980s, some still regarded the extinct Heath Hen, the Greater Prairie-Chicken, the Lesser Prairie-Chicken, and the Attwater's Prairie-Chicken as four separate subspecies within *T. cupido* (Johnsgard 1983). The scientific name of the Lesser Prairie-Chicken, *T. pallidicinctus*, has been recognized by the American Ornithologists Union since the first edition of "The Checklist of North and Middle American Birds" in 1886. However, the standardized common name of the species was originally Lesser Prairie Hen, which was then changed to Lesser Prairie Chicken in 1910 and finally to Lesser Prairie-Chicken in 1983.

Despite earlier suggestions that the Lesser Prairie-Chicken was a subspecies of the Greater Prairie-Chicken, recent genetic evidence indicates that it is a distinct species with recent separation primarily driven by sexual selection associated with their lek-based mating systems and different landscapes occupied by the two

species (Ellsworth et al. 1994, 1995; Chapter 5, this volume). However, hybridization of Greater and Lesser Prairie-Chickens has resulted in fertile offspring in captivity (Crawford 1978), and hybridization does occur under natural conditions in contact zones where the species distributions overlap (Bain and Farley 2002, Hagen and Giesen 2005). Genetic studies conducted on Lesser Prairie-Chickens indicate that the population at the southwestern most extent of their distribution in New Mexico and west Texas has lower genetic diversity than populations in northeastern Texas, Oklahoma, and Kansas (Van Den Bussche et al. 2003, Hagen et al. 2010, Corman 2011). Reductions in diversity may be evidence of reduced connectivity via isolation by distance among populations (Chapter 5, this volume). Further, current populations in Colorado and southwest Kansas are becoming increasingly isolated due to ongoing population declines.

Life History

The phenology of Lesser Prairie-Chickens is generally considered in context of a breeding season and a nonbreeding season. Lesser Prairie-Chickens are a promiscuous species with a lek mating breeding system. Leks are a distinct, key habitat feature where males congregate, defend small territories, and communally display in an attempt to attract females for mating. Chronologically, the breeding season consists of the lekking period, the nesting period, and the brood-rearing period. During the nonbreeding season, most birds stay close to leks but are more widely distributed across the area than during the breeding season when birds stay near leks as activity centers.

Leks are typically located in areas of sparse vegetation and located higher, even if only slightly, than the surrounding area. Females may visit more than one lek, so clusters of leks within an area are important for population persistence on a local scale (Applegate and Riley 1998). Females are also more likely to visit established leks rather than newly formed or "satellite" leks (Haukos and Smith 1999). Male mating success is often skewed because a subset of only a few males will dominate mating at the lek (Behney et al. 2012). The affinity of Lesser Prairie-Chickens to display sites has also made lek sites an important asset for purposes of inventory and monitoring (e.g., McRoberts et al. 2011; Timmer et al. 2013; McDonald et al. 2014a,b). Ecologically, the importance of established leks is further highlighted by the fact that most birds rarely move farther than 4.8 km from the nearest lek, regardless of time of year (Riley et al. 1994, Woodward et al. 2001, Hagen and Giesen 2005, Kukal 2010, Boal et al. 2014, Grisham et al. 2014). Furthermore, Lesser Prairie-Chickens are a resident species that generally occupies the same areas during both breeding and nonbreeding seasons (Taylor and Guthery 1980, Hagen and Giesen 2005). Thus, most conservation planning, habitat management, and estimation of habitat availability center around lek locations.

Lekking activity starts as early as February with female visitation to leks occurring at a discrete period, peaking in mid-April (Haukos and Smith 1999, Behney et al. 2012). Males will remain on leks into June to mate with females that lose their first clutch and return to the lek prior to renesting. Males also return to leks for a short period during fall (late September–early October), during which they will engage in some lekking behaviors but not mate. At leks, males engage in elaborate advertisement displays to attract females. Courtship displays include body posturing in which the males droop their wings and spread their feathers, erect their pinnae overhead, erect the tail feathers, and enlarge eye combs while inflating their esophageal air sacs to produce a "booming" vocalization (Figure 1.1; Hagen and Giesen 2005). Males stamp their feet rapidly while moving about with their heads extended forward and down. Display behaviors are intended to attract females to mate

and, somehow, they work. When a female selects a male, copulations appear to occur primarily at the lek; mating is the extent of paternal investment by the male for recruitment of the next generation.

Lesser Prairie-Chickens have uniparental care of offspring, and egg laying, incubation, and brood rearing are completed by the female alone. Nesting usually occurs from mid-April through May, but renesting may continue through June (Hagen and Giesen 2005). Females typically nest within 3.2 km of the nearest lek site (not necessarily the lek of mating) and usually select for nesting habitats with similar vegetation structure across their range, with some differences reported among ecoregions (Chapter 6, this volume). In a shallow nest bowl, females lay an average clutch size of 10–12 eggs. After about 26 days of incubation (Sutton 1968), eggs hatch synchronously and precocial chicks depart the nest with the female within 24 hours (Hagen and Giesen 2005). The female cares for the brood throughout the summer until September when brood breakup begins and the species moves into the nonbreeding season phase of the annual cycle. Overall reproductive success varies among populations and among years within populations. Demographic rates, the factors driving them, and population-level implications are more closely examined elsewhere (Chapter 4, this volume).

The food habitats of Lesser Prairie-Chickens, especially for chicks, are poorly understood. The species has been reported to eat a variety of invertebrates, seeds, acorns, leaves, and buds in natural habitats (Riley et al. 1993, Jamison et al. 2002, Hagen et al. 2005) and may visit agriculture fields to forage on waste grains during the winter (Copelin 1963, Ahlborn 1980). Invertebrates, particularly grasshoppers and treehoppers, appear to be especially important for young Lesser Prairie-Chickens and adults at certain times of the year (Davis et al. 1979). Lesser Prairie-Chickens are thought to acquire sufficient water through dietary sources, but birds will use free water when available in all seasons (Hagen and Giesen 2005). More recently, Boal et al. (2013) documented the use of bird-friendly wildlife water guzzlers by the species.

Lesser Prairie-Chickens are subject to predation throughout the year (Chapter 8, this volume). Lesser Prairie-Chickens probably experience relatively constant year-round predation threat from resident mammalian predators such as coyotes (*Canis latrans*) and foxes (*Vulpes* spp.), but females are likely more susceptible to predation when attending ground nests during incubation. Some birds of prey are resident and present across the distribution of Lesser Prairie-Chickens, but numbers of raptors may increase during migratory periods and winter when a lack of territoriality allows for greater raptor densities and increased predation risk. In contrast, predation risk from snakes and ravens is primarily limited to the breeding season and is mainly focused on eggs and chicks.

Lesser Prairie-Chicken populations are characterized by considerable fluctuations in abundance based on reproductive success or lack thereof in response to environmental conditions (Chapter 4, this volume). Individuals have a relatively short life span and high annual mortality. Campbell (1972) estimated a 5-year maximum life span for males, but Pruett et al. (2011) estimated generation time for Lesser Prairie-Chickens as 1.9 years in Oklahoma and 2.7 years in New Mexico. Most recently, the average natural life span or generation time has been estimated at 1.95 years (Van Pelt et al. 2013).

Populations and Occupied Range

Once presumed to be distributed widely across the Southern Great Plains, the Lesser Prairie-Chicken currently occupies a substantially reduced portion of its perceived historic range (Figure 1.2). However, considerable uncertainty prevails in regard to "historical" occupied range given the paucity of documented records of Lesser Prairie-Chickens until the late 1800s following initial settlement of the region. Uncertainty about historic distributions is affected by confusion regarding species identification in early sight records and questionable interpretation of locations of anecdotal sightings (Chapter 2, this volume). Additionally, sightings of individuals during extensive dispersal events in years of poor resources following years of good resources and therefore high reproduction success may have been incorrectly deduced as representing extant population presence (Chapter 2, this volume). Much of the landscape included in the historical range delineation was short-grass prairie dominated by buffalo grass (*Buchloë dactyloides*) and blue grama (*Bouteloua gracilis*), which do not

provide habitat for Lesser Prairie-Chickens (Van Pelt et al. 2013). In addition, the southern extent of the range includes arid ecoregions incapable of providing Lesser Prairie-Chicken habitat. The U.S. Fish and Wildlife Service (2014) estimated an 85% decline between historical "occupied" range of the species and recent delineations of occupied range (Table 1.1).

McDonald et al. (2014a,b) and Van Pelt et al. (2013) have adopted an ecoregional approach to defining populations of Lesser Prairie-Chickens in the Southern Great Plains (Figure 1.2). Extant populations of the species were stratified into four principal ecoregions of the Southern Great Plains: the Short-Grass Prairie/CRP (Conservation Reserve Program) Mosaic Ecoregion of northwestern Kansas; the Sand Sagebrush (*Artemisia filifolia*) Prairie Ecoregion of southwestern Kansas and southeastern Colorado; the Mixed-Grass Prairie Ecoregion of northwest Oklahoma, northeast Texas Panhandle, and south-central Kansas; and the Sand Shinnery Oak (*Quercus havardii*) Prairie Ecoregion of eastern New Mexico and western Texas (Figure 1.2, Table 1.2). Population density, abundance trends, limiting factors, available habitat, reproductive potential, conservation priorities, and land use differ profoundly among the four ecoregions. Each ecoregion consists of different vegetative communities (Chapter 6, this volume) and experiences different environmental conditions (Chapter 12, this volume). In general, though, it appears that Lesser Prairie-Chickens select habitats more on the basis of structure than the composition of specific vegetation species (Chapter 6, this volume). Here, the populations and specific characteristics of habitat and ecology of Lesser Prairie-Chickens are addressed separately for each ecoregion (Chapters 14 to 17, this volume). Environmental conditions within the current range of the Lesser Prairie-Chicken vary extensively among ecoregions and interannually within regions. The temperature, precipitation, growing-season range, and extreme and unpredictable fluctuations required Lesser Prairie-Chickens to develop unique adaptive strategies to persist in frequently inhospitable conditions. As an example, regular concerns that the species was nearing extinction have been expressed multiple times during the past 130 years (Chapter 2, this volume). A dividing line can be considered at ~100th–101st meridians separating the extant occupied range of Lesser Prairie-Chickens based

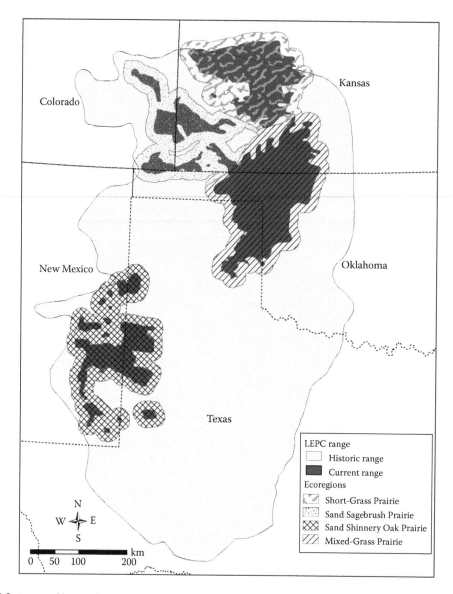

Figure 1.2. Presumed historical range and current distribution of Lesser Prairie-Chickens (LEPC) in the Southern Great Plains.

on average annual rainfall, likelihood of sufficient annual rainfall for quality habitat, ranking of potential limiting factors, relative impacts of anthropogenic factors, approaches to conservation, population demography, and application of management practices.

Estimates of historical population size prior to 1960 should be considered indeterminate at best and primarily anecdotal. Lack of access to private lands, limited roads, and challenges in locating leks create great difficulties for counting birds and leks, estimating densities, and determining potentially occupied habitat for extrapolation.

For example, the Texas Game, Fish, and Oyster Commission (1945) speculated that as many as 2 million Lesser Prairie-Chickens were present in Texas prior to 1900, but had declined to 12,000 birds by 1937. Standardized methods for generating population estimates were not available. Starting in the 1960s, states developed a variety of approaches to monitor Lesser Prairie-Chicken populations (Chapter 4, this volume). Hagen et al. (2010) compiled population estimates by states for the period of 2000–2011 for an estimated maximum population size of ~44,500 birds, assuming that estimates for each

TABLE 1.1

Presumed extent of historical and current occupied range of Lesser Prairie-Chickens in five states of the Southern Great Plains.

State	No. of counties		Occupied area		% decline
	Historical	Current	Historical	Current	
Colorado	6	4	32,821.1 km² (12,672.3 mi²)	4,456.4 km² (1,720.6 mi²)	86.4
Kansas	38	35	76,757.4 km² (29,636.2 mi²)	34,479.6 km² (13,312.6 mi²)	55.1
New Mexico	12	7	52,571.2 km² (20,297.9 mi²)	8,570.1 km² (3,308.9 mi²)	83.7
Oklahoma	22	9	68,452.1 km² (26,429.5 mi²)	10,969.1 km² (4,235.2 mi²)	84.0
Texas	34	21	236,396.2 km² (91,273.1 mi²)	12,126.5 km² (4,682.1 mi²)	94.9
Total	112	76	466,998.0 km² (180,308.9 mi²)	70,601.7 km² (27,259.5 mi²)	84.9

SOURCE: U.S. Fish and Wildlife Service (2014).

TABLE 1.2

Area, population estimates, and population goal for the defined ecoregions for Lesser Prairie-Chicken populations in the Southern Great Plains.

Ecoregion	Estimated area (km²)	Population estimate (90% confidence intervals)			Population goal
		2012	2013	2014	
Sand Shinnery Oak Prairie	27,675	3,379 (1,705–10,500)	1,840 (1,019–4,069)	1,292 (640–2,955)	8,000
Sand Sagebrush Prairie	15,975	2,460 (998–5,049)	2,017 (971–3,477)	477 (241–839)	10,000
Mixed-Grass Prairie	39,600	9,489 (4,507–18,066)	4,173 (2,175–8,212)	7,372 (3,621–15,050)	24,000
Short-Grass Prairie and Conservation Reserve Program Mosaic	37,350	19,895 (9,763–36,747)	10,708 (4,109–21,305)	13,273 (7,063–27,118)	25,000
Total	120,600	35,223 (20,328–71,183)	18,747 (10,351–35,774)	22,415 (13,406–44,882)	67,000

SOURCES: Van Pelt et al. (2013), McDonald et al. (2014a,b).

state were taken simultaneously (Table 1.3). Starting in 2012, a statistically designed aerial survey was initiated (McDonald et al. 2014a). Aerial surveys produce point estimates for each ecoregion with associated confidence intervals (Table 1.2). As additional years are added to the survey, data are pooled across all previous survey years to estimate the probabilities of detection of clusters of Lesser Prairie-Chickens (McDonald et al. 2014b). Based on the pooled dataset, population estimates are provided for the survey year and estimates for previous years are updated, and initial estimates may change as information from future surveys are added to

TABLE 1.3

Population estimates of Lesser Prairie-Chickens by state for 2000–2012.

State	Population estimate (no. of birds)
Colorado	<1,500 (in 2000)
Kansas	19,700–31,100 (in 2006)
New Mexico	6,130 (in 2011)
Oklahoma	<3,000 (in 2000)
Texas	1,254–2,649 (in 2010–2011)
Total	<45,000

SOURCES: Hagen et al. (2010), U.S. Fish and Wildlife Service (2014).

the analyses. The total number of Lesser Prairie-Chickens in 2014 was estimated as 22,415 birds (McDonald et al. 2014b; Table 1.2). Van Pelt et al. (2013) established a range-wide population goal of 67,000 birds, based on an annual spring average over a 10-year time frame (Table 1.2).

Status and Limiting Factors

Until 2014, the Lesser Prairie-Chicken was considered an upland game bird, with each state responsible for conservation and management within state boundaries (Chapter 3, this volume). A variety of regulatory measures and authorities existed in each state where the species occurs. Concern for Lesser Prairie-Chickens is not a recent occurrence; management of the species at the state level started as early as 1861 in Kansas and 1912 in Oklahoma. However, concern for the species increased dramatically over the last two decades due to continued population declines and increasing potential threats. State and federal agencies engaged in active conservation efforts, but continued population declines, most recently exacerbated by severe drought, resulted in a determination by the U.S. Fish and Wildlife (2014) that the Lesser Prairie-Chicken would receive federal protection as a threatened species under the Endangered Species Act of 1973. Included in the listing was a 4(d) special rule recognizing the significant conservation planning efforts made by state and federal wildlife agencies within the range of the Lesser Prairie-Chicken (Chapter 3, this volume).

A host of threats are faced by Lesser Prairie-Chicken populations. Identified potential threats include conversion of native prairie to row crop agriculture, unmanaged livestock grazing, shrub eradication, altered fire regimes, invasive vegetation species, energy development, disease, roads, houses and other infrastructure associated with human presence, collision with fences and power lines, predation, drought, hybridization, and changing climatic conditions (USFWS 2014). Determining population-level influences of any single factor is challenging. Further, the impact of any one factor is probably variable depending on ecoregion. The one concept that virtually all researchers and managers agree on is that the loss of quality usable space with the appropriate structure and food resources is the primary threat and limiting factor for populations of Lesser Prairie-Chickens. Several opportunities provide reasons to be optimistic, however. Among avian conservation efforts, grassland birds have experienced the greatest benefit from state and federal programs designed to facilitate conservation on private lands (North American Bird Conservation Initiative 2013). Given that ~85% of grasslands in the United States are in private ownership (North American Bird Conservation Initiative 2013), engagement of private landowners is a key tool for conservation efforts that can reverse population declines of not only Lesser Prairie-Chickens but populations of other sensitive species of grassland birds as well.

SCOPE OF THIS VOLUME

This volume is divided into four parts: Historical and Legal Perspectives, Ecology, Emerging Issues, and finally Conservation and Management. Part I includes our introduction to the species and is followed by Randy Rodgers' comprehensive historical overview of Lesser Prairie-Chickens and then by William Van Pelt's review of the legal status of the species. The three introductory chapters provide the necessary context for the remainder of the volume.

Part II focuses on the life history and environmental aspects that influence Lesser Prairie-Chickens. Edward Garton and his coauthors present one of the first examinations of the population dynamics of the species. Randy DeYoung and Damon Williford continue with the theme of population ecology, but do so by assessing the genetic aspects of conservation. David Haukos and Jennifer Zavaleta examine multiscale aspects of habitats across the species distribution. Lesser Prairie-Chickens are traditionally considered an upland game bird, and David Haukos and his coauthors provide a historical perspective and overview of harvest management. Clint Boal examines predator–prey relationships of Lesser Prairie-Chickens and is followed by Markus Peterson's review of the influence of parasites and disease on the species.

Dwayne Elmore and David Dahlgren open Part III with their chapter that focuses attention on the challenges of conservation in a landscape comprised of both public and private lands. The direct and indirect influences of anthropogenic activities, especially energy development, are assessed by Anne Bartuszevige and Alex Daniels. The effects of current and future climate change on wildlife populations are a major contemporary issue, which Blake Grisham and his coauthors examine in the context of Lesser Prairie-Chicken ecology. Closing Part III, in their chapter, Patricia McDaniel and Betty Williamson provide an insightful case history on conservation efforts for Lesser Prairie-Chickens.

Part IV consists of four chapters with a focus on specific conservation and management as applied to the major ecoregions within the range of Lesser Prairie-Chickens. At the northern end of the species distribution, David Dahlgren and coauthors discuss the ecology of Lesser Prairie-Chickens on the short-grass prairies and the Conservation Reserve Program (CRP) landscape of northwestern Kansas and eastern Colorado. Moving south, David Haukos and coauthors examine populations inhabiting the Sand Sagebrush Prairie Ecoregion of southwestern Kansas and southeastern Colorado. Farther east and south in the Lesser Prairie-Chicken distribution, Donald Wolfe and his coauthors focus their attention on Lesser Prairie-Chickens in the mixed-grass prairies of northwest Oklahoma,

northeast Texas Panhandle, and south-central Kansas. Last, at the southwestern extent of the species distribution, Blake Grisham and his coauthors review what is known of Lesser Prairie-Chickens in the Sand Shinnery Oak Prairie Ecoregion of eastern New Mexico and western Texas. Finally, Christian Hagen and Dwayne Elmore summarize the general information and insights of the volume with thoughtful predictions of future conservation success.

Given the considerable attention and controversy surrounding the conservation and management of Lesser Prairie-Chickens, we included representatives from state and federal agencies, nongovernmental organizations, and private landowners as authors to limit future debate over inconsequential issues and instead stimulate conservation planning and implementation for the betterment of the species. Our goal was to compile the state of existing knowledge into a single useful resource summarizing the history, ecology, and conservation of Lesser Prairie-Chickens and how these factors are consistent or differ among four ecoregions of the Southern Great Plains where the species still persists. In his book *Grassland Grouse and Their Conservation*, Paul Johnsgard (2002) titled a chapter "Only the Silence Persists: the Heath Hen and New England's Scrub Oak Barrens." The chapter was a sad, poignant tale of extinction brought about by human actions and the inability or desire to engage in appropriate conservation efforts. Our hope is that this volume in the series *Studies in Avian Biology* may play some role in ensuring that a similar chapter is never needed for the Lesser Prairie-Chicken.

ACKNOWLEDGMENTS

We would be ill prepared to author our chapter without firsthand knowledge gained from field research focused on Lesser Prairie-Chickens. Our research efforts would not have been possible without the willingness of all the landowners that allowed us access to private lands. Additionally, we thank the many graduate students and technicians, past and present, who have worked with us on our research focused on Lesser Prairie-Chickens, and without whom, little of our work could have been accomplished. We thank S. Carleton and M. Rice for providing reviews of this chapter and M. Houts for preparing the range map. Last, we thank Brett Sandercock for his support and for ultimately making our book possible.

LITERATURE CITED

Ahlborn, C. G. 1980. Brood-rearing habitat and fall-winter movements of Lesser Prairie-Chickens in eastern New Mexico. M.S. thesis, New Mexico State University, Las Cruces, NM.

Applegate, R. D., and T. Z. Riley. 1998. Lesser Prairie-Chicken management. Rangelands 20:13–15.

Bain, M. R., and G. H. Farley. 2002. Display by apparent hybrid prairie chickens in a zone of geographic overlap. Condor 104:683–687.

Behney, A. C., B. A. Grisham, C. W. Boal, H. A. Whitlaw, and D. A. Haukos. 2012. Sexual selection and mating chronology of Lesser Prairie-Chickens. Wilson Journal of Ornithology 124:96–105.

Boal, C. W., P. K. Borsdorf, and T. Gicklhorn. 2013. Assessment of Lesser Prairie-Chicken use of wildlife water guzzlers. Bulletin of the Texas Ornithological Society 46:10–18.

Boal, C. W., B. Grisham, D. A. Haukos, J. C. Zavaleta, and C. Dixon. 2014. Lesser Prairie-Chicken nest site selection, microclimate, and nest survival in association with vegetation response to a grassland restoration program. U.S. Geological Survey Open-File Report 2013-1235, Reston, VA.

Brennan, L. A., and W. P. Kuvlesky, Jr. 2005. Invited paper: North American grassland birds: an unfolding conservation crisis. Journal of Wildlife Management 69:1–13.

Campbell, H. 1972. A population study of Lesser Prairie-Chickens in New Mexico. Journal of Wildlife Management 36:689–699.

Committee on the Status of Endangered Wildlife in Canada. 2009. COSEWIC assessment and status report on the Greater Prairie-Chicken *Tympanuchus cupido pinnatus* in Canada. Committee on the Status of Endangered Wildlife in Canada. Ottawa. vi + 28pp. <www.sararegistry.gc.ca/status/status_e.cfm> (16 August 2014).

Copelin, F. F. 1963. The Lesser Prairie-Chicken in Oklahoma. Technical Bulletin No. 6, Oklahoma Department of Wildlife Conservation, Oklahoma City, OK.

Corman, K. S. 2011. Conservation and landscape genetics of Texas Lesser Prairie-Chickens: population structure and differentiation, genetic variability, and effective size. M.S. thesis, Texas A&M University–Kingsville, Kingsville, TX.

Crawford, J. A. 1978. Morphology and behavior of Greater × Lesser Prairie-Chicken hybrids. Southwestern Naturalist 23:591–596.

Davis, C. A., T. Z. Riley, H. R. Suminski, and M. J. Wisdom. 1979. Habitat evaluation of Lesser Prairie-Chickens in eastern Chaves County, New Mexico. Final Report to Bureau of Land Management, Roswell, Contract YA-512-CT6-61. Department of Fishery and Wildlife Sciences, New Mexico State University, Las Cruces, NM.

Ellsworth, D. L., R. L. Honeycutt, and N. J. Silvy. 1995. Phylogenetic relationships among North American grouse inferred from restriction endonuclease analysis of mitochondrial DNA. Condor 97:492–502.

Ellsworth, D. L., R. L. Honeycutt, N. J. Silvy, K. D. Rittenhouse, and M. H. Smith. 1994. Mitochondrial-DNA and nuclear-gene differentiation in North American prairie grouse (Genus *Tympanuchus*). Auk 111:661–671.

George, T. L., and D. S. Dobkin (editors). 2002. Effects of habitat fragmentation on birds in western landscapes: contrasts with paradigms from the eastern United States. Studies in Avian Biology 25.

Grisham, B. A., P. K. Borsdorf, C. W. Boal, and K. K. Boydston. 2014. Nesting ecology and nest survival of Lesser Prairie-Chickens on the Southern High Plains of Texas. Journal of Wildlife Management 78:857–866.

Hagen, C. A., and K. M. Giesen. 2005. Lesser Prairie-Chicken (*Tympanuchus pallidicinctus*). No. 364 in A. Poole (editor), The Birds of North America Online, Cornell Lab of Ornithology, Ithaca, NY. <http://bna.birds.cornell.edu.bnaproxy.birds.cornell.edu/bna/species/364> (20 January 2014).

Hagen, C. A., G. C. Salter, J. C. Pitman, R. J. Robel, and R. D. Applegate. 2005. Lesser Prairie-Chicken brood habitat in sand sagebrush: invertebrate biomass and vegetation. Wildlife Society Bulletin 33:1080–1091.

Hagen, C. A., J. C. Pitman, B. K. Sandercock, D. H. Wolfe, R. J. Robel, R. D. Applegate, and S. J. Oyler-McCance. 2010. Regional variation in mtDNA of the Lesser Prairie-Chicken. Condor 112:29–37.

Haukos, D. A., and L. M. Smith. 1999. Effects of lek age on age structure and attendance of Lesser Prairie-Chickens (*Tympanuchus pallidicinctus*). American Midland Naturalist 142:415–420.

Jamison, B. E., R. J. Robel, J. S. Pontius, and R. D. Applegate. 2002. Invertebrate biomass: associations with Lesser Prairie-Chicken habitat use and sand sagebrush density in southwestern Kansas. Wildlife Society Bulletin 30:517–526.

Johnsgard, P. A. 1983. The grouse of the world. University of Nebraska Press, Lincoln, NE.

Johnsgard, P. A. 2002. Grassland grouse and their conservation. Smithsonian Institution Press, Washington, DC.

Kukal, C. A. 2010. The over-winter ecology of Lesser Prairie-Chicken (*Tympanuchus pallidicinctus*) in the northeast Texas Panhandle. M.S. thesis, Texas Tech University, Lubbock, TX.

McDonald, L., K. Adachi, T. Rintz, G. Gardner, and F. Hornsby. [online]. 2014b. Range-wide population size of the Lesser Prairie-Chicken: 2012, 2013, and 2014. Western EcoSystems Technology, Inc., Laramie, WY. <http://www.wafwa.org/documents/LPC-aerial-survey-results-2014.pdf> (12 August 2014).

McDonald, L., G. Beauprez, G. Gardner, J. Griswold, C. Hagen, D. Klute, S. Kyle, J. Pitman, T. Rintz, and B. Van Pelt. 2014a. Range-wide population size of the Lesser Prairie-Chicken: 2012 and 2013. Wildlife Society Bulletin 38:536–546.

McRoberts, J. T., M. J. Butler, W. B. Ballard, H. A. Whitlaw, D. A. Haukos, and M. C. Wallace. 2011. Detectability of Lesser Prairie-Chicken leks: a comparison of surveys from aircraft. Journal of Wildlife Management 75:771–778.

North American Bird Conservation Initiative. 2013. The state of the birds 2013 report on private lands. U.S. Department of Interior, Washington, DC.

Noss, R. F., and A. Y. Cooperrider. 1994. Saving nature's legacy: protecting and restoring biodiversity. Island Press, Washington, DC.

Pruett, C. L., J. A. Johnson, L. C. Larson, D. H. Wolfe, and M. A. Patten. 2011. Low effective population size and survivorship in a grassland grouse. Conservation Genetics 12:1205–1214.

Ridgway, R. 1873. A new variety of prairie chicken. Bulletin of the Essex Institute 5:199.

Ridgway, R. 1885. Some emended names of North American birds. Proceedings of the United States National Museum 8:354–356.

Riley, T. Z., C. A. Davis, M. A. Candelaria, and R. Suminski. 1994. Lesser Prairie-Chicken movements and home ranges in New Mexico. Prairie Naturalist 26:183–186.

Riley, T. Z., C. A. Davis, and R. A. Smith. 1993. Autumn and winter foods of the Lesser Prairie-Chicken (*Tympanuchus pallidicinctus*) (Galliformes: Tetraonidae). Great Basin Naturalist 53:186–189.

Samson, F. B., and F. L. Knopf. 1994. Prairie conservation in North America. BioScience 44:418–421.

Samson, F. B., and F. L. Knopf. 1996. Prairie conservation: preserving North America's most endangered ecosystem. Island Press, Washington, DC.

Samson, F. B., F. L. Knopf, and W. R. Ostlie. 2004. Great Plains ecosystems: past, present, and future. Wildlife Society Bulletin 32:6–15.

Savage, C. 2004. Prairie: a natural history. Greystone Books, Vancouver, BC, Canada.

Sutton, G. M. 1968. The natal plumage of the Lesser Prairie Chicken. Auk 85:69.

Svedarsky, W. D., R. L. Westemeier, R. J. Robel, S. Gough, and J. E. Toepfer. 2000. Status and management of the Greater Prairie-Chicken in North America. Wildlife Biology 6:277–284.

Taylor, M. A., and F. S. Guthery. 1980. Status, ecology, and management of Lesser Prairie Chicken. USDA Forest Service General Technical Report RM-77. Rocky Mountain Forest and Range Experiment Station, Fort Collins, CO. 15 p.

Texas Game, Fish, and Oyster Commission. 1945. Principal game birds and mammals of Texas: their distribution and management. Austin, TX.

Timmer, J. M., M. J. Butler, W. B. Ballard, C. W. Boal, and H. A. Whitlaw. 2013. Abundance and density of Lesser Prairie-Chicken leks in Texas. Wildlife Society Bulletin 37:741–749.

U.S. Fish and Wildlife Service (USFWS). 1967. Native fish and wildlife: endangered species. Federal Register 32:6.

U.S. Fish and Wildlife Service (USFWS). 2006. Endangered and threatened wildlife and plants; 90-day finding on a petition to list the Columbian Sharp-tailed Grouse as threatened or endangered. Federal Register 71:67318–67325.

U.S. Fish and Wildlife Service (USFWS). 2014. Endangered and threatened wildlife and plants; special rule for the Lesser Prairie-Chicken. Federal Register 79:20074–20085.

Van Den Bussche, R. A., S. R. Hoofer, D. A. Wiedenfeld, D. H. Wolfe, and S. K. Sherrod. 2003. Genetic variation within and among fragmented populations of Lesser Prairie-Chickens (*Tympanuchus pallidicinctus*). Molecular Ecology 12:675–683.

Van Pelt, W. E., S. Kyle, J. Pitman, D. Klute, G. Beauprez, D. Schoeling, A. Janus, and J. Haufler. 2013. The Lesser Prairie-Chicken range-wide conservation plan. Western Association of Fish and Wildlife Agencies, Cheyenne, WY.

Woodward, A. J. W., S. D. Fuhlendorf, D. M. Leslie, and J. Shackford. 2001. Influence of landscape composition and change on Lesser Prairie-Chicken (*Tympanuchus pallidicinctus*) populations. American Midland Naturalist 145:261–274.

PART ONE

Historical and Legal Perspectives

CHAPTER TWO

A History of Lesser Prairie-Chickens*

Randy D. Rodgers

Abstract. Scattered encounters with Lesser Prairie-Chickens (*Tympanuchus pallidicinctus*) on the Southern Great Plains were first reported by Euro-American explorers in the mid-1800s. By the 1870s, Euro-Americans had extirpated the American bison (*Bison bison*), introduced cattle, and begun agricultural cultivation in the region. Lesser Prairie-Chicken populations initially increased with changes on the landscape that came with early settlements, with populations sufficiently abundant to support both subsistence and market hunting. The proportion of prairies converted to cropland steadily increased from the late 1800s to the early 1900s. Habitat loss, combined with drought and high winds, precipitated the Dust Bowl, an ecological disaster during the 1930s in the heart of the species' range. By the late 1930s, Lesser Prairie-Chicken populations were reported to be near extirpation in Kansas, New Mexico, and Colorado, with greatly reduced numbers in Texas and Oklahoma. The disaster of the Dust Bowl prompted many of the first landscape conservation efforts to benefit Lesser Prairie-Chickens, including the eventual establishment of the Cimarron and Comanche National Grasslands. Lesser Prairie-Chicken populations generally recovered during the 1940s–1960s, but new threats to the species' habitats emerged during that period. Conversion of grasslands to croplands continued, and development of center-pivot irrigation systems made it possible to farm previously nonarable sand prairie, sand sagebrush (*Artemisia filifolia*) prairie, and sand shinnery oak (*Quercus havardii*) prairie, all prime habitats for Lesser Prairie-Chicken. Oil and gas development and associated infrastructure encroached on and degraded Lesser Prairie-Chicken habitats throughout the species' range, most intensively in the Permian Basin of west Texas and adjacent New Mexico. The encroachment of invasive trees on grasslands, facilitated by a century of fire suppression, reached levels intolerable to Lesser Prairie-Chickens and caused further habitat loss within the range of the species. Extensive herbicide treatments of both sand sagebrush and sand shinnery oak prairies have degraded the quality of these remaining habitats. By the late 1900s, only 10%–15% of the historical range remained occupied by Lesser Prairie-Chickens. A petition to list the species as "threatened" under the Endangered Species Act was submitted to the U.S. Fish and Wildlife Service (USFWS) in 1995. The USFWS determined that the listing was "warranted but precluded" by higher priorities, leaving the Lesser Prairie-Chicken as a "candidate" species. The change in conservation status prompted new research and conservation activity for the species, including land acquisitions. Considerable effort and resources were channeled through the Conservation Reserve Program

* Rodgers, R. D. 2016. A history of Lesser Prairie-Chickens. Pp. 15–38 *in* D. A. Haukos and C. W. Boal (editors), Ecology and conservation of Lesser Prairie-Chickens. Studies in Avian Biology (no. 48), CRC Press, Boca Raton, FL.

(CRP) of the U.S. Department of Agriculture. Lesser Prairie-Chicken populations expanded their range and numbers substantially in Kansas in response to native species grasslands established through the CRP. Much of the conservation activity was monitored through the Lesser Prairie-Chicken Interstate Working Group, which produced three documents that culminated with a range-wide conservation plan put into action in early 2014. Continuing loss and degradation of Lesser Prairie-Chicken habitats due to energy development and prolonged drought prompted the USFWS to list the species as "threatened" on May 12, 2014. The listing decision was vacated by a judicial decision in September 2015, creating considerable regulatory uncertainty regarding the legal status of the species.

Key Words: human settlement, Southern Great Plains, taxonomy, *Tympanuchus pallidicinctus*.

Native peoples who inhabited southwestern sections of the Great Plains prior to the arrival of Euro-Americans would have lived closely with Lesser Prairie-Chickens (*Tympanuchus pallidicinctus*) and their habitats. Anyone who has personally viewed the spectacular spring displays of these birds, then or now, could not help but be captivated. Unfortunately, no records exist describing the historical status of these birds or their relevance to the culture of Native American societies. Records from the early stages of European exploration and settlement are also scarce. Thus, my "history" of Lesser Prairie-Chickens in North America is, by default, a history of the dramatic changes that the species has experienced since its prehistoric range was explored, settled, and ultimately transformed by peoples of European descent.

My historical account is not comprehensive. Instead, I have attempted to paint a broad-brush picture of Euro-Americans' earliest known encounters with Lesser Prairie-Chickens and how the species has reacted to the many drastic alterations of the land wrought at the hands of modern man. I provide an overview and examples of major land use changes and conservation actions that have affected and continue to impact Lesser Prairie-Chickens. More extensive and detailed ecological information is provided elsewhere in this volume. My chapter concludes with a discussion of many of the conservation actions that have been directed toward assuring that the unthinkable prospect of extinction of this species does not become reality.

PRESETTLEMENT

The members of the Spanish party led by Francisco de Coronado in 1541 were the first Europeans to traverse the Llano Estacado (Staked Plains or Southern High Plains) south of the Canadian River in what is now the Texas Panhandle and eastern New Mexico, but records from the expedition make no mention of birds that can be recognized as Lesser Prairie-Chickens. Despite the regular passage of travelers along the Santa Fe Trail by the 1820s, through what probably constituted Lesser Prairie-Chicken habitat, there are no credible reports of the species. A detailed account of records of the species from the mid-1800s through the 1920s has been assembled by (Hubbard et al., unpubl. manuscript) the Museum of Southwestern Biology at the University of New Mexico in Albuquerque; much of the following borrows from that work.

With the exception of probable reports by James Abert (in Galvin 1970:50–53) and his contingent in 1845, none of the other expeditions that explored the region along the Canadian River reported prairie chickens. As part of John C. Fremont's third expedition to the West (1845–1846), Abert was ordered to explore along the upper Canadian River and then eastward. Abert and his party left Bent's Fort in southeastern Colorado traveling south for 35 days through southeastern Colorado, northeastern New Mexico, and the Texas Panhandle before encountering on September 21 prairie chickens that "started from beneath our feet" from a "grove" of sand shinnery oak (*Quercus havardii*), probably in what is now Wheeler County, Texas. Proceeding east, he reported a "large flock" five days later, likely in either Ellis or Dewey counties in present-day Oklahoma.

Naturalist Samuel W. Woodhouse (in Tomer and Brodhead 1996:146 and 178–179) recorded prairie chickens near the Cimarron River in present-day Major County, Oklahoma on August 14, 1849. Based on Sutton (1967:133–136), those birds could have been either Lesser or Greater Prairie-Chickens (*T. cupido*), although Lessers are more likely. Captain R. B. Marcy indicated having seen prairie chickens in 1852 in what is now southwestern Oklahoma (Cotton and Kiowa counties) and the adjacent Texas Panhandle (possibly Brisco or Hall counties) as his expedition searched for the source of the Red River (Marcy 1866).

Captain John Pope and his U.S. Army command collected co-type specimens of Lesser Prairie-Chicken in the southern part of the Llano Estacado during the spring of 1854 (Pope 1854). The collection site was probably in the vicinity of Sulphur Spring or Big Spring in present-day Martin County or Howard County, Texas. Other expeditions that explored the Llano Estacado south of the Canadian River failed to report prairie chickens.

Lesser Prairie-Chickens were first recorded in Kansas in 1872 when a hunting party commanded by Colonel Richard I. Dodge took 32 "grouse" in October on the small tributaries of the Cimarron River, likely in present-day Ford and Clark counties (Fleharty 1995:27). No records of the species are known from Colorado prior to 1900, though sand sagebrush (*Artemisia filifolia*) habitats certainly were present along both the Arkansas and Cimarron River drainages (Giesen 2000). Lesser Prairie-Chickens were first scientifically described in 1873 as a subspecies of Greater Prairie-Chicken but were later assigned species status in 1885 (Ridgway 1885).

The fact that so many travelers through the range of Lesser Prairie-Chickens failed to record the species, combined with the limited reports of those noted earlier, suggests that the presettlement distribution of the species was patchy and perhaps limited to isolated habitats (Hubbard et al., unpubl. manuscript). However, a lack of reports might also be attributed to the early explorers' preoccupation with larger game species or to the secretive nature of the species during summer, when most early travelers were present in the region.

SETTLEMENT

With the exception of the Kansas Territory, most of the presumed range of Lesser Prairie-Chickens remained unsettled by Europeans prior to the American Civil War. Earlier settlement in the Kansas Territory was at least partly facilitated by the passage of the Federal Preemption Act of 1841, which offered a "first option to buy" to individuals who were already living on federal lands. Settlement was accelerated when President Abraham Lincoln signed the Homestead Act of 1862, which offered 65 ha (160 acres) to anyone aged 21 years or older who would pay a small filing fee and live for five years on the property. Larger parcels were consolidated through a variety of means, legal and otherwise.

Euro-American incursions onto Indian lands had already triggered years of retaliations from the Indians of the Southern Great Plains. Incursions were supposed to be ended by the Treaty of Medicine Lodge in 1867, which guaranteed the Indians of the Southern Plains exclusive hunting grounds south of the Arkansas River. But in the early 1870s, professional bison-hide hunters violated the Treaty and entered the Texas Panhandle from western Kansas. Violation of treaty terms precipitated further wars between the Indians and Euro-Americans that were essentially over by early 1875. Thereafter, hide hunters decimated the remaining herds of American bison (*Bison bison*) and the Southern Great Plains was opened to settlement in several stages.

Consequential Euro-American settlement first occurred in northern portions of the Lesser Prairie-Chicken range, with the last area to be settled in the early 1900s in the driest parts of the Llano Estacado of what is now eastern New Mexico. The Kansas Territory was opened to settlement in 1854, with the first settlers being farmers. Most settlers did not fare well with early attempts to grow crops ill adapted to the Kansas climate. It was not until the 1870s, when Mennonite farmers from Russia first introduced wheat, that farming became well established. With increased popularity and new technology, extensive areas of wheat were being harvested by 1880. The first cattle to be grazed in southwestern Kansas were driven out of southern Texas by the Barton Brothers Company in 1872. Early cattle ranchers did not own much land—typically only small tracts obtained through the Homestead Act for their buildings and corrals. Ranchers grazed cattle on unoccupied lands surrounding their deeded lands. The well-known cattle drives from Texas first reached Dodge City, Kansas, a few years later; the first permanent cattle ranches in southwest Kansas were established in the early 1880s.

With the decimation of the bison in the Texas Panhandle, the first domestic livestock brought into the region in the mid-1870s were sheep. However, cattle ranching quickly became dominant in the late 1870s. Charles Goodnight and James Adair drove cattle from Colorado and established the JA Ranch in 1876 in Palo Duro Canyon as the first cattle operation established in the Texas Panhandle, and it grew to encompass 526,316 ha (1.3 million acres). Many other ranches soon followed, including the XIT, which once encompassed 1,214,575 ha (3 million acres). The first attempts at farming in the Texas Panhandle occurred in the early 1880s but were unsuccessful in drought conditions. Successful farming did not take hold in the panhandle until the relatively wet period of the early 1900s, when wheat and cotton became viable in the region.

In what is now Oklahoma, cattle grazing was established in the late 1870s and early 1880s. Large consortiums leased grazing rights from the Cherokee tribe in what was known as the Cherokee Outlet. The largest were the Box Ranch and the Comanche Pool, which operated in both Kansas and Indian Territory (Oklahoma). The Homestead Act unfolded somewhat differently in the Oklahoma Territory, where it was first applied in 1889. From 1889 through 1895, a series of land "runs" took place mainly in the West where "unassigned lands" not assigned to Indian reservations were opened to settlement. By 1905, "surplus" Indian holdings had also been opened to settlement. Examples in the Lesser Prairie-Chicken range included the sparsely populated Cheyenne–Arapaho reservation on the western border of the Territory, which was opened in 1892, as well as the Kiowa–Comanche–Apache and Wichita–Caddo reservations, which opened in 1901. The pattern of settlement resulted in smaller land holdings in some areas, and eventually, a greater density of barbed-wire cross-fencing than elsewhere in the Lesser Prairie-Chicken range. One result of this pattern is that fence collisions have been documented as a significant source of Lesser Prairie-Chicken mortality in northwestern Oklahoma (Patten et al. 2005, Wolfe et al. 2007), but collisions appear to be less common in the remainder of the range (Chapter 17, this volume).

The Federal Timber Culture Act of 1873 likely also had a notable impact on the Lesser Prairie-Chicken. The Act offered 65 ha (160 acres) to any claimant who would plant 16 ha (40 acres) of trees and maintain them for a period of 10 years. In drier parts of the species' range, impacts of the law may have been limited to some additional fragmentation of the prairies and acceleration of the rate of Euro-American settlement. However, in eastern sections of the species' range, where precipitation was greater, plantings provided a source of seeds, which along with fire suppression, resulted in serious invasion of trees into remaining prairies. Tree establishment resulted in major fragmentation and the eventual loss of much Lesser Prairie-Chicken habitat, particularly in the sand prairies south of the Arkansas River in central Kansas (Rodgers 2003).

Settlement of the dry westernmost sections within the range of Lesser Prairie-Chickens mainly occurred after federal actions in the early 1900s made farming and ranching more viable in these areas. The Enlarged Homestead Act of 1909 doubled the size of the individual parcels that could be obtained to 129 ha (320 acres). The Stock Raising Homestead Act of 1916 further increased the size to 259 ha (640 acres). Some of the earliest attempts to settle western sections of the Llano Estacado were by farmers, but they quickly switched to ranching when farming failed.

Abundant evidence indicates that Lesser Prairie-Chicken populations initially increased and remained at relatively high levels during the first few decades of Euro-American settlement (1870s–1920s) due to several factors acting simultaneously. Initial population increases have been primarily attributed to the increased food resources made available by the relatively primitive and scattered nature of grain production during that period (Bent 1932, Baker 1953, Jackson and DeArment 1963). Also at that time, predator populations may have been greatly suppressed by farmers and ranchers who perceived both avian and mammalian predators as threats to subsistence agriculture and livestock production. In the interim period after bison were extirpated and before cattle became abundant, the grazing rest afforded to the grasslands of the Southern and Central Plains could have temporarily enhanced the quality of the habitat available to Lesser Prairie-Chickens. Additionally, Hubbard et al. (unpubl. manuscript) speculated that the species' populations may have responded to the waning of the Little Ice Age as it oscillated and diminished in the latter half of the 1800s.

Lesser Prairie-Chicken populations evidently increased to high densities by the late 1870s and 1880s. Local populations were sufficient to support extensive market hunting, with many carcasses shipped on ice to northeastern U.S. markets (Judd 1905, Ligon 1927, Bent 1932). Jackson and DeArment (1963:733) acknowledged the "unrestricted slaughter" of the species and noted that "railways ran specials for sportsmen to such towns as Higgins in Lipscomb County (Texas) and placed iced cars on sidings for preservation of the kill." Colvin (in Bent 1932) reported counting 15,000–20,000 birds in a single field in Seward County, Kansas, in 1904, where the birds were exploiting maize (*Zea mays*) and milo (*Sorghum bicolor*). Hubbard et al. (unpubl. manuscript) suggested that such population levels could have led to the species' occupancy of southeastern Colorado, where its presence was not confirmed until 1914 (Lincoln 1918). Litton (1978) speculated that up to 2 million birds were present in Texas prior to 1900; however, this figure has been questioned by Davis et al. (2008) and others because it implies a density exceeding 8 birds/km^2 (20 birds/mile^2) over the entire historical presumed range in Texas, including vast regions that were primarily short-grass prairie, which is not generally considered habitat for Lesser Prairie-Chickens. In any case, it is clear that Lesser Prairie-Chickens were abundant through the first two decades of the 1900s (Bent 1932, Baker 1953, Bailey and Niedrach 1965, Oberholzer 1974, Crawford 1980). The species was sufficiently abundant (even in the driest sections of eastern New Mexico) that early farmers and ranchers in the region relied heavily on Lesser Prairie-Chickens for subsistence. One undated (but prior to 1928) report by M. S. Murray indicated that Lesser Prairie-Chickens were so numerous they damaged homesteaders' crops in present-day Roosevelt County, New Mexico (J. P. Hubbard et al., unpubl. data).

A series of mostly winter incursions of Lesser Prairie-Chickens into southern and eastern Texas occurred during this period of abundance (e.g., Oberholzer 1974) and are probably the reason Bent (1932) considered the species to be migratory. Hubbard et al. (unpubl. manuscript) found no evidence of regular migrations by the species; however, they did report numerous extralimital expansions from the 1870s through the early 1900s

into central and eastern Kansas, southwestern and central Nebraska, and even southwestern Missouri. The authors suggested such dispersals could have been triggered by grain crop failures that resulted in autumn and winter food shortages at times when prairie chicken populations were high.

Even during early periods of abundance, Lesser Prairie-Chickens were apparently not common on the short-grass prairies encompassed within the boundaries of what is now presumed to be their historical range. Citing literature references and their interviews of many "old timers," Duck and Fletcher (1944) indicated that Lesser Prairie-Chickens were not present on the short-grass prairies of the High Plains in western Oklahoma but were common in shrublands on sandy soils scattered across the High Plains. When interviewed during the 1930s, a number of the earliest settlers of Throckmorton and Young counties in north Texas indicated that prairie chickens were confined to the sandy grasslands further north during the breeding season (Jackson and DeArment 1963). Schwilling (1955:3) interviewed long-time resident and hunter Frank Schulman of Garden City, Kansas, which was geographically located between extensive hilly, sand sagebrush prairie to the south and flat, fine soil, short-grass prairie to the north. Schulman indicated that "the large numbers, often spoken of, were found only in the rough sandhill-sagebrush areas along the Arkansas and Cimarron Rivers as well as sagebrush areas along streams and rivers further north." He noted "[They] never had many chickens on the flatlands." Similar accounts have been recorded by early occupants of eastern New Mexico.

LAND ABUSE, ECOLOGICAL DISASTER, AND THE FEDERAL RESPONSE

As the proportion of cropland within the range of the Lesser Prairie-Chicken increased and gradually exceeded the area occupied by grasslands and shrublands, the quantity of habitat available to the species declined (Jackson and DeArment 1963, Crawford 1980). The quality of the remaining prairie was also degraded as cattle numbers increased and grasslands were overgrazed. Clean-tillage farming left the landscape vulnerable to the severe drought and abnormally high winds that began in 1933 and lasted through 1939. More than 350 dust storms struck the High Plains from 1933 to 1937 (Hartman and McDonald 1988). The region that became known as the "Dust Bowl" broadly overlapped the Lesser Prairie-Chicken range. Many of the dust storms were so intense that they blackened the sky, sickened and killed both people and livestock, and no doubt directly killed much wildlife, including Lesser Prairie-Chickens.

The adversity of the Dust Bowl period was such that Lesser Prairie-Chicken populations may have been near extirpation by the late 1930s in New Mexico (Lee 1950), Colorado (Bailey and Niedrach 1965), and Kansas (Baker 1953). Baker (1953) reported the comments of Edward Gebhard of Meade, who believed that only two small flocks were left in Kansas (one on the 75 mile² [19,425 ha] XI Ranch in Meade County). Monitored populations at the Davison Ranch, near Arnett, Oklahoma, declined substantially in the 1930s (Davison 1940), and by 1937, the Texas population was estimated to contain only 12,000 birds. All states responded to the population decline by closing hunting seasons (Copelin 1963; Chapter 7, this volume).

The ecological disaster of the Dust Bowl and other land abuses precipitated a federal land conservation response. The Bankhead–Jones Farm Tenant Act of 1937 authorized the federal government to acquire lands that were greatly damaged in the Dust Bowl region and rehabilitate them for a variety of purposes. In Morton County, Kansas, for example, the government had purchased 43,320 ha (107,000 acres) by 1939. George Atwood, who was hired to oversee the work there, made the statement, "You could stand here and look all the way to Colorado without seeing a living piece of vegetation except for an occasional clump of sagebrush" (Hartman and McDonald 1988:13). The U.S. Soil Conservation Service was formed in 1935 (renamed the Natural Resources Conservation Service in 1994) was charged with replanting and restoring these areas. The 43,796 ha (108,222 acres) acquired in southwestern Kansas became the Cimarron National Grassland in 1960 and is administered by the U.S. Forest Service (USFS). The 187,600 ha (463,570 acres) now comprising the Comanche National Grassland in southeastern Colorado and the 12,666 ha (31,298 acres) of the Black Kettle National Grassland in western Oklahoma were acquired, restored, and administered in the same time frame using the same political process. The Cimarron and the Comanche National Grasslands still support small but important populations of Lesser Prairie-Chickens at the time of this writing.

In 1934, the Taylor Grazing Act was passed by Congress to control the use of 708,200 km² (175 million acres) of western rangelands and, as a result, the U.S. Grazing Service was established. The Grazing Service was combined with the General Land Office in 1946 to become the Bureau of Land Management (BLM). In 1976, the Federal Land Policy and Management Act (BLM's Organic Act) repealed outdated land laws and established a standard of multiple use with balanced land stewardship that would sustain the health, diversity, and productivity of these public lands. The BLM, through its Roswell and Carlsbad field offices, currently administers 603,239 ha (1.49 million acres) in New Mexico, of which 181,781 ha (449,000 acres) are considered occupied by Lesser Prairie-Chickens.

POPULATION RECOVERY AND DECLINE

Based on maps provided by Aldrich (1963), Taylor and Guthery (1980) estimated that the species' occupied range had decreased by 92% and the population had declined by 97% since Euro-American settlement. Few historical records exist of the original range or numbers of Lesser Prairie-Chickens, and the estimates of range contraction and population decline are difficult to verify. However, it is clear that the occupied range and abundance of the species was severely diminished by the late 1900s (Chapter 4, this volume).

Generally, Lesser Prairie-Chicken numbers increased during the period from the 1940s through the early 1960s, but recovery was not uniform across the species' remaining range or without periodic interruption (Crawford 1980). Duck and Fletcher (1944) estimated the 1940 population in Oklahoma to be ~15,000 birds, but the species' population in the state may have been reduced to as few as 2,500 in the 1950s (Summars 1956), at least partly as a result of renewed severe droughts that occurred through 1957. Copelin (1963) indicated the Oklahoma population again increased as much as fivefold by 1960. In New Mexico, populations were reported to have peaked at between 40,000 and 50,000 birds between 1949 and 1961 (Sands 1968). Some of the positive population numbers in the late 1950s into the early 1960s can almost certainly be attributed to the grassland restoration that resulted from the Conservation Reserve component of the Federal Soil Bank Program (Jackson and DeArment 1963, Sands 1968). The Soil Bank was created under Title 1 of the Agricultural Act of 1956, and the program was primarily active during the following five years. The Soil Bank Program was relatively short-lived, but the Conservation Reserve component provided a valuable model for the current Conservation Reserve Program (CRP), which was implemented almost three decades later. Jackson and DeArment (1963:737) referred to the grasslands produced by the Soil Bank Program as "a bright spot in an otherwise not very promising picture" for Lesser Prairie-Chickens.

Many human-induced changes to the landscape were continuing to eat away at the habitat base of the Lesser Prairie-Chicken even as populations recovered somewhat after the 1930s. Despite ongoing grassland restoration occurring at the National Grasslands and as a result of the Soil Bank Program, additional native grasslands were steadily being converted to cultivation. However, much of the best habitat for the species was too rough, too sandy, or too arid to be successfully cropped. Habitat loss continued with the advent of center-pivot irrigation as a new technology for agricultural production.

Frank Zybach had been a mechanical innovator since he was a young farm boy in Nebraska, but none of his inventions had proven economically successful. After attending a demonstration of a manually moved circular irrigation system in 1947, the inventor set out to find a better way. By 1948, his first prototype of a self-propelled, center-pivot irrigation system was running in a field near Strasburg, Colorado. Zybach obtained a patent for the system in 1952 and eventually sold the manufacturing rights to a small company in the little town of Valley, just west of Omaha, Nebraska. The Valley Manufacturing Company sold only six sprinkler systems in 1955, their first year (Ashworth 2006). Those systems were the vanguard of an innovation that would rapidly transform vast expanses of rough grazing land into what today seems to be endlessly repeating circles of irrigated corn, alfalfa, and other water-intensive crops in the core of the historical range of Lesser Prairie-Chickens.

Many center pivots replaced existing row flood irrigation systems and caused no additional loss of Lesser Prairie-Chicken habitat. Elsewhere, wherever inexpensive grassland overlaid sufficient supplies of groundwater, mostly in the Ogallala Aquifer, the agricultural economics of converting that land to center-pivot irrigated cropland were obvious. Transformations of prairie chicken habitat occurred rapidly in many parts of southwest Kansas through the Texas Panhandle and even into eastern New Mexico.

Lesser Prairie-Chickens initially took advantage of the grain provided by center-pivot irrigation, but the food resource has not compensated for the overwhelming habitat loss that these populations have suffered. Waddell and Hanzlick (1978) documented habitat loss in south-central and southwest Kansas, where center-pivot irrigation was established during the 1960s. Based on satellite imagery obtained in the mid-1970s, they determined that sand prairie in south-central Kansas and sand sagebrush prairie in southwest Kansas were being lost at annual rates of 5% and 6%, respectively. In one 749 km^2 (289 mile^2) tract that in 1974 consisted mainly of sand sagebrush prairie south and west of Garden City, Kansas, 44 sections (15%) had lost their grass by 1976. At the observed rate, they predicted complete conversion by 1987. Public hearings began in 1984 and led to a designation in 1986 of an Intensive Groundwater Use Control Area by the Kansas Division of Water Resources that halted additional water right allocations, or the prediction would likely have been accurate. About 20% of the sand sagebrush prairie in the tract remains today (R. Rodgers, unpubl. data).

One of the remnants of sand sagebrush prairie in that same southwest Kansas tract is the Sandsage Bison Range (SBR) located just south of Garden City and operated by Kansas Department of Wildlife, Parks, and Tourism. The site was established in 1916 when President Woodrow Wilson granted 1,223 ha (3,022 acres) for the establishment of a game preserve. The area now totals 1,522 ha (3,760 acres) and is bounded on the west and south by center-pivot irrigated cropland and by Garden City to the north. Monitoring of Lesser Prairie-Chicken numbers through the use of spring lek surveys was begun at SBR in 1977 during the period when pivot development was displacing Lesser Prairie-Chickens onto remaining sand sagebrush habitats. The density of breeding prairie chickens at SBR was estimated at almost 15 birds/km^2 (38 birds/mile^2) in the spring of 1977 and by 1980, the breeding density at SBR was 19.3 birds/km² (50 birds/mile²). Densities steadily declined to the point where no leks remained in 1996, though occasional use by the species still occurs (R. Rodgers, pers. obs.).

A notable spin-off effect of center-pivot irrigation, and its capacity to produce huge quantities of feed grains and forage in otherwise semiarid landscapes, has been the expansion of confined animal feeding operations. Cattle feedlots and industrial-scaled hog farms and dairies have often been developed on lands that previously provided habitat for Lesser Prairie-Chickens. Such lands are relatively inexpensive to acquire and tend to be sparsely populated, thus reducing the impact of odor or groundwater contamination on nearby towns or communities.

Just as the presence of groundwater facilitated the spread of center-pivot irrigated croplands and feeding operations, the development of other underground resources has dealt another heavy blow to Lesser Prairie-Chicken habitats. Nowhere within the range of the species is the impact of oil and gas infrastructure more evident than in the Permian Basin of Texas and adjacent New Mexico. Here, fossil fuel resources underlie an area roughly 402 km (250 miles) in width and 482 km (300 miles) long that overlap much of the Llano Estacado. Some Permian oil fields lie north and east of Lubbock, Texas, but the majority of the fossil fuel–producing region is to the west and south of this city and extends into four southeastern counties of New Mexico.

The first producing oil wells in the Permian Basin were drilled in the early 1920s, but it was a deep test well drilled in 1928 that triggered a major expansion of the oil and gas industry in the region. Production peaked in the 1970s and >14.9 billion total barrels of oil had been produced from the Basin by 1993. Not only did production occur over a wide area, but it was also highly intensive. Using Google Earth (Google, Mountain View, California) to view the region revealed extensive areas with 20–30 well pads per section and up to 60 wells on some sections. Additional areas of intense oil and gas development in the historical range of the Lesser Prairie-Chicken occur in the northeastern Texas Panhandle and parts of northwest Oklahoma. Other areas of substantial, but less dense, oil and gas infrastructure are scattered throughout the species' range, most notably in the Mississippian Lime formation that underlies the Red Hills region of northern Oklahoma and southern Kansas.

Direct habitat loss from roads, well pads, pipelines, compressor stations, and other production facilities has had major impacts on Lesser Prairie-Chickens. Further, oil and gas infrastructure has also fragmented the remaining habitat and degraded its quality by increasing erosion, facilitating the spread of invasive weeds, and reducing habitat patch size. Oil field traffic probably also disturbs Lesser Prairie-Chickens, potentially interfering with breeding activity and effectively decreasing the area of available habitat (Crawford and Bolen 1976, Braun et al. 2002). Transmission lines associated with this infrastructure provide perches for raptors that potentially increase mortality rates of prairie chickens (Bidwell et al. 2003), and have further fragmented the landscape.

A better understanding of the impacts of oil and gas development, and other infrastructure, resulted from research in southwestern Kansas. A 6-year radiotelemetry study revealed that Lesser Prairie-Chickens were less likely to nest near anthropogenic features such as buildings, oil pumps, power lines, roads, and compressor stations (Pitman et al. 2005). The field project, combined with earlier studies, demonstrated how Lesser Prairie-Chicken populations could be completely lost from intensive oil and gas production fields, even though much of the original vegetation remains. Project results also revealed how even widely scattered oil and gas infrastructure can still have negative impacts on Lesser Prairie-Chicken populations.

The spread of agriculture, energy developments, and other human infrastructures have directly destroyed or rendered much Lesser Prairie-Chicken habitat unusable, but indirect effects of landscape changes have also proved to be destructive. Colonization of the prairies by trees likely began with the arrival of the earliest settlers, who attempted to control range fires. Even the early pattern of clean-tilled cropland and a system of roads scattered across the prairies must have limited the spread of wildfires. Overstocking of cattle and confinement to fenced pastures reduced the amount and stature of the dry grasses that could fuel wildfire and reduced the competitive advantage of native grasses, making prairies less resistant to invasive woody species. As human developments increased in both extent and density, active fire suppression was added to the fire limitations already noted.

Fire suppression favored establishment and spread of native and introduced trees that early settlers planted for timber products, shade, windbreaks, and as reminders of their Eastern origins. America's love of trees was evident in The Timber Culture Act of 1873, in the designation in 1908 of 1,214 km² (469 mile²) of sand sagebrush prairie along the Arkansas River to become the Kansas National Forest (a failed endeavor), and in the thousands of kilometers of windbreaks that were planted and nurtured as a response to the calamity of the Dust Bowl. In mesic parts of the range of Lesser Prairie-Chicken, trees were established from planted seeds or trees that survived in ravines or other natural fire refuges. Even in the drier parts of the range, mesquite beans were spread in the manure of cattle. Where enough moisture was available, the absence of fire also allowed woody plants already present on prairies to gradually transform these habitats into savanna forests and even closed canopy woodlands.

The slow-moving expansion of invasive trees did not seem to be recognized as a serious threat to Lesser Prairie-Chickens until the late 1900s, after it had progressed to a stage where large tracts of former prairie and shrubland were being rapidly lost. By that time, the encroachment of trees onto open habitats was occurring at an exponential rate (Archer 1994, Fuhlendorf et al. 1996), and was strongly linked with the decline of Lesser Prairie-Chicken populations in western Oklahoma and the northeastern Texas Panhandle (Fuhlendorf et al. 2002).

Even as the quantity of Lesser Prairie-Chicken habitat rapidly diminished in the second half of the 1900s, the quality of some remaining habitats also deteriorated. Overstocking cattle on semiarid rangelands decreased the prevalence of tall and midgrasses, in some cases to the point of near absence. Grazing management has gradually improved on both public and private lands as more stockmen have increased their knowledge of range management, but restoration of overgrazed lands through better grazing management can take decades or longer. Historical grazing pressure probably also increased the dominance of sand shinnery oak and sand sagebrush on many tracts, leading some range managers to use chemicals or other quick solutions to increase grass production.

Efforts to improve cattle–forage production on semiarid shrublands by reducing competition from shrubs have sometimes had the unintended effect of damaging the habitat's ability to support Lesser Prairie-Chickens. In New Mexico, the BLM treated 35,295 ha (87,215 acres) of sand shinnery oak habitat with the herbicide tebuthiuron over 25 years between 1969 and 1993. Researchers working over three decades have pointed out potential pitfalls of herbicide treatments on sand shinnery oak prairies. Sand shinnery oak can be completely eliminated at high application rates, reducing food availability and the ability of the habitat to provide suitable microclimates necessary for critical life stages of Lesser Prairie-Chickens (Doerr and Guthery 1983, Olawsky and Smith 1991, Bell et al. 2010). Haukos (2011) summarized the consensus view of researchers that the goal of tebuthiuron treatments for habitat restoration should seek to create a codominant community of shrubs and native grasses interspersed in patchy mosaics, which was the case for the 10,121 ha (25,000 acre) Weaver Ranch near Milnesand, New Mexico. If herbicide treatments are attempted, Haukos (2011) recommended tebuthiuron should be applied at low rates, should not be applied to sand shinnery oak habitats in poor range condition, and should always be followed by extended grazing deferment with a duration dependent on the subsequent response and climatic conditions.

A cautious approach has typically been recommended relative to herbicide use for habitat improvement, but such ideals have not always been followed in actual practice. One such example occurred on the Cimarron National Grasslands in southwest Kansas from 1978 to 1983. In 1978, 1,174 ha (2,900 acres) of sand sagebrush prairie were treated with moderate rates of the herbicide 2,4-D in accordance with a wildlife management plan (USFS, Kansas Fish and Game Commission) that recommended treatment in alternating strips and small blocks with the aim of creating a habitat mosaic beneficial to wildlife. Apparently unsatisfied with early results, the USFS deviated from that plan. From 1979 through 1983, an additional 14,771 ha (36,500 acres) of sand sagebrush prairie were treated in large blocks with more than double the previously recommended rate of 2,4-D. In their evaluation of this project, Rodgers and Sexson (1990) found that avian species richness, abundance, and diversity all sharply declined, but only ≥4 years after treatment when the structure of the dead sand sagebrush began to significantly deteriorate. Rodgers and Sexson (1990) contended that the near-total kill of sand sagebrush over such extensive tracts would probably prove detrimental to Lesser Prairie-Chickens. During the two decades (1985–2004) following extensive spraying on the Cimarron National Grasslands, the species' populations averaged only 46% of those of the prior two decades (1965–1984), and population numbers have continued to decline. The amount and degree of herbicide-induced habitat alteration of sand sagebrush or sand shinnery oak prairies that has occurred on private lands since the 1960s is unclear but has likely had a significant negative impact on Lesser Prairie-Chicken populations.

EARLY STATE CONSERVATION EFFORTS

The earliest effort by a state to conserve Lesser Prairie-Chicken habitat occurred in New Mexico when funds obtained through the Pittman–Robertson Federal Aid in Wildlife Restoration Act of 1937 were used to acquire farms and ranches that had failed during the Great Depression. Additional lands for the species were acquired in New Mexico in 1967, 1972, and 1991. In all, these lands became a collection of Prairie Chicken Areas (PCA) administered by the New Mexico Department of Game and Fish (NMDGF). New Mexico's 29 PCAs total ~109 km² (27,000 acres) ranging from 0.12 to 31.6 km² (29 to 7,800 acres) in size, mostly in 2.6–5.1 km² (642–1,260) units. The PCAs are scattered throughout the state's Lesser Prairie-Chicken range and have been primarily managed through livestock exclusion or light grazing. Tracts acquired in 1991 from the BLM and the U.S. Department of Agriculture (USDA) Farmers Home Administration have not been fenced to exclude livestock.

Within the historical range of the Lesser Prairie-Chicken in northwestern Oklahoma, the Cooper, Beaver River, and Packsaddle Wildlife Management Areas (WMAs) together total >21,457 ha (53,000 acres), much of which consists of sand sagebrush or sand shinnery oak prairies. These areas once held significant numbers of Lesser Prairie-Chickens, but they are now either absent or present only in low numbers. Today, the Cooper WMA is bordered by wind turbines and the Packsaddle WMA has extensive ongoing natural gas development to the degree that Lesser Prairie-Chickens may never again occur on the property.

Located in the northeast corner of the Texas Panhandle, the Gene Howe WMA was purchased by the Texas Parks and Wildlife Department in the early 1950s. About 972 ha (2,400 acres) of the 2,383 ha (5,889 acre) area consists of sand sagebrush and midgrass habitats that still harbor some Lesser Prairie-Chickens.

In south-central Kansas, most of the 2,314 ha (5,718 acre) Pratt Sandhills WMA was purchased in 1969. The wildlife management area supported substantial numbers of Lesser Prairie-Chickens on sand prairie habitats through the early 1990s, but it was encircled by center-pivot irrigated cropland on surrounding private lands in the 1970s and also suffered from tree invasion. No Lesser Prairie-Chicken leks have been found on the area since 1999. Similarly, the species was also extirpated from the Sandsage Bison Range south of Garden City, Kansas. No WMAs in Colorado support Lesser Prairie-Chickens.

Monitoring

All surveys historically used by the five states have relied on spring surveys of male Lesser Prairie-Chickens displaying at lek sites. However, interstate differences in terrain, available funding, personnel, access to private lands, and politics produced a hodgepodge of survey methods, lack of consistency in timing and locations of where surveys were conducted, and insufficient sampling of the geographic areas occupied by the species. Structural problems with sampling methodology made it difficult to make comparisons among states or to derive statistically valid population trends for the suitable habitat areas that still remain.

The first steps toward monitoring of Lesser Prairie-Chicken populations in the five states were undertaken in the mid-1900s. In 1941, two study areas were set up in the northeastern Texas Panhandle and monitored in 1942 using a lek-listening survey that provided a total male index (Jackson and DeArment 1963). Listening-based surveys of these study areas were resumed in 1952 and continued through 1986 (Sullivan et al. 2000). For the remainder of the 1900s, Texas relied on annual monitoring of four selected leks on each of these two study areas. Starting in 1969, one lek in each of six southwestern panhandle counties was also monitored annually, with three more study areas added in the late 1990s. This survey method for monitoring relied on an index of males per lek to provide some indication of population trends (Sullivan et al. 2000).

In New Mexico, the BLM began monitoring Lesser Prairie-Chickens on its lands in 1971 by attempting to conduct spring counts on all known leks. The annual effort put into these surveys was, however, variable. Starting in 2003, the Carlsbad, New Mexico, district of the BLM also conducted general roadside listening surveys through sand shinnery oak and sand sagebrush habitats. The NMDGF first conducted roadside surveys in 1998 and established 29 routes scattered throughout the species' range by 1999, as well as monitoring at 29 sites designated as Prairie Chicken Areas (Davis et al. 2008).

The Kansas Forestry, Fish, and Game Commission (now Kansas Department of Wildlife, Parks, and Tourism) began monitoring of Lesser Prairie-Chickens in 1964 with the establishment of three 51.8 km^2 (12,800 acre) survey areas—one each in Finney, Meade, and Morton counties. Each of these fixed areas consisted of a 16.1 km (10 mile) driving route and the land extending 1.6 km (1 mile) on either side. Survey routes are monitored annually using spring listening surveys designed to locate all the leks on each area and obtain flush counts from each lek. Roadside survey methodology has allowed annual population density indices and long-term trends to be derived. More survey units were gradually added over time, and currently, there are 15 routes spread throughout the species' range in Kansas, roughly in proportion to the area occupied by the different types of Lesser Prairie-Chicken habitats found in the state. Similar surveys have continued at the Sandsage Bison Range (1,294 ha [5 mile^2], since 1977) and Pratt Sandhills Wildlife Area (3,625 ha [14 mile^2], since 1980).

The oldest method used to monitor Lesser Prairie-Chicken populations in Oklahoma consisted of annual counts of a selection of known leks. In 1982, a series of lek density detection routes were established. Surveys were patterned after methods used in Kansas, but no attempts were made to locate and flush the leks detected. Effort dedicated to these Oklahoma surveys has varied greatly over time, which has limited their value in estimating population density and for generating population trends.

In Colorado, the first attempts at surveying Lesser Prairie-Chickens occurred in 1959, but counts were haphazard and inconsistent in methodology and application. Surveys have been regularly conducted since 1976, but the methodologies

used and the annual survey effort have varied over time. More intensive surveys occurred from 1980 to 1990 when biologists attempted to obtain three separate counts of all previously known leks and survey adjacent areas for others, but this degree of effort ended as a result of personnel cuts. In 1998, an intensive effort involving 16 biologists was made to check all historical leks and survey all suitable habitats within the known or suspected range of the species in Baca and Prowers counties (Giesen 2000).

Translocations

Attempts at establishing new Lesser Prairie-Chicken populations through translocation have occurred, but all were technically unsophisticated and released only small numbers of birds. The earliest such efforts occurred in New Mexico in the 1930s and 1940s but were unsuccessful (Snyder 1967). The first translocation in Colorado was in 1961, when eight birds trapped in Colorado were released at a site southwest of Campo. In 1968, 27 birds trapped in Kansas were released south of Hugo. Between 1972 and 1975, 103 Kansas-trapped birds were released on the Comanche National Grasslands in Baca County. Thirty-two Colorado-trapped birds were released in Pueblo County in 1988 and 1989 and another 77 Kansas-trapped birds were released in the same vicinity in 1993 and 1994. None of these Colorado translocations succeeded (Giesen 2000). A small number (<30) of Lesser Prairie-Chickens were trapped in Wheeler County, Texas, from 1982 through 1984; they were fitted with transmitters and released in Dickens County, Texas. The Texas releases also failed, as have translocation attempts in Oklahoma (Horton 2000). Though specific information is unavailable, Lesser Prairie-Chickens were apparently translocated in 1934 to the relatively arid and privately owned Hawaiian island of Ni'ihau, where they reportedly persisted for many years (Fisher 1951, Pratt et al. 1987, Pyle and Pyle 2009).

EARLY IMPLEMENTATION OF THE CONSERVATION RESERVE PROGRAM

Mostly as a result of huge international grain sales, U.S. farm commodity prices increased sharply in the late 1960s through the early 1970s, and land values increased along with commodities.

Economics of row-crop agriculture led to yet another round of the breaking and cultivation of prairies, some of which had been suitable habitat for Lesser Prairie-Chickens. When grain markets crashed in the mid-1970s, many farmers who had borrowed to purchase high-priced land and equipment found themselves financially over-extended. Overproduction of grains kept prices low and resulted in the farm crisis of the late 1970s and early 1980s. One of the ways the U.S. Congress confronted this crisis was through the establishment of the CRP created by the Federal Food Security Act (Farm Bill) of 1985. Similar in design to the Soil Bank Program (1956–1961), the CRP sought to pay farmers to remove their least productive lands from crop production and establish perennial cover to prevent soil erosion during 10-year contract periods.

Initially, wildlife conservation was considered a side benefit of the CRP and not a primary goal. Administration of the CRP was eventually handed to the USDA's Agricultural Stabilization and Conservation Service (now the Farm Service Agency), but early decisions on implementation of CRP fell to the USDA Soil Conservation Service (now the Natural Resources Conservation Service). Within general guidelines, state Soil Conservation Service officials made decisions as to what types of perennial covers could be established on CRP contract lands. Of the five states where Lesser Prairie-Chickens were present, Kansas and Colorado opted to primarily use mixtures of native warm-season grasses. In Oklahoma and Texas, monocultures of introduced warm-season grasses were the predominant vegetation established on CRP contract lands. Both introduced and native grasses were seeded in eastern New Mexico (Rodgers and Hoffman 2005). Forbs and legumes were not included in the early CRP seed mixtures used within the species' range in any of the five states.

Farmland enrollment in the CRP on the Southern and Central High Plains was successful, and by the late 1980s, several million acres of CRP grasslands had been established within the original range of the Lesser Prairie-Chicken. Recognition of the potential for CRP grasslands to benefit Lesser Prairie-Chickens and other prairie grouse was slow. Virtually all previous population studies of the species had been conducted in prairies dominated by either sand sagebrush or sand shinnery oak, and many biologists believed the Lesser Prairie-Chicken to be an obligate shrubland species. This viewpoint changed after reports that the species had substantially increased its numbers and range by the late 1990s in west-central Kansas where native grass CRP stands had been established (Rodgers 1999). Most stands were seeded in the late 1980s and typically required ≥4 years to mature. It apparently took another 3–5 years for prairie chickens to utilize and show demographic responses to the creation of large expanses of new habitat. Research later conducted in this zone of expansion showed these CRP native grass stands to be highly favored for nesting (Fields et al. 2006).

A positive but much smaller response was also observed in eastern New Mexico. Although consisting of mixed native species, the CRP grass stands originally established in southeast Colorado were dominated by shorter species that provided insufficient height and structural diversity to significantly benefit Lesser Prairie-Chickens. Monocultures of exotic bluestem grasses (Bothriochloa spp.) seeded in northwest Oklahoma and the Texas Panhandle apparently offered little habitat value to the species because no population increase or range expansion occurred (Rodgers and Hoffman 2005). Decisions to seed exotic grass monocultures not only constituted lost opportunities to benefit wildlife, but also have proven difficult to undo. Wherever exotic bluestems have been established, they have been nearly impossible to eliminate and proven to be aggressive invaders that are likely to further diminish the habitat quality of remaining native grasslands.

THE PETITION TO LIST AND GROWING THREATS

In October 1995, the U.S. Fish and Wildlife Service (USFWS) was petitioned by the Biodiversity Legal Foundation of Boulder, Colorado, to list the Lesser Prairie-Chicken as threatened under provisions of the Endangered Species Act of 1973. The action occurred after a sharp decline in the species' populations had occurred throughout its range but before any positive population response to the CRP was evident. The 90-day response by the USFWS did not come until July 8, 1997, when it was published in the Federal Register (62 FR 36482). The delay was due to budgetary constraints and a moratorium that had been placed on listing activities. The initial response indicated that further investigation was warranted. The 12-month finding published on June 9, 1998 (63 FR 31400) stated that listing the species as threatened was "warranted but precluded" indicating that scientific evidence supported listing the species, but that the agency had higher priority species to address given limited resources. The Lesser Prairie-Chicken has subsequently been considered a "candidate" species, effectively elevating its status in attracting funding for conservation-oriented management and research. In the USFWS ranking hierarchy of "candidate" species, the Lesser Prairie-Chicken was initially given a priority rating of 8, essentially two steps away from being listed as threatened (Chapter 3, this volume).

Energy Production

While Lesser Prairie-Chicken populations generally recovered from the 1990s decline and increased in New Mexico and Kansas, additional serious and immediate threats to the species' remaining habitats were developing throughout the five-state range (Chapter 11, this volume). The most conspicuous of the threats was posed by the development of the wind power industry. Wind power generation was mostly regarded as a California phenomenon prior to 2000, but rapid expansion of the industry on the windy High Plains and elsewhere was spurred with the help of legislation for new Production Tax Credits that was passed and regularly extended by the U.S. Congress. Other factors pushing wind power expansion included steadily improving turbine efficiency, increasing demand for electricity, upgrades to the nation's electrical grid in both scale and the capacity to manage variable power sources, and a loss in favor of coal-based power generation due to the production of carbon dioxide as the main greenhouse gas driving global warming. During the 2000s, representatives from established and start-up wind power companies were scrambling across the Western Great Plains to secure rights for wind power development.

Areas offering the best and most easily developed wind resources were often the same lands with remaining habitat for Lesser Prairie-Chickens. Grazing lands have remained unbroken for cultivation largely because their rough topography was unsuitable for farming. Sites with high elevation or topographic variation tend to have consistent or accelerated air flows, thus increasing the value of the wind resource. Compared to farms, ranches generally tend to be held in larger tracts with relatively few owners, which simplifies the legal, political, and logistical processes needed to develop a wind production facility. No better example of this dilemma can be found than in northwest Oklahoma, where, in the mid-2000s, more than three-quarters of the known Lesser Prairie-Chicken leks were located on lands considered to have high potential for wind power development. Much of that land has subsequently been developed. Negative effects of wind power infrastructure for the species were predicted to be comparable to the effects of the infrastructure of oil and gas facilities previously documented by Pitman et al. (2005).

The mid-2000s also brought an almost frenetic growth back to the domestic oil and gas industry. Underlying the economic growth were the price increases that began in 2002 after the terrorist attacks of September 11, 2001, and that were sustained by ongoing conflicts in the Middle East, and continuing growth in world oil demand, particularly in China. Major technological advances have also facilitated the location and extraction of oil and gas. New technologies include improvements in seismic detection of oil-bearing formations, carbon dioxide injection into older oil fields, the development of horizontal drilling techniques, and hydraulic fracturing. New oil and gas infrastructure has sprung up in many parts of the remaining range of Lesser Prairie-Chickens, but this boom has been particularly intense in the species' remaining range in the northeast Texas Panhandle and adjacent northwest Oklahoma.

High oil prices and U.S. dependency on imported oil have also been the driving forces for alternative biofuels, particularly in the development of the ethanol industry. The biofuel industry was kick-started by a subsidy of 40 cents/gal provided in the Federal Energy Policy Act of 1978. Since that time, subsidies have reached as much as 60 cents/gal. Other federal legislation has sought to encourage the use of ethanol by providing tax breaks on the purchase of vehicles that could use fuels containing up to 85% ethanol. The most notable growth in the ethanol industry was triggered by the establishment of national ethanol production targets that seemed to guarantee the viability of the industry well into the future.

The Federal Energy Policy Act of 2005 established targets of 18.2 billion liters (4 billion gallons) of ethanol by 2006 and 31.9 billion liters (7.5 billion gallons) by 2012. A national demand for ethanol helped fuel a sharp increase in the price of corn beginning in 2006 and that increase was further accelerated by the long-term target of 153 billion liters (36 billion gallons) for annual production of biofuels by 2022, as established by the Energy Independence and Security Act of 2007. The price of wheat and sorghum was also driven up along with the price of corn. High commodity prices, coupled with the safety net of federal crop insurance unencumbered by prohibitions against converting native prairie to cultivation, resulted in yet another drive to convert grasslands to croplands. An undesirable outcome that is relevant to Lesser Prairie-Chickens is the fact that high commodity prices substantially reduced farmer interest in enrollment in the CRP.

Current energy policy with subsidies for the ethanol industry has been called into question in the U.S. Congress. The science of ethanol production has shown that little or no additional energy can be obtained from corn-derived ethanol than is used to produce it (Patzek 2004). Conversion of native prairies to grain production, made worse by the ethanol industry's demand for corn, has also been shown to release vast amounts of carbon dioxide into the atmosphere, which accelerates climate change (Hill et al. 2008). High commodity prices have driven up the price of food not just domestically but globally. Davis et al. (2008) suggested that ethanol production from cellulosic feedstock, if managed appropriately, might contribute to the habitat base for Lesser Prairie-Chickens but that form of production has not become economically viable at this point in time. All of these factors increase risk to population persistence of Lesser Prairie-Chickens (Chapter 12, this volume). Citing an increasing list of emerging threats, the USFWS candidate species' priority rating was elevated to level 2 in December 2008 which was one step away from a listing decision.

THE CONSERVATION RESPONSE

Given that the vast majority of Lesser Prairie-Chicken habitat is on privately owned land, the pros and cons of a federal listing of the species as "threatened" have been hotly debated, even among advocates who are equally passionate about conserving the species (Chapter 10, this volume). The debate continues even today, with the primary concern of many stakeholders being their own welfare rather than that of the species. Controversy continues as to whether federal listing is a real or only a perceived threat to private enterprise. What cannot be debated is that the 1995 petition to list stimulated a broad-based response to conserve the Lesser Prairie-Chicken, with a goal of keeping the species common enough that there would be no need for listing. The conservation response has been diverse in both its sources and in its applications, and nearly all of these efforts have involved extensive collaboration among the federal government, state governments, NGOs, and other private entities.

Harvest

Modern regulated hunting has never been scientifically implicated as a factor in the decline of Lesser Prairie-Chicken populations, but the 1995 petition to list the species, along with low populations during the early and mid-1990s, prompted state wildlife agencies to reconsider their hunting seasons. New Mexico had regulated hunting seasons of varying structure from 1948 through 1995, with a peak harvest estimate of ~4,000 birds in 1987. In Oklahoma, a 9-day season was in place at least since the late 1960s, with a peak harvest estimated at around 9,500 birds in 1982. Lesser Prairie-Chickens currently retain the status of a game bird in both states, but the hunting season was closed in 1996 in New Mexico and in 1997 in Oklahoma. Texas began regulated hunting of the species in 1967 with a 2-day season. The highest modern harvest in Texas was estimated at 1,388 birds in 1987 (limited harvest estimates prior to 1987), but annual harvest had dropped to a few hundred birds by the mid-1990s. New regulations begun in 2005 precluded hunting of Lesser Prairie-Chickens except on properties participating in a special management program administered by Texas Parks and Wildlife Department, at which point annual harvest dropped to just a handful of birds. Texas ended the limited hunting opportunity in 2009. Kansas opened its first modern hunting season for Lesser Prairie-Chicken with a 2-day season in 1970 and season length was gradually lengthened through the remainder of the decade. Season length and structure have been modified several times since 1970 with a peak estimated harvest of ~6,000 birds in 1982. The population growth and expansion that occurred as a result of the CRP allowed Kansas to continue to conservatively provide harvest opportunity for the species, with annual harvests typically being a few hundred birds. With the smallest overall population and range, Lesser Prairie-Chickens have not been hunted in Colorado since 1917 and were listed as "threatened" by the state of Colorado in 1973. Harvest management for Lesser Prairie-Chickens is considered in greater detail by Haukos et al. (Chapter 7, this volume).

Lesser Prairie-Chicken Interstate Working Group

The first meeting specifically called to discuss the status of the Lesser Prairie-Chicken occurred in Amarillo, Texas, in February 1995. Biologists at that meeting reported on the status of the species in their respective states and explored potential common causes for the species' decline since 1990. Drought, exacerbated by prior habitat losses, was generally considered the proximate cause for population decline during that period. In 1996, staff with the Wildlife Management Institute set up and conducted a small meeting in Texas with a goal of facilitating a cooperative conservation effort among the five states and federal agencies with responsibilities for Lesser Prairie-Chickens. In a letter sent to the USFWS on October 10, 1996, the heads of the five state wildlife agencies pledged to work together for the conservation of the species. The draft letter functioned as a charter for the formation of the Lesser Prairie-Chicken Interstate Working Group (LPCIWG). Members of the LPCIWG included representatives from the five state wildlife agencies, the USFS, BLM, USFWS, Natural Resources Conservation Service, and Wildlife Management Institute. Other participants in the LPCIWG eventually included the Western Governors' Association (WGA) and a number of NGOs (Mote et al. 1999).

By February 1999, the LPCIWG had produced an "Assessment and Conservation Strategy for the Lesser Prairie-Chicken" (Mote et al. 1999). The assessment document was the first effort to list potential threats to the species but also indicated that much remained to be learned. The authors recognized that successful conservation would have to involve private landowners while preserving their abilities to derive income from their land

(Chapter 10, this volume). The document outlined the legal implications of listing on federal agencies, created a series of prioritized goals (the Conservation Strategy), and made general habitat recommendations.

Even before publication of the first LPCIWG document, the first of 12 "Ranch Conversations" was held in Buffalo, Oklahoma, on January 14, 1999, with the assistance of the High Plains Resource Conservation and Development Council. Members of the LPCIWG recognized that ranchers not only controlled most of the lands where Lesser Prairie-Chickens remained, but also that they held many of the same values regarding land and wildlife as did conservationists. The meetings offered presentations that covered the basics of Lesser Prairie-Chicken life history, population distribution and numbers, and threats to the species. However, the meetings were primarily geared toward obtaining ranchers' input into how best the species could be conserved and toward cultivating relationships that would serve conservation goals. By February 2000, four Ranch Conversations had been held in New Mexico and two each in Texas, Oklahoma, Kansas, and Colorado. Ranchers who attended the meetings and indicated an interest in Lesser Prairie-Chickens were later contacted by federal or state biologists. Many participants became important partners in subsequent on-the-ground conservation efforts.

The Ranch Conversations were an extension of a new environmental doctrine of the Western Governors' Association (WGA). The "Enlibra" doctrine advocated obtaining solutions through collaboration rather than confrontation, using incentives where possible rather than mandates, and dealing with issues over appropriate geographic scales rather than within discrete political boundaries. The High Plains Partnership for Species at Risk (HPP) was one such example where this doctrine was to be applied. The HPP aspired to bring a diverse group of partners together, from Montana to Texas, to work toward community-based, voluntary approaches for reversing the decline of species, with the WGA serving as a coordinating entity. The WGA worked directly with the LPCIWG to facilitate this type of approach. The HPP did not persist, but the LPCIWG has continued and can claim many accomplishments.

With assistance from many sources, the LPCIWG was able to better define its estimation of the outer boundaries of the presettlement breeding range by combining species records, expert opinion, and the known distribution of vegetation communities that likely provided suitable habitat for the species. By 2007, they had determined that ~14% of the ~456,000 km^2 (113 million acres) original range was still occupied by the species. The estimate of 14% was somewhat greater than the earlier estimate of 8% made by Taylor and Guthery (1980), largely due to the recent CRP-induced range expansion that had occurred in Kansas.

Annual meetings of the LPCIWG have been held at many venues within the historical range of the Lesser Prairie-Chicken. A primary objective of these meetings has been the exchange of information not only among members, but circulation of information to resource professionals who could influence the management of the landscapes where Lesser Prairie-Chickens still occur. Participation by active landowners, USDA officials, university extension staff, NGOs, and many others was invited and welcomed. In addition to presentations, associated field trips helped broaden participants' collective knowledge of the range of habitat types and configurations that Lesser Prairie-Chickens could utilize. The meetings helped lay the groundwork for many cooperative conservation efforts that have followed.

In an effort to broaden the educational effort and build greater support for range-wide conservation action, in 2004 the LPCIWG made a decision to develop a video that would introduce Lesser Prairie-Chickens to a wider audience, and focus on habitats, threats, and conservation opportunities across the five-state range. Coordination, production, and financing of the documentary were taken on by the Kansas Department of Wildlife and Parks. More than 5,000 DVDs of the resulting 40-min video, "The Lesser Prairie-Chicken: Echoes of the Prairie," had been distributed to ranchers, conservationists, developers, state and federal officials, schools, and others in the five states by early 2007. Many more found their way to federal officials, politicians, and NGOs in Washington, DC.

Nine years after publication of the initial "Assessment and Conservation Strategy for the Lesser Prairie-Chicken," the LPCIWG produced the much more comprehensive "Lesser Prairie-Chicken Conservation Initiative" (LPCCI, Davis et al. 2008). The 114-page document provided a thorough review of the scientific literature on the species, updated the status and range of the species, and documented the conservation efforts to that point in time. The document was the first to establish state population goals and a range-wide goal of ~80,000 birds. The LPCCI outlined 39 strategies that would facilitate the conservation of the species, although it did not specifically define mechanisms to fund these strategies. The LPCCI also contained documents that established the LPCIWG as one of the technical groups associated with the Western Association of Fish and Wildlife Agencies (WAFWA) Grassland Initiative. Perhaps most importantly, the LPCCI called upon energy production industries, which have been responsible for past and continuing habitat loss, to take a major role in financing the conservation efforts for Lesser Prairie-Chickens.

Recent Conservation Actions

Programs directed at Lesser Prairie-Chicken conservation by the federal government began with the determination of candidate status for the species in 1998 and have been occurring simultaneously, and in cooperation with work by the states. Initial efforts to maintain existing habitats were through the development of prelisting voluntary agreements between the USFWS and willing partners. Candidate Conservation Agreements seek mutual cooperation between the USFWS and other federal or private entities to identify threats to a candidate species and design, implement, and monitor appropriate conservation measures. For example, a Candidate Conservation Agreement has been established among the USFWS, the BLM, and private companies relative to oil and gas leasing in New Mexico. Under this agreement, companies must minimize habitat fragmentation, restore damage from earlier development, and provide a mechanism whereby companies can financially contribute to the species' conservation. Candidate Conservation Agreements with Assurances have similar goals but provide nonfederal landowners with added incentives to engage in voluntary conservation practices. A Candidate Conservation Agreement with Assurances offers assurances that if the landowner implements and maintains agreed-upon conservation practices, they will not be subject to additional restrictions should the species become listed under provisions of the Endangered Species Act.

Since the species was designated as candidate for listing in 1998, biologists with the USFWS Partners for Fish and Wildlife program have worked directly with landowners and land managers (many first contacted through the Ranch Conversations) to improve Lesser Prairie-Chicken habitat. Ongoing efforts have included educational, technical, and financial assistance aimed at improving range management and, ultimately, habitat conditions in rangelands managed for livestock production. The goal has been accomplished in part through direct contacts, workshops, range tours, and range evaluations. Partners for Fish and Wildlife biologists have helped promote applicable conservation practices offered through the USDA. Considerable emphasis has been placed on removing invasive trees from prairies. In Texas, tree control has mainly taken the form of removing honey mesquite (*Prosopis glandulosa*) from shrubland habitats in the Permian Basin region. In Kansas and Oklahoma, tree removal has focused mainly on eastern red-cedar (*Juniperus virginiana*). Control of cedars has been accomplished through both mechanical means and prescribed fire. Establishing the use of prescribed fire for periodic habitat maintenance has also been a priority in Kansas and Oklahoma, where tree invasion is a serious threat to Lesser Prairie-Chickens. In Kansas, for example, Partners for Fish and Wildlife efforts to promote prescribed fire have been greatly facilitated by partnerships developed with organizations such as the Comanche Pool Prairie Resource Foundation, Kansas Grazing Lands Coalition, The Nature Conservancy, and other state and federal agencies.

Seeing the threat posed to Lesser Prairie-Chicken habitats by the rapid growth of the wind power industry, the USFWS began a series of meetings with representatives of the industry in the early 2000s. These meetings also involved members of the LPCIWG and interested NGOs. Over time, this informational exchange has been further facilitated by The Nature Conservancy and by staff of the Playa Lakes Joint Venture. Along with other partners, The Nature Conservancy staff based in Kansas organized a conference in Kansas City in March of 2003 that helped focus the wind industry on the likely negative impacts of wind power on Great Plains wildlife (Manes et al. 2004). Geographic Information System specialists from involved agencies and groups have since created many informational tools aimed at helping energy developers voluntarily avoid siting of energy infrastructure in habitats that are critical to the sustainability of the Lesser Prairie-Chicken and other declining species.

The Lesser Prairie-Chicken has been designated a species of special concern for the Cimarron and Comanche National Grasslands. Additional emphasis on maintaining the species' habitat has been applied on the National Grasslands primarily through changes in grazing rotations and cattle stocking rates. Stocking rate reductions, and even temporary destocking, have been critical steps taken to maintain grassland health in the face of recent severe drought. Prescribed burns have also been conducted to benefit the species.

In New Mexico, the Pecos District of the BLM approved in 2008 a Special Status Species Resource Management Plan Amendment that provided greater protection to the Lesser Prairie-Chicken and dune sagebrush lizard (*Sceloporus arenicolus*), which occupies the same habitat. The planning effort established both a Core Management Area and the Primary Population Area for Lesser Prairie-Chickens where no new oil and gas leases will be issued on ~157,895 ha (390,000 acres) of federal mineral holdings. The plan also established a 23,482 ha (58,000 acres) Area of Critical Environmental Concern within the Core Management Area. The Area of Critical Environmental Concern is located ~72 km (45 miles) east of Roswell in eastern Chavez County, New Mexico, and has combined the management of ~12,146 ha (30,000 acres) of federal land, 4,453 ha (11,000 acres) of state land, and 4,049 ha (10,000 acres) of private land. Most of the private land within the Area of Critical Environmental Concern was purchased through the work of various NGOs. For example, The Conservation Fund was able to facilitate the BLM's purchase of 729 ha (1,800 acres) of a property known as the Sand Ranch. The Conservation Fund also leveraged funds from the Richard King Mellon Foundation to purchase 17,004 ha (42,000 acres) of federal and state grazing permits on the Area of Critical Environmental Concern. As a result, no grazing currently occurs in the area. Lands within the Area of Critical Environmental Concern are also closed to all future oil and gas leasing. These efforts have been aimed at maintaining the integrity of the best remaining Lesser Prairie-Chicken habitat under its control, but the BLM has also launched an ambitious program of habitat restoration in association with the Restore New Mexico Initiative. One aspect of this effort has focused on reclaiming lands historically disturbed by oil and gas development. As of 2014, ~328 ha (810 acres) of well pads and abandoned roads have been reclaimed, and abandoned power lines have been removed within the Resource Management Plan Amendment boundaries, thereby reducing fragmentation of surrounding habitat. Additionally, mesquite has been treated on >107,692 ha (266,000 acres) of habitat within the Resource Management Plan Amendment Area.

Also in New Mexico, The Nature Conservancy purchased the 7,490 ha (18,500 acre) Creamer Ranch in 2005 to establish the Milnesand Prairie Preserve in Roosevelt County. Another 3,725 ha (9,200 acres) of the adjacent Johnson Ranch were added in 2009. The sand shinnery oak prairies are located near Milnesand in the heart of New Mexico's Lesser Prairie-Chicken range. A cow–calf operation is used to manage the preserve's pastures in a rotational grazing system. With assistance from the Grasslans Charitable Foundation and the Weaver Ranch, The Nature Conservancy's Preserve also hosts many of the activities, such as the Milnesand Prairie-Chicken Festival, which has been held annually since 2001. The popular festival occurs over 2 days each spring, during which participants not only view Lesser Prairie-Chicken lek activities, but they also have the opportunity to go on various tours featuring activities such as local birds and insect identification (Chapter 13, this volume).

Across the Texas–New Mexico state line, east of Milnesand, The Nature Conservancy manages the >2,915 ha (7,200 acre) Yoakum Dunes Preserve in Yoakum, Cochran, and Terry counties. Land acquisition for this preserve began in 2007 with the purchase of 2,429 ha (6,000 acres); additional purchases have continued through 2013. Preserve lands have been augmented by adjacent lands acquired by the Texas Parks and Wildlife Department and Conoco-Phillips Petroleum Company. Taken together, these lands total ~4,453 ha (11,000 acres).

In 1999, The Nature Conservancy acquired the 6,883 ha (17,000 acre) Smoky Valley Ranch preserve in Logan County, Kansas. There are ~1,822 ha (4,500 acres) of habitat currently suitable for Lesser Prairie-Chickens on the Preserve and another 1,781 ha (4,400 acres) have been identified as having potential to provide Lesser Prairie-Chicken habitat. The latter habitats are being restored with a regime of moderate rest rotation livestock grazing, expedited by season-long grazing deferments. The Smoky Valley Ranch lies within the zone where the Lesser Prairie-Chicken range overlaps the range of the Greater Prairie-Chicken, possibly as a result of CRP-related population expansions. Given the importance of CRP stands and other privately owned grasslands in this area, The Nature Conservancy is working with surrounding landowners to maintain the integrity of the habitat base that supports this population.

In 2012, The Nature Conservancy established a permanent conservation easement on 11,943 ha (29,500 acres) in Cheyenne County, Colorado. The ranch consists of a contiguous area of sand-sagebrush prairie with leks of Lesser Prairie-Chickens that are known to be active. The Nature Conservancy also holds another 1,700 ha (4,200 acre) conservation easement with active leks in Prowers County, Colorado.

The Oklahoma Department of Wildlife Conservation recently acquired several substantial new tracts of land to benefit Lesser Prairie-Chickens. The 1,376 ha (3,400 acre) Cimarron Bluff WMA in Harper County and the 1,538 ha (3,800 acre) Cimarron Hills WMA in Woods County were purchased in 2006 and 2007, respectively, with state Legacy funds and State Wildlife Grant funds. Two other areas, including the 1,903 ha (4,700 acre) Packsaddle WMA Expansion in Ellis County and the 2,389 ha (5,900 acre) Beaver River Expansion in Beaver County were acquired in 2011 and 2012, respectively. Both of these properties were, in part, purchased with funds provided by Oklahoma Gas and Electric Company to help offset the impacts of their wind farm located in Harper and Woodward counties. The same source of funding has allowed Oklahoma Department of Wildlife Conservation to obtain 30-year leases on 1,619 ha (4,000 acres) of School Lands to manage for Lesser Prairie-Chickens. Management practices include prescribed grazing, prescribed fire, and removal of upland trees.

No federal program has offered more potential to build the habitat base available to Lesser Prairie-Chickens than the Conservation Reserve Program. Unfortunately, the potential benefits to the species were mostly unrealized in four of the five states in the first decade of the program

(Rodgers and Hoffman 2005). The population growth and expansion observed in Kansas as a result of its native grass CRP stands clearly showed that quality habitat for Lesser Prairie-Chickens could be restored. Equally important were the numerous opportunities made available to further enhance the quality and better target the distribution of CRP stands. Nowhere else in the range of the Lesser Prairie-Chicken were these opportunities better utilized than in western Kansas.

A positive working relationship between the Natural Resources Conservation Service and the Kansas Fish and Game Commission (now Kansas Department of Wildlife, Parks, and Tourism) had been carefully cultivated since the early 1980s. As a consequence, Natural Resources Conservation Service officials in Kansas often quickly became aware of emerging wildlife science and were better able to apply this knowledge to programs for which they had responsibility. Additional motivation to improve CRP stands for wildlife came when the 1996 Federal Agriculture Improvement and Reform Act (i.e., Farm Bill) elevated wildlife conservation to equal status as one of the primary purposes of the program. When initial CRP contracts began to expire in the late 1990s, Natural Resources Conservation Service officials responsible for technical aspects of the CRP were aware of the importance of invertebrates (mainly insects) in the diet of young grassland birds (Hill 1985, Erpelding et al. 1987), and that insect diversity and availability were greater when broad-leaved plants (forbs and legumes) were present in restored habitats (Rands 1985). As a result, CRP seeding mixtures for new stands were improved as broad-leaved perennial species became required components. Forbs and legumes were not included in the early CRP seeding mixtures, and inter-seeding with broad-leaved plants became a condition for reenrollment of existing CRP stands. The Kansas Department of Wildlife, Parks, and Tourism initiated its own inter-seeding program in four counties, where many large blocks of CRP stands had been reenrolled prior to the initiation of this practice. The state-run effort was technically aimed at improving population numbers of Ring-necked Pheasants (*Phasianus colchicus*), but was primarily focused in Gove and Logan counties where Lesser Prairie-Chicken populations had already shown remarkable growth. Through this program, a diverse mixture of perennial forbs and legumes were seeded at no cost to the landowner

on ~12,146 ha (30,000 acres). Combined, these federal and state efforts resulted in the quality of hundreds of thousands of acres of existing CRP stands being enhanced within the species' range in Kansas.

Kansas was also the first state to establish CRP Conservation Priority Areas intended specifically to benefit Lesser Prairie-Chickens. Potential enrollees whose lands were within a Conservation Priority Area were given an important competitive advantage toward acceptance in the program, which provided an opportunity to direct CRP leases into areas of particular conservation concern. Relative to Lesser Prairie-Chickens, enrollment was done somewhat crudely at first by simply selecting key counties where the species was present and interest in CRP enrollment was high. Periodically revised, the designation of Conservation Priority Areas became far more refined as the CRP matured, with small watersheds becoming the geographical units of choice. The fine-tuned targeting of CRP enrollments only became possible with a much improved knowledge of the distribution of Lesser Prairie-Chickens in Kansas that was obtained through a series of listening surveys conducted by Kansas Department of Wildlife, Parks, and Tourism staff during the springs of 1999 through 2004.

The success of efforts in Kansas to establish and enhance CRP stands for Lesser Prairie-Chicken populations became well known among conservationists throughout the species' range, eventually helping to change how the CRP was administered in the other states. Kansas' success was conveyed through many channels, but notably by LPCIWG presentations, field tours, and presentations at the national conference on the CRP in Fort Collins, Colorado, in 2004 (Rodgers and Hoffman 2005). Kansas State Farm Services Agency officials and Kansas Department of Wildlife, Parks, and Tourism collaborated to train field staff by explaining the benefits of forbs and legumes, and of prescribed fire, thus improving CRP stand management in western Kansas. One particular high point in this relationship was coordinated by staff of the Playa Lakes Joint Venture and conducted in the field by Kansas Department of Wildlife, Parks, and Tourism staff. High-level Farm Services Agency staff from Washington, DC, and the five states with Lesser Prairie-Chickens, toured western Kansas CRP stands over the course of two days in October 2009. This tour, along with other interorganizational communications, resulted in an increased allocation of CRP acres that could be targeted to benefit the species through the State Acres for Wildlife Enhancement practice, first offered through the Continuous Signup of the CRP in January of 2008. Each state, except Kansas, established large geographic areas within their Lesser Prairie-Chicken range into which all or part of their initial State Acres for Wildlife Enhancement allocation could be enrolled. In Kansas, the Kansas Department of Wildlife, Parks, and Tourism worked with Geographic Information Specialists to carefully focus their second State Acres for Wildlife Enhancement area allocation into core habitat areas where USDA data indicated that the most immediate CRP contract expirations were about to occur. This type of targeting has since become standard in efforts to conserve the Lesser Prairie-Chicken.

In 2010, the Natural Resources Conservation Service began its Lesser Prairie-Chicken Initiative (LPCI), which combined and modified facets of previously existing programs into one package intended to improve the species' habitat on working lands. The goals of the Lesser Prairie-Chicken Initiative were to help transform expiring CRP stands into well-managed grazing lands, improve grazing lands sustainability by developing grazing plans that would optimize grass production for habitat quality and profitability, and remove invasive woody species from grasslands. With the help of partners, additional staff has been added in critical locations through the Lesser Prairie-Chicken range to offer technical assistance and cost sharing for a variety of practices available through the Environmental Quality Incentives Program and Wildlife Habitat Incentive Program. During the first 3 years of the LPCI, ~$19M in assistance was provided to >600 producers to improve habitat quality on ~283,400 ha (700,000 acres) of land in high-priority focal areas within the remaining range of the Lesser Prairie-Chicken. In addition, >666 km (410 miles) of fence that posed a significant threat to prairie chickens was either removed or marked to reduce collision mortalities (Wolfe et al. 2007).

Building upon much of the aforementioned work and acknowledging the fact that most of the remaining habitat for the species is on private land, the LPCIWG, in association with WAFWA and the Ecosystem Management Research Institute, recently completed the five-state Lesser Prairie-Chicken Range-wide Conservation Plan (Van Pelt et al. 2013). The plan recognizes that it is crucial to work with private landowners in a proactive manner to assure persistence of the species. The Range-wide Conservation Plan specifically designates key focal (core) habitat areas, and connecting corridors between them, that must be protected. A Geographic Information System (GIS)-based system called the Crucial Habitat Assessment Tool has been the primary mechanism for selecting focal areas and corridors within each of four ecoregions in the Lesser Prairie-Chicken range. The four ecoregions include (1) Mixed-Grass Prairie of south-central Kansas, northwest Oklahoma, and northeast Texas Panhandle; (2) Sand Sagebrush Prairie of southwest Kansas, southeast Colorado, and the Oklahoma Panhandle; (3) Sand Shinnery Oak Prairie of eastern New Mexico and the western South Plains of Texas (Llano Estacado); and (4) Short-Grass Prairie/CRP Mosaic of west-central Kansas. Using helicopter-based surveys conducted in the springs of 2012 and 2013 (McDonald et al. 2014), along with survey data from earlier years, the Range-wide Conservation Plan set a benchmark goal of maintaining an average of 67,000 birds across the species' range during the next decade. Subpopulation goals of 24,000 birds in the Mixed-Grass Prairie, 10,000 birds in the Sand Sagebrush Prairie, 8,000 in the Sand Shinnery Oak Prairie, and 25,000 birds in the Short-Grass Prairie/CRP Mosaic ecoregions were also established.

The Range-wide Conservation Plan established a voluntary, incentive-based mechanism to encourage improved habitat management practices on private land and discourage industrial development within the focal areas and connecting corridors. The Range-wide Conservation Plan maintains that much of the financial responsibility for conserving what remains of Lesser Prairie-Chicken habitat should be borne by the energy industries whose activities threaten the species. To make that goal possible, the Range-wide Conservation Plan created an in-lieu fee program, administered by the states and WAFWA, intended to provide a convenient means for industry to finance appropriate off-site mitigation practices if development in focal areas and corridors proves to be unavoidable. Public input for the Range-wide Conservation Plan was provided at numerous informational meetings that were conducted throughout the range of the species to convey the need for, and mechanisms of, the plan.

All of the programs and initiatives that have been crafted to benefit Lesser Prairie-Chickens have depended on the cumulative knowledge gained from research conducted on the species. Universities within the five states have particularly stepped up to meet the challenge of filling critical gaps in our knowledge since the petition to list the species was submitted in 1995. A listing of such studies is not possible here, but many have already been cited or will be cited elsewhere in this volume. Much valuable research has recently been conducted by scientists associated with the University of Oklahoma's G. M. Sutton Avian Research Center, Texas Tech University, Oklahoma State University, Kansas State University, Colorado State University, New Mexico State University, Texas A&M University, the University of New Mexico, Auburn University, and Fort Hays State University. Valuable knowledge has also been gained from studies by state biologists and through funding of independent research by the Grasslans Charitable Foundation in New Mexico. In addition to research, universities have provided expertise in conveying to landowners and the general public information to increase appreciation of Lesser Prairie-Chickens, understand their needs, and maintain their remaining habitat. Extension specialists with Oklahoma State University have particularly excelled in this regard. The work of a variety of GIS specialists has also been critical in identifying threats, providing resources that developers can use to minimize their potential impacts on the Lesser Prairie-Chicken, and in selecting where the best remaining opportunities for conservation of the species occur.

Many actions of NGOs have already been cited, but the importance of nongovernment conservation advocacy and educational efforts for Lesser Prairie-Chicken conservation should not be underestimated. Their important work has been conducted at all levels, on ranches, and in the halls of Congress. The Nature Conservancy and the Playa Lakes Joint Venture have done much to move the wind power and electrical transmission industries toward appropriate location of their infrastructure. The Playa Lakes Joint Venture, the North American Grouse Partnership, the Wildlife Management Institute, and the Teddy Roosevelt Conservation Partnership, among others, have worked with politicians and USDA officials in Washington, DC. to positively impact policies and obtain needed funding. Organizations such as the Rocky Mountain Bird Observatory and the Environmental Defense Fund have organized outreach activities and provided input into the development and implementation of conservation programs. Pheasants Forever has placed biologists in strategically located Natural Resources Conservation Service offices to help facilitate the delivery of valuable Farm Bill programs and practices, particularly the Lesser Prairie-Chicken Initiative.

Though much has been accomplished toward conservation of the Lesser Prairie-Chicken, a mix of daunting opposition and hopeful cooperation clouds the future. Elements within the oil and gas industry and a loosely knit coalition of county governments in Kansas have chosen to actively campaign against even the voluntary conservation mechanisms proposed in the Range-wide Conservation Plan. In contrast, utilities such as Oklahoma Gas and Electric, Westar Energy in Kansas, and Clean Line Energy Partners (based in Texas) have demonstrated a clear desire to minimize or mitigate potential negative impacts of new infrastructure. A small start-up company known as Common Ground Capital is even seeking to harness industrial financing in the establishment of permanent conservation easements for large ranches with high-quality habitat for Lesser Prairie-Chickens.

Overshadowing all of the current initiatives is the specter of climate change. Years of extreme drought and high temperatures, coupled with ongoing habitat loss, have driven down the breeding population of the Lesser Prairie-Chicken to ~17,600 birds by the spring of 2013, likely fewer than remained after the ecological disaster of the Dust Bowl in the 1930s. With the potential for the climate on the Southern and Central Plains to become hotter, drier, and even more unstable, and with substantial agricultural barriers that could prevent the species from shifting northward, it is clear that efforts to conserve Lesser Prairie-Chickens will remain a priority for the foreseeable future (Chapter 12, this volume).

In May 2014, citing recent population declines and continued threats to habitat quantity and quality, the U.S. Fish and Wildlife Service listed Lesser Prairie-Chickens as threatened under provisions of the Endangered Species Act (USFWS 2014). The listing became operational on May 12, 2014, forever changing conservation and management of Lesser Prairie-Chickens. Unfortunately, additional uncertainty for future conservation and management of the species was created when the listing decision was vacated by judicial decision in September 2015.

ACKNOWLEDGMENTS

I offer special thanks to D. Swepston, who went to a great deal of effort to locate and forward much valuable information used in this chapter, and to D. Haukos, who provided guidance, patient encouragement, and much reference material. G. Miller, P. McDaniel, D. Davis, D. Wolfe, H. Whitlaw, K. Giesen, R. Horton, and G. Beauprez also were extremely helpful in locating and providing historical information. Many others provided information regarding specific efforts to conserve Lesser Prairie-Chickens. I also thank S. Manes, B. Crouch, H. Parman, K. Mote, D. Rideout, T. Riley, M. Bain, C. Hise, S. Gallagher, T. Toombs, D. Elmore, C. Kowaleski, A. Chappell, M. Smith, G. Kramos, J. Hughes, K. Fitzgerald, J. Ungerer, R. Winkler, M. Houts, W. Heck, J. Pitman, N. Silvy, and C. Nichols for their valuable help.

LITERATURE CITED

Aldrich, J. W. 1963. Geographic orientation of American Tetraonidae. Journal of Wildlife Management 27:529–545.

Archer, S. 1994. Woody plant encroachment into southwestern grasslands and savannas: rates, patterns, and proximate causes. Pp. 13–68 in M. Varva, W. A. Laycock, and R. D. Pieper (editors), Ecological implications of livestock herbivory in the west. Society for Range Management, Denver, CO.

Ashworth, W. 2006. Ogallala blue: water and life on the high plains. The Countryman Press, Woodstock, VT.

Bailey, A. M., and R. J. Niedrach. 1965. Birds of Colorado, Vol. 1. Denver Museum of Natural History, Denver, CO.

Baker, M. F. 1953. Prairie Chickens of Kansas. University of Kansas Museum of Natural History and Biological Survey of Kansas, Miscellaneous Publication 5. University of Kansas Museum of Natural History and Biological Survey of Kansas, Lawrence, KS.

Bell, L. A., S. D. Fuhlendorf, M. A. Patten, D. H. Wolfe, and S. K. Sherrod. 2010. Lesser Prairie-Chicken hen and brood habitat use on sand shinnery oak. Rangeland Ecology and Management 63:478–486.

Bent, A. C. 1932. Life histories of North American gallinaceous birds. U.S. National Museum Bulletin 162. Dover Publications, New York, NY.

Bidwell, T. G., S. Fuhlendorf, B. Gillen, S. Harmon, R. Horton, R. Manes, R. Rodgers, S. Sherrod, and D. Wolfe. 2003. Ecology and management of the Lesser Prairie-Chicken in Oklahoma. Oklahoma State University Extension Circular E-970. Oklahoma Cooperative Extension Unit, Stillwater, OK.

Braun, C. E., O. O. Oedekoven, and C. L. Aldridge. 2002. Oil and gas development in western North America: effect on sagebrush steppe avifauna with particular emphasis on Sage Grouse. Transactions of the North American Wildlife and Natural Resources Conference 67:337–349.

Copelin, F. F. 1963. The Lesser Prairie Chicken in Oklahoma. Oklahoma Wildlife Conservation Department Technical Bulletin 6. Oklahoma Wildlife Conservation Department, Oklahoma City, OK.

Crawford, J. A. 1980. Status, problems, and research needs of the Lesser Prairie Chicken. Pp. 1–7 in P. A. Vohs and F. L. Knopf (editors), Proceedings of the Prairie Grouse Symposium. Oklahoma State University, Stillwater, OK.

Crawford, J. A., and E. G. Bolen. 1976. Effects of lek disturbance on Lesser Prairie Chickens. Southwestern Naturalist 21:238–240.

Davis, D. M., R. E. Horton, E. A. Odell, R. D. Rodgers, and H. A. Whitlaw. [online]. 2008. Lesser Prairie-Chicken Conservation Initiative. Lesser Prairie-Chicken Interstate Working Group, Colorado Division of Wildlife, Fort Collins, CO. <http://www.wafwa.org/documents/LPCCI_FINAL.pdf> (May 2008).

Davison, V. E. 1940. An 8-year census of Lesser Prairie Chickens. Journal of Wildlife Management 4:55–62.

Doerr, T. B., and F. S. Guthery. 1983. Effects of tebuthiuron on Lesser Prairie-Chicken habitat and foods. Journal of Wildlife Management 47:1138–1142.

Duck, L. G., and J. B. Fletcher. 1944. A Survey of the Game and Furbearing Animals of Oklahoma. Oklahoma Game and Fish Commission Bulletin 3. Oklahoma Game and Fish Commission, Oklahoma City, OK.

Erpelding, R., R. O. Kimmel, and D. J. Lockman. 1987. Foods and feeding behavior of young Gray Partridge in Minnesota. Pp. 17–30 in R. O. Kimmel, J. W. Schultz, and G. J. Mitchell (editors), Perdix IV: gray Partridge Workshop, September 29–October 4, 1986, Regina, SK, Canada. Minnesota Department of Natural Resources, Madelia, MN.

Fields, T. L., G. C. White, W. C. Gilgert, and R. D. Rodgers. 2006. Nest and brood survival of Lesser Prairie-Chickens in west central Kansas. Journal of Wildlife Management 70:931–938.

Fisher, H. I. 1951. The avifauna of Niihau Island, Hawaiian Archipelago. Condor 53:31–42.

Fleharty, E. D. 1995. Wild animals and settlers on the great plains. University of Oklahoma Press, Norman, OK.

Fuhlendorf, S. D., F. E. Smeins, and W. E. Grant. 1996. Simulation of a fire-sensitive ecological threshold: a case study of Ashe juniper on the Edwards Plateau of Texas, USA. Ecological Modeling 90:245–255.

Fuhlendorf, S. D., A. J. W. Woodward, D. M. Leslie, Jr., and J. S. Shackford. 2002. Multi-scale effects of habitat loss and fragmentation on Lesser Prairie-Chicken populations of the US Southern Great Plains. Landscape Ecology 17:617–628.

Galvin, J. (editor). 1970. Through the country of the Comanche Indians in the fall of the year 1845 (1845 diary of J. W. Abert). John Howell Books, San Francisco, CA.

Giesen, K. M. 2000. Population status and management of Lesser Prairie-Chicken in Colorado. Prairie Naturalist 32:137–148.

Hartman, J., and M. McDonald. 1988. The cornerstone of Kansas. Kansas Wildlife and Parks Magazine 45:9–15.

Haukos, D. A. [online]. 2011. Use of tebuthiuron to restore sand shinnery oak grasslands of the Southern High Plains. Pp. 103–124 in M. Naguib and A. El-Ghany Hasaneen (editors), Herbicides—mechanisms and mode of action. InTech, Rijeka, Croatia <http://www.intechopen.com/books/herbicides-mechanisms-and-mode-of-action/use-of-tebuthiuron-to-restore-sand-shinnery-oak-grasslands-of-the-southern-high-plains> (22 December 2013).

Hill, D. A. 1985. The feeding ecology and survival of pheasant chicks on arable farmland. Journal of Applied Ecology 22:645–654.

Hill, J., S. Polasky, E. Nelson, D. Tilman, H. Huo, L. Ludwig, J. Neumann, H. Zheng, and D. Bonta. 2008. Climate change and health costs of air emissions from biofuels and gasoline. Proceedings of National Academy of Sciences 106:2077–2082.

Horton, R. E. 2000. Distribution and abundance of Lesser Prairie-Chickens in Oklahoma. Prairie Naturalist 32:189–195.

Hubbard, J. P., C. M. Milensky, and C. Dove. Emended type locality and historic status of the Lesser Prairie-Chicken. Unpublished manuscript, Division of Birds, Museum of Southwestern Biology, University of New Mexico, Albuquerque, NM.

Jackson, A. S., and R. DeArment. 1963. The Lesser Prairie Chicken in the Texas Panhandle. Journal of Wildlife Management 27:733–737.

Judd, S. D. 1905. The grouse and Wild Turkeys of the United States and their economic values. USDA, Biological Survey Bulletin 24. U.S. Department of Agriculture, Washington, DC.

Lee, L. 1950. Kill analysis for the Lesser Prairie Chicken in New Mexico, 1949. Journal of Wildlife Management 14:475–477.

Ligon, J. S. 1927. Wildlife in New Mexico: its conservation and management. New Mexico State Game Commission, Santa Fe, NM.

Lincoln, F. C. 1918. Notes on some species new to the Colorado list of birds. Auk 35:236–237.

Litton, G. W. 1978. The Lesser Prairie Chicken and its management in Texas. Texas Parks and Wildlife Department Booklet 7000-25. Texas Parks and Wildlife Department, Austin, TX.

Manes, R., S. A. Harmon, B. K. Overseer, and R. D. Applegate. March 19–20, 2004. Wind energy and wildlife in the Great Plains: identification of concerns and ways to alleviate them. Proceedings of the Great Plains Wind Power and Wildlife Workshop, Kansas City, MO.

Marcy, R. B. 1866. Thirty years of army life on the border. Harper & Brothers, New York, NY.

McDonald, L., G. Beauprez, G. Gardner, J. Griswold, C. Hagen, D. Klute, S. Kyle, J. Pitman, T. Rintz, and B. Van Pelt. 2014. Range-wide population size of the Lesser Prairie-Chicken: 2012 and 2013. Wildlife Society Bulletin 38:536–546.

Mote, K. D., R. D. Applegate, J. A. Bailey, K. M. Giesen, R. Horton, and J. L. Sheppard. 1999. Assessment and conservation strategy for the Lesser Prairie-Chicken (Tympanuchus pallidicinctus). Kansas Department of Wildlife and Parks, Emporia, KS.

Oberholzer, H. C. 1974. The bird life of Texas, Vol. 1. University of Texas Press, Austin, TX.

Olawsky, C. D., and L. M. Smith. 1991. Lesser Prairie-Chicken densities on tebuthiuron-treated and untreated sand shinnery oak rangelands. Journal of Range Management 44:364–368.

Patten, M. A., D. H. Wolfe, E. Shochat, and S. K. Sherrod. 2005. Habitat fragmentation, rapid evolution, and population persistence. Evolutionary Ecology Research 7:1–15.

Patzek, T. W. 2004. Thermodynamics of the corn–ethanol biofuel cycle. Critical Reviews in Plant Sciences 23:519–567.

Pitman, J. C., C. A. Hagen, R. J. Robel, T. M. Loughlin, and R. D. Applegate. 2005. Location and success of Lesser Prairie-Chicken nests in relation to vegetation and human disturbance. Journal of Wildlife Management 69:1259–1269.

Pope, J. 1854. Report of the exploration of a route for the Pacific railroad near the thirty-second parallel of north latitude from the Red River to the Rio Grande. Report to Jefferson Davis, Secretary of War, Washington, DC.

Pratt, H. D., P. L. Bruner, and D. G. Berrett. 1987. A field guide to the birds of Hawaii and the tropical Pacific. Princeton University Press, Princeton, NJ.

Pyle, R. L., and P. Pyle. [online]. 2009. The birds of the Hawaiian Islands: occurrence, history, distribution, and status. B.P. Bishop Museum, Honolulu, HI. Version 1. <http://hbs.bishopmuseum.org/birds/rlp-monograph/> (31 December 2013).

Rands, M. R. W. 1985. Pesticide use on cereals and the survival of Grey Partridge chicks: a field experiment. Journal of Applied Ecology 22:49–54.

Ridgway, R. 1885. Some emended names of North American birds. Proceedings of the U.S. National Museum 3:354–356.

Rodgers, R. D. 1999. Recent expansion of Lesser Prairie-Chickens to the northern margin of their historic range. Proceedings of the Prairie Grouse Technical Council Meeting 23:18–19.

Rodgers, R. D. 2003. Tree invasion. Kansas Wildlife and Parks 60:17–24.

Rodgers, R. D., and R. W. Hoffman. 2005. Prairie grouse population response to Conservation Reserve Program grasslands: an overview. Pp. 120–128 in A. W. Allen and M. W. Vandever (editors), The Conservation Reserve Program—Planting for the Future: proceedings of a National Conference, June 6, 2004, Fort Collins, CO. USGS Biological Resources Division, Scientific Investigation Report 2005-5145, Reston, VA.

Rodgers, R. D., and M. L. Sexson. 1990. Impacts of extensive chemical control of sand sagebrush on breeding birds. Journal of Soil and Water Conservation 45:494–497.

Sands, J. L. 1968. Status of the Lesser Prairie Chicken. Audubon Field Notes 22:454–456.

Schwilling, M. D. 1955. A Study of the Lesser Prairie Chicken in Kansas. Job Completion Report. Kansas Forestry, Fish, and Game Commission, Pratt, KS.

Snyder, W. A. 1967. Lesser Prairie Chicken. Pp. 121–128 in W. S. Huey (editor), New Mexico wildlife management. New Mexico Department of Game and Fish, Santa Fe, NM.

Sullivan, R. M., J. P. Hughes, and J. E. Lionberger. 2000. Review of the historical and present status of the Lesser Prairie-Chicken (*Tympanuchus pallidicinctus*) in Texas. Prairie Naturalist 32:177–188.

Summars, V. C. 1956. Lesser Prairie-Chicken census. Federal Aid in Wildlife Restoration Project W-062-R-1. Oklahoma Game and Fish Department, Oklahoma City, OK.

Sutton, G. M. 1967. Oklahoma Birds. University of Oklahoma Press, Norman, OK.

Taylor, M. A., and F. S. Guthery. 1980. Status, ecology, and management of the Lesser Prairie Chicken. USDA Forest Service General Technical Report RM-77. USDA Forest Service, Rocky Mountain Forest and Range Experiment Station, Fort Collins, CO.

Tomer, J. A., and M. J. Brodhead (editors). 1996. A naturalist in Indian territory (1849–1850 diary of S. W. Woodhouse). University of Oklahoma Press, Norman, OK.

U.S. Fish and Wildlife Service (USFWS). 2014. Endangered and threatened wildlife and plants; special rule for the Lesser Prairie-Chicken. Federal Register 79:20074–20085.

Van Pelt, W. E., S. Kyle, J. Pitman, D. Klute, G. Beauprez, D. Schoeling, A. Janus, and J. Haufler. 2013. The Lesser Prairie-Chicken range-wide conservation plan. Western Association of Fish and Wildlife Agencies, Cheyenne, WY.

Waddell, B., and B. Hanzlick. 1978. The vanishing sandsage prairie. Kansas Fish and Game 35:17–23.

Wolfe, D. H., M. A. Patten, E. Shochat, C. L. Pruett, and S. K. Sherrod. 2007. Causes and patterns of mortality in Lesser Prairie-Chickens *Tympanuchus pallidicinctus* and implications for management. Wildlife Biology 13(Supplement 1):95–104.

CHAPTER THREE

Legal Status of the Lesser Prairie-Chicken*

William E. Van Pelt

Abstract. The Lesser Prairie-Chicken (*Tympanuchus pallidicinctus*) is a nonmigratory avian species, traditionally managed by individual states as an upland game bird. Concern for the species' conservation has been longstanding, with state management occurring as early as 1861 in Kansas, and recommendation for hunting season closure in Oklahoma in 1912. Population management efforts through harvest regulations, land management, and land purchases were initiated by state wildlife management agencies as early as the 1940s. Regardless of conservation efforts, populations and distributions of Lesser Prairie-Chickens have diminished across their historical range, with recent estimates of only 17% of their historical range being currently occupied. Population declines and range contractions have been primarily attributed to habitat loss, degradation, and fragmentation due to anthropogenic activities and the loss of natural ecological drivers. In 1995, the U.S. Fish and Wildlife Service (USFWS) was initially petitioned to provide federal protection to the Lesser Prairie-Chicken. In 2008, the species listing priority was increased by the USFWS based upon their perception of increased threats caused by energy development. In response, five core states within the range of Lesser Prairie-Chickens developed and submitted a conservation plan to the USFWS, with specific population and habitat objectives to increase and conserve range-wide populations of Lesser Prairie-Chickens over a decade. Continuing population declines resulted in a determination by the USFWS on March 27, 2014, that the Lesser Prairie-Chicken would receive federal protection as a threatened species under the Endangered Species Act of 1973. Included in the listing was a 4(d) special rule recognizing the significant conservation planning efforts made by state and federal wildlife agencies within the range of the Lesser Prairie-Chicken. The 4(d) special rule provides that take will not be prohibited if it is (1) incidental to activities conducted by a participant enrolled in, and operating in compliance with, the Lesser Prairie-Chicken Interstate Working Group's Lesser Prairie-Chicken Range-wide Conservation Plan; (2) incidental to the conditioned conservation practices carried out in accordance with a conservation plan developed by Natural Resources Conservation Service in connection with the Lesser Prairie-Chicken

* Van Pelt, W. E. 2016. Legal status of the Lesser Prairie-Chicken. Pp. 39–46 in D. A. Haukos and C. W. Boal (editors), Ecology and conservation of Lesser Prairie-Chickens. Studies in Avian Biology (no. 48), CRC Press, Boca Raton, FL.

Initiative; or (3) incidental to activities conducted during the continuation of routine agricultural practices on cultivated lands that are in row crop, seed-drilled untilled crop, hay, or forage production. However, the listing of the Lesser Prairie-Chicken was vacated by a judicial decision in September 2015. If conservation of the Lesser Prairie-Chicken can facilitate long-term recovery, a strong

and mutually respective partnership will be necessary between states, federal, and nongovernmental conservation entities and private landowners. The goal of my chapter is to examine the current status of the species at both a state and a national level.

Key Words: authority, permit, regulation, regulatory, rule, state-listed, statute, threatened.

Lesser Prairie-Chickens (*Tympanuchus pallidicinctus*) and their associated habitats have diminished across their historical range since the 19th century (Crawford and Bolen 1976, Taylor and Guthery 1980a). Recent estimates of current occupied range total ~80,000 km^2 (30,900 mile^2), or ~17% of the estimated area of their historical range (Figure 1.2), although boundaries of this estimated range include many areas that are unlikely to be occupied, including riparian corridors, forests, and desert. The reduction in occupied range has been primarily attributed to habitat loss and fragmentation (U.S. Fish and Wildlife Service 2012). Two primary elements have contributed to habitat losses. In the western portion of the range, habitat loss is associated with conversion of native prairie to cropland (Bent 1932, Copelin 1963, Jackson and DeArment 1963, Crawford and Bolen 1976, Taylor and Guthery 1980b), and in the eastern part of the range, habitat loss is attributed to tree invasion (Woodward et al. 2001, Fuhlendorf et al. 2002). Degradation of remaining habitat has been due to fire suppression (Woodward et al. 2001, Jones 2009), grazing management practices (Jackson and DeArment 1963, Taylor and Guthery 1980a, Riley et al. 1992), and herbicide spraying for shrub control, all of which can reduce the quality of Lesser Prairie-Chicken habitat (Jackson and DeArment 1963, Peterson and Boyd 1998, Thacker et al. 2012; Chapter 6, this volume). In addition to habitat loss and degradation factors, existing habitat has been fragmented by oil and gas development (Hunt 2004), and possibly by effects of wind energy development (Pruett et al. 2009). In addition, Lesser Prairie-Chicken populations have been influenced by collision mortalities with fences and utility lines (Wolfe et al. 2007, Hagen 2010), prolonged drought (Merchant 1982, Lyons et al. 2011, Grisham 2012), and climate change (U.S. Department of Agriculture,

Natural Resources Conservation Service 2012, U.S. Fish and Wildlife Service 2012). Specifics related to these impacts are provided throughout this volume. However, it is important to review how regulatory measures have contributed to the legal status of the species from both state and federal perspectives.

As a resident species of upland game bird, Lesser Prairie-Chickens have been historically managed as a state trust species, with each state responsible for local populations of the species within state boundaries. States have independently opened and closed hunting seasons based on anecdotal observations and perceptions of relative abundance without quantitative estimates based on systematic population surveys and trends (Chapter 6, this volume). For example, in 1912, the Oklahoma State Game and Fish Warden Report recommended a season closure of 4–5 years "as an experiment to determine whether or not it is possible for this bird to increase in numbers or survive in a settled country" (Doolin 1912:157–159). Historically, collaboration among states for management has been limited and was primarily for the purpose of translocation of individual birds (Chapter 2, this volume). In addition to managing population through setting hunting seasons, states also began to purchase and manage lands by controlling grazing, by supplemental planting of food plants used by Lesser Prairie-Chickens, and by restricting harvest of Lesser Prairie-Chickens as early as the 1940s (U.S. Department of the Interior 1966). Ultimately, however, the primary role of state wildlife agencies in relation to Lesser Prairie-Chickens has been to regulate take. To manage harvest and incidental take, a variety of regulatory measures and authorities exist in each state where the species occurs.

In 1966, the Bureau of Sports Fisheries and Wildlife listed the Lesser Prairie-Chicken as rare

(U.S. Department of Interior 1966), which was defined as "a rare species or subspecies is one that, although not presently threatened with extinction, is in such small numbers throughout its range that it may be endangered if its environment worsens. Close watch of its status is necessary." The agency attributed low numbers of birds to poor range conditions due to the severe drought and poor grazing practices. The report also stated that Lesser Prairie-Chickens were "almost extinct in the 1930s and populations fluctuate markedly." Although considered, Lesser Prairie-Chickens were not included in the first list of species receiving federal protection under the Endangered Species Act in 1974. Following a formal petition to the U.S. Fish and Wildlife Service (USFWS) for federal protection in 1995 and subsequent reviews, the Lesser Prairie-Chicken received federal protection in 2014 as a threatened species under the Endangered Species Act, with 4(d) rules (U.S. Fish and Wildlife Service 1997, 2014). However, the listing of the Lesser Prairie-Chicken was vacated by a judicial decision in September 2015. In the following, I provide an overview of regulatory measures and authorities for five states in the Great Plains and then describe the current status under the federal listing and rules.

STATE STATUS OF LESSER PRAIRIE-CHICKENS

Colorado

Colorado Parks and Wildlife is responsible for the management and conservation of wildlife resources, including conservation and management of state-listed threatened, endangered, and species of concern, within state borders as defined and directed by state laws (i.e., Colorado Revised Statutes, Title 33, Article 1). Under Title 33 Article 1-101, the legislative declaration states: "It is the policy of the State of Colorado that the wildlife and their environment are to be protected, preserved, enhanced and managed for the use, benefit, and enjoyment of the people of this state and its visitors. It is further declared to be the policy of this state that there shall be provided a comprehensive program designed to offer the greatest possible variety of wildlife-related recreational opportunity to the people of this state and its visitors and that, to carry out such program and policy, there shall be a continuous operation of planning, acquisition, and development of wildlife

habitats and facilities for wildlife-related opportunities." The Lesser Prairie-Chicken was listed as a state threatened species in 1973 and has not been hunted in Colorado since 1937 (Chapter 6, this volume).

Oil and gas well permits are issued by the Colorado Oil and Gas Conservation Commission (COGCC). As of April 2009, the 1200 series of COGCC rules address oil and gas development threats to the Lesser Prairie-Chicken. The rules require producers to use online resources to identify sensitive wildlife habitat and areas of restricted surface occupancy. Currently, sensitive habitats for Lesser Prairie-Chickens are defined as production areas that include 80% of the nesting and brood-rearing habitat surrounding leks that have been active at least once in the preceding 10 years. Restricted surface occupancy areas for Lesser Prairie-Chickens are defined as areas within 1.0 km (0.6 mile) of leks that have been active at least once in the preceding 10 years. Under COGCC rules, potential oil and gas wells identified within the priority areas have mandated consultation with Colorado Parks and Wildlife, where best management practices are provided to industry to minimize impacts to Lesser Prairie-Chickens.

Kansas

The Kansas Department of Wildlife, Parks and Tourism (KDWPT) manages Lesser Prairie-Chickens under the authorities in Kansas Statutes Annotated (KSA) 32-702, which states: "It shall be the policy of the state of Kansas to protect, provide and improve outdoor recreation and natural resources in this state and to provide for the wise management and use of the state's natural resources, thus contributing to and benefiting the public's health and its cultural, recreational and economic life. For these purposes, the secretary, the commission and the department are hereby vested with the duties and powers hereinafter set forth."

Harvest of Lesser Prairie-Chickens is regulated in Kansas through bag limits and seasons (Chapter 6, this volume). Recent population studies in southeast Kansas and agency monitoring have indicated that harvest is an insignificant source of mortality in Kansas (Fields 2004; Hagen et al. 2006, 2007; J. Pitman, KDWPT, unpubl. data), and removing harvest as a source of mortality will not result in a significant increase in population growth (Hagen et al. 2009). The KDWPT closed the harvest of Lesser Prairie-Chickens in 2014 as a result of the federal listing decision (kdwpt. state.ks.us/news/Hunting/Hunting-Regulations/ Maps/Greater-Prairie-Chicken-Unit-Map).

In 2009, KDWPT was petitioned to list the Lesser Prairie-Chicken as either a threatened or endangered species at the state level. The Threatened and Endangered Task Committee reviewed the petition and recommended not listing the species at that time; the recommendation was supported by the Wildlife and Parks Commission.

The KDWPT also has regulatory authority over some developments pursuant to KSA 32-957 to 963, 32-1009 to 1012, and 32-1033 of the Kansas Nongame and Endangered Species Conservation Act. The KDWPT conducts environmental reviews and permits activities that are publicly funded or require some other type of state or federal permit. The KDWPT requires mitigation when those reviews indicated expected impacts to state-listed species. The Lesser Prairie-Chicken is not currently a state-listed species in Kansas, but it shares similar habitats with long-nose snakes (*Rhinocheilus lecontei*) as a state-listed species of reptile in a substantial portion of its range, including areas south of the Arkansas River. Thus, the Lesser Prairie-Chicken is being provided with indirect protections in those areas through the Kansas Nongame and Endangered Species Conservation Act.

New Mexico

The New Mexico Department of Game and Fish (NMDGF) manages Lesser Prairie-Chickens under the statutory authority of the Wildlife Conservation Act (Chapters 17-2-37 through 17-2-46) of New Mexico Statutes Annotated 1978: "It is the purpose of this act and the policy of the state of New Mexico to provide an adequate and flexible system for the protection of the game and fish of New Mexico and for their use and development for public recreation and food supply, and to provide for their propagation, planting, protection, regulation and conservation to the extent necessary to provide and maintain an adequate supply of game and fish within the state of New Mexico." Hunting seasons for Lesser Prairie-Chicken have been closed in New Mexico since 1996 (Chapter 6, this volume).

In 1997, NMDGF was petitioned to investigate the status of the Lesser Prairie-Chicken for listing as either a state threatened or endangered species.

In 2006, the NMDGF found that the prospects for survival and recruitment of the Lesser Prairie-Chicken were not jeopardized to a degree that warranted a change in classification as threatened or endangered under the Wildlife Conservation Act. The NMDGF recommendation regarding the Lesser Prairie-Chicken investigation in response to the listing was brought before the State Game Commission in November 2006. The motion to accept the Final Listing Investigation Report and recommendation that the Lesser Prairie-Chicken not be listed under the Wildlife Conservation Action was carried unanimously.

Oklahoma

The Oklahoma Department of Wildlife Conservation (ODWC) manages Lesser Prairie-Chickens under the authority provided by the Oklahoma Wildlife Conservation Code, Title 29, Oklahoma Statutes, §29-3-101. The ODWC is governed by the Wildlife Conservation Commission that has defined functions, powers, and duties within Title 29, Oklahoma Statutes, §29-3-103, A., which states: "The Wildlife Conservation Commission shall constitute an advisory, administrative and policymaking board for the protection, restoration, perpetuation, conservation, supervision, maintenance, enhancement, and management of wildlife in this state as provided in the Oklahoma Wildlife Conservation Code...." The mission of the ODWC is to manage Oklahoma's wildlife resources and habitat to provide scientific, educational, aesthetic, economic, and recreational benefits for present and future generations of hunters, anglers, and others who appreciate wildlife. The Lesser Prairie-Chicken is considered an upland game bird in Oklahoma, but there has not been an open season since 1997 (Chapter 6, this volume).

Texas

The Texas Parks and Wildlife Department (TPWD) manages Lesser Prairie-Chickens under the authority of the Texas Statute Parks and Wildlife Code, Title 2 Chapters 11 and 12, and PWC Title 5 Chapters 61 and 64. The code states that "The mission of TPWD is to manage and conserve the natural and cultural resources of Texas and to provide hunting, fishing, and outdoor recreation opportunities for the use and enjoyment of present and future generations." Texas Statute, Parks and Wildlife Code Title 2; Chapter 12 subchapter A Sec. 12.011 establishes the responsibility for protecting the state's fish and wildlife resources, which authorizes TPWD to provide comments and recommendations on projects to minimize impacts to fish and wildlife. Lesser Prairie-Chickens are considered an upland game bird in Texas, but harvest and hunting seasons for the species were suspended indefinitely in 2009 (Chapter 6, this volume).

HISTORICAL FEDERAL ACTIONS

Due to ongoing declines in population numbers of Lesser Prairie-Chickens, range contractions relative to their historical occurrence, and presumed increases in the scope and intensity of identified impacts, the U.S. Fish and Wildlife Service was petitioned to list the Lesser Prairie-Chicken as threatened by the Biodiversity Legal Foundation in 1995 (USFWS 1997). After review, the USFWS issued a finding in 1998 that the species warranted listing, but was precluded because of actions needed for other higher priority species (USFWS 2012). The USFWS assigned the Lesser Prairie-Chicken a listing priority number of 8 on a 12-point scale where 1 indicates the highest need for action and 12 is the lowest risk. The listing was revised to a priority number 2 in 2008 because the USFWS considered the threat of wind development and associated development of transmission lines within the occupied range had increased significantly since the previous analysis (USFWS 2012). On December 11, 2012, the USFWS expressed concerns that a number of existing and expanding threats are currently outside of the regulatory authority of the states to control, and proposed listing the Lesser Prairie-Chicken as threatened with a final listing decision scheduled for no later than September 30, 2013 (USFWS 2012). Publication of the proposed rule opened a 90-day comment period that closed on March 11, 2013.

Public comments received by the USFWS during the comment period expressed concerns regarding the sufficiency and accuracy of data related to the listing proposal for the species and the positive impacts of conservation programs on Lesser Prairie-Chicken populations. Conservation initiatives included state and federal programs that had enrolled millions of acres in partner programs such as the Natural Resource Conservation Service's Lesser Prairie-Chicken Initiative. Public comments indicated that some Lesser Prairie-Chicken populations were stable, but identified a need for concerted efforts to address the declines in other ecoregions (Chapters 14 to 17, this volume).

On May 6, 2013, the USFWS announced the publication of a proposed special rule under the authority of section 4(d) of the Endangered Species Act. A comment period on the proposed listing rule was opened to provide an opportunity for the public to simultaneously provide comments on the proposed listing rule with a proposed special rule, and a draft range-wide conservation plan for the Lesser Prairie-Chicken prepared by the five state wildlife agencies in collaboration with the Western Association of Fish and Wildlife Agencies. The comment period was open for about a month from May 6 to June 20, 2013.

On July 9, 2013, the USFWS announced a 6-month extension of the final listing determination based on their finding of substantial disagreement regarding the sufficiency or accuracy of the available data relevant to their determination regarding the proposed listing rule. The Service reopened the comment period to solicit additional information. The second comment period closed on August 8, 2013.

On December 11, 2013, the Service reopened the comment period, to solicit comments on a revised proposed special 4(d) rule and the December 11, 2012, proposed listing rule as a result of endorsing the Western Association of Fish and Wildlife Agencies' Lesser Prairie-Chicken Range-wide Conservation Plan. The third comment period closed on January 10, 2014. However, the endorsed version of the Western Association of Fish and Wildlife Agencies' Lesser Prairie-Chicken Range-wide Conservation Plan (Van Pelt et al. 2013) was not available on the USFWS websites, as stated in the December 11, 2013, revised proposed special 4(d) rule. Subsequently, the USFWS reopened the comment period on January 29, 2014, to allow the public the opportunity to have access to this range-wide plan and submit comments on the revised proposed special rule and the December 11, 2012, proposed listing rule. The final comment period closed on February 12, 2014.

CURRENT STATUS

On March 27, 2014, the USFWS announced the listing determination of threatened species status for the Lesser Prairie-Chicken under the Endangered

Species Act of 1973, as amended (Act) (U.S. Fish and Wildlife Service 2014). The final rule implemented the federal protections provided by the Act for the Lesser Prairie-Chicken. Critical habitat is prudent but not determinable at the time of listing. In addition, the USFWS published a final special rule under section 4(d) of the Act for the Lesser Prairie-Chicken. Under section 4(d) of the Act, the Secretary of the Interior may publish a special rule that modifies the standard protections for threatened species with special measures tailored to the conservation of the species that are determined to be necessary and advisable. The 4(d) special rule does not remove or alter in any way the consultation requirements under section 7 of the Act. Under the 4(d) special rule, the Service provides that all of the prohibitions under 50 CFR 17.31 and 17.32 will apply to the Lesser Prairie-Chicken, except those noted in the rule itself. The final 4(d) special rule provides that take is incidental to activities conducted by a participant enrolled in, and operating in compliance with, the Lesser Prairie-Chicken Interstate Working Group's Lesser Prairie-Chicken Range-wide Conservation Plan will not be prohibited. The Service included the provision in the final 4(d) special rule in recognition of the significant conservation planning efforts of the five state wildlife agencies within the range of the Lesser Prairie-Chicken (Van Pelt et al. 2013).

The final 4(d) special rule also stated that take of the Lesser Prairie-Chicken will not be prohibited provided the take is incidental to conditioned conservation practices that are carried out in accordance with a conservation plan or related activities developed by the Natural Resources Conservation Service (NRCS) in connection with the Lesser Prairie-Chicken Initiative. Conservation activities for Lesser Prairie-Chickens needed to be consistent with the provisions of the November 22, 2013, conference opinion that was developed in coordination with the USFWS. Conditioned conservation practices are NRCS standard conservation practices to which the USFWS and NRCS have added specific requirements in the form of conservation measures so that when the measure is followed, impacts to the Lesser Prairie-Chicken will be avoided or minimized.

Last, the final 4(d) special rule determined that take of Lesser Prairie-Chicken will not be prohibited, provided the take is incidental to activities that are conducted during the continuation of routine agricultural practices on cultivated lands that are in row crop, seed-drilled untilled crop, hay,

or forage production. Agricultural lands must meet the definition of cropland as defined in 7 CFR 718.2 and must have been cultivated by tilling, planting, or harvest within the 5-year period preceding the proposed routine agricultural practice that might result in incidental take. Thus, this particular provision does not include coverage for incidental take occurring during any new conversion of grasslands into agriculture. The Lesser Prairie Chicken began receiving full protection under the Endangered Species Act on May 12, 2014.

Shortly after the decision to list the Lesser Prairie-Chicken, five different lawsuits were filed against the listing decision and the process for making the decision. The first case was heard in July 2015 and the listing of the Lesser Prairie-Chicken was vacated by a judicial decision on September 1, 2015. The decision functionally removed the ESA protections for the Lesser Prairie-Chicken until which time the decision is successfully appealed or the USFWS makes another decision warranting listing the species. Because this case has not concluded, and its final action could influence other lawsuits, all other cases are pending. If conservation of the Lesser Prairie-Chicken is to lead to successful recovery in the long term, a strong and mutually respective partnership will be necessary between the state, federal, and nongovernmental agencies and private landowners. The foundation of that partnership is embedded under Section 6 (a–b) of the Endangered Species Act. The relevant section of the Act clearly directs the USFWS to cooperate to the maximum extent practicable with state fish and wildlife agencies, and provides the authority for the USFWS to carry that partnership forward. Agreement on a range-wide conservation plan across five states will provide a solid foundation for future conservation efforts for Lesser Prairie-Chicken in the Great Plains.

ACKNOWLEDGMENTS

I thank the Lesser Prairie-Chicken Interstate Working Group for being committed to conserving the species under their charge by looking outside the box and developing a new paradigm for conserving wildlife by state wildlife agencies. I also thank the directors from the five wildlife agencies for their leadership and stalwart support for drafting a scientifically driven conservation plan. Last, I thank the U.S. Fish and Wildlife Service for their willingness to use the inherent flexibility built into the Endangered Species Act to conserve a species and for maintaining the key role of state wildlife agencies in managing wildlife under their trust.

LITERATURE CITED

Bent, A. C. 1932. Life histories of North American gallinaceous birds. U.S. National Museum Bulletin 162. Dover Publications, New York, NY.

Copelin, F. F. 1963. The Lesser Prairie-Chicken in Oklahoma. Oklahoma Department of Wildlife Technical Bulletin 6. Oklahoma Wildlife Conservation Department, Oklahoma City, OK.

Crawford, J. A., and E. G. Bolen. 1976. Effects of land use on Lesser Prairie-Chickens in Texas. Journal of Wildlife Management 40:96–104.

Doolin, J. B. 1912. Field forest and stream in Oklahoma. Annual report of the state game and fish warden. Co-operative Publishing Co., Gutherie, OK.

Fields, T. L. 2004. Breeding season habitat use of Conservation Reserve Program (CRP) land by Lesser Prairie-Chickens in west central Kansas. M.S. thesis, Colorado State University, Fort Collins, CO.

Fuhlendorf, S. D., A. J. Woodward, D. M. Leslie, Jr., and J. S. Shackford. 2002. Multiscale effects of habitat loss and fragmentation on Lesser Prairie-Chicken populations. Landscape Ecology 17:601–615.

Grisham, B. N. 2012. The ecology of Lesser Prairie-Chickens in shinnery oak–grassland communities in New Mexico and Texas with implications toward habitat management and future climate change. Ph.D. dissertation, Texas Tech University, Lubbock, TX.

Hagen, C. A. 2010. Impacts of energy development on prairie grouse ecology: a research synthesis. Transactions of the 75th North American Wildlife and Natural Resources Conference 75:96–103.

Hagen, C. A., B. A. Grisham, C. W. Boal, and D. A. Haukos. 2006. A meta-analysis of Lesser Prairie-Chicken nesting and brood rearing habitats: implications for habitat. Wildlife Society Bulletin 37:750–758.

Hagen, C. A., J. C. Pitman, B. K. Sandercock, R. J. Robel, and R. D. Applegate. 2007. Age-specific survival and probable causes of mortality in female Lesser Prairie-Chickens. Journal of Wildlife Management 71:518–525.

Hagen, C. A., B. K. Sandercock, J. C. Pitman, R. J. Robel, and R. D. Applegate. 2009. Spatial variation in Lesser Prairie-Chicken demography: a sensitivity analysis of population dynamics and management alternatives. Journal of Wildlife Management 73:1325–1332.

Hunt, J. L. 2004. Investigation into the decline of the Lesser Prairie-Chicken (Tympanuchus pallidicinctus Ridgway) in southeastern New Mexico. Ph.D. dissertation, Auburn University, Auburn, AL.

Jackson, A. S., and R. DeArment. 1963. The Lesser Prairie-Chicken in the Texas Panhandle. Journal of Wildlife Management 27:733–737.

Jones, R. S. 2009. Seasonal survival, reproduction, and use of wildfire areas by Lesser Prairie-Chickens in the northeastern Texas panhandle. M.S. thesis, Texas A&M University, College Station, TX.

Lyons, E. K., R. S. Jones, J. P. Leonard, B. E. Toole, R. A. McCleery, R. R. Lopez, M. J. Peterson, and N. J. Silvy. 2011. Regional variation in nesting success of Lesser Prairie-Chickens. Pp. 223–232 in B. K. Sandercock, K. Martin, and G. Segelbacher (editors), Ecology, conservation, and management of grouse. Studies in Avian Biology (no. 39), University of California Press, Berkeley, CA.

Merchant, S. S. 1982. Habitat-use, reproductive success, and survival of female Lesser Prairie-Chickens in two years of contrasting weather. M.S. thesis, New Mexico State University, Las Cruces, NM.

Peterson, R. S., and C. S. Boyd. 1998. Ecology and management of sand shinnery communities: a literature review. USDA Forest Service, General Technical Report. USDA Forest Service, Rocky Mountain Forest and Range Experiment Station, Fort Collins, CO.

Pruett, C. L., M. A. Patten, and D. H. Wolfe. 2009. Avoidance behavior of prairie grouse: implications for wind and energy development. Conservation Biology 23:1253–1259.

Riley, T. Z., C. A. Davis, M. Ortiz, and M. J. Wisdom. 1992. Vegetative characteristics of successful and unsuccessful nests of Lesser Prairie-Chickens. Journal of Wildlife Management 56:383–387.

Taylor, M. A., and F. S. Guthery. 1980a. Status, ecology, and management of the Lesser Prairie-Chicken. USDA Forest Service, General Technical Report RM-77. USDA Forest Service, Rocky Mountain Forest and Range Experiment Station, Fort Collins, CO.

Taylor, M. A., and F. S. Guthery. 1980b. Fall–winter movements, ranges, and habitat use of Lesser Prairie-Chickens. Journal of Wildlife Management 44:521–524.

Thacker, E. T., R. L. Gillen, S. A. Gunter, and T. L. Springer. 2012. Chemical control of sand sagebrush: implications for Lesser Prairie-Chicken habitat. Rangeland Ecology and Management 65: 516–522.

U.S. Department of Agriculture, Natural Resources Conservation Service. 2012. USDA conservation program contributions to Lesser Prairie-Chicken conservation in the context of projected climate change. CEAP Conservation Insight Conservation Effects Assessment Project. USDA Natural Resources Conservation Service. <http://www.nrcs.usda.gov/Internet/FSE_DOCUMENTS/stelprdb1080286.pdf> (8 August 2014).

U.S. Department of the Interior. 1966. Rare and endangered fish and wildlife of the United States. Resource Publication 34. Compiled by Committee on Rare and Endangered Wildlife Species, Bureau of Sport Fisheries and Wildlife, Washington, DC.

U.S. Fish and Wildlife Service. 1997. Endangered and threatened wildlife and plants; 90-day finding for a petition to list the Lesser Prairie-Chicken as threatened. Federal Register 62:36482–36484.

U.S. Fish and Wildlife Service. 2012. Endangered and threatened wildlife and plants; listing the Lesser Prairie-Chicken as a threatened species. Federal Register 77238:73827–73888.

U.S. Fish and Wildlife Service. 2014. Endangered and threatened wildlife and plants; special rule for the Lesser Prairie-Chicken. Federal Register 79:20074–20085.

Van Pelt, W. E., S. Kyle, J. Pitman, D. Klute, G. Beauprez, D Schoeling, A. Janus, and J. Haufler. 2013. The Lesser Prairie-Chicken range-wide conservation plan. Western Association of Fish and Wildlife Agencies, Cheyenne, WY.

Wolfe, D. H., M. A. Patten, E. Shochat, C. L. Pruett, and S. K. Sherrod. 2007. Causes and patterns of mortality in Lesser Prairie-Chickens *Tympanuchus pallidicinctus* and implications for management. Wildlife Biology 13(Suppl 1):95–104.

Woodward, A. J., S. D. Fuhlendorf, D. M. Leslie, Jr., and J. Shackford. 2001. Influence of landscape composition and change on Lesser Prairie-Chicken (*Tympanuchus pallidicinctus*) populations. American Midland Naturalist 145:261–274.

PART TWO

Ecology

CHAPTER FOUR

Population Dynamics of the Lesser Prairie-Chicken*

Edward O. Garton, Christian A. Hagen, Grant M. Beauprez, Sean C. Kyle, James C. Pitman, Doug D. Schoeling, and William E. Van Pelt

Abstract. We conducted a comprehensive analysis of Lesser Prairie-Chicken (*Tympanuchus pallidicinctus*) populations throughout the species' range in the central United States by accumulating and analyzing survey counts of birds at 504 individual leks and 28 lek routes over a 49-year period from 1964 to 2012. Counts from a relatively small number of initial lek routes were expanded through time to provide a combined total of 633 counts of lek routes through 2012. Our compilation of datasets represent the best long-term database from which to estimate annual rates of population change and reconstruct unbiased estimates of minimum abundance of Lesser Prairie-Chicken in four ecoregions and range-wide. Population reconstructions projected abundance increasing from ~50,000 to 200,000 birds in the late 1960s, stabilizing to ~150,000 birds for nearly two decades, and then declining to ~25,000 birds from the late 1980s to the mid-1990s. Populations rebounded to ~80,000 birds in 2008 but then declined to 37,000 birds five years later in 2012. We considered a comprehensive set of 26 models, but the best models for describing changes in population growth rates were density-dependent models of both the Gompertz and Ricker type, including a time trend in carrying capacity or quasi-equilibrium abundance. Of the populations in four ecoregions, two populations exhibited declining trends through time of 1%–2.3% annually (Mixed-Grass Prairie and Sand Sagebrush Prairie); one population was stable (Sand Shinnery Oak Prairie); and the fourth population increased by ~8% per year during 2000–2012 (Short-Grass Prairie). Estimates of historical movement rates were based on population genetics and indicated patterns of movement among closely spaced populations within ecoregions. We forecasted population persistence across the four ecoregions and the range-wide metapopulation using a hierarchy of best models applied to 2012 estimated population size (n = 34,440 birds at leks). Multimodel projections indicated that the probability of individual ecoregion populations and range-wide population persisting at levels above effective population sizes (N_e) of 50 and 500 in the short and long term are significantly increased if low rates of movement among ecoregions are maintained (ca. 1 in 10,000 per generation). Thus, conserving adequate habitat in each ecoregion and maintaining connectivity among populations with movement are paramount for population persistence. Conservation actions that promote successful movement across unsuitable habitat will require concerted efforts to maintain or increase Lesser Prairie-Chicken habitat and population viability throughout the species' range.

Key Words: density-dependent, density-independent, effective population size, Gompertz, lek counts, metapopulation, N_e, PVA, Ricker, stochastic process.

* Garton, E. O., C. A. Hagen, G. M. Beauprez, S. C. Kyle, J. C. Pitman, D. D. Schoeling, and W. E. Van Pelt. 2016. Population dynamics of the Lesser Prairie-Chicken. Pp. 49-76 in D. A. Haukos and C. W. Boal (editors), Ecology and conservation of Lesser Prairie-Chickens. Studies in Avian Biology (no. 48), CRC Press, Boca Raton, FL.

onservation of imperiled species requires a fundamental understanding of population biology, which necessitates answers to several fundamental ecological questions: what is the current population status and long-term trends, does the species exhibit regular cyclical patterns or lag effects in abundance, what are the effects and magnitude of density dependence, has carrying capacity changed, is the population spatially structured as a metapopulation, and what do patterns of population dynamics forecast for future population viability? Fundamental information on population dynamics is needed to assist conservationists in establishing population goals for recovery, identifying thresholds that should initiate more intensive management, recognizing normal ebb and flow of population size, and identifying populations that may require more intensive action.

Answering the ecological questions can be challenging and an absence of answers has hindered conservation efforts to successfully manage species of conservation concern. Generally, the demographic and movement data required to adequately address the set of questions are unavailable for many species of wildlife (Beissinger and Westphal 1998), and lack of data may lead to misguided conservation actions (Reed et al. 2002). Ideally, species conservation would be based in detailed analyses of vital rates (e.g., reproductive success, age at maturity, annual survival), but such detail is rarely available across the entire distribution of a species (Beissinger and Westphal 1998). Thus, combining regional demographic patterns and trends based on extensive count data with fine-scale demographic data from intensive population studies at local field sites provides a solid framework for building a Strategic Habitat Conservation approach.

Conservation concern for the Lesser Prairie-Chicken (*Tympanuchus pallidicinctus*) dates back to the Dust Bowl era in the 1930s when the species was thought to be near extinction because of catastrophic habitat conditions at the time (U.S. Department of Interior 1966). In May 2014, the U.S. Fish and Wildlife Service (USFWS) listed the Lesser Prairie-Chicken as a threatened species under the Endangered Species Act because of long-term declining population trends and ongoing threats across the five-state range (USFWS 2014). However, the listing rule was vacated by judicial decision in September 2015, creating considerable

regulatory uncertainty at the time of this volume. The Lesser Prairie-Chicken currently occupies four principal ecoregions: Sand Sagebrush (*Artemisia filifolia*) Prairie of southwestern Kansas and southeastern Colorado; Short-Grass Prairie/Conservation Reserve Program (CRP) landscape of northwestern Kansas and eastern Colorado; portions of Mixed-Grass Prairie of northwest Oklahoma, northeast Texas panhandle, and south-central Kansas; and Sand Shinnery Oak (*Quercus havardii*) Prairie of eastern New Mexico and western Texas.

The current distribution of Lesser Prairie-Chicken is ~16% of the anecdotal historical range at the time of European settlement (Hagen and Giesen 2004, Van Pelt et al. 2013). Since 2000, there have been shifts in the species' distribution within portions of historical range (Hagen and Giesen 2005). In Kansas, portions of the historical distribution have been reoccupied, and the current distribution now extends beyond the known historical range (Van Pelt et al. 2013, Figure 1.2). Shifts in the distribution of Lesser Prairie-Chickens have been largely attributed to the planting and maintenance of U.S. Department of Agriculture CRP fields that provide the necessary vegetation structure for breeding birds throughout the short-grass prairie in Kansas (Rodgers and Hoffman 2005). Population trends in this region are concomitant with observed changes in the species distribution (Van Pelt et al. 2013).

Population trends have been monitored using spring lek counts since the 1940s, and lek counts have provided the basis for estimating population trends used to assess species status. Survey effort and methods have varied over time, but data from lek monitoring comprise the best available datasets to assess long-term trends (Davis et al. 2008). Lek counts include the number of active leks and the number of birds counted per lek (usually male biased) and is often summarized as average birds per active lek. Patterns of lek attendance, sex ratio of birds during the breeding season, and probability of detection of leks are not well understood (Behney et al. 2012). Sampling issues and variation in survey methods suggest that estimation of population size (N) from lek count information requires a set of assumptions (Walsh et al. 2004, Garton et al. 2011).

Until recently, a unified approach has not been used to assess population trends of lek mating grouse across broad geographic ranges (Garton et al. 2011). However, at least three range-wide assessments are available for population dynamics and persistence of Greater Sage-Grouse (*Centrocercus urophasianus*, Connelly et al. 2004, WAFWA 2008, Garton et al. 2011). Similar issues of data consistency and variation in sampling effort were common to all three studies. Using population reconstruction from annual counts at leks, density-independent models (Dennis et al. 1991, Staples et al. 2004) and density-dependent models of population growth can be fit to lek monitoring data (Dennis and Taper 1994, Dennis et al. 2006, Garton et al. 2011). The models provide estimates of population growth (λ_t), carrying capacity (or quasi-equilibrium population size), and long-term trends in abundance (Garton et al. 2011). The tools of multimodel inference enable forecasts of future abundance and estimates of the probability of quasi-extinction from a suite of models that are each weighted according to the fit of the model to the data from lek monitoring programs (Garton et al. 2011).

Our primary goal was to provide a unified assessment of Lesser Prairie-Chicken population dynamics across four ecoregions (Van Pelt et al. 2013) and the entire species distribution using the general methods described by Garton et al. (2011). Our objectives were to: (1) assess long-term (1965–2012) changes in Lesser Prairie-Chicken populations by ecoregion using lek count data, (2) reconstruct population abundance through time using lek count data and population estimates from 2012 aerial survey data, (3) assess the potential of alternative ecological models for describing the dynamics of each population, and (4) use the best models to project the likely future probabilities of population persistence for Lesser Prairie-Chickens. Our secondary objectives were to estimate effective range-wide population size (N_e), dispersal rates between ecoregions based on genetic characteristics, and correlative relationships among demographic rates in different ecoregions.

METHODS

Study Area

We analyzed count data from lek monitoring surveys across the four ecoregions and distribution currently occupied by the Lesser Prairie-Chicken, as defined by Van Pelt et al. (2013). The four ecoregions that subdivide the range provide biologically meaningful units for the analysis of population data and for planning conservation efforts (Figure 1.2). The Mixed-Grass Prairie Ecoregion is the largest of the four regions, extending from southwest Kansas through western Oklahoma and into the northeast panhandle of Texas. The Short-Grass Prairie/CRP Ecoregion is contained within western Kansas and north of the Arkansas River and estimated to host the majority of the current population (McDonald et al. 2014). The Sand Shinnery Oak Prairie Ecoregion is the third-largest geographic area, occupying portions of northwest Texas and eastern New Mexico. The Sand Sagebrush Prairie Ecoregion extends from southeast Colorado along the Arkansas River into Kansas and Oklahoma (Figure 1.2).

Field Methods

Field methods for monitoring of spring populations of Lesser Prairie-Chickens varied among the five states. Generally, the monitoring techniques can be categorized into two methods: counts of birds attending individual leks or counts of birds at leks detected along a set survey route. In both cases, counts were a tabulation of birds counted annually with the difference being that the first method summarizes the data specifically to a point on the map, and the second method summarizes data across multiple points within an area. Although the fundamentals are the same, these two different reporting types are problematic in regard to reporting standard statistics such as average males per lek, because one method is point based whereas the other is area based. Thus, we report average number of males per lek or lek route.

Monitoring Effort

A regional difference in the lek survey data is that Kansas, Oklahoma (post-2001), and New Mexico use "flush counts," which are a total count of all birds encountered at a lek without any attempt to ascertain sex of birds counted. Alternatively, Texas and Colorado have used the high count of males from repeated counts within a survey year to estimate population trend. Fortunately, observers in Texas also recorded females and individuals of unknown gender, which provided an opportunity to understand the degree to which flush counts of birds of both sexes might affect trend analyses. Oklahoma used counts of males as an index of population trend until 2001, but thereafter "flush counts" of leks were implemented, which may include some unknown number of females.

Nearly all males attend leks every day, whereas females attend leks only for short periods of time and visit multiple leks over the course of several days. Thus, females are not likely to measurably influence the data regardless of whether data are collected by lek or flush counts. Nevertheless, lek data provide annual counts over time that are suitable for population reconstruction; counts yield unbiased estimates of annual abundance and rates of population change suitable for modeling patterns of population dynamics (Garton et al. 2011). A brief summary of monitoring and data summarization by state are presented later, and a detailed summary of lek monitoring techniques by state is given by Davis et al. (2008).

Data Summary

New Mexico lek counts were started in 1971 and have been conducted by the U.S. Department of Interior Bureau of Land Management, the New Mexico Department of Game and Fish, and The Nature Conservancy biologists. Spatial information for localities of each lek observation was updated each year and led to a large number of "lek sites" that were effectively the same group of birds monitored over time with small changes in lek sites. To develop a time series of lek count data, we spatially aggregated the 9,017 lek count observations into discrete geographic units. Thus, we buffered each recorded lek location (n = 1,101) by 500 m, which is half the mean (1,003 m) nearest-neighbor distance among leks, and in most instances a reasonable distance to detect satellite activity around primary leks. Aggregation at a 500-m scale resulted in 404 lek complex polygons. Next, we spatially joined all lek observation data to these 404 lek complexes. We used the peak daily count for each lek complex during a year as the basis for estimating annual trends. An additional 703 lek counts were conducted in the autumn (September–November) and another 88 counts without a date recorded, all of which were excluded from the analysis. Last, 709 lek detections were "estimated" counts based on auditory or an incomplete census of the lek, and these observations were also excluded. The remaining 7,517 lek counts provide the basis for the New Mexico portion of trend analysis and population reconstruction in the Sand Shinnery Oak Ecoregion.

Lek data from the other states were summarized from the raw counts because naming conventions

and spatial data were recorded such that discrete time series could be generated for each lek or lek route. Kansas surveys routes for leks twice during the breeding season, with the highest count of the two surveys used as the metric to index the population. Each route was treated as an individual count unit. Colorado counts all known leks (and satellite leks) multiple times per year and uses the high count as the metric to index the population. Each lek or lek complex with a main lek and satellite locations is treated as an individual count unit. Texas counts leks at established study areas, where each area is treated as an individual count unit. Oklahoma counted individual leks and satellite locations for a 24-year period (1968–1991), and then began counting birds at leks using a route system similar to Kansas. For the early period, lek complexes were treated as individual count units, but in the more recent period, routes were treated as individual count units.

Range-wide aerial surveys were conducted in 2012 to estimate the abundance of leks and a minimum population size of Lesser Prairie-Chicken (McDonald et al. 2014). Data were obtained using helicopter surveys over a stratified random sample of spatial blocks corrected for visibility bias based on distance from the helicopter and different views on either side of the helicopter. The minimum population size for 2012 served as the baseline from which to reconstruct populations to assess trend and conduct stochastic population projections. The surveys recorded both lek and nonlek observations where <3 nondisplaying birds were observed from the aircraft. Detection probabilities were similar between the two groups, and lek and nonlek observations comprised 74% and 26% of the population estimate, respectively. Sex of birds could not be determined for either group, and for the purposes of our analyses, we assumed that lek observations were mostly males. Thus, the minimum population estimate is primarily an estimate for male Lesser Prairie-Chickens.

Statistical Analyses

Population Reconstruction

Analysis began for populations in each ecoregion with reconstruction of population dynamics with the approach pioneered in Garton et al. (2011), with a few minor modifications. We took the 2012 aerial survey estimate of Lesser Prairie-Chickens as a base count and reconstructed populations backward in time from 2012 using estimated annual rates of change based on ground counts (θ) as described in Garton et al. (2011). Our approach begins with the best available recent population estimate from the 2012 aerial survey and uses paired counts in previous, successive years through time to estimate the annual finite rates of population change λ_i (or lambda), the continuous or instantaneous analog r_i, and the reciprocal θ_i (or theta), which expressed abundance in the previous year as a proportion of the current year estimate. Paired counts were used to estimate rates of change and thetas for a particular pair of years. We used only ground lek counts that were conducted in both years to produce unbiased estimates through time even if the number of leks counted differed between years. Each of the three estimates represents a ratio of counts in paired years for which the statistic is a weighted ratio estimator with straightforward variance and confidence intervals (Scheaffer et al. 1996).

Sampling effort devoted to counting leks varied considerably among years, with number of leks counted in particular years varying from 70 to >150 in a 3-year interval. It was necessary to standardize estimates and remove any bias due to variable sample size of leks or lek complexes that were surveyed. Thus, we treated the number of birds counted in the 2012 aerial survey as the baseline abundance. We then reconstructed estimates of earlier abundance by applying ratio estimators based on ground counts through time (Scheaffer et al. 1996:200). To estimate the finite rate of change from males counted from the ground at each lek, lek complex, or lek route surveyed in both 2011 and 2012, we treated each count as cluster samples of individual males in successive years and the pair of counts as a ratio estimator. The ratio of males counted in a pair of successive years served as an estimate of the finite rate of change for males at that lek or lek complex in that 1-year interval. These ratios were combined across leks within a population for each year to estimate the finite rate of change, $\lambda(t)$, as the ratio estimator:

$$\lambda(t) = \frac{\sum_{i=1}^{n} M_i(t+1)}{\sum_{i=1}^{n} M_i(t)} \qquad (4.1)$$

where $M_i(t)$ = number of males counted at lek i in year t, across n leks counted in both years t

and t + 1. Note that while the number of leks or lek complexes counted varied from 1 year to the next, the estimator was unbiased by sample size because we did not use counts for leks or lek sites counted in only one of the paired years. Precision (variance and SE) of finite rates of change was estimated by treating this finite rate of change, $\lambda(t)$, as a standard ratio estimator (Scheaffer et al. 1996:200):

$$Var(\lambda_t) = \frac{fpc}{n\overline{M}(t)^2} \frac{\sum_{i=1}^{n} \left[M_i(t+1) - \lambda(t)M(t) \right]^2}{n-1}$$

$$(4.2)$$

where fpc was assumed to be 1.0 and $\overline{M}(t) = \left(\sum_{i=1}^{n} M_i(t) \right)/n$.

An index to the relative size of the population during the previous year, $\theta(t)$, was estimated in an analogous manner from the paired samples as the reciprocal of $\lambda(t)$:

$$\theta(t) = \frac{\sum_{i=1}^{n} M_i(t)}{\sum_{i=1}^{n} M_i(t+1)} \qquad (4.3)$$

with analogous variance:

$$Var(\theta_t) = \frac{fpc}{n\overline{M}(t+1)^2} \frac{\sum_{i=1}^{n} \left[M_i(t) - \theta_t M_i(t+1) \right]^2}{n-1}$$

$$(4.4)$$

We then calculated an index of population size by first considering the estimated number of males present in the 2012 aerial survey as a minimum estimate of breeding male population size for each ecoregion ($\hat{N}_{2012}$). We then reconstructed the previous year's minimum male abundance index ($\hat{N}_{2011}$) by multiplying the 2012 abundance by the ratio estimator of the relative number of males attending the same leks in 2011 compared to 2012 ($\hat{N}_{t-1} = \hat{N}_t \hat{\theta}_{t-1}, \hat{N}_{2011} = \hat{N}_{2011} \hat{\theta}_{2011}$). Using successive estimates of θ_t to reconstruct population abundance simply required multiplying the current population estimate ($\hat{N}_t$) by θ_{t-1} during the previous interval repeatedly backward through time. As such, the estimates of abundance backward through time are each a product of estimators for which a variance and standard error are readily estimated as a product of variances following methods of Goodman (1962; see Garton et al. 2011:302). The variance of previous year's population indices clearly involve the variance of a series of products of $\hat{\theta}_t$, which can be combined as follows:

$$\hat{Var}\left(\prod_{i=1}^{k}\hat{\theta}_i\right)=\prod_{i=1}^{k}\left(\hat{Var}\left(\hat{\theta}_i\right)+\hat{\theta}_i^2\right)-\prod_{i=1}^{k}\hat{\theta}_i^2\ldots \quad (4.5)$$

It has been shown that multiple counts are often required to ascertain a reasonable estimate of numbers of males attending leks (Beck and Braun 1980, Connelly et al. 2003, Walsh et al. 2004, Garton et al. 2011). We treated the estimates obtained by the 2012 aerial surveys for each ecoregion and range-wide as minimum estimates because each estimate was based on a single aerial survey. The approach was reasonable for populations in the Mixed-Grass Prairie, Short-Grass Prairie/CRP, and Sand Shinnery Oak Prairie Ecoregions. One exception was the Sand Sagebrush Prairie Ecoregion, where ground counts of leks gave a higher estimate of abundance than the estimates obtained from the aerial surveys (McDonald et al. 2014). Here, we substituted the estimate from ground counts for the estimate from aerial surveys. We made use of the 2012 aerial survey estimates of minimum population size for three of the ecoregions as the basis for reconstructing earlier population sizes because each estimate was conducted as a true probability sample of the spatial regions delineated for populations of Lesser Prairie-Chicken in each ecoregion. Thus, we capitalized on the strengths of each approach by using paired sequential ground counts to estimate unbiased, annual rates of change and thetas combined with the unbiased minimum population size baseline estimate for 2012 to estimate annual historical minimum population sizes.

Fitting Population Growth Models

Previous analyses for Greater Sage-Grouse demonstrated that incorporating potential time period differences can improve models of population dynamics (Garton et al. 2011). Thus, we explored three approaches to time period for Lesser Prairie-Chicken in a preliminary analysis: (1) two periods consisting of the 21-year period from 1965 to 1985 versus a later 27-year period of 1986 to 2012 when laws affecting pumping water for irrigation in Kansas and CRP were implemented; (2) three periods consisting of the first two periods above with an additional categorical variable (Period 3) after 1997 when additional conservation attention was focused on the species, culminating in the 2007 listing as a candidate species; and (3) two periods categorized as prior to 1997 and following 1997. Preliminary analyses of reconstructed population estimates were conducted for long-term datasets from the Sand Sagebrush, Sand Shinnery Oak, and Mixed-Grass Prairie Ecoregions. Model selection based on Akaike's Information Criteria (AIC) values for these models demonstrated that the third approach was the best fit to the count data, and a two-category classification of periods was used for the full analysis of potential population models.

We fit a suite of stochastic population growth models, including two density-independent models to the time series of reconstructed population indices for each population following Garton et al. (2011): (1) exponential growth with process error (EGPE; Dennis et al. 1991), and (2) exponential growth with differing mean rates of change between the two time periods (Period 1 = 1965–1996, Period 0 = 1997–2012). We also fit 24 density-dependent models consisting of all combinations of four factors: (1) Ricker-type density dependence in population growth (Ricker; Dennis and Taper 1994), or Gompertz-type density dependence in population growth (Gompertz; Dennis et al. 2006); (2) presence or absence of a time delay in the effect density has on population growth rate (no delay, 1-year delay, or 2-year delay); (3) a period effect (period, as described earlier); and (4) time trend in population-carrying capacity (year).

Specifically, let $N(t)$ be the observed population index at time t, $Y(t) = \log[N(t)]$, and the annual growth rate $r(t) = Y(t + 1) - Y(t)$. The global stochastic model incorporating Ricker-type density dependence was

$$r(t) = a + b \times N(t - \Delta) + c \times \text{Year}$$
$$+ d \times \text{Period} + E(t) \quad (4.6)$$

and the analogous model for Gompertz-type density dependence was

$$r(t) = a + b \times \ln\left(N(t-\Delta)\right) + c \times \text{Year}$$

$$+ d \times \text{Period} + E(t) \qquad (4.7)$$

where $Y(t) = \log[N(t)]$, and the annual growth rate $r(t) = Y(t + 1) - Y(t)$. The global statistical model incorporated a difference in time periods by setting Period = 1 if Year = 1965–1996 and Period = 0 if Year = 1997–2012. $E(t)$ represented environmental and demographic (i.e., process) variation in realized growth rates and was a normally distributed random deviate with mean = 0 and variance = σ^2. These models yielded five parameters (i.e., a, b, c, d, and σ^2) that were estimated via maximum likelihood using the indices to past abundance data estimated from the population reconstruction.

The main difference between the different models is that the Ricker model assumes that growth rates are a linear function of population size, whereas the Gompertz model assumes growth rates are a linear function of the natural log of population size. Density-dependent models such as Gompertz and Ricker provide an objective approach to estimate a quasi-equilibrium value, which is defined as the population size at which the growth rate is 0 and can be considered as an estimate of carrying capacity. The carrying capacity represents a threshold in abundance below which population size tends to increase and above which population size tends to decrease.

Parameter Estimation

We estimated the carrying capacity of the range-wide species population based on the best stochastic growth models fit to the full dataset available for each ecoregion (Garton et al. 2011:303, Equation 15.10). From the set of base models, several plausible scenarios for population growth can be realized. Models involving time trends (+ Year) and period differences (+ Period) can be interpreted as inferring that the carrying capacity is changing over time or between periods, where negative slopes or lower values indicate reductions in carrying capacity. For example, the parameter estimates from the Ricker model with a time trend

(Year) and period effect (Period) can be used to estimate a carrying capacity (K) as follows:

$$\hat{K} = -\hat{b}^{-1}\left(\hat{a} + \hat{c}\text{Year} + \hat{d}\text{Period}\right) \qquad (4.8)$$

The hat ($\wedge$) notation over a parameter indicates that this value was the maximum likelihood estimate for that parameter when fit to the past abundance data. When parameters b and c are set to 0, these models reduce to the EGPE model (Dennis et al. 1991) with a declining or increasing trend in abundance, depending on a, and including period simply allows for differing carrying capacities between two discrete time periods.

We fit 26 candidate models to each set of observed abundance data with the statistical computing program R (ver. 2.6.1; R Foundation for Statistical Computing, Vienna, Austria) and mixed procedure of Program SAS (ver. 9.2; SAS Institute, Cary, NC). The stochastic growth models treat annual rates of change (r_t) as mixed effects of fixed effects (year and period) and random effects (reconstructed population index with or without log transformation and time lag). Annual rates of change (r_t) were consistently described well by a normal distribution. We used Akaike's Information Criteria corrected for small sample size (AIC_c) to rank the fit of each candidate model to the time series data (Burnham and Anderson 2002). Likewise, we followed Akaike (1973), Buckland et al. (1997), and Burnham and Anderson (2002:75) in calculating AIC_c weights (w_i), which we treated as relative likelihoods for a model given the data. For a given analysis unit, we report a 95% confidence set of models based on the best model using the sum of model weights ≥ 0.95 (Burnham and Anderson 2002). Our approach reduced the number of models reported for all analysis units to those models with some potential for explaining the data, but did not necessarily drop all models with $\Delta AIC_c > 2$ or 3.

Stochastic Population Projections

For each population, we projected 4,000 replicate abundance trajectories for 30 years and 100 years post-2012 using the following expression:

$$N(t+1) = N(t) + e^{\hat{r}(t)} \qquad (4.9)$$

where $\hat{r}(t)$ was the stochastic growth rate calculated using maximum likelihood parameter estimates for the given model. For example, to project populations based on the Gompertz model with no time lag, a time trend in carrying capacity and a difference between periods, we used

$$N(t+1) = N(t) \times e^{\hat{a} + \hat{b}\,N(t) + \hat{c}\,\text{Year} + \hat{d}\,\text{Period} + E(t)} \quad (4.10)$$

where $N(0)$, the initial abundance for population projections was taken as the final index of population size recorded in 2012, Period = 0 indicated that future growth would be analogous to what occurred from 1997 to 2012, and $E(t)$ was a random deviate drawn from a normal distribution with mean 0 and SD equal to the square root of maximum likelihood estimate of mean squared error remaining from mixed model. The replicate time series were then used to calculate the probability that the population would decline below a quasi-extinction threshold corresponding to minimum counts of 85 and 852 males at leks. The probability of quasi-extinction was the proportion of replications in which population abundance fell below the quasi-extinction threshold at some point during the time horizon (30 or 100 years). Thresholds of 85 and 852 were chosen to approximate the standard 50/500 rule as thresholds for effective population size (N_e; Franklin 1980, Soule 1980), expressed in terms of breeding males counted at leks and mean adult sex ratio at leks. In other words, forecasting the future probability of a local population declining below effective population size of 50 breeding adults ($N_e = 50$ corresponding to an index based on minimum males counted at leks of ≤ 85) identifies populations at short-term risk for extinction (Franklin 1980, Soule 1980). Similarly, a local population declining below effective population size of 500 breeding adults ($N_e = 500$ corresponding to an index based on minimum males counted at leks of 852 or less) identifies populations at long-term risk for extinction (Franklin 1980, Soule 1980).

Based on our comparison of AIC_c values, most populations had multiple candidate models that could be considered a competing best model by scoring within the 95% set (usually $\Delta AIC_c < 3$). To incorporate model selection uncertainty into forecasts of population viability, we projected future population abundances using each of the 26 models and used model averaging (Burnham and Anderson 2002:159) to generate an overall (i.e., based on all fitted models) estimate of the probability of quasi-extinction. Generally, a "model-averaged" prediction can be obtained by calculating the predicted value of probability of quasi-extinction or other parameters of interest for each model and by taking a weighted average of the predictions where the weights are the relative likelihoods of each model,

$$\hat{Pr}(\text{Extinction}) = \sum_{i=1}^{R} \left\langle \hat{Pr}\left(\text{Extinction}|\text{Model}_i\right) \times w_i \right\rangle$$

$$(4.11)$$

Note that the probability of extinction under a particular model is conditional on that model and its maximum likelihood parameter estimates. To assess the precision of these model-averaged probabilities of quasi-extinction, we calculated a weighted variance for these probabilities of extinction (Krebs 1999:276), similar to the variance of a mean for grouped data (Remington and Schork 1970:46)

$$\hat{Var}\left[\hat{Pr}(\text{Extinction})\right]$$

$$= \sum_{i=1}^{R} w_i^2 \times \left[\hat{Pr}(\text{Extinction}) - \hat{Pr}(\text{Extinction}|\text{Model}_i)\right]^2$$

$$(4.12)$$

Metapopulation Analyses

We analyzed the viability of the metapopulations of Lesser Prairie-Chicken populations similar to analyses for individual populations, but with two exceptions (Garton et al. 2011). First, instead of basing population projections on all 26 models, we used only the highest-ranked AIC_c models across all four populations—Ricker + Year and Gompertz + Year—both of which were density-dependent models with carrying capacity changing through time. Second, the metapopulation model required two additional inputs: dispersal rates between the four populations in each ecoregion and an estimate of the degree to which populations experience correlated dynamics.

Information on dispersal movements is lacking for Lesser Prairie-Chickens (Pitman et al. 2006). However, Hagen et al. (2010) conducted tests of mtDNA genetic differentiation (Φ_{ST}) among Lesser Prairie-Chickens across the range and indicated minimal difference between populations, except in the Sand Shinnery Oak Ecoregion. We built upon these measures of genetic difference to estimate relative dispersal rates between populations using the methods of Hudson et al. (1992).

We estimated the numbers of migrants per generation (Nm) to estimate dispersal rate between pairs of populations (d_{ij}) where $Nm = 0.5$ ($1/ \Phi_{ST} - 1$) and Φ_{ST} is an estimate of Φ_{ST} (Hudson et al. 1992). When estimates of Φ_{ST} were negative, we treated them as zero (Table 4.1). To estimate d_{ij}, we averaged all pairwise Φ_{ST} values between two given ecoregions (i and j, Tables 4.1 and 4.2). We then divided Nm between a pair of ecoregions (i, j) by the sum of the average effective populations (N_e) in the two ecoregions (i, j). From the population reconstruction of each ecoregion, we calculated the average population size for the entire time period of the reconstruction. We then multiplied our estimated ratio of N_e/N_c (0.585; see Results) by the average population size to generate estimates of average N_e for each region to estimate d_{ij} between each pair of ecoregions.

Correlated dynamics among populations were modeled by including a covariance in the random deviates used to portray environmental stochasticity. We obtained estimates of covariance by correlating the residuals of the fitted models used to project population abundances. We used simulations, analogous to parametric bootstraps given earlier, conducted in SAS using custom scripts for metapopulation projections, and we used the same approaches that we applied to populations in each ecoregion to evaluate the probability of persistence and other metrics of the range-wide metapopulation. Dispersal movements among populations were simulated at the beginning of each time step during late winter or early spring, and prior to projection of annual changes in abundance associated with seasonal reproduction and survival during the year.

Data Considerations and Limitations

Another key issue in analyzing the lek data concerns the magnitude of sampling error in lek counts of Lesser Prairie-Chickens. Sampling error could invalidate assumptions of linear mixed models and inflate the estimates of process error, leading to stochastic forecasts of future population viability that are excessively conservative. We evaluated this question by analyzing each reconstructed population time series using a Gompertz state-space model approach that simultaneously estimates observation and process error (Dennis et al. 2006). We found that the population reconstruction time series provide unbiased estimates of process with sampling error from combining counts at tens to hundreds of leks approaching zero. Four estimates of sampling error (τ^2) were zero for three populations in different ecoregions and range-wide. Only the short-term dataset (2001–2012) for the Short-Grass Prairie/CRP Ecoregion indicated a nonzero sampling error estimate ($\tau^2 = 0.0025$). Thus, we were able to estimate parameters and likelihoods for models including observation error within a single error term combining both process error from stochastic environmental and demographic processes and sampling error (Garton et al. 2011). Consequently, forecasts from these models of probability of persistence will be slightly conservative, implying that the probability of persistence was at least as large as our estimates or slightly larger. The assumption of no measurement error in estimates of reconstructed population size or its logarithmic transformation is likely incorrect, but most authorities and many studies verify that the linear models are robust to violations of this assumption (Zar 1984:268).

A second issue concerns the effects of the severe drought throughout the prairie region in the summer of 2011. We hypothesized that the resulting low lek counts in 2012 would skew the analysis sufficiently that analyses should be conducted with data from that final year eliminated from the lek counts. The potential bias was examined in two ways, conceptually and statistically. All of the candidate models evaluated for annual rates of change of lek counts were stochastic models incorporating and estimating the magnitude of random variation about average annual rates of change. Eliminating a large negative value would lead to underestimation of the level of random process variation, with resulting overly optimistic estimates of the probability of persistence. On that basis, lek count data from 2012 should not be eliminated unless the value is an extreme value with high leverage on the coefficients, which was not the case.

TABLE 4.1

Genetic distance (Φ_{ST}) of Lesser Prairie-Chicken ecoregions in the lower quadrant (adapted from Hagen et al. 2010) and estimated number of migrants per generation (N_m) in the upper quadrant (applying Hudson et al.'s [1992] estimator to Φ_{ST}).

	Sampled populations by County and State per Hagen et al. (2010)									
	Gove, KS	Ness, KS	Finney, KS	Kearny, KS	Baca, CO	Prowers, CO	Comanche, KS	Beaver, OK	Harper, OK	Roosevelt, NM
Ecoregion[a]	SGPR	SGPR	SSPR	SSPR	SSPR	SSPR	MGPR	MGPR	MGPR	SOPR
SGPR	—	0[b]	11	9	18	3	5	11	4	2
SGPR	-0.038[a]	—	246	47	0	4	19	0	10	3
SSPR	0.043	0.002	—	0	0	20	13	39	10	3
SSPR	0.051	0.011	-0.007	—	49	2,000	10	19	18	2
SSPR	0.027	-0.008	-0.032	0.010	—	5	18	0	9	3
SSPR	0.149	0.117	0.024	0.000	0.087	—	34	19	0	3
MGPR	0.093	0.026	0.038	0.049	0.028	0.014	—	0	0	24
MGPR	0.042	-0.010	0.013	0.025	-0.002	0.025	-0.003	—	20	6
MGPR	0.110	0.048	0.045	0.028	0.054	-0.039	-0.012	0.025	—	4
SOPR	0.168	0.128	0.127	0.170	0.126	0.151	0.020	0.081	0.124	—

[a] SGPR, Short-Grass Prairie/CRP Ecoregion; SSPR, Sand Sagebrush Prairie Ecoregion; MGPR, Mixed-Grass Prairie Ecoregion; SOPR, Sand Shinnery Oak Prairie Ecoregion.
[b] Negative Φ_{ST} values were treated as zeros for the purposes of estimating Nm.

TABLE 4.2

Average estimates of reconstructed population size (N) for the entire time series, minimum effective population size (N_e), number of migrants per generation (N_m), correlations in demographic rates λ, and estimated dispersal rates among four ecoregions of Lesser Prairie-Chickens[a].

Estimated[b]	SGPR	SSPR	MGPR	SOPR
N	20,069	34,317	41,777	5,105
N_e	11,751	20,092	24,461	2,989
Average Φ_{ST}			Nm	
SGPR	—	10	9	3
SSPR	0.050	—	15	3
MGPR	0.053	0.033	—	6
SOPR	0.148	0.143	0.075	—
Correlations in λ		Dispersal rate		
SGPR	—	0.000120	0.000098	0.000078
SSPR	0.536	—	0.000133	0.000052
MGPR	0.221	0.302	—	0.000089
SOPR	0.519	0.223	0.298	—

[a] SGPR, Short-Grass Prairie/CRP Ecoregion; SSPR, Sand Sagebrush Prairie Ecoregion; MGPR, Mixed-Grass Prairie Ecoregion; SOPR, Sand Shinnery Oak Prairie Ecoregion.

[b] N = average population size from population reconstructions of each ecoregion, and N_e = N × 0.585. Dispersal rate was estimated as $N_{m_i}/\Sigma(N_{e_{all}})$, and each rate assumed 2.5-year mean generation time.

A third concern is that the final year estimates form the starting points for all future projections of population abundance and as such could skew estimates of the probability of quasi-extinction assessed as the number of simulations falling below a specified minimum abundance of 50 or 500 adults under the classic 50:500 rule for short-term and long-term persistence (Franklin 1980, Soule 1980). We evaluated the same 26 models previously applied to Greater Sage-Grouse (Garton et al. 2011), but used maximum likelihood methods to estimate parameters and information theoretic measures of model probability for Lesser Prairie-Chickens (Burnham and Anderson 2002). Fortunately, the most strongly supported models of population dynamics included various density-dependent models of the Gompertz or Ricker form, which proved completely unaffected in trials testing whether the 2011 or 2012 lek estimates were treated as the initial abundance (N_0) for future projections (Garton 2013).

Estimating Effective Population Size

A critical input in future forecasts of population viability consists of thresholds representing critical values of population abundance from a genetic perspective. We used the widely accepted effective population size (N_e) values of 50 for short-term persistence and 500 for long-term persistence (Franklin 1980, Soule 1980). Elsewhere, Pruett et al. (2011) investigated effective population sizes for populations of Lesser Prairie-Chicken in two portions of the range from a demographic and genetic perspective and provided a framework to estimate N_e range-wide.

We employed the effective population size (N_e) equation of Nunney and Elam (1994) to estimate a demographic N_e for Lesser Prairie-Chicken (Pruett et al. 2011). We summarized survival and nest success estimates from recent radiotelemetry studies across the range to estimate N_e (1980–2012, Table 4.3). We bootstrapped (1,000 replications) of survival and reproductive rates to estimate an average N_e/N_c (Table 4.3). Additionally, we modified the Pruett et al. (2011) input parameter for male breeding success from 20% to 35% because Behney et al. (2012) revealed that more Lesser Prairie-Chicken males were successful in copulating than previously thought (Hagen and Giesen 2005). Thus, we used the average breeding success ($\bar{x} = 35\%, 95\%$ CI $= 24\% - 43\%$) from four leks reported by Behney et al. (2012). The N_e employed assumes that both sexes are equally surveyed in the population, and because Lesser Prairie-Chicken monitoring is male biased, we

TABLE 4.3
Parameter estimates for annual survival, nest success, and other demographic parameters (SD based on 1,000 bootstrap samples) and calculations for estimating demographic effective population size using the minimal method.

Parameters	Definition	Estimate	SD	Calculation	Source[a]
r	Sex ratio	0.50	0.000	N/A	1, 12
A_f	Adult female life span	2.55	0.828	$A_f = 1/(1 - V_f)$	9
A_m	Adult male life span	2.33	0.343	$A_m = 1/(1 - V_m)$	9
T	Generation time	2.44	0.445	$T = M - 1 + (A_f + A_m)/2$	9
M	Age at first breeding	1.00	0.000	N/A	1, 12
I_{af}	Variance in female life span	0.61	0.107	V_f	3, 4, 7, 8
I_{am}	Variance in male life span	0.57	0.061	V_m	4, 6, 8, 10
I_{bf}	Variance in female reproduction	1.00	0.478	$I_{bf} = (1 - \alpha_f)/\alpha_f$	10
α_f	Female breeding success	0.50	0.062	P	2, 3, 4, 8, 10, 11
I_{bm}	Variance in male reproduction	2.36	0.549	$I_{bm} = \{r/[(1 - r)\alpha_f)]\}/$ $(1 - \alpha_m)/\alpha_m$	9
α_m	Proportion of males breeding	0.35	0.063	b	1

SOURCES: After Nunney and Elam (1994), Pruett et al. (2011).

[a] (1) Behney et al. (2012), (2) Davis (2009), (3) Fields (2004), (4) Grisham (2013), (5) Hagen and Giesen (2004), (6) Hagen et al. (2005), (7) Hagen et al. (2009), (8) Lyons et al. (2009), (9) Nunney and Elam (1994), (10) Patten et al. (2005), (11) Pitman et al. (2005), and (12) Pruett et al. (2011).

estimated a correction factor based on two data sources: (1) Texas lek counts using proportion of males (0.83) to total count (n = 70 count-years) and (2) Oklahoma counts using the proportion of males (0.67) in the count where both flush count and male counts were recorded from 1999 to 2001. We combined these estimates to calculate that an adjustment of 75% is required for aerial and ground counts to reflect count composition averaging 75% males. Using the sex ratio correction and the equations of Nunney and Elam (1994), we estimated that ground or aerial counts of birds at leks of 85 birds (primarily males) represent $N_e = 50$ and counts of 852 birds represent $N_e = 500$. Following Garton et al. (2011), a significant likelihood of extirpation was defined as a result of >50% of simulated population forecasts falling below the respective N_e for either 30 or 100 years.

RESULTS

Lesser Prairie-Chicken monitoring and descriptive range-wide population trends summarized over 5-year periods followed a widespread pattern of sporadic sampling for the first decade (mid-1960s to mid-1970s), more frequent but inconsistent sampling across regions for the next two decades (mid-1970s to mid-1990s), and sustained increased sampling effort since 1996. Range-wide, the number of leks counted each year increased from a handful in the 1960s to dozens in the middle period to ≥100 per year in the past 18 years.

Mixed-Grass Prairie Ecoregion

Survey effort increased by over fourfold from an average of three routes in 1965–1969 to 14 routes in 2005–2012 and from zero lek surveys to an average of nine lek surveys during the same time periods. The proportion of active routes and leks decreased 38% and 52%, respectively, over the assessment period (Table 4.4). Population trends as indicated by average number of males per lek declined 46% from 1965–1969 to 1995–1999 but then recovered during 2000–2007 (Table 4.4). Average rates of change (λ) were >1.0 for three of the nine analysis periods; however, standard errors were large for all estimates of λ, making inferences difficult from these 5-year averages. We used the 2012 minimum population estimate of 8,076 males (95% CI = 969–15,183) to reconstruct minimum population estimates for males back to 1965 (Figure 4.1). The population increased from ~2,400 males in 1965 to exceed 50,000 birds from late 1960s to late 1980s, when a

TABLE 4.4

Lesser Prairie-Chicken monitoring and population trends, 1965–2012,
summarized over 5-year periods for the Mixed-Grass Prairie Ecoregion.

Parameter[a]	Years								
	05–12[b]	00–04	95–99	90–94	85–89	80–84	75–79	70–74	65–69
Routes counted per year	14	12	6	3	3	4	3	3	3
Average males per route	24	61	38	24	126	202	274	257	187
Leks counted per year	9	10	10	12	8	0	0	0	0
Average males per lek	2	6	4	6	8				
% active routes	62	84	87	100	100	100	100	100	100
% active leks	38	58	48	61	72				
Annual rate of change (λ)	0.834	0.946	1.466	0.879	0.938	0.980	1.002	0.997	1.230
SE (λ)	0.118	0.080	0.278	0.206	0.128	0.069	0.078	0.145	0.134

[a] Averaged over years for each period.
[b] Eight years of data in this period.

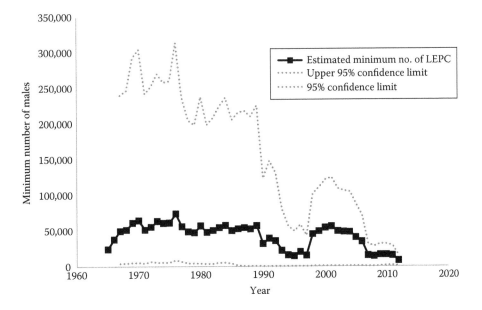

Figure 4.1. Estimated minimum number of Lesser Prairie-Chickens (LEPC) attending leks 1965–2012 (95% CI) in the Mixed-Grass Prairie Ecoregion based on population reconstruction.

precipitous drop from 1989 to 1990 was followed by a continuous decline through 1997. The population peaked again to ~56,000 birds in 2001 and remained at relatively high levels until 2007.

We found considerable model uncertainty in describing the best stochastic process for the Mixed-Grass Prairie Ecoregion of Lesser Prairie-Chicken, with nine models having a ΔAIC_c <3 (Table 4.5). The top-ranked model was a Gompertz + Year with no lag effects and

a declining time trend in abundance of -1.0% per year (r_t = 22.902 − 0.232 lnN_t − 0.010 year, σ = 0.272, r^2 = 0.18). A parametric bootstrap based on the Gompertz model with declining time trend revealed a minimal chance of the population declining below N_e = 50. A decline below N_e = 500 was more likely (28% probability) within 30 years (Table 4.6), but the possibility did not rise to the level of statistical significance. If the observed trend were to continue for

TABLE 4.5
Candidate model set (contains 95% of model weight) and model statistics for estimating population trends and persistence
probabilities for Lesser Prairie-Chickens in the Mixed-Grass Prairie Ecoregion of the central United States, 1965–2012.

| | Model statistics[a] | | | | |
Model	r^2	K	ΔAIC_c	w_i	Σw_i
Gompertz + Year	0.175	5	0.0[b]	0.145	0.145
Ricker + Year	0.170	5	0.3	0.125	0.270
Gompertz t-2 + Year	0.165	5	0.6	0.108	0.378
Gompertz t-1 + Year	0.152	5	1.3	0.076	0.453
Ricker t-2 + Year	0.147	5	1.6	0.066	0.519
Ricker t-1 + Year	0.146	5	1.6	0.064	0.583
EGPE[c]	0.000	3	2.2	0.048	0.632
Gompertz + Year, Period	0.176	6	2.5	0.041	0.673
Ricker + Year, Period	0.170	6	2.9	0.035	0.708
Gompertz t-2+ Period	0.119	5	3.1	0.031	0.739
Gompertz t-2 + Year, Period	0.166	6	3.1	0.030	0.769
Ricker t-2+ Period	0.105	5	3.8	0.021	0.791
Gompertz + Period	0.105	5	3.8	0.021	0.812
Ricker + Period	0.105	5	3.8	0.021	0.834
Gompertz t-1 + Year, Period	0.153	6	3.8	0.021	0.855
Gompertz t-1 + Period	0.101	5	4.1	0.019	0.874
Ricker t-2 + Year, Period	0.147	6	4.1	0.018	0.892
Ricker t-1 + Period	0.098	5	4.2	0.018	0.910
Ricker t-1 + Year, Period	0.146	6	4.2	0.018	0.928
Gompertz t-2	0.043	4	4.5	0.015	0.943
Period	0.035	4	4.9	0.013	0.956

[a] Model fit described by the coefficient of determination (r^2), the number of parameters (K), the difference in Akaike's Information Criterion corrected for small sample size (ΔAIC_c), AIC_c weights (w_i), and cumulative w_i (Σw_i).
[b] $AIC_c = 11.7$ for the top-ranked model.
[c] EGPE = model of exponential growth with process error.

100 years, there would be a 39% chance of the population declining below $N_e = 50$ and a 75% probability of it declining below $N_e = 500$.

Sand Sagebrush Prairie Ecoregion

Survey effort increased by nearly threefold from an average of three routes in 1965–1969 to eight routes in 2005–2012 and from zero lek surveys to an average of 64 lek surveys during the same time periods. The proportion of active routes and leks decreased 29% and 57%, respectively, over the assessment period (Table 4.7). Population trends as indicated by average number of males per route and lek declined by 81% and 66%, respectively (Table 4.7). Average λ was >1.0 for three of the nine analysis

periods; however, standard errors were large for all estimates of λ, making inference difficult from the 5-year averages. We used the 2012 minimum population estimate of 3,005 males (95% CI = 1485–4511) to reconstruct minimum population estimates for males back to 1965 (Figure 4.2). The population increased from ~19,000 males in 1966 to peak at >85,000 males from 1970 to 1975, followed by a continuous decline through 1997. The population increased post-1997 and peaked again at ~20,000 birds in the early 2000s, but subsequently declined to its lowest levels from 2007 to 2012.

The top-ranked model describing rates of change for the Sand Sagebrush Prairie Ecoregion of Lesser Prairie-Chicken was the Gompertz; adding a time trend (Year effect) also resulted in a strongly

TABLE 4.6

Multimodel forecasts of probability (weighted mean percentage and SE) of number of birds attending leks counted in four ecoregions and the range-wide population of Lesser Prairie-Chickens declining to abundances below quasi-extinction levels representing N_e = 50 and N_e = 500 total breeding adults within 30 or 100 years from 2012.

Population	Prob ($<N_e$) in 30 years (%)		Prob ($<N_e$) in 100 years (%)	
	N_e = 50[a]	N_e = 500	N_e = 50	N_e = 500
Mixed-Grass Prairie Ecoregion				
Probability	14.7	28.3	39.0	75.2
SE	6.5	7.7	9.0	8.3
Sand Sagebrush Prairie Ecoregion				
Probability	37.6	76.1	81.6	81.8
SE	9.7	7.4	7.7	7.7
Sand Shinnery Oak Prairie Ecoregion				
Probability	2.3	19.2	9.4	38.0
SE	1.0	5.7	4.4	7.7
Short-Grass Prairie/CRP Ecoregion				
Probability	0.0	0.1	13.7	17.8
SE	0.0	0.0	3.6	3.4
Range-wide				
Probability	4.8	15.2	18.5	83.5
SE	3.7	6.5	7.7	7.3

[a] N_e50 = 85 birds counted and N_e500 = 852 birds counted at leks based on range-wide estimates and minimum method of Nunney and Elam (1994) and Pruett et al. (2011).

TABLE 4.7

Lesser Prairie-Chicken monitoring and population trends 1965–2012 summarized over 5-year periods for the Sand Sagebrush Ecoregion.

Parameter[a]	Years								
	05–12[b]	00–04	95–99	90–94	85–89	80–84	75–79	70–74	65–69
Routes counted per year	8	7	7	7	6	6	4	3	3
Average males per route	25	48	27	37	61	109	133	167	86
Leks counted per year	64	49	65	29	48	46	16	0	0
Average males per lek	3	6	4	4	9	8	9		
% active routes	71	89	82	91	94	100	100	100	100
% active leks	38	63	51	89	83	65			
Annual rate of change (λ)	0.878	0.931	1.210	0.827	0.948	0.885	0.984	1.037	1.316
SE (λ)	0.102	0.089	0.124	0.133	0.098	0.106	0.121	0.241	

[a] Averaged over years for each period.
[b] Eight years of data in this period.

supported model (Table 4.8). A Gompertz + Year model indicated an annual decline of -2.3% in abundance (r_t = 48.8085 − 0.3041 $\ln N_t$ − 0.023 year, σ = 0.2578, r^2 = 0.30). A parametric bootstrap based on the Gompertz model with declining time trend indicated a relatively low likelihood (38%) that the Sand Sagebrush population would decline below N_e = 50, but a significant likelihood (76%) that it would decline below N_e = 500 in 30 years (Table 4.6). If the trend were to continue for

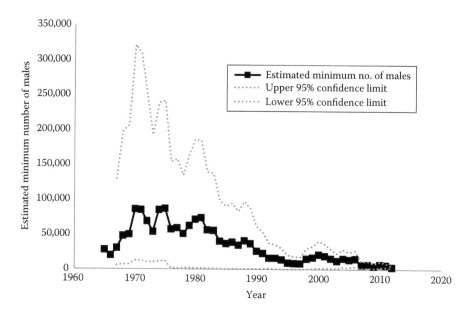

Figure 4.2. Estimated minimum number of Lesser Prairie-Chickens attending leks 1964–2012 (95% CI) in the Sand Sagebrush Prairie Ecoregion based on population reconstruction.

TABLE 4.8

Candidate model set (contains 95% of model weight) and model statistics for estimating population trends and persistence probabilities for Lesser Prairie-Chicken in the Sand Sagebrush Prairie Ecoregion, 1965–2012.

	Model statistics[a]				
Model	r^2	K	ΔAIC_c	w_i	Σw_i
Gompertz + Year	0.295	5	0.0[b]	0.348	0.348
Gompertz + Year, Period	0.315	6	1.3	0.186	0.535
Ricker + Year	0.274	5	1.3	0.180	0.715
Ricker + Year, Period	0.302	6	2.1	0.119	0.834
Ricker t-1 + Year	0.228	5	4.2	0.044	0.878
Gompertz t-1 + year	0.219	5	4.7	0.033	0.911
Gompertz t-2 + Year	0.215	5	4.9	0.030	0.941
Ricker t-1 + Year, Period	0.240	6	6.1	0.017	0.958

[a] Model fit described by the coefficient of determination (r^2), the number of parameters (K), the difference in Akaike's Information Criterion corrected for small sample size (ΔAIC_c), AIC_c weights (w_i), and cumulative w_i (Σw_i).
[b] $AIC_c = 16.8$ for the top-ranked model.

100 years, there was a strong likelihood (>80%) that the Sand Sagebrush population would decline below $N_e = 50$ and 500.

Sand Shinnery Oak Prairie Ecoregion

Survey effort along routes increased from one route in 1969–1974 to two routes in 2005–2012 and from 12 lek surveys to an average of 119 lek surveys during the same time periods. The proportion of active routes and leks decreased 69% and 66%, respectively, over the assessment period (Table 4.9). The average number of males per lek was relatively stable over the assessment period, but average number of males per route declined by 60%. Average values of rate of change were $\lambda > 1.0$ for four of the eight analysis periods; however, standard errors were large for all estimates of λ, making inference

TABLE 4.9

Lesser Prairie-Chicken monitoring and population trends 1965–2012 summarized over 5-year periods for the Sand Shinnery Oak Prairie Ecoregion.

Parameter[a]	Years								
	05–12[b]	00–04	95–99	90–94	85–89	80–84	75–79	70–74	65–69
Routes counted per year	2	3	1	0	0	0	1	1	ND
Average males per route	40	28	19				96	107	ND
Leks counted per year	119	121	90	37	83	34	35	12	ND
Average males per lek	9	7	2	4	10	11	8		ND
% active routes	57	42	28	56	80	79	68	87	ND
% active leks	69	67	100	ND	ND	ND	100	100	ND
Annual rate of change (λ)	0.941	1.166	1.221	0.916	0.956	1.318	1.055	0.956	ND
SE (λ)	0.131	0.113	0.239	0.250	0.208	0.266	0.209	0.072	ND

[a] Averaged over years for each period.
[b] Eight years of data in this period.

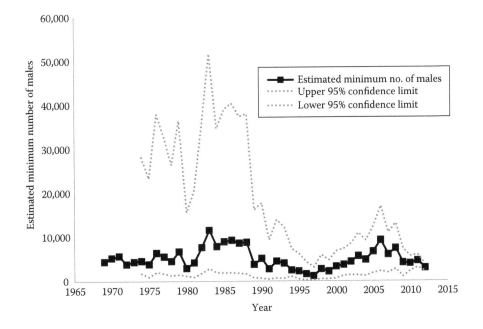

Figure 4.3. Estimated minimum number of Lesser Prairie-Chickens attending leks 1969–2012 (95% CI) in the Sand Shinnery Oak Ecoregion based on population reconstruction.

difficult from the 5-year averages. We used the 2012 minimum population estimate of 2,946 males (95% CI = 2,164–3,727) to reconstruct minimum population estimates for males back to 1969 (Figure 4.3). The population was relatively stable from 1969 to 1975 and then peaked to ~11,000 males in 1983 and maintained at high levels until 1988, followed by a continuous decline to the lowest numbers in 1997 (1,100 males). The population post-1997 peaked again to ~9,000 birds in 2006, but subsequently declined to 2,900 males by 2012.

The best model to describe a stochastic process for the Sand Shinnery Oak Prairie Ecoregion of Lesser Prairie-Chicken was the Gompertz, but the Ricker model was also strongly supported (Table 4.10). The Gompertz model without Year or Period trends suggested a relatively stable population fluctuating around a constant carrying capacity (r_t =

TABLE 4.10

Candidate model set (contains 95% of model weight) and model statistics for estimating population trends and persistence probabilities for Lesser Prairie-Chicken in the Sand Shinnery Oak Prairie Ecoregion, 1969–2012.

Model	Model statistics[a]				
	r^2	K	ΔAIC_c	w_i	Σw_i
Gompertz	0.147	4	0.0[b]	0.234	0.234
Ricker	0.143	4	0.2	0.214	0.447
EGPE[c]	0.000	3	2.1	0.083	0.530
Gompertz + Year	0.154	5	2.2	0.078	0.608
Ricker + Year	0.149	5	2.4	0.069	0.677
Gompertz + Period	0.148	5	2.5	0.067	0.745
Ricker + Period	0.143	5	2.7	0.06	0.805
Gompertz + Year, Period	0.192	6	3.0	0.052	0.857
Ricker + Year, Period	0.167	6	4.3	0.028	0.885
Gompertz t-2	0.041	4	4.8	0.021	0.906
Ricker t-2	0.037	4	5.0	0.02	0.926
Gompertz t-1	0.014	4	5.9	0.012	0.938
Ricker t-1	0.008	4	6.2	0.011	0.949
Period	0.002	4	6.4	0.009	0.958

[a] Model fit described by the coefficient of determination (r^2), the number of parameters (K), the difference in Akaike's Information Criterion corrected for small sample size (ΔAIC_c), AIC_c weights (w_i), and cumulative w_i (Σw_i).

[b] $AIC_c = 39.1$ for the top-ranked model.

[c] EGPE = model of exponential growth with process error.

$2.522 - 0.3008 \ln N_t$, $\sigma = 0.3686$, $r^2 = 0.15$). A parametric bootstrap based on the Gompertz model with a stable time trend indicated virtually no chance (2.3%) that the Sand Shinnery Oak Prairie population will decline below $N_e = 50$, and a low likelihood (19%) that it will decline below $N_e = 500$ in 30 years (Table 4.6). If the trend were to continue for 100 years, there is a low likelihood (<38%) that the Lesser Prairie-Chicken population in the Sand Shinnery Oak Prairie ecoregion will decline below $N_e = 50$ and 500.

Short-Grass Prairie/CRP Ecoregion

Prior to the early 2000s, the Short-Grass Prairie/CRP Ecoregion was thought to be unoccupied by Lesser Prairie-Chickens (Hagen and Giesen 2004). Anecdotal reports from landowners and hunters led to exploratory surveys to determine the species status in this region (Rodgers and Hoffman 2005). Subsequently, three formal survey routes were established, with the first survey conducted in 2001 (Table 4.11). Population trend, indicated by average number of males per route, increased slightly over the two time periods. Average estimates of the rate of change were $\lambda > 1.0$ for both analysis periods; however, standard errors were large for all estimates of λ, making inference difficult from these 5-year averages. We used the 2012 minimum population estimate of 20,413 males (95% CI = 8,549–32,277) to reconstruct minimum population estimates for males back to 2001 (Figure 4.4). The population increased from ~19,000 males in 2001 and peaked at >28,000 males in 2011.

The top-ranked models describing a stochastic process of Lesser Prairie-Chicken population fluctuation for the Short-Grass Prairie Ecoregion was the Ricker + Year, as well as a model without a Year effect (Table 4.12). The Ricker + Year effect indicated an annual increase of 7.7% ($r_t = 152.42 - 0.000075\ N_t + 0.0767$ year, $\sigma = 0.101$, $r^2 = 0.81$). A parametric bootstrap based on the Ricker model trend indicated a low likelihood (0%) that Lesser Prairie-Chicken numbers will decline below $N_e = 50$ or $N_e = 500$ in 30 years in the Short-Grass Prairie/CRP Ecoregion (Table 4.6).

TABLE 4.11
Lesser Prairie-Chicken monitoring and population trends 1965–2012
summarized over 5-year periods for the Short-Grass Prairie/CRP Ecoregion.

Parameter[a]	Years								
	05–12[b]	00–04	95–99	90–94	85–89	80–84	75–79	70–74	65–69
Routes counted per year	3	1	ND[c]	ND	ND	ND	ND	ND	ND
Average males per route	92	80	ND	ND	ND	ND	ND	ND	ND
Leks counted per year	0	0	ND	ND	ND	ND	ND	ND	ND
% active routes	100	100	ND	ND	ND	ND	ND	ND	ND
Annual rate of change (λ)	1.02	1.04	ND	ND	ND	ND	ND	ND	ND
SE (λ)	0.098	ND	ND	ND	ND	ND	ND	ND	ND

[a] Averaged over years for each period.
[b] Eight years of data in this period.
[c] No detections due to lack of survey.

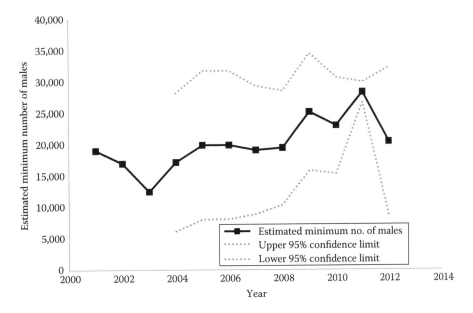

Figure 4.4. Estimated minimum number of Lesser Prairie-Chickens attending leks 2001–2012 (95% CI) in the Short-Grass Prairie/CRP Ecoregion based on population reconstruction.

If the trend were to continue for 100 years, there remains a low likelihood (<20%) that the Short-Grass Prairie population will decline below $N_e = 50$ and 500.

Range-Wide Population

Survey effort along routes increased more than fourfold from an average of six routes in 1965–1974 to 26 routes in 2005–2012 and from 12 lek surveys to an average of 191 lek surveys during these same time periods. The proportion of active leks decreased 40% over the assessment period (Table 4.13). Average estimates of population change were $\lambda > 1.0$ for four of nine analysis periods; however, standard errors were large for all estimates of λ, making inference difficult from these 5-year averages. We used the 2012 minimum population estimate of 34,440 males (95% CI = 24,010–44,869) to reconstruct minimum population estimates for males back to 1965 (Figure 4.5). The population increased from 1965 to 1970, peaking at 186,000 males. The population maintained estimated levels

TABLE 4.12
Candidate model set (contains 95% of model weight) and model statistics for estimating population trends and persistence probabilities for Lesser Prairie-Chicken in the Short-Grass Prairie/CRP Ecoregion, 2001–2012.

Model	Model statistics[a]				
	r^2	K	ΔAIC_c	w_i	Σw_i
Ricker + Year	0.811	5	0.0[b]	0.323	0.323
Ricker	0.613	4	0.2	0.295	0.619
Gompertz	0.575	4	1.0	0.194	0.812
EGPE[c]	0.000	3	2.0	0.119	0.931
Gompertz + Year	0.700	5	4.2	0.040	0.971

[a] Model fit described by the coefficient of determination (r^2), the number of parameters (K), the difference in Akaike's Information Criterion corrected for small sample size (ΔAIC_c), AIC_c weights (w_i), and cumulative w_i (Σw_i).
[b] $AIC_c = -1.3$ for the top-ranked model.
[c] EGPE = model of exponential growth with process error.

TABLE 4.13
Lesser Prairie-Chicken monitoring and population trends 1965–2012 summarized over 5-year periods for range-wide metapopulation consisting of populations in four ecoregions.

Parameter[a]	Years								
	05–12[b]	00–04	95–99	90–94	85–89	80–84	75–79	70–74	65–69
Routes counted per year	26	23	14	9	9	10	7	7	6
Average males per route	32	53	31	34	83	142	180	195	137
Leks counted per year	191	180	165	78	139	80	51	12	0
Average males per lek	7	6	3	4	10	10	8	7	ND
% active routes	69	84	85	94	95	100	100	100	100
% active leks	49	48	37	55	81	71	73	87	ND
Annual rate of change (λ)	0.920	1.012	1.273	0.842	0.913	0.953	1.000	0.983	1.226
SE (λ)	0.071	0.060	0.128	0.100	0.067	0.082	0.092	0.118	0.187

[a] Averaged over years for each period.
[b] Eight years of data in this period.

of >100,000 males until 1989, after which it steadily declined to an all-time low of 25,000 males in 1997. The population post-1997 peaked again at ~92,000 birds in 2006, but subsequently declined to 34,440 males in 2012.

The best model to describe a stochastic process for the range-wide population of Lesser Prairie-Chickens, ignoring any movement among ecoregions, was the Ricker + Year + Period, but Gompertz + Year + Period was also strongly supported as a model (Table 4.14). The Ricker + Year + Period model indicated a declining annual trend of -1.7% over the assessment period ($r_t = 34.901 - 0.232\ N_t - 0.0174$ year $- 0.266$ period, $\sigma = 0.1402$, $r^2 = 0.43$). A parametric bootstrap based on the Ricker model with declining time trend indicated a low likelihood (4.8%) that

the range-wide population will decline below $N_e = 50$ and a low likelihood (15.2%) that it will decline below $N_e = 500$ in 30 years (Table 4.6). If the trend were to continue for 100 years, there is a low likelihood (<18.5%) that the range-wide population will decline below $N_e = 50$ but a stronger likelihood (83.5%) that it will decline below $N_e = 500$.

Metapopulation Analysis

Evaluating the most strongly supported models across all four ecoregions implied that the two simple density-dependent models without lag or period effects but with a strong time trend in carrying capacity constituted the entire 95% model set (Gompertz + Year and Ricker + Year, Table 4.15).

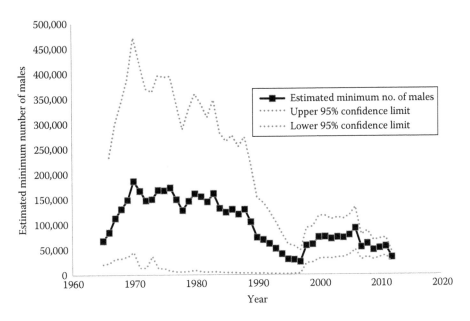

Figure 4.5. Estimated range-wide (in the central United States) minimum number of Lesser Prairie-Chickens attending leks 1964–2012 based on population reconstruction.

TABLE 4.14

Candidate model set (contains 95% of model weight) and model statistics for estimating population trends and persistence probabilities for the range-wide population of Lesser Prairie-Chickens, 1965–2012.

Model	Model statistics[a]				
	r^2	K	ΔAIC_c	w_i	Σw_i
Ricker + Year, Period	0.428	6	0.0[b]	0.240	0.240
Gompertz + Year, Period	0.394	6	1.3	0.124	0.364
EGPE[c]	0.000	3	1.5	0.112	0.476
Gompertz t-1 + Year, Period	0.348	6	3.0	0.054	0.530
Ricker t-1 + Year, Period	0.348	6	3.0	0.053	0.583
Ricker + Year	0.241	5	3.1	0.051	0.634
Gompertz t-2 + Year, Period	0.341	6	3.3	0.047	0.681
Gompertz t-2 + Year	0.230	5	3.4	0.043	0.724
Ricker t-2 + Year, Period	0.332	6	3.6	0.040	0.764
Ricker t-1 + Year	0.220	5	3.7	0.038	0.802
Gompertz t-1 + Year	0.218	5	3.8	0.037	0.839
Ricker t-2 + Year	0.212	5	3.9	0.033	0.872
Gompertz + Year	0.198	5	4.4	0.027	0.899
Gompertz	0.033	4	5.6	0.015	0.914
Ricker	0.018	4	5.9	0.013	0.927
Ricker t-1	0.020	4	5.9	0.013	0.940
Gompertz t-1	0.014	4	6.0	0.012	0.952

[a] Model fit described by the coefficient of determination (r^2), the number of parameters (K), the difference in Akaike's Information Criterion corrected for small sample size (ΔAIC_c), AIC_c weights (w_i), and cumulative w_i (Σw_i).

[b] AIC_c = -22.1 for the top-ranked model.

[c] EGPE = model of exponential growth with process error.

TABLE 4.15
Highest ranked models across the range of the Lesser Prairie-Chicken, 1964–2012,
based on up to 48 years of lek surveys for four ecoregions.

Model	Three long-term ecoregion datasets[a]			All four ecoregion datasets[b]		
	AIC_c	ΔAIC_c	w_i	AIC_c	ΔAIC_c	w_i
Gompertz + Year	-315.3	0.0	0.651	-340.1	2.3	0.243
Ricker + Year	-313.4	1.9	0.252	-342.3	0.0	0.754
Gompertz + Year, Period	-310.8	4.5	0.069	NA[c]	NA	NA
Ricker + Year, Period	-308.2	7.1	0.019	NA	NA	NA
Gompertz t-2 + Year	-304.8	10.5	0.003	-321.6	20.7	0.004
EGPE[c]	-304.0	11.3	0.002	-330.9	11.4	0.003
Gompertz t-1 + year	-303.1	12.2	0.001	-320.2	22.1	0.002
Ricker t-1 + Year	-303.1	12.2	0.001	-320.4	21.9	0.002
Ricker t-2 + Year	-301.5	13.8	0.001	-318.2	24.1	0.001
Gompertz t-2	-294.7	20.6	0	-315.5	26.9	0.006
Ricker t-2	-293.9	21.4	0	-315.2	27.2	0.007
Gompertz t-1	-292.5	22.8	0	-313.6	28.7	0.007
Ricker t-1	-292.1	23.2	0	-313.2	29.1	0.006

[a] Mixed-Grass Prairie, Sand Sagebrush Prairie, Sand Shinnery Oak Prairie Ecoregions.
[b] Adds Short-Grass Prairie/CRP Ecoregion to three mentioned earlier.
[c] EGPE = model of exponential growth with process error.

We estimated the correlation in rates of population change among the four ecoregions from correlations among residuals of the Gompertz + Year models fit to each population (Table 4.2). Correlations with residuals from the Ricker + Year model were similar, and we used the same set of correlations to model stochastic variation in rates of change for each of the four ecoregions constituting the metapopulation.

Pairwise Φ_{ST} values among the four ecoregions show the pattern one would expect with values near zero among samples from the same populations, and larger values between the more isolated Sand Shinnery Oak Prairie Ecoregion and the other three more closely spaced ecoregions (Tables 4.1 and 4.2). The estimator of Hudson et al. (1992) yielded reasonable estimates of numbers of migrants among these ecoregions (Table 4.2). Assuming an average generation length of 2.5 years and average effective population sizes estimated from population reconstruction over the past 45 years, estimates of annual movement rates of Lesser Prairie-Chickens between ecoregions vary from a low of <1 in 10,000 birds moving between the isolated Sand Shinnery Oak Prairie Ecoregion and the other more closely spaced ecoregions and a high of >1 in 10,000 birds moving among the three more closely spaced ecoregions in the northern half of the range (Table 4.2).

Multimodel forecasts of probability of persistence for Lesser Prairie-Chickens in individual ecoregions (Table 4.6) were similar to metapopulation estimates (Table 4.16), but the estimates for metapopulations were almost always greater. All ecoregions show no short- (30 years) or long-term (100 years) probability of extinction exceeding 50%. A few estimates approach the threshold of 50% but variance around the estimates make inference difficult. The predictions differ substantially from forecasts of significant likelihood of extinction with abundance $<N_e = 50$ or $<N_e = 500$ for the Mixed-Grass Prairie Ecoregion, Sand Sagebrush Prairie Ecoregion, and Range-wide population under simpler models assuming no movement among populations in different ecoregions (Table 4.6).

DISCUSSION

Our analyses have provided new insights into the population dynamics of Lesser Prairie-Chickens in four discrete ecoregions and their range-wide metapopulation structure. Density-dependent models of population dynamics best described long-term fluctuations in bird numbers across the species range. Lesser Prairie-Chicken populations

TABLE 4.16

Multimodel metapopulation forecasts of probability (weighted mean percentage and SE) of number of birds counted in four ecoregions and the range-wide population of Lesser Prairie-Chickens declining to abundances below quasi-extinction levels representing $N_e = 50$ and $N_e = 500$ total breeding adults within 30 or 100 years from 2012.

Population	Prob ($<N_e$) in 30 years (%)		Prob ($<N_e$) in 100 years (%)	
	$N_e = 50$[a]	$N_e = 500$	$N_e = 50$	$N_e = 500$
Mixed-Grass Prairie Ecoregion				
Probability	12.1	36.6	37.7	49.9
SE	15.3	46.4	47.9	39.7
Sand Sagebrush Prairie Ecoregion				
Probability	37.7	49.7	37.8	49.9
SE	47.9	39.8	47.7	39.7
Sand Shinnery Oak Prairie Ecoregion				
Probability	0.0	13.1	1.2	44.9
SE	0.0	14.3	1.5	40.7
Short-Grass Prairie/CRP Ecoregion				
Probability	24.5	34.3	37.7	37.7
SE	31.1	43.6	47.9	47.9
Range-wide				
Probability	0.0	0.9	0.2	34.8
SE	0.0	1.2	0.2	44.2

[a] $N_e50 = 85$ birds counted and $N_e500 = 852$ birds counted at leks based on range-wide estimates and minimum method of Nunney and Elam (1994) and Pruett et al. (2011).

apparently fluctuate around a dynamic carrying capacity and were not adequately characterized by a single-trend line. However, the Mixed-Grass Prairie and Sand Sagebrush Prairie Ecoregions had declining trends in both carrying capacity and abundance. The Sand Shinnery Oak Prairie Ecoregion has been relatively stable, with a population fluctuating around a carrying capacity of ~4,500 birds. The stochastic variation was greatest in the Sand Shinnery Oak Prairie Ecoregion, which might be expected given more frequent episodes of severe drought and extreme weather events that negatively impact components of reproduction (Grisham et al. 2013). Surprisingly, even the relatively short time period (11 years) during which the Short-Grass Prairie/CRP Ecoregion had been monitored, it too exhibited strong density dependence, and grew by 8% annually. Combined, range-wide carrying capacity stabilized in the latter part of the assessment period and is projected to increase modestly (6%) by the year 2042. None of the populations exhibited lag effects in density dependence. Lack of lag effects in prairie chickens was surprising

compared to Greater Sage-Grouse, where the best models for six of seven Management Zones included either 1- or 2-year lag effects (Garton et al. 2011). Lag effects might be more likely in sage-grouse than prairie chickens because sage grouse are a long-lived species with stronger patterns of age-specific variation in productivity and survival (Patten et al. 2005, Hagen et al. 2009, Connelly et al. 2011). However, the Sand Shinnery Oak Prairie Ecoregion population of prairie chickens has a life history strategy more akin to Greater Sage-Grouse with lower productivity and higher annual survival (Patten et al. 2005, Grisham 2013), but did not demonstrate lag effects in density dependence. We conclude that yearling and second-year Lesser Prairie-Chickens are more likely to successfully breed than the same age classes of Greater Sage-Grouse and thereby minimize potential lag effects of annual production.

Our metapopulation analyses provided two salient pieces of inference. First, modest estimates of dispersal rates among ecoregions afforded some level of "rescue effect" for Mixed-Grass Prairie and

Sand Sagebrush Prairie Ecoregions. In both cases, the probability of persistence was significantly improved when dispersal was accounted for in the stochastic metapopulation models. Consequently, the long-term probability of persistence increased markedly for a range-wide population versus a range-wide population without movement among ecoregions. Our findings emphasize the importance of conservation actions or policy designed to increase the quality and quantity of available habitat in core areas and ensure connectivity among them (Van Pelt et al. 2013).

Reconstructed populations show an increase in Lesser Prairie-Chicken populations in the last half of the 1960s, followed by long-term declines through the 1980s until the mid-1990s when population numbers apparently stabilized. Stabilization was coincident with improvements in multiple habitat factors, including cessation of conversion of grasslands to center-pivot irrigation in Kansas, maturation of CRP grasslands, and increased conservation efforts as the species was petitioned for listing under Endangered Species Act in 1995. One striking observation about the pattern of change range-wide is that the period of substantial long-term decline and fairly dramatic 5-year periodicity in abundance from 1970 to 1990 is synchronous with the pattern observed in Greater Sage-Grouse in the Great Plains Management Zone (Garton et al. 2011:301, Figure 15.2e). Moreover, the best models describing Greater Sage-Grouse dynamics in the Great Plains were also Gompertz density-dependent models with and without declining carrying capacities through time (Garton et al. 2011:315, Table 15.12). In both cases, we are struck by the modest synchrony of periodicity in lek counts overlaid on a long-term decline lasting about 25 years followed by 15 years of an apparently stabilized population that does not exhibit predictable fluctuations.

The coefficient of variation (CV) for the ground survey estimate in the Sand Sagebrush Prairie Ecoregion (0.27) was lower than estimates obtained from aerial surveys (0.77), which was the case for most of the CVs obtained from ground counts and aerial surveys in 2012. Greater precision underscores the importance of ground counts and their potential to provide precise estimates of the minimum number of birds present at leks, if conducted within a probability sampling framework like the methods used for the aerial surveys. Adopting a robust sampling design for future lek counts would benefit population estimates for Lesser Prairie-Chickens (Garton et al. 2011).

Estimates of short- and long-term persistence for Lesser Prairie-Chickens based on population reconstructions from ≥ 45 years of lek counts provide optimism for potential reverses of recent declines when assessed as a metapopulation. The fact that 68% of the current range-wide population shows stable to increasing year coefficients implies that carrying capacities have been and will continue to be increasing. However, the other 32% of the range in the Sand Sagebrush Prairie and Mixed-Grass Prairie Ecoregions will require targeted conservation efforts to reverse current population trends. Treating the Sand Sagebrush Prairie Ecoregion as an independent population forecasts significant likelihoods of extirpation in all but 30-year $N_e = 50$ scenarios. Similarly, the Mixed-Grass Prairie Ecoregion had a significant likelihood of extinction for the 100-year $N_e = 500$ scenarios, but the variation was too great to be conclusive.

Potential Trajectories

A critical assumption of our metapopulation forecasts for the next 30 or 100 years is that dispersal among ecoregions will occur at the same rate as we estimated from genetic studies. The Φ_{ST} we used was based on mtDNA or maternally inherited DNA, and the estimates represent historic dispersal rates coupled with successful breeding. Current levels of habitat fragmentation may restrict movement to levels lower than occurred historically and lower than what we assumed in our analyses. Thus, future probabilities of persistence may lie between forecasts from the isolated independent population models and the range-wide metapopulation model if there are no future changes in connectivity and extent and quality of necessary habitats.

A series of factors may affect connectivity in the future, namely, anthropogenic developments from fossil fuel extraction, wind energy development, biofuel development, and urban development, encroachment by invasive woody plants, and continued conversion of grasslands to agricultural production. Energy development and transmission can affect space use, habitat selection, and movement patterns of Lesser Prairie-Chickens (Pitman et al. 2005, Hagen 2010, Hagen et al. 2011, Pruett et al. 2011). The mechanisms that alter Lesser Prairie-Chicken behavior relative to these developments are not well understood, but the evidence is growing for displacement and avoidance of such features. Similarly, the encroachment of invasive woody plants, such as eastern red-cedar (*Juniperus virginiana*) or honey mesquite (*Prosopis glandulosa*), has the potential to serve as a barrier to movement. Although not directly measured in Lesser Prairie-Chicken populations, both Greater Sage-Grouse and Greater Prairie-Chicken (*T. cupido*) exhibit avoidance of otherwise suitable habitat with as little as 5% cover of invasive woody plants (Doherty et al. 2008). Recent work provides growing evidence that encroachment by invasive woody plants is negatively related to population growth or occupancy of all three species of grouse (Fuhlendorf et al. 2002, McNew et al. 2012, Baruch-Mordo et al. 2013). Invasive woody plant encroachment is often the result of land use changes, primarily changes in use of prescribed fire or livestock grazing management (Fuhlendorf et al. 2002). Thus, proactive approaches to encourage reintroducing fire to the landscape in a strategic manner could assist in enhancing connectivity both within and among these ecoregions.

One particular concern is the role that CRP plays in the trajectory of Lesser Prairie-Chicken in the Short-Grass Prairie/CRP Ecoregion (Rodgers and Hoffman 2005, Fields et al. 2006), and the potential for this population to serve as a future source population for other nearby regions, especially if there are significant amounts of prairie conversion and losses of CRP acreage to uses incompatible with Lesser Prairie-Chicken requirements. Early projections suggested that as much as 14% of the species' range would be lost to conversion of CRP to cropland between 2008 and 2012 (U.S. Fish and Wildlife Service 2012). However, these losses have not occurred and >80% of CRP contracts that expired have been retained as grasslands either through reenrollment in the program or maintained as grasslands for other uses (C.A. Hagen, unpubl. data). Notwithstanding, voluntary conservation actions and policy to maintain and manage existing CRP fields as grasslands should remain a high priority range-wide, but especially in the Short-Grass Prairie/CRP Ecoregion. Otherwise, significant losses of these grasslands will likely affect our forecasts of carrying capacity and population persistence range-wide.

Conservation Implications

Prairie grouse appear to be a group of area-sensitive species (Ryan et al. 1998, Fuhlendorf et al. 2002, McNew et al. 2012), and their population dynamics are largely driven by components of reproductive output such as nest success and chick survival (Wisdom and Mills 1997, Peterson et al. 1998, Hagen et al. 2009). Our analyses indicated that populations consistently exhibited density dependence, but with geographic variation in the trends of carrying capacity and abundance. Our metapopulation analyses reaffirmed the importance of maintaining connectivity within and among ecoregions. Armed with new knowledge, we can begin to strategically target conservation actions in each ecoregion that will simultaneously improve conditions for reproduction and connectivity.

Consistent with the Range-wide Conservation Strategy for Lesser Prairie-Chickens (Van Pelt et al. 2013), our work emphasizes the importance of avoiding energy development and agricultural production that would further fragment or disconnect focal areas from one another. While our finding should apply range-wide, reducing habitat loss is especially important in portions of the Mixed-Grass Prairie and Sand Shinnery Oak Prairie Ecoregions where policies on oil- and gas-well density and placement are less restrictive (Van Pelt et al. 2013). Additionally, an aggressive strategy is needed to address the encroachment of cedars in the Mixed-Grass Ecoregion and mesquite in the Sand Shinnery Oak Prairie Ecoregion to improve nesting and brood rearing at a site scale and connectivity at a landscape scale. Creation of larger areas of suitable habitat will provide a myriad of benefits, and perhaps provide a stronger buffer to severe and extended droughts. Similarly, in the Short-grass Prairie/CRP Ecoregion, a strategy for retaining CRP grasslands as grasslands and managing habitat conditions for nesting and brood rearing will increase the likelihood that abundance and carrying capacity will at a minimum be maintained, but optimally those parameters will continue to increase. Much of the potentially suitable habitat in the Sand Sagebrush Prairie Ecoregion has been converted to center-pivot irrigation agriculture, and reversing or stabilizing of population declines will likely require a strategic approach to habitat restoration. The Range-wide Conservation Plan and U.S. Department of Agriculture Natural

Resources Conservation Service Lesser Prairie-Chicken Initiative contain new, innovative conservation strategies with a framework for implementation, which have the potential to positively affect future trajectories relative to projections based on models of current and historical stochastic processes describing abundance of Lesser Prairie-Chickens.

ACKNOWLEDGMENTS

We thank numerous biologists and volunteers from over the years for their diligence in collecting these data, often under challenging field conditions. We recognize the many landowners who have granted access to private lands to make many of the survey counts possible. M. East of the Natural Heritage New Mexico Program provided data and invaluable assistance in data retrieval for lek counts in New Mexico. D. Klute of Colorado Parks and Wildlife provided useful comments on the manuscript and insights on the count data from Colorado. Our work was supported in part from a National Fish and Wildlife Foundation and Pheasants Forever grant to Oregon State University. L. Wait provided suggestions on analyzing genetic patterns and dispersal that were helpful. Comments of A. Gregory and an anonymous reviewer greatly improved the quality of this manuscript.

LITERATURE CITED

Akaike, H. 1973. Information theory as an extension of the maximum likelihood principle. Pp. 37–52 in B. N. Petrov and F. Csaki (editors), Second International Symposium on Information Theory. Akademiai Kiado, Budapest, Hungary.

Baruch-Mordo, S., J. S. Evans, J. P. Severson, D. E. Naugle, J. D. Maestas, J. M. Kiesecker, M. J. Falkowski, C. A. Hagen, and K. P. Reese. 2013. Saving Sage-Grouse from trees: a proactive solution to reducing a key threat to candidate species. Biological Conservation 167:233–241.

Beck, T. D. I., and C. E. Braun. 1980. The strutting ground count: variation, traditionalism, management needs. Proceedings of the Western Association of Fish and Wildlife Agencies 60:558–566.

Behney, A. C., B. A. Grisham, C. W. Boal, H. A. Whitlaw, and D. A. Haukos. 2012. Sexual selection and mating chronology of Lesser Prairie-Chickens. Wilson Journal of Ornithology 124:96–105.

Beissinger, S. R., and M. I. Westphal. 1998. On the use of demographic models of population viability in endangered species management. Journal of Wildlife Management 62:821–841.

Buckland, S. T., K. P. Burnham, and N. H. Augustine. 1997. Model selection: an integral part of inference. Biometrics 53:603–618.

Burnham, K. P., and D. R. Anderson. 2002. Model selection and multimodel inference: a practical information-theoretic approach. Springer, New York, NY.

Connelly, J. W., C. A. Hagen, and M. A. Schroeder. 2011. Characteristics and dynamics of Greater Sage-Grouse populations. Pp. 53–68 in S. T. Knick and J. W. Connelly (editors), Ecology and conservation of Greater Sage-Grouse: a landscape species and its habitats. Studies in Avian Biology (no. 38), University of California Press, Berkeley, CA.

Connelly, J. W., S. T. Knick, M. A. Schroeder, and S. J. Stiver. 2004. Conservation assessment of Greater Sage-Grouse and sagebrush habitats. Western Association of Fish and Wildlife Agencies, Cheyenne, WY.

Connelly, J. W., K. P. Reese, and M. A. Schroeder. 2003. Monitoring of Greater Sage-Grouse habitats and populations. College of Natural Resources Experiment Station Bulletin 80. University of Idaho, Moscow, ID.

Davis, D. M. 2009. Nesting ecology and reproductive success of Lesser Prairie-Chickens in shinnery oak–dominated rangelands. Wilson Journal of Ornithology 121:322–327.

Davis, D. M., R. E. Horton, E. A. Odell, R. D. Rodgers, and H. A. Whitlaw. [online]. 2008. Lesser Prairie-Chicken conservation initiative. Lesser Prairie-Chicken Interstate Working Group, Colorado Division of Wildlife, Fort Collins, CO. <http://www.wafwa.org/documents/LPCCI_FINAL.pdf> (12 December 2013).

Dennis, B., P. L. Munholland, and J. M. Scott. 1991. Estimation of growth and extinction parameters for endangered species. Ecological Monographs 61:115–143.

Dennis, B., J. M. Ponciano, S. R. Lele, and M. Taper. 2006. Estimating density dependence, process noise and observation error. Ecological Monographs 76:323–341.

Dennis, B., and M. L. Taper. 1994. Density dependence in time series observations of natural populations: estimation and testing. Ecological Monographs 64:205–224.

Doherty, K. E., D. E. Naugle, B. L. Walker, and J. M. Graham. 2008. Greater Sage-Grouse winter habitat selection and energy development. Journal of Wildlife Management 72:187–195.

Fields, T. L. 2004. Breeding season habitat use of conservation reserve program (CRP) land by Lesser Prairie-Chickens in west central Kansas. M.S. thesis, Colorado State University, Fort Collins, CO.

Fields, T. L., G. C. White, W. C. Gilgert, and R. D. Rodgers. 2006. Nest and brood survival of Lesser Prairie-Chickens in west central Kansas. Journal of Wildlife Management 70:931–938.

Franklin, I. R. 1980. Evolutionary change in small populations. Pp. 135–139 in M. E. Soule and B. A. Wilcox (editors), Conservation biology: an ecological–evolutionary perspective. Sinauer Associates, Sunderland, MA.

Fuhlendorf, S. D., A. J. W. Woodward, D. M. Leslie, Jr., and J. S. Shackford. 2002. Multi-scale effects of habitat loss and fragmentation on Lesser Prairie-Chicken populations. Landscape Ecology 17:617–628.

Garton, E. O. 2013. An assessment of population dynamics and persistence of Lesser Prairie-Chickens. Final Report to Western Association of Fish and Wildlife Agencies, Moscow, ID.

Garton, E. O., J. W. Connelly, J. S. Horne, C. A. Hagen, A. Moser, and M. A. Schroeder. 2011. Greater Sage-Grouse population dynamics and probability of persistence. Pp. 293–382 in S. T. Knick and J. W. Connelly (editors), Ecology and conservation of Greater Sage-Grouse: a landscape species and its habitats. Studies in Avian Biology (no. 38), University of California Press, Berkeley, CA.

Goodman, L. A. 1962. The variance of the product of K random variables. Journal of American Statistical Association 57:54–60.

Grisham, B. A. 2013. The ecology of Lesser Prairie-Chickens in shinnery oak–grassland communities in New Mexico and Texas with implications toward habitat management and future climate change. Ph.D. dissertation, Texas Tech University, Lubbock, TX.

Grisham, B. A., C. W. Boal, D. A. Haukos, D. M. Davis, K. K. Boydston, C. Dixon, and W. R. Heck. 2013. The predicted influence of climate change on Lesser Prairie-Chicken reproductive parameters. PLoS One 8:e68225.

Hagen, C. A. 2010. Impacts of energy development on prairie grouse ecology: a research synthesis. Transactions of the North American Wildlife and Natural Resources Conference 75:98–105.

Hagen, C. A., and K. M. Giesen. [online]. 2004. Lesser Prairie-Chicken (Tympanuchus pallidicinctus). Birds of North America 364. <http://bna.birds.cornell.edu/bna/species/364/articles/introduction> (15 November 2013).

Hagen, C. A., J. C. Pitman, T. L. Loughin, B. K. Sandercock, R. J. Robel, and R. D. Applegate. 2011. Potential impacts of anthropogenic features on Lesser Prairie-Chicken habitat use. Pp. 63–75 in S. T. Knick and J. W. Connelly (editors), Ecology and conservation of Greater Sage-Grouse: a landscape species and its habitats. Studies in Avian Biology (no. 38), University of California Press, Berkeley, CA.

Hagen, C. A., J. C. Pitman, B. K. Sandercock, R. J. Robel, and R. D. Applegate. 2005. Age-specific variation in apparent survival rates of male Lesser Prairie-Chickens. Condor 107:78–86.

Hagen, C. A., J. C. Pitman, B. K. Sandercock, R. J. Robel, and R. D. Applegate. 2009. Spatial variation in Lesser Prairie-Chicken demography: a sensitivity analysis of population dynamics and management alternatives. Journal of Wildlife Management 73:1325–1332.

Hagen, C. A., J. C. Pitman, B. K. Sandercock, D. H. Wolfe, R. J. Robel, R. D. Applegate, and S. J. Oyler-McCance. 2010. Regional variation in mtDNA of the Lesser Prairie-Chicken. Condor 112:29–37.

Hudson, R. R., M. Slatkin, and W. P. Maddison. 1992. Estimation of levels of gene flow from DNA sequence data. Genetics 132:583–589.

Krebs, C. J. 1999. Ecological methodology, 3rd edn. Addison Wesley Educational Publishers, Menlo Park, CA.

Lyons, E. K., B. A. Collier, N. J. Silvy, R. R. Lopez, B. E. Toole, R. S. Jones, and S. J. Demaso. 2009. Breeding and non-breeding survival of Lesser Prairie-Chickens Tympanuchus pallidicinctus in Texas, USA. Wildlife Biology 15:89–96.

McDonald, L., G. Beauprez, G. Gardner, J. Griswold, C. Hagen, D. Klute, S. Kyle, J. Pitman, T. Rintz, and B. Van Pelt. 2014. Range-wide population size of the Lesser Prairie-Chicken: 2012 and 2013. Wildlife Society Bulletin 38:536–546.

McNew, L. B., T. J. Prebyl, and B. K. Sandercock. 2012. Effects of rangeland management on site occupancy dynamics of prairie chickens in a protected prairie preserve. Journal of Wildlife Management 76:36–47.

Nunney, L., and D. R. Elam. 1994. Estimating the effective population size of conserved populations. Conservation Biology 8:175–184.

Patten, M. A., D. H. Wolfe, E. Shochat, and S. K. Sherrod. 2005. Effects of microhabitat and microclimate on adult survivorship of the Lesser Prairie-Chicken. Journal of Wildlife Management 69:1270–1278.

Peterson, M. J., W. T. Grant, and N. J. Silvy. 1998. Simulation of reproductive stages limiting productivity of the endangered Attwater's Prairie-Chicken. Ecological Modelling 111:283–295.

Pitman, J. C., C. A. Hagen, R. J. Robel, T. M. Loughin, and R. D. Applegate. 2005. Location and success of Lesser Prairie-Chicken nests in relation to human disturbance. Journal of Wildlife Management 69:1259–1269.

Pitman, J. C., B. E. Jamison, C. A. Hagen, R. J. Robel, T. M. Loughin, and R. D. Applegate. 2006. Brood break-up and dispersal of juvenile Lesser Prairie-Chickens in Kansas. Prairie Naturalist 38:86–99.

Pruett, C. L., J. A. Johnson, L. C. Larsson, D. H. Wolfe, and M. A. Patten. 2011. Low effective population size and survivorship in a grassland grouse. Conservation Genetics 12:1205–1214.

Reed, J. M., L. S. Mills, J. B. Dunning, E. S. Menges, K. S. McKelvey, R. Frye, S. R. Beissinger, M.-C. Anstett, and P. Miller. 2002. Emerging issues in population viability analyses. Conservation Biology 16:7–19.

Remington, R. D., and M. A. Schork. 1970. Statistics with application to the biological and health sciences. Prentice-Hall, Englewood Cliffs, NJ.

Rodgers, R. D., and R. W. Hoffman. 2005. Prairie grouse population response to conservation reserve grasslands: an overview. Pp. 120–128 in A. W. Allen and M. W. Vandever (editors), The Conservation Reserve Program—Planting for the Future: proceedings of a National Conference. USGS Biological Resources Division, Scientific Investigation Report 2005-5145. USGS Biological Resources Division, Fort Collins, CO.

Ryan, M. R., L. W. Burger, D. P. Jones, and A. P. Wywialowski. 1998. Breeding ecology of Greater Prairie-Chickens (Tympanuchus cupido) in relation to prairie landscape configuration. American Midland Naturalist 140:111–121.

Scheaffer, R. L., W. Mendenhall, III, and R. L. Ott. 1996. Elementary survey sampling. Wadsworth Publishing, Belmont, CA.

Soulé, M. E. 1980. Thresholds for survival: maintaining fitness and evolutionary potential. Pp. 151–169 in M. E. Soule and B. A. Wilcox (editors), Conservation biology: an ecological–evolutionary perspective. Sinauer Associates, Sunderland, MA.

Staples, D. F., M. L. Taper, and B. Dennis. 2004. Estimating population trend and process variation for PVA in the presence of sampling error. Ecology 85:923–929.

United States Department of the Interior (USDI). 1966. Rare and endangered fish and wildlife of the United States. Resource Publication 34, Compiled by Committee on Rare and Endangered Wildlife Species, Bureau of Sport Fisheries and Wildlife, Washington, DC.

U.S. Fish and Wildlife Service. 2012. Endangered and Threatened Wildlife and RWPts; Listing the Lesser Prairie-Chicken as a Threatened Species. Federal Register 77:73827–73888.

U.S. Fish and Wildlife Service. 2014. Endangered and threatened wildlife and plants; special rule for the Lesser Prairie-Chicken. Federal Register 79:20074–20085.

Van Pelt, W. E., S. Kyle, J. Pitman, D. Klute, G. Beauprez, D. Schoeling, A. Janus, and J. Haufler. 2013. The Lesser Prairie-Chicken range-wide conservation plan. Western Association of Fish and Wildlife Agencies, Cheyenne, WY.

Walsh, D. P., G. C. White, T. E. Remington, and D. C. Bowden. 2004. Evaluation of the lek-count index for Greater Sage-Grouse. Wildlife Society Bulletin 32:56–68.

Western Association of Fish and Wildlife Agencies (WAFWA). 2008. Greater Sage-Grouse population trends and analysis of lek databases. Unpublished report. Western Association of Fish and Wildlife Agencies, Cheyenne, WY.

Wisdom, M. J., and L. S. Mills 1997. Sensitivity analysis to guide population recovery: prairie chicken as an example. Journal of Wildlife Management 61:302–312.

Zar, J. H. 1984. Biostatistical analysis. Prentice-Hall Inc., Englewood Cliffs, NJ.

CHAPTER FIVE

Genetic Variation and Population Structure in the Prairie Grouse*

IMPLICATIONS FOR CONSERVATION
OF THE LESSER PRAIRIE-CHICKEN

Randy W. DeYoung and Damon L. Williford

Abstract. The Lesser Prairie-Chicken (*Tympanuchus pallidicinctus*) is one of three species of prairie grouse in genus *Tympanuchus* in subfamily Tetraoninae. The prairie grouse experienced range expansion and divergence dating to climatic and habitat shifts that occurred during the Pleistocene glaciations. Ecological niche modeling suggests that repeated episodes of range shifts, expansion, and contraction during the late Pleistocene probably led to periods of population isolation followed by secondary contact and opportunities for hybridization. Taxonomic placement within the genus has been complicated because the species are of recent origin and lineage sorting is incomplete. Despite the low taxonomic resolution afforded by neutral mitochondrial DNA sequence data, the Lesser Prairie-Chickens, Greater Prairie-Chickens (*T. cupido*), and Sharp-tailed Grouse (*T. phasianellus*) display diagnostic morphological and behavioral differentiation attributed to sexual selection associated with their lek mating system. Prairie grouse have declined in abundance and geographic range during the past century due to changes in land use practices that have resulted in the fragmentation and loss of native prairie habitat. Population declines have been most severe for Lesser and Greater Prairie-Chickens, which occupy <15% of their historical range. Unlike populations of Greater Prairie-Chickens, however, most extant populations of Lesser Prairie-Chickens have retained relatively high levels of neutral genetic diversity, though peripheral and isolated sites display increased genetic differentiation or evidence of reduced connectivity among sites. If current trends continue, populations will become increasingly fragmented and isolated, begin to lose genetic diversity, and become increasingly susceptible to stochastic events; similar events have already occurred in relict populations of Greater Prairie-Chickens (*T. c. pinnatus*) in Wisconsin and Illinois, the Attwater's Prairie-Chicken (*T. c. attwateri*) in Texas, and the now extinct Heath Hen (*T. c. cupido*). The fate of all prairie grouse is intimately tied to habitat. Preservation and restoration of habitat, including dispersal corridors, will be critical to maintain a sufficiently large effective population size to ensure the long-term persistence of these grassland specialists.

Key Words: conservation, dispersal, ecological niche modeling, habitat, metapopulation, pinnated grouse, population genetics, prairie grouse, taxonomy.

* DeYoung, R. W. and D. L. Williford. 2016. Genetic variation and population structure in the prairie grouse: Implications for conservation of the Lesser Prairie-Chicken. Pp. 77–97 in D. A. Haukos and C. W. Boal (editors), Ecology and conservation of Lesser Prairie-Chickens. Studies in Avian Biology (no. 48), CRC Press, Boca Raton, FL.

Grouse (Tetraoninae) are a subfamily of Family Phasianidae in the Order Galliformes (American Ornithologists' Union 1998, Sangster et al. 2012). Eight genera and 19 species of grouse and ptarmigan are currently recognized (American Ornithologists' Union 1998, Sangster et al. 2012), including *Falcipennis* (Siberian and Spruce Grouse), *Dendragapus* (Dusky and Sooty Grouse), *Lagopus* (Ptarmigan and Red Grouse), *Tetrao* (Capercaillie and Black Grouse), *Tetrastes* (Hazel and Severtzov's Grouse), *Bonasa* (Ruffed Grouse), *Centrocercus* (Greater and Gunnison's Sage-Grouse), and *Tympanuchus* (prairie grouse). The history of taxon revision within the clade reflects the often difficult process of reconciliation between morphology and genetic data. Grouse have alternately been treated as a distinct family, Tetraonidae (Ridgway and Friedmann 1946, Wetmore 1960), or as a subfamily of Phasianidae (Holman 1964, American Ornithologists' Union 1998). Recent phylogenetic studies based on nuclear and mitochondrial DNA indicate that grouse are nested within Phasianidae (Gutíerrez et al. 2000, Kaiser et al. 2007, Wang et al. 2013). Short (1967) and Johnsgard (1983) recognized six genera, whereas Madge and McGowan (2002) recognized nine genera.

The grouse subfamily dates to at least the Pliocene and diverged from Meleagridinae (or Meleagrididae, the turkeys) about 6.3 million years ago (Drovetski 2003). Grouse originated in the Nearctic region and experienced up to three separate radiations, corresponding to the ancestors of prairie grouse, forest grouse, and ptarmigan, species adapted to tundra and alpine habitats (Drovetski 2003). Speciation events for grouse correlate with major climactic changes during the Pliocene and more recent Pleistocene glaciations, similar to many other avian taxa in North America (Johnson and Cicero 2004). For instance, *Centrocercus*, *Dendragapus*, and *Tympanuchus* diverged 1.4–1.9 million years before present (bp), which dates to a period of major glacial activity (Drovetski 2003, Galla 2013, J. A. Johnson, unpubl. data). Glacial activity and climate-induced range shifts prompted a series of vicariant events that led to speciation (Avise 2000).

Genetic data support the vicariance hypothesis for most speciation events within the grouse family, with the exception of *Tympanuchus*, where neutral genetic markers have been unable to fully resolve relationships within the genus (Ellsworth et al. 1994, Drovetski 2003, Oyler-McCance et al. 2010,

Figure 5.1). Phylogenies based on mitochondrial DNA (mtDNA) have failed to recover Lesser Prairie Chickens (*Tympanuchus pallidicinctus*), Greater Prairie-Chickens (*T. cupido*), and Sharp-tailed grouse (*T. phasianellus*) as reciprocally monophyletic lineages. Many individuals in each species of *Tympanuchus* share mtDNA haplotypes with individuals of the other two species despite obvious morphological, behavioral, and ecological differences (Ellsworth et al. 1994, Palkovacs et al. 2004, Spaulding et al. 2006).

GENETIC AND MORPHOLOGICAL DIFFERENTIATION WITHIN TYMPANUCHUS

Our review is focused on the Lesser Prairie-Chicken. However, comparatively few genetic studies are available for the Lesser Prairie-Chicken, in contrast to other species of grouse with long-standing conservation concerns, such as the Greater Prairie-Chicken or the Greater Sage-Grouse (*C. urophasianus*). Similarities in the behavior, life history, and recent demographic history of *Tympanuchus* suggest that population studies of Greater Prairie-Chickens and Sharp-tailed Grouse are directly relevant to Lesser Prairie-Chickens. Furthermore, the recent radiation of *Tympanuchus* dictates that the evolutionary history and conservation status of the Lesser Prairie-Chicken can be fully understood only in context with the other members of the genus.

Collectively, *Tympanuchus* are referred to as "prairie grouse" due to their adaptation to tall-, mixed-, and short-grass habitats. In addition, prairie chickens are sometimes referred to as "pinnated grouse" because the males have elongated feathers on the sides of the neck that are erected to reveal air sacs during displays; the erect feathers resemble pinnae (external ears) and are termed "pinna feathers" (Hagen and Giesen 2004, Johnson et al. 2011). Species and subspecies of *Tympanuchus* were first described on the basis of minor morphology (including plumage and skeletal size), behavior, and geographic range (Table 5.1). Three subspecies are recognized for Greater Prairie-Chickens, including the extinct Heath Hen (*T. c. cupido*), the Greater Prairie-Chicken (*T. c. pinnatus*), and the endangered Attwater's Prairie-Chicken (*T. c. attwateri*). The Sharp-tailed Grouse are morphologically distinctive from other races of *Tympanuchus* and were once considered a separate genus (Johnsgard 1983, Madge and McGowan 2002). The Sharp-tailed Grouse

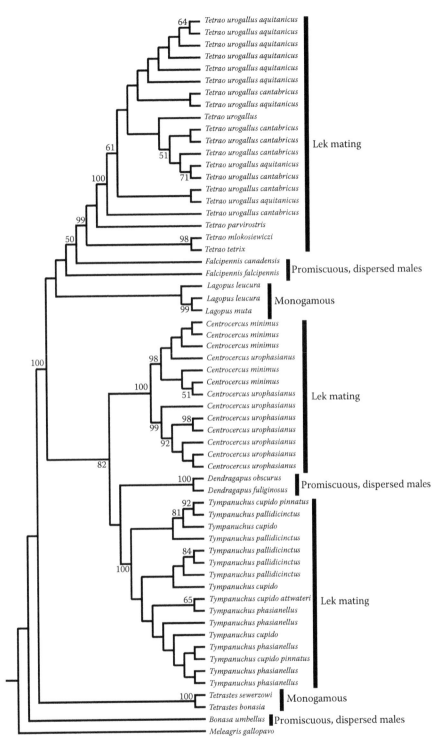

Figure 5.1. Maximum likelihood phylogeny of grouse mtDNA control region sequences (1,022 bp) based on the GTR + I + Γ nucleotide substitution model. Numbers represent the bootstrap support for each node based on 1,000 bootstrap replicates; sequence data were downloaded from GenBank. The prevailing mating system with each clade is indicated. Lek mating species do not form well-supported groups due to the influence of sexual selection on diversification within lineages. (After Oyler-McCance et al. 2010.)

TABLE 5.1

Morphological variation in the genus Tympanuchus, including number of subspecies, body size, plumage and tail characteristics, gular sacs, and behavior.

Taxon (no. subspecies)[a]	Size and mass	Characteristic plumage	Tail	Gular sacs, comb	Behavior[d]
T. cupido (3)[b]	Length: 41–47 cm M: 990 g F: 770 g	Elongated and pointed feathers that form pinnae in M and F, barred markings	Short, rounded	Yellow-orange; yellow	Habitat: native prairie; tri-syllabic, droning "booming" calls during lek display
T. pallidicinctus (0)	Length: 38–41 cm M: 790 g F: 700 g	Pinnae similar to Greater Prairie-Chicken, shorter in females, barred markings on breast	Short, rounded	Reddish-purple; yellow, conspicuous	Habitat: grasslands, low shrubs; vocalizations shorter and higher pitched than Greater Prairie-Chicken, resemble "hooting"
T. phasianellus (7)[c]	Length: 38–48 cm M: 950 g F: 815 g	No pinnae, V-shaped markings on breast	Pointed, whitish	Purple; yellow, inconspicuous	Habitat: variable, includes open and early successional forest, up to 50% woody cover

SOURCES: Johnsgard (1983), Madge and McGowan (2002).

[a] Distinctive morphological features shared by Tetraoninae include the following: Originated in Holarctic and most species are adapted to cool climates. Grouse have short, feathered tarsi without spurs; feathers extend to the base of the toes in some taxa. Feathers cover the nostrils and both sexes possess a flesh-like comb above the eye, which is most conspicuous in males during mating behaviors. Most species perform conspicuous mating displays.

[b] Only two subspecies are extant, the Heath Hen is extinct.

[c] Six extant, one subspecies is extinct.

[d] All Tympanuchus have a lek-breeding, promiscuous mating system.

has an extensive geographic range, where plumage and minor morphology vary extensively; seven subspecies are recognized, including one extinct form and six extant subspecies. The Lesser Prairie-Chicken was formerly considered a geographic race of the Greater Prairie-Chicken and displays little geographic variation in morphology or behavior over its comparatively restricted geographic range (Figure 1.2, Johnsgard 1983, Hagen and Giesen 2004).

A lack of taxonomic resolution within *Tympanuchus* has been attributed to either incomplete lineage sorting or interspecific hybridization, as formerly isolated groups came into secondary contact after the Pleistocene glaciations (Ellsworth et al. 1994). Data from several different mtDNA genes imply that species in genus *Tympanuchus* are of recent origin. Coalescent analyses of mtDNA sequences have dated the divergence of Greater and Lesser Prairie-Chickens to ~10,000–80,000 years bp, corresponding to the Wisconsin glaciation (Johnson 2008). Molecular methods dated the divergence of Sharp-tailed Grouse from Lesser and Greater Prairie-Chickens at ca. 300,000 years bp (Spaulding et al. 2006, J. A. Johnson, unpubl. data; 95% Bayesian highest posterior density [HPD] 200,000–550,000 years bp). A recent multilocus analysis that included autosomal loci estimated divergence between Greater and Lesser Prairie-Chickens at about 170,000 years bp (95% HPD = 80,000–290,000 years bp; Galla 2013, J. A. Johnson, unpubl. data, Galla and Johnson, in review). The multilocus divergence estimates date to the last interglacial or perhaps the Illinoisan glacial period. Similarly, analyses of divergence dates of Sharp-tailed Grouse from other races of Prairie-Chickens produced dates of ca. 500,000 years bp (Spaulding et al. 2006, J. A. Johnson, unpubl. data). The disparity in divergence dates among mitochondrial and autosomal genes is probably due to different rates of molecular evolution and the limited precision of molecular dating methods. Overall, the genetic data support the hypothesis that *Tympanuchus* experienced population expansion and diversification during the Pleistocene and comparatively rapid evolution of morphological traits (Spaulding 2007, Oyler-McCance et al. 2010).

The extent of morphological and behavioral diversity in *Tympanuchus* without corresponding genetic differentiation is anomalous relative to the rest of Tetraonidae. Many species of grouse form well-defined, monophyletic groups of mtDNA haplotypes during phylogenetic analyses (Figure 5.1). However, sexual selection may have played an important role in the recent, rapid diversification within *Tympanuchus* (Spaulding 2007, Oyler-McCance et al. 2010). Grouse exhibit three different types of mating systems (Emlen and Oring 1977): social monogamy (*Lagopus*), promiscuity with dispersed males (or exploded leks, *Bonasa*, *Dendragapus*, and *Falcipennis*), and promiscuity in a classic lek mating system (*Tympanuchus* and *Centrocercus*, Johnsgard 1983, Madge and McGowan 2002). Nonlekking species of grouse usually form monophyletic groups, whereas lekking species display shallow mtDNA gene trees and share mtDNA haplotypes within genera (Figure 5.1). Therefore, the recent and comparatively rapid diversification within *Tympanuchus* appears to have been influenced by sexual selection, where highly skewed mating success that is typical of lek mating systems has helped drive morphological divergence (Emlen and Oring 1977).

Variation in plumage, behavior, and other minor morphology without corresponding differentiation at neutral genetic loci or reciprocal monophyly is a common pattern observed among recently diverged avian taxa that have large ancestral population sizes (Zink 2004). The lack of correspondence between morphology and genetic data is a persistent and vexing issue for avian conservation. Species with broad, continuous geographic ranges often display clinal variation in plumage, body size, song, and other physical characters, adding further complexity to taxonomic identification. Nonetheless, discrete patterns of morphological variation may correlate with adaptive variation that is not detectable via neutral genetic markers due to different rates of character evolution, especially in noncontinuous populations (Winker 2010). Divergence at neutral genetic loci appears minimal within *Tympanuchus*, but species in the genus differ in morphology and behavior, including plumage coloration, body size, habitat selection, and vocalizations (Table 5.1). Many of these physical and behavioral traits are likely to have a genetic basis and are at least partially subject to sexual selection.

Recent divergence within *Tympanuchus* is also supported by the observation that reproductive isolation within the genus appears incomplete and is primarily a function of allopatry among separate geographic ranges. Hybridization occurs infrequently among Greater Prairie-Chicken × Sharp-tailed Grouse (0.3%–3% of offspring) where the species overlap, but the F_1 hybrids are fertile (Johnsgard 1983). Populations of the Greater and Lesser Prairie-Chickens have been allopatric in recent times, but ranges probably overlapped prior to European colonization and during Pleistocene range shifts. Historically, the occurrence of hybridization between Lesser and Greater Prairie-Chickens has been difficult to estimate. Recent range expansions in portions of western Kansas have resulted in range overlap and observations of individuals that exhibit intermediate behavior or coloration, or some mixture of physical traits from both Lesser and Greater Prairie-Chickens (Bain and Farley 2002). The hybrid zone in Kansas has not been subject to detailed study using genetic markers, but preliminary data indicate some degree of introgression between the two species (S. J. Oyler-McCance, unpubl. data).

EFFECTS OF HISTORICAL AND RECENT CHANGES IN HABITAT

Like all grouse, *Tympanuchus* have experienced historical and recent habitat changes and subsequent shifts in geographic range, each of which have affected population differentiation, structure, and genetic variation (Ellsworth et al. 1994, Drovetski 2003, Johnson 2008). Both Greater and Lesser Prairie-Chickens exhibit strong habitat selection for open grasslands to the degree that shrub encroachment may limit their geographic distribution (Fuhlendorf et al. 2002, Johnson et al. 2011). In contrast, Sharp-tailed Grouse appear more tolerant of woody cover and rely on woody vegetation for forage and cover during winter (Johnsgard 1983). Within Tetraoninae, prairie and arid grassland specialists occur only in North America. Pleistocene era changes in amount and geographic of distribution of suitable habitat probably influenced speciation within *Tympanuchus* (Drovetski 2003), as sexual selection resulted in morphological and behavioral divergence (Spaulding 2007, Johnson 2008, Oyler-McCance et al. 2010).

The planet has experienced 10 full glacial cycles during the past 3 million years, with a full cycle comprised of a glacial period lasting 60,000–90,000 years and shorter, warmer interglacial periods of 10,000–40,000 years (Lowe and Walker 1997). Animal and plant species were displaced as ice sheets advanced and changes in temperatures and rainfall patterns led to the fragmentation of species' ranges (Hewitt 2000, 2004). Therefore, the amount and distribution of suitable habitat for *Tympanuchus* species may have varied extensively throughout the past ca. 3 million years of glacial cycles. The fossil record for prairie grouse is sparse, and it is difficult to visualize the magnitude of range shifts, their geographic location, and the degree of allopatry among species of *Tympanuchus*.

Ecological niche models are a recent tool that can illustrate the probable magnitude of climate-induced range shifts during the Pleistocene (Figure 5.2, Phillips et al. 2006, Phillips and Dudík 2008). The maximum entropy approach estimates the geographic distribution of a species based on presence data as the distribution that is closest to uniform, or the distribution of maximum entropy, under the constraint that environmental predictors and functions thereof are close to the average (Phillips et al. 2006). One of the advantages of the maximum entropy method is that it requires only presence data. We used collection locations of museum specimens of prairie grouse as presence data, obtained through the ORNIS database (www.ornisnet.org). We removed duplicate records for specimens collected from localities <1 km apart and records with only country-, state-, province-, or county-level data. The final dataset consisted of 268 records for the Sharp-tailed Grouse, 199 for the Greater Prairie-Chicken, and 54 for the Lesser Prairie-Chicken. We downloaded three raster datasets of 19 bioclimatic variables from WorldClim (www.worldclim.org): contemporary climate (1950–2000, Hijmans et al. 2005), the Last Glacial Maximum (21,000 years bp, Braconnot et al. 2007), and the Last Interglacial Period (130,000 years bp, Otto-Bliesner et al. 2006). We constructed ecological niche models for each species of prairie grouse based on bioclimatic variables from the contemporary range using the subsampling method, with 30% of the presence records randomly selected and removed to test model performance. Each analysis consisted of 25 replicates of 5,000 iterations each, using the default settings for other parameters. The model trained on contemporary bioclimatic variables was then projected to the climatic conditions of the Last Glacial Maximum and the Last Interglacial Period.

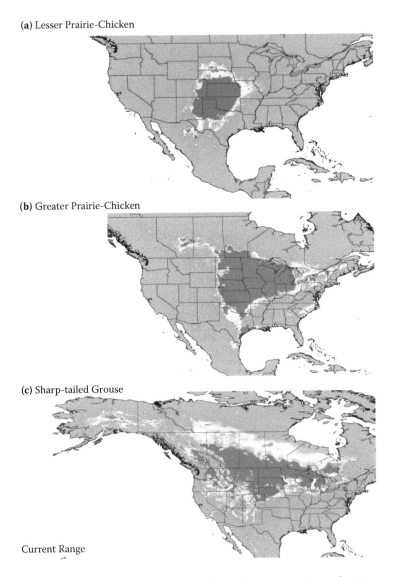

(a) Lesser Prairie-Chicken

(b) Greater Prairie-Chicken

(c) Sharp-tailed Grouse

Current Range

Figure 5.2. Ecological niche models based on the collection localities of museum specimens of prairie grouse (*Tympanuchus*). Warmer colors indicate areas of greater climatic suitability for each species. Niche models were generated using the computer program MaxEnt. (From Phillips et al. 2006, Phillips and Dudík 2008.)

Predicted range shifts were most extreme during the Last Glacial Maximum; current and historical ranges are almost completely allopatric (Figure 5.2), and some were widely separated (Figure 5.3). Predicted ranges during the Last Interglacial period resembled contemporary ranges and are consistent with the recent multilocus estimates of divergence that date the split between the Lesser and Greater Prairie-Chickens to ca. 170,000 years bp (Figure 5.4). Fossil and pollen data largely support the ecological niche model predictions, as tundra, steppe, and cold-adapted forests existed immediately south of the ice sheets (Jackson et al. 2000, Williams et al. 2000), while the southwestern United States and northwestern México were cooler and wetter during the Last Glacial Maximum (Metcalfe et al. 2000, Thompson and Anderson et al. 2000). Fossil and pollen data indicate that Florida, and perhaps the expanded Atlantic and Gulf Coasts, had a milder climate with savanna forests and open woodlands (Russell et al. 2009).

The temporal and spatial details of range expansion can affect genetic diversity within

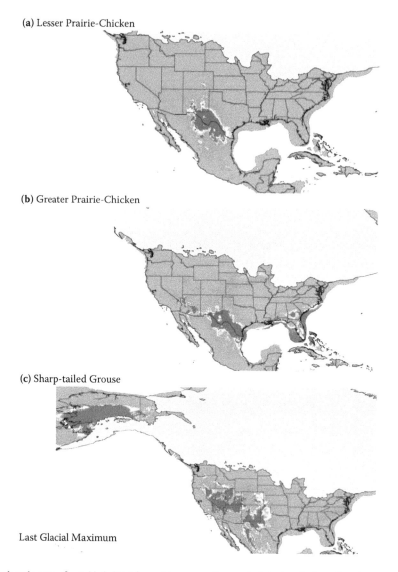

(a) Lesser Prairie-Chicken

(b) Greater Prairie-Chicken

(c) Sharp-tailed Grouse

Last Glacial Maximum

Figure 5.3. Predicted range of suitable habitat for prairie grouse (*Tympanuchus*) during the Last Glacial Maximum (ca. 21,000 years before present) derived from ecological niche modeling. Warmer colors indicate areas of greater climatic suitability for each species. Stippled areas display the geographic extent of the glacial ice. Note the expanded shoreline and land area of North America as a result of a lower sea level during the Last Glacial Maximum. Niche models were generated using the computer program MaxEnt. (From Phillips et al. 2006, Phillips and Dudík 2008.)

populations and differentiation among populations (Excoffier et al. 2009). For instance, coalescent models of range distribution and size revealed that changes in the size or geographic location of a species' range may have disparate effects on patterns of neutral genetic diversity depending on the timing and spatial extent of range shifts (Arenas et al. 2012). Although we cannot predict the relative rapidity of a range shift, the proportion and source of stocks during expansion, or the relative proportion of habitat available during the range shifts, these factors all influence genetic diversity and differentiation. Therefore, it is probable that genetic differentiation within *Tympanuchus* was influenced by chance events during climate-induced changes in population size due to expansion or contraction and to the location of the geographic range of each species, and not strictly to isolation events.

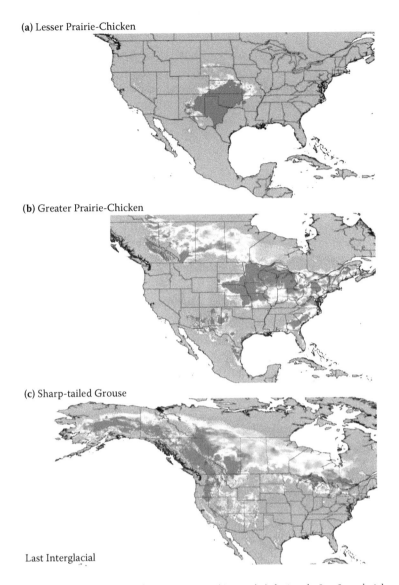

(a) Lesser Prairie-Chicken

(b) Greater Prairie-Chicken

(c) Sharp-tailed Grouse

Last Interglacial

Figure 5.4. Predicted range of suitable habitat for prairie grouse (*Tympanuchus*) during the Last Interglacial period (ca. 130,000 years before present) derived from ecological niche modeling. Warmer colors indicate areas of greater climatic suitability for each species. Niche models were generated using the computer program MaxEnt. (From Phillips et al. 2006, Phillips and Dudík 2008.)

More recently, the abundance and geographic range of the Lesser Prairie-Chicken have declined by more than 90% during the past century (Taylor and Guthery 1980, Hagen and Giesen 2004). The loss of habitat coincided with changes in land use practices throughout the range of prairie chickens, especially the conversion of native prairies to large-scale row crop agriculture or improved pasture (Crawford and Bolen 1976a,b; Hagen and Giesen 2004; Brennan and Kuvlesky 2005). Elsewhere, fire suppression and overgrazing have allowed shrubs to invade formerly continuous grasslands, a process complemented by the intentional planting of trees for windbreaks and shade. The prolonged droughts during the "Dust Bowl" era, the 1950s, and more recent droughts in the 1990s have severely affected populations of Lesser Prairie-Chickens (Sutton 1967, Hagen and Giesen 2004). The current range of the Lesser Prairie-Chicken has not only contracted to a fraction of the original range but has been divided into at least two widely separated regional populations by agricultural and urban development (Figure 1.2, Hagen and Giesen 2004, Corman 2011). Populations in New Mexico and the southwestern panhandle region of Texas are isolated from contiguous populations in Oklahoma, Kansas, Colorado, and the northeastern portion of the Texas Panhandle.

GENETIC DIVERSITY WITHIN AND AMONG POPULATIONS

Changes in the amount and distribution of habitat can have lasting effects on genetic variation within populations. The association of habitat changes and long-term population trends indicates that some formerly contiguous populations of Lesser Prairie-Chicken have become isolated (Figure 5.5a,b), and these isolated populations may be prone to extirpation (Fuhlendorf et al. 2002, Pruett et al. 2011). The effects of recent habitat changes on genetic diversity have been well documented for populations of the Greater Prairie-Chicken (Bouzat et al. 1998a,b; Westemeier et al. 1998; Bellinger et al. 2003; Johnson et al. 2003, 2004), including the extinction of the Heath Hen (Johnson and Dunn 2006) and the current, perilous existence of the remaining individuals of the Attwater's Prairie-Chicken (Johnson and Dunn 2006, Johnson et al. 2007, Bollmer et al. 2011). Similar trends in habitat loss have contributed to an increase in genetic structure and loss of diversity in Sage Grouse (Oyler-McCance et al. 2005a,b). Some populations of Greater Prairie-Chickens have also lost adaptive diversity, including variation at genes that control immune function (Bollmer et al. 2011, Eimes et al. 2011). Accumulation of inbreeding in isolated populations has resulted in reduced egg viability and poor hatching success (Westemeier et al. 1998), and lower survival of inbred progeny (Hammerly et al. 2013).

Few genetic studies have been conducted on Lesser Prairie-Chickens relative to the Greater Prairie-Chicken. However, isolated and peripheral populations of Lesser Prairie-Chickens display lower neutral genetic diversity than observed in continuous populations (Bouzat and Johnson 2004, Hagen et al. 2010, Corman 2011, Pruett et al. 2011). Region-wide, populations of the Lesser Prairie-Chicken in New Mexico and adjacent sites in the southwest panhandle region of Texas have less mtDNA diversity than sites in Oklahoma, Kansas, and the northeastern panhandle region of Texas (Van den Bussche et al. 2003, Hagen et al. 2010, Corman 2011). It is unclear if differences in mtDNA diversity are the result of recent events, or if the observed pattern of mtDNA variation was influenced by historical processes (Van den Bussche et al. 2003, Corman 2011, Pruett et al. 2011). Populations in New Mexico and adjacent habitat in Texas are at the periphery of the historic range of the species. The central-marginal hypothesis predicts that populations at the edge of a species' range tend to have lower diversity than populations near the core (Eckert et al. 2008, but see Bush et al. 2011). Alternatively, the overall process and the relative rapidity of changes in a species' geographic range can influence the maintenance of genetic variation in remnant populations (Excoffier et al. 2009, Arenas et al. 2012). One or more events that have affected the New Mexico population during the past century, such as the Dust Bowl, droughts, or expansion of agriculture or energy development, may have contributed to loss of neutral variation via reduction in the effective population size.

Despite the severe decline in overall geographic range and census size, most extant populations of the Lesser Prairie-Chicken display comparatively high levels of neutral genetic variation (Van den Bussche et al. 2003, Bouzat and Johnson 2004, Hagen et al. 2010, Corman 2011, Pruett et al. 2011, S. J. Oyler-McCance, unpubl. data). No historical specimens of Lesser Prairie-Chickens in museums or other natural history collections have been analyzed yet. However, genetic variation in most populations of the Lesser Prairie-Chicken exceeds that of many populations of Greater Prairie-Chickens that have experienced genetic bottlenecks and have persisted at low census sizes for decades (Bouzat et al. 1998a,b; Bellinger et al. 2003; Johnson et al. 2004, 2007). Genetic variation at mtDNA and microsatellite loci in most populations of Lesser Prairie-Chickens is comparable to levels observed in historical, prebottleneck samples of Greater Prairie-Chickens (Johnson et al. 2004, 2007). Few studies have focused on adaptive traits in Lesser Prairie-Chickens, but there is some evidence that a recent increase in anthropogenic mortality in Oklahoma populations has selected for life history traits that affect clutch size and renesting (Patten et al. 2005).

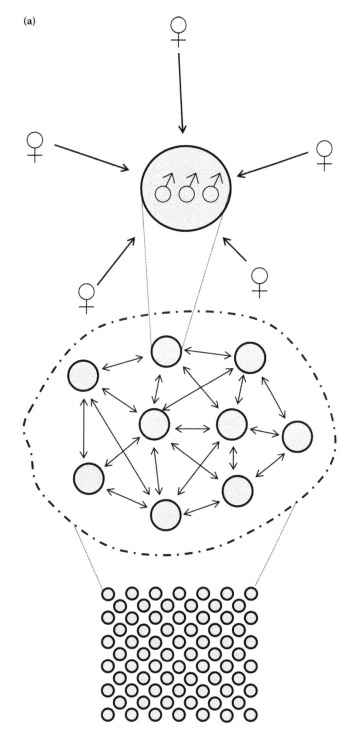

Figure 5.5. (a) A conceptual representation of population structure in Lesser Prairie-Chickens. Male relatives remain near the natal area and form leks composed of related individuals (depicted by circles), while females are the primary dispersers and facilitate gene flow (arrows) among local populations; the overall population is structured as a metapopulation.

(Continued)

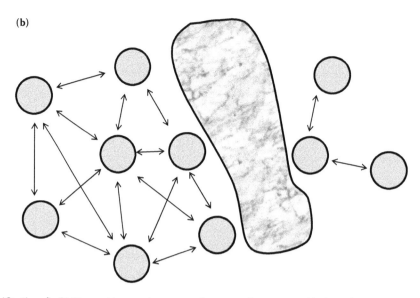

(b)

Figure 5.5. (*Continued*) (b) Dispersal in a continuous population may be interrupted by loss of suitable habitat or disturbance (shaded area), resulting in the isolation of peripheral leks. Population structure may resemble a stepping stone or continent-island model, or populations may become isolated.

It is unclear why most populations of Lesser Prairie-Chickens have retained more neutral variation overall than relict populations of Greater Prairie-Chickens. Many of the genetically depauperate populations of Greater Prairie-Chickens experienced severe declines in abundance during the 1950s to the 1960s and have remained small and isolated since that time (Bouzat et al. 1998a,b; Bellinger et al. 2003; Johnson et al. 2004, 2007). Populations lose genetic diversity during bottlenecks in two stages (Nei et al. 1975, Hedrick 2005). First, rare alleles are lost as population numbers are reduced, but individual heterozygosity may remain relatively high for several generations. The imbalance between population levels of allelic diversity and individual heterozygosity is termed "heterozygosity excess" and reflects disruption in the equilibrium between mutation and drift that occurs in large, stable populations (Cornuet and Luikart 1996). Thus, individuals will temporarily have greater levels of heterozygosity than expected based on the number of alleles in the population until the population reaches a new equilibrium or experiences further loss of variation due to genetic drift. Unless a bottle-necked population recovers rapidly, additional variation will be lost through genetic drift as long as the effective number of breeders remains small. Thus, the genetic effects of reductions in population size may not yet be observed in most populations of Lesser Prairie-Chickens and may not be detectable until a new equilibrium is reached between mutation and drift—a process that may take several generations. Recent studies suggest that the effective population size for several current populations of Lesser Prairie-Chickens may be low enough to result in genetic drift (Corman 2011, Pruett et al. 2011).

It is possible that remaining populations of Lesser Prairie-Chickens have maintained an effective population size sufficient to limit loss of genetic diversity due to skewed mating success or gene flow, at least until recently. Animal mtDNA is maternally inherited and the expected effective size for mtDNA in an idealized population is one-fourth relative to estimates based on neutral markers (Avise 2000). Therefore, one would expect to observe changes in mtDNA diversity and increased levels of population differentiation among isolated populations before the effects become detectable via nuclear genetic markers, such as DNA microsatellites (Avise 2004). However, lek mating systems, characterized by promiscuous mating, skewed male mating success, and female-biased dispersal (Figure 5.5a), are a significant departure from an idealized, random mating population. Lek mating systems, such as the classic lek mating system of prairie grouse, can produce a pattern of gene dynamics that cause the effective size for mtDNA to be nearly as high as for biparentally inherited markers (Chesser and Baker 1996, Johnson et al. 2003).

Female mating at leks was long thought to be skewed toward one or a few males, but several recent studies have found evidence for multiple mating in grouse, where females may mate with >1 male (Hess et al. 2012, Bird et al. 2013, R. DeYoung, unpubl. data). The occurrence of multiple mating increases the effective number of breeders and may slow the rate of genetic drift in local populations (Sugg and Chesser 1994). Differences in clutch size, number of nesting attempts, and variance in reproductive success can also influence the effective size of populations (Stiver et al. 2008, Pruett et al. 2011). Immigration from one or more adjacent populations may erase the signatures of a population bottleneck, depending on the number of immigrants and their reproductive success (Keller et al. 2001). For instance, Greater Sage-Grouse (*Centrocercus urophasianus*) have retained genetic diversity in fragmented areas at the northern extent of the range due to high rates of dispersal (Bush et al. 2011).

A pessimistic view is that demographic collapse may have occurred in many portions of the former range, and populations were extirpated before any loss of genetic diversity could be detected (Lande 1988). Alternatively, habitat alteration or disturbance may have prompted birds from a wide geographic area to congregate on the small amount of remaining suitable habitat (Figure 5.6). Rapid range contraction can maintain high genetic diversity in remnant populations and can produce a signal of low genetic differentiation among populations that persists for many generations (Arenas et al. 2012). Last, genetic sampling of Lesser Prairie-Chickens may simply be biased to larger leks due to ease of sample acquisition, and peripheral sites or populations that have experienced a loss of diversity may be underrepresented.

POPULATION STRUCTURE AND DISPERSAL

Male Lesser Prairie-Chickens gather at communal lek sites to display and compete for breeding opportunities with females (Johnsgard 1983). Lek structure differs among species of grouse, where leks of some species contain male relatives, while others do not (Gibson et al. 2005, Bush et al.

2011). Genetic data have confirmed that Lesser Prairie-Chicken leks often contain male relatives, while females are unrelated and appear to be the primary dispersers (Bouzat and Johnson 2004, Corman 2011). Populations structured around leks connected by dispersal might conform to a type of metapopulation (Segelbacher et al. 2003, Figures 5.5a and 5.6). The metapopulation is a conceptual model for populations distributed in an array of habitat patches, with some level of migration among patches (Levins 1969, Hanski 1998). The metapopulation is a flexible model that may be extended to a wide range of spatial scales, habitat configurations, extinction and recolonization scenarios, and dispersal rates, all of which have important consequences for population structure and, ultimately, population persistence (Hanski and Gilpin 1997). Unlike some northern populations, grouse that inhabit southerly regions of North America do not display cyclic population dynamics (Williams et al. 2004). Nonetheless, prairie grouse display high rates of population turnover due to life history and the influence of extrinsic variables, such as weather and habitat quality.

Lesser Prairie-Chickens have a r-selected life history strategy, where clutch sizes average about 10–12 eggs, and annual adult survival rates range from about 30% to 45%; maximum observed longevity is 5 years (Hagen and Giesen 2004). In addition, recruitment, harvest rates, and population trends are influenced by annual variation in precipitation in the arid grassland that provide habitat for Lesser Prairie-Chickens (Brown 1978, Giesen 2000). For instance, estimates of population density for Lesser Prairie-Chickens may vary more than sevenfold among years (Hagen and Giesen 2004). Therefore, the frequency and magnitude of environmental variation may play a large role on the maintenance of genetic variation in a metapopulation system (Banks et al. 2013). In the absence of dispersal, environmentally mediated fluctuations in population size can act as serial bottlenecks, where genetic diversity is lost through genetic drift during the low phase (Nei et al. 1975, Wright 1978).

Range-wide, the genetic data for populations of Lesser Prairie-Chicken fit a pattern of isolation by distance, which suggests a history of connectivity among populations (Hagen et al. 2010), and similar results have been found for other species of grouse (Johnson et al. 2003,

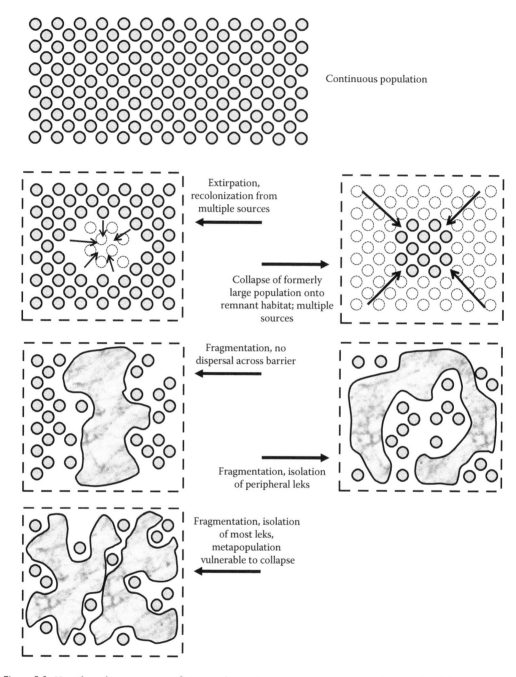

Continuous population

Extirpation, recolonization from multiple sources

Collapse of formerly large population onto remnant habitat; multiple sources

Fragmentation, no dispersal across barrier

Fragmentation, isolation of peripheral leks

Fragmentation, isolation of most leks, metapopulation vulnerable to collapse

Figure 5.6. Hypothetical representation of metapopulation structure in a continuous population, where leks composed of male relatives are connected by the dispersal of females throughout the range (top panel). Changes in habitat may result in the temporary or permanent extirpation of populations from former range, complex recolonization scenarios, and fragmentation, or isolation of leks (middle panels). If leks become isolated, the metapopulation may collapse if small populations become extirpated through stochastic events (bottom panel). Unsuitable habitat is indicated by the shaded polygons; filled circles indicate active leks, while empty circles represent abandoned or extirpated leks.

Segelbacher et al. 2003, Oyler-McCance et al. 2005b). Historical accounts include the suggestion that Lesser Prairie-Chickens made large seasonal movements, though these reports have been difficult to verify and no such movements are known to occur presently (Hagen and Giesen 2004). Most activity of Lesser Prairie-Chickens occurs within 4–5 km of a lek, but occasional, long-distance seasonal movements >30 km have been documented (Woodward et al. 2001, Hagen 2003). Juveniles tend to move greater distances than adults (Copelin 1963, Campbell 1972); natal dispersal for males may be <3 km, whereas natal dispersal of females is 2–25 km (Copelin 1963, Pitman 2003). Spatial autocorrelation of microsatellite DNA loci indicated that males at leks were not independent within 7.5–10 km, a spatial area of 44.2–78.5 km² (Corman 2011). Estimates of spatial autocorrelation based on population genetics agree with field estimates of habitat area derived by radiotelemetry (32–72 km², Taylor and Guthery 1980; 25.2–61.9 km², Giesen 1991) and may provide a means to test hypotheses about metapopulation structure.

Metapopulation theory predicts that species with high rates of population turnover must be good dispersers or the metapopulation will collapse (Harrison and Hastings 1996). Immigration and emigration are difficult to track empirically and may be hard to discern using genetic data for recently diverged populations, where dispersal is unbalanced or asymmetrical, or for non-equilibrium populations (Waples and Gagiotti 2006). Most information on dispersal in populations of Lesser Prairie-Chickens was derived from banding or telemetry studies, where not all individuals can be marked or recovered, and the reproductive status of dispersers is often unknown. Lesser Prairie-Chickens have the potential for long-distance dispersal, but their ability to make dispersal movements over unsuitable habitat is not known. Dispersal events may be infrequent or irregular and not observable by short-term studies (Fedy et al. 2008). Furthermore, the handling process and attachment methods of radio-marking might be a handicap for some galliform birds (Guthery and Lusk 2004). Hagen et al. (2006) observed no effect of modern radio-marking techniques on the survival of adult male prairie-chickens, but the effects of radio-marking on movements and survival of females or juveniles are unknown.

PREDICTING THE FUTURE: NEEDS AND TRENDS

Genetic variation is important for long-term persistence of populations and for the preservation of their evolutionary potential (Frankham 1995). If current population trends continue, isolated populations of the Lesser Prairie-Chicken will lose diversity through genetic drift and become vulnerable to extirpation through stochastic events. Local populations may collapse or begin to display signs of inbreeding (Westemeier et al. 1998, Briskie and Mackintosh 2004). Recent estimates of effective number of breeders for several remnant populations are sufficiently low (<100 birds) to result in genetic drift (Corman 2011, Pruett et al. 2011). Furthermore, there is some evidence for recent selection on life history traits that affect reproduction in populations of Lesser Prairie-Chickens in Oklahoma (Patten et al. 2005). Life history changes with increased clutch size and a greater probability of renesting do not seem to be negative, but any selective pressure that results from anthropogenic mortality should be viewed with caution.

Loss of genetic variation is a cause for concern, but the plight of the Lesser Prairie-Chicken is primarily a problem of habitat loss, both amount and spatial extent (Silvy et al. 2004). Concerns about habitat loss are paramount because loss of genetic variation, small population size, amount of habitat, and stochastic events operate in a synergistic, not isolated, manner (Soulé and Mills 1998). Habitat is the key to the maintenance of populations and long-term success of active management, including translocation programs (Bouzat et al. 2009). Insufficient habitat remains to support the long-term viability of most populations, and habitat restoration efforts will be critical to the long-term persistence of prairie grouse. Recently, the plight of the Lesser Prairie-Chicken has prompted their consideration, and subsequent listing, as a threatened species under the Endangered Species Act (U.S. Fish and Wildlife Service 2013, 2014), although the listing rule was vacated by judicial decision in September 2015. Most potential habitat for Lesser Prairie-Chickens occurs on private lands, and the development and implementation of incentive programs may be the only means to reverse the loss of habitat (Hagen et al. 2004, Riley 2004, Silvy et al. 2004). Unfortunately, habitat issues may be exacerbated regardless of active management because climate change may increase the frequency of droughts and severe weather known to affect recruitment in Lesser Prairie-Chickens (Grisham et al. 2013).

Genetic restoration or "genetic rescue" via translocations into inbred populations is a useful management tool with several successful applications, including Greater Prairie-Chickens (Westemeier et al. 1998, Bouzat et al. 2009, Bateson et al. 2014) and the Florida panther (*Puma concolor coryi*) (Hedrick and Fredrickson 2010). It may provide little benefit, however, to release birds if the habitat is not sufficient to support a viable population (Bouzat et al. 2009). Translocation programs for grouse have met with mixed outcomes (Hagen et al. 2004). Notable success stories include the restoration of normal hatching success in populations of Greater Prairie-Chickens via translocation of wild birds into isolated populations (Westemeier et al. 1998). However, the potential for successful management of upland birds via translocations may depend on the availability of wild stocks. In fact, the most successful restoration program for upland game birds in North America involved the dramatic recovery of the Wild Turkey (*Meleagris gallopavo*). Sustaining populations were produced only when wild stocks were released, and Leopold (1944) attributed the poor performance of captive stocks to selection or behavioral changes that occurred during the transition to captivity. Restoration efforts using captive stocks for the Attwater's Prairie-Chicken and the Masked Bobwhite (*Colinus virginianus ridgwayi*) have largely failed to produce sustaining populations. Releases of captive stocks have aided in the maintenance of wild birds, but the populations depend on continuing releases (Hernandez et al. 2006, Pratt 2010). Propagation and release methods for captive galliforms continue to improve, but habitat is not always available. It is possible that the highly dedicated biologists in charge of the restoration programs will someday unlock the secrets to the successful use of captive stocks for restoration of upland birds. However, unless techniques improve, one must take the pragmatic view that captive breeding programs can prevent extinction but may never fully restore wild populations. Once wild populations are extirpated, they may be gone for good.

An additional concern for active management programs, such as translocations or the release of captive stocks, is that the immigrants may erase locally adapted gene complexes (Bouzat et al. 2009). A more complicated problem involves hybridization and introgression between species of *Tympanuchus*. Narrow hybrid zones are relatively common in areas of secondary contact among formerly allopatric taxa and not typically of management concern. However, introgression could dilute remnant genetic stocks and complicate the conservation status of threatened or endangered populations (Rhymer and Simberloff 1996). For instance, managers could face the dilemma of whether to interfere with hybridization if the process threatened to dilute unique lineages (Latch et al. 2006, Gutiérrez et al. 2007).

At present, available information is insufficient to evaluate many key hypotheses about metapopulation structure, effects of habitat alteration, and other processes that are likely to influence the maintenance of genetic variation in remnant populations of Lesser Prairie-Chickens. Successful management of the Lesser Prairie-Chicken depends on the preservation and restoration of habitat. The maintenance of sufficient habitat is critical to ensure effective population size remains large enough to offset genetic drift and sufficient dispersal and gene flow to maintain genetic variation (Segelbacher et al. 2003, Bush et al. 2011). A better understanding of the dispersal process may help to guide management actions, including the factors that cue dispersal events as well as choice of post-dispersal range. Without additional insights into dispersal and gene flow, it may be difficult to understand local population dynamics and structure, as well as response to disturbance, habitat loss or gain, and climate change. Furthermore, an analysis of museum specimens, if sufficient tissues are available, may aid in evaluating historical trends in genetic diversity and population structure. For instance, the analysis of museum specimens may help to determine whether the lower neutral diversity in New Mexico and the southwest Texas Panhandle region is a result of recent or historical events. Analyses of historical specimens could also help to determine if adaptive diversity has been lost.

The establishment of monitoring programs that combine mark-resight, telemetry, and genetic data could greatly improve our understanding of population trends. Genetic data will be most useful in conjunction with demographic data for long-term monitoring of populations because current levels of genetic diversity may not reflect recent trends due to the lag time for the effects of population isolation, bottlenecks, and genetic drift to occur (Landguth et al. 2010). With the continuing development of high-throughput DNA sequencing technology, the detection of neutral and adaptive variation at the DNA sequence level is becoming increasingly feasible. Such analyses could reveal additional details about genetic differentiation within *Tympanuchus* and other grouse and will provide useful information necessary for management decisions.

ACKNOWLEDGMENTS

The authors were supported by the Caesar Kleberg Wildlife Institute during manuscript preparation. Their foray into the population genetics of prairie grouse began in 2007 with funds from the Texas Parks and Wildlife Department. K. Corman, H. Whitlaw, W. Ballard, C. Boal, M. Wallace, S. DeMaso, L. Brennan, R. Perez, and many unpaid volunteers were instrumental in the Texas grouse research. The authors are grateful to J. A. Johnson and S. J. Oyler-McCance for sharing unpublished data, to J. A. Johnson and an anonymous reviewer for insightful comments on prior drafts.

LITERATURE CITED

American Ornithologists' Union. 1998. Check–List of North American Birds, 7th edn. American Ornithologists' Union, Washington, DC.

Arenas, M., N. Ray, M. Currat, and L. Excoffier. 2012. Consequences of range contractions and range shifts on molecular diversity. Molecular Biology and Evolution 29:207–218.

Avise, J. C. 2000. Phylogeography: the history and formation of species. Harvard University Press, Cambridge, MA.

Avise, J. C. 2004. Molecular markers, natural history, and evolution, 2nd edn. Sinauer Associates, Sunderland, MA.

Bain, M. R., and G. H. Farley. 2002. Display by apparent hybrid prairie chickens in a zone of geographic overlap. Condor 104:683–687.

Banks, S. C., G. J. Cary, A. L. Smith, I. D. Davies, D. A. Driscoll, A. M. Gill, D. B. Lindenmayer, and R. Peakall. 2013. How does ecological disturbance influence genetic diversity? Trends in Ecology and Evolution 28:670–678.

Bateson, Z. W., P. O. Dunn, S. D. Hull, A. E. Henschen, J. A. Johnson, and L. A. Whittingham. 2014. Genetic restoration of a threatened population of Greater Prairie-Chickens. Biological Conservation 174:12–19.

Bellinger, M. R., J. A. Johnson, J. Toepfer, and P. Dunn. 2003. Loss of genetic variation in Greater Prairie-Chickens following a population bottleneck in Wisconsin, U.S.A. Conservation Biology 17:717–724.

Bird, K. L., L. A. Cameron, J. E. Carpenter, C. A. Paszkowski, M. S. Boyce, and D. W. Coltman. 2013. The secret sex lives of sage grouse: multiple paternity and intraspecific nest parasitism revealed through genetic analysis. Behavioral Ecology 24:29–38.

Bollmer, J. L., E. A. Ruder, J. A. Johnson, J. A. Eimes, and P. O. Dunn. 2011. Drift and selection influence geographic variation at immune loci of prairie chickens. Molecular Ecology 20:4695–4706.

Bouzat, J. L., H. H. Cheng, H. A. Lewin, R. L. Westemeier, J. D. Brawn, and K. N. Paige. 1998a. Genetic evaluation of a demographic bottleneck in the Greater Prairie-Chicken. Conservation Biology 12:836–843.

Bouzat, J. L., and K. Johnson. 2004. Genetic structure among closely spaced leks in a peripheral population of Lesser Prairie-Chickens. Molecular Ecology 13:499–505.

Bouzat, J. L., J. A. Johnson, J. E. Toepfer, S. A. Simpson, T. L. Esker, and R. L. Westemeier. 2009. Beyond the beneficial effects of translocations as an effective tool for the genetic restoration of isolated populations. Conservation Genetics 10:191–201.

Bouzat, J. L., H. A. Lewin, and K. N. Paige. 1998b. The ghost of genetic diversity past: historical DNA analysis of the Greater Prairie-Chicken. American Naturalist 152:1–6.

Braconnot, P., B. Otto–Bliesner, S. Harrison, S. Joussaume, J.–Y. Peterchmitt, A. Abe-Ouchi, M. Crucifix et al. 2007. Results of PMIP2 coupled simulations of the Mid–Holocene and Last Glacial Maximum—Part 1: experiments and large scale features. Climate of the Past 3:261–277.

Brennan, L. A., and W. P. Kuvlesky, Jr. 2005. North American grassland birds: an unfolding conservation crisis? Journal of Wildlife Management 69:1–13.

Briskie, J. V., and M. Mackintosh. 2004. Hatching failure increases with severity of population bottlenecks in birds. Proceedings of the National Academy of Sciences of the USA 101:558–561.

Brown, D. E. 1978. Grazing, grassland cover, and gamebirds. Transactions of the North American Wildlife Conference 43:477–485.

Bush, K. L., C. K. Dyte, B. J. Moynahan, C. L. Aldridge, H. S. Sauls, A. M. Battazzo, B. L. Walker, K. E. Doherty, J. Tack, J. Carlson, D. Eslinger, J. Nicholson, M. S. Boyce, D. E. Naugle, C. A. Paszkowski, and D. W. Coltman. 2011. Population structure and genetic diversity of Greater Sage-Grouse (*Centrocercus urophasianus*) in fragmented landscapes at the northern edge of their range. Conservation Genetics 12:527–542.

Campbell, H. 1972. A population study of Lesser Prairie Chickens in New Mexico. Journal of Wildlife Management 36:689–699.

Chesser, R. K., and R. J. Baker. 1996. Effective sizes and dynamics of uniparentally and diparentally inherited genes. Genetics 144:1225–1235.

Copelin, F. F. 1963. The Lesser Prairie Chicken in Oklahoma. Oklahoma Wildlife Conservation Department Technical Bulletin 6, Oklahoma City, OK.

Corman, K. S. 2011. Conservation and landscape genetics of Texas Lesser Prairie-Chickens: population structure and differentiation, genetic variability, and effective size. M.S. thesis, Texas A&M University–Kingsville, Kingsville, TX.

Cornuet, J. M., and G. Luikart. 1996. Description and power analysis of two tests for detecting recent population bottlenecks from allele frequency data. Genetics 144:2001–2014.

Crawford, J. A., and E. G. Bolen. 1976a. Effects of land use on Lesser Prairie Chickens in Texas. Journal of Wildlife Management 40:96–104.

Crawford, J. A., and E. G. Bolen. 1976b. Effects of lek disturbances on Lesser Prairie Chickens. Southwestern Naturalist 21:238–240.

Davis, D. M., R. E. Horton, E. A. Odell, R. D. Rodgers, and H. A. Whitlaw. 2008. Lesser Prairie-Chicken conservation initiative. Lesser Prairie-Chicken Interstate Working Group. Unpublished Report. Colorado Division of Wildlife, Fort Collins, CO.

Drovetski, S. V. 2003. Plio-Pleistocene climatic oscillations, Holarctic biogeography and speciation in an avian subfamily. Journal of Biogeography 30:1173–1181.

Eckert, C. G., K. E. Samis, and S. C. Lougheed. 2008. Genetic variation across species' geographical ranges: the central-marginal hypothesis and beyond. Molecular Ecology 17:1170–1188.

Eimes, J. A., J. L. Bollmer, L. A. Whittingham, J. A. Johnson, C. van Oosterhout, and P. O. Dunn. 2011. Rapid loss of MHC class II variation in a bottlenecked population is explained by drift and loss of copy number variation. Journal of Evolutionary Biology 24:1847–1856.

Ellsworth, D. L, R. L. Honeycutt, N. J. Silvy, K. D. Rittenhouse, and M. H. Smith. 1994. Mitochondrial-DNA and nuclear-gene differentiation in North American prairie grouse (Genus *Tympanuchus*). Auk 111:661–671.

Emlen, S. T., and L. W. Oring. 1977. Ecology, sexual selection, and the evolution of mating systems. Science 197:215–223.

Excoffier, L., M. Foll, and R. J. Petit. 2009. Genetic consequences of range expansions. Annual Review of Ecology, Evolution, and Systematics 40:481–501.

Fedy, B. C., K. Martin, C. Ritland, and J. Young. 2008. Genetic and ecological data provide incongruent interpretations of population structure and dispersal in naturally subdivided populations of White-tailed Ptarmigan (*Lagopus leucura*). Molecular Ecology 17:1905–1917.

Frankham, R. 1995. Conservation genetics. Annual Reviews in Genetics 29:305–327.

Fuhlendorf, S. D., A. J. W. Woodward, D. M. Leslie, Jr., and J. S. Shackford. 2002. Multi-scale effects of habitat loss and fragmentation of Lesser Prairie-Chicken populations of the US Southern Great Plains. Landscape Ecology 17:617–628.

Galla, S. J. 2013. Exploring the evolutionary history of North American prairie grouse (Genus: *Tympanuchus*) using multi-locus coalescent analyses. M.S. thesis, University of North Texas, Denton, TX.

Gibson, R. M., D. Pikes, K. S. Delaney, and R. K. Wayne. 2005. Microsatellite DNA analysis shows that Greater Sage-Grouse leks are not kin groups. Molecular Ecology 14:4453–4459.

Giesen, K. M. 1991. Population inventory and habitat use by Lesser Prairie-Chickens in southeast Colorado. Federal Aid in Wildlife Restoration Report W-152-R. Colorado Division of Wildlife, Fort Collins, CO.

Giesen, K. M. 2000. Population status and management of Lesser Prairie-Chicken in Colorado. Prairie Naturalist 32:137–148.

Grisham, B. A., C. W. Boal, D. M. Davis, K. K. Boydstun, C. Dixon, and W. R. Heck. 2013. The predicted influence of climate change on Lesser Prairie-Chicken reproductive parameters. PLoS One 8:e68225.

Guthery, F. S., and J. J. Lusk. 2004. Radiotelemetry studies: are we radio-handicapping Northern Bobwhites? Wildlife Society Bulletin 32:194–201.

Gutiérrez, R. J., G. F. Barrowclough, and J. G. Groth. 2000. A classification of the grouse (Aves: Tetraoninae) based on mitochondrial DNA sequences. Wildlife Biology 6:205–211.

Gutiérrez, R. J., M. Cody, S. Courtney, and A. B. Franklin. 2007. The invasion of Barred Owls and its potential effect on the Spotted Owl: a conservation conundrum. Biological Invasions 9:181–196.

Hagen, C. A. 2003. A demographic analysis of Lesser Prairie-Chicken populations in southwestern Kansas: survival, population viability, and habitat use. Ph.D. dissertation, Kansas State University, Manhattan, KS.

Hagen, C. A., and K. M. Giesen. [online]. 2004. Lesser Prairie-Chicken (*Tympanuchus pallidicinctus*). Birds of North America 364. <http://bna.birds.cornell.edu/bna/species/364/articles/introduction> (20 January 2014).

Hagen, C. A., B. E. Jamison, K. M. Giesen, and T. Z. Riley. 2004. Guidelines for managing Lesser Prairie-Chicken populations and their habitats. Wildlife Society Bulletin 32:69–82.

Hagen, C. A., J. C. Pitman, B. K. Sandercock, D. H. Wolfe, R. J. Robel, R. D. Applegate, and S. J. Oyler-McCance. 2010. Regional variation in mtDNA of the Lesser Prairie-Chicken. Condor 112:29–37.

Hagen, C. A., B. K. Sandercock, J. C. Pitman, R. J. Robel, and R. D. Applegate. 2006. Radiotelemetry survival estimates of Lesser Prairie-Chickens in Kansas: are there transmitter biases? Wildlife Society Bulletin 34:1064–1069.

Hammerly, S. C., M. E. Morrow, and J. A. Johnson. 2013. A comparison of pedigree- and DNA-based measures for identifying inbreeding depression in the critically endangered Attwater's Prairie-Chicken. Molecular Ecology 22:5313–5328.

Hanski, I. 1998. Metapopulation dynamics. *Nature* 396:41–49.

Hanski, I., and M. E. Gilpin (editors). 1997. Metapopulation biology: ecology, genetics and evolution. Academic Press, San Diego, CA.

Harrison, S., and A. Hastings. 1996. Genetic and evolutionary consequences of metapopulation structure. Trends in Ecology and Evolution 11:180–183.

Hedrick, P. W. 2005. Genetics of populations, 3rd edn. Jones & Bartlett, Sudbury, MA.

Hedrick, P. W., and R. Fredrickson. 2010. Genetic rescue guidelines with examples from Mexican wolves and Florida panthers. Conservation Genetics 11:615–626.

Hernandez, F., W. P. Kuvlesky, Jr., R. W. DeYoung, L. A. Brennan, and S. A. Gall. 2006. Recovery of rare species: case study of Masked Bobwhite. Journal of Wildlife Management 70:617–631.

Hess, B. D., P. O. Dunn, and L. Whittingham. 2012. Females choose multiple mates in the lekking Greater Prairie-Chicken. Auk 129:133–139.

Hewitt, G. M. 2000. The genetic legacy of the Quaternary ice ages. Nature 405:907–925.

Hewitt, G. M. 2004. The structure of biodiversity—Insights from molecular phylogeography. Frontiers in Zoology 1:4.

Hijmans, R. J., S. E. Cameron, J. L. Parra, P. G. Jones, and A. Jarvis. 2005. Very high resolution interpolated climate surfaces for global land areas. International Journal of Climatology 25:1965–1978.

Holman, J. A. 1964. Osteology of gallinaceous birds. Quarterly Journal of the Florida Academy of Science 22:230–252.

Jackson, S. T., R. S. Webb, K. H. Anderson, J. T. Overpeck, T. Webb III, J. W. Williams, and B. C. S. Hansen. 2000. Vegetation and environment in eastern North America during the Last Glacial Maximum. Quaternary Science Reviews 19:489–508.

Johnsgard, P. A. 1983. The grouse of the world. University of Nebraska Press, Lincoln, NE.

Johnson, J. A. 2008. Recent range expansion and divergence among North American prairie grouse. Journal of Heredity 99:165–173.

Johnson, J. A., M. R. Bellinger, J. E. Toepfer, and P. Dunn. 2004. Temporal changes in allele frequencies and low effective population size in Greater Prairie-Chickens. Molecular Ecology 13:2617–2630.

Johnson, J. A., and P. O. Dunn. 2006. Low genetic variation in the Heath Hen prior to extinction and implications for the conservation of prairie chicken populations. Conservation Genetics 7:37–48.

Johnson, J. A., P. O. Dunn, and J. L. Bouzat. 2007. Effects of recent population bottlenecks on reconstructing the demographic history of prairie chickens. Molecular Ecology 16:2203–2222.

Johnson, J. A., M. A. Schroeder, and L. A. Robb. [online] 2011. Greater Prairie-Chicken (*Tympanuchus cupido*). Birds of North America 036. <http://bna.birds.cornell.edu/bna/species/036 doi:10.2173/bna.36> (3 March 2014).

Johnson, J. A., J. E. Toepfer, and P. O. Dunn. 2003. Contrasting patterns of mitochondrial and microsatellite population structure in fragmented populations of Greater Prairie-Chickens. Molecular Ecology 12:3335–3347.

Johnson, N. K., and C. Cicero. 2004. New mitochondrial DNA data affirm the importance of Pleistocene speciation in North American birds. Evolution 58:1122–1130.

Kaiser, V. B., M. Van Tuinen, and H. Ellegren. 2007. Insertion events of CR1 retrotransposable elements elucidate the phylogenetic branching order in galliform birds. Molecular Biology and Evolution 24:338–347.

Keller, L. F., K. J. Jeffery, P. Arcese, M. A. Beaumont, W. M. Hochachka, J. N. M. Smith, and M. W. Bruford. 2001. Immigration and the ephemerality of a natural population bottleneck: evidence from molecular markers. Proceedings of the Royal Society of London B 268:1387–1394.

Lande, R. 1988. Genetics and demography in biological conservation. Science 241:1455–1460.

Landguth, E. L., S. A. Cushman, M. K. Schwartz, K. S. McKelvey, M. Murphy, and G. Luikart. 2010. Quantifying the lag time to detect barriers in landscape genetics. Molecular Ecology 19:4179–4191.

Latch, E. K., L. A. Harveson, J. S. King, M. D. Hobson, and O. E. Rhodes, Jr. 2006. Assessing hybridization in wildlife populations using molecular markers: a case study in Wild Turkeys. Journal of Wildlife Management 70:485–492.

Leopold, A. S. 1944. The nature of heritable wildness in turkeys. Condor 46:133–197.

Levins, R. 1969. Some demographic and genetic consequences of environmental heterogeneity for biological control. Bulletin of the Entomological Society of America 15:237–240.

Lowe, J. J., and M. J. C. Walker. 1997. Reconstructing quaternary environments, 2nd edn. Routledge, London, U.K.

Madge, S., and P. McGowan. 2002. Pheasants, partridges, and grouse: a guide to the pheasants, partridges, quails, grouse, guineafowl, buttonquails, and sandgrouse of the world. Princeton University Press, Princeton, NJ.

Metcalfe, S. E., S. L. O'Hara, M. Caballero, and S. J. Davies. 2000. Records of Late Pleistocene–Holocene climatic change in Mexico—A review. Quaternary Science Reviews 19:699–721.

Nei, M., T. Maruyama, and R. Chakraborty. 1975. The bottleneck effect and genetic variability in populations. Evolution 29:1–10.

Otto–Bliesner, B. L., S. J. Marshall, J. T. Overpeck, G. H. Miller, A. Hu, and the CAPE Last Interglacial Project members. 2006. Simulating Arctic climate warmth and icefield retreat in the last interglaciation. Science 311:1751–1753.

Oyler-McCance, S. J., J. St. John, and T. W. Quinn. 2010. Rapid evolution in lekking grouse: implications for taxonomic definitions. Ornithological Monographs 67:114–122.

Oyler-McCance, S. J., J. St. John, S. E. Taylor, A. D. Apa, and T. W. Quinn. 2005a. Population genetics of Gunnison Sage-Grouse: implications for management. Journal of Wildlife Management 69:630–637.

Oyler-McCance, S. J., S. E. Taylor, and T. W. Quinn. 2005b. A multilocus population genetic survey of the Greater Sage-Grouse across their range. Molecular Ecology 14:1293–1310.

Palkovacs, E. P., A. J. Oppenheimer, E. Gladyshev, J. E. Toepfer, G. Amato, T. Chase, and A. Caccone. 2004. Genetic evaluation of a proposed introduction: the case of the Greater Prairie-Chicken and the extinct Heath Hen. Molecular Ecology 13:1759–1769.

Patten, M. A., D. H. Wolfe, E. Shochat, and S. K. Sherrod. 2005. Habitat fragmentation, rapid evolution and population persistence. Evolutionary Ecology Research 7:235–249.

Phillips, S. J., R. P. Anderson, and R. E. Schapire. 2006. Maximum entropy modeling of species geographic distributions. Ecological Modeling 190:231–259.

Phillips, S. J., and M. Dudík. 2008. Modeling of species distributions with MaxEnt: new extensions and a comprehensive evaluation. Ecography 31:161–175.

Pitman, J. C. 2003. Lesser Prairie-Chicken nest site selection and nest success, juvenile gender determination and growth, and juvenile survival and dispersal in southwestern Kansas. M.S. thesis, Kansas State University, Manhattan, KS.

Pratt, A. C. 2010. Evaluation of the reintroduction of Attwater's Prairie-Chickens in Goliad County, Texas. M.S. thesis, Texas A&M University–Kingsville, Kingsville, TX.

Pruett, C. L., J. A. Johnson, L. C. Larsson, D. H. Wolfe, and M. A. Patten. 2011. Low effective population size and survivorship in a grassland grouse. Conservation Genetics 12:1205–1214.

Rhymer, J. M., and D. Simberloff. 1996. Extinction by hybridization and introgression. Annual Review of Ecology and Systematics 27:83–109.

Ridgway, R., and H. Friedmann. 1946. The birds of North and Middle America. Bulletin of the United States National Museum, no. 50, part 10. Smithsonian Institute Press, Washington, D.C.

Riley, T. Z. 2004. Private-land habitat opportunities for prairie grouse through federal conservation programs. Wildlife Society Bulletin 32:83–91.

Russell, D. A., F. J. Rich, V. Schneider, and J. Lynch–Stieglitz. 2009. A warm thermal enclave in the Late Pleistocene of the south–eastern United States. Biological Reviews 84:173–202.

Sangster, G., J. M. Collinson, P.-A. Crochet, A. G. Knox, D. T. Parkin, and S. C. Voiter. 2012. Taxonomic recommendations for British birds: eighth report. Ibis 154:874–883.

Segelbacher, G., J. Höglund, and I. Storch. 2003. From connectivity to isolation: genetic consequences of population fragmentation in Capercaillie across Europe. Molecular Ecology 12:1773–1780.

Short, L. L. 1967. A review of the genera of grouse (Aves: Tetraonidae). American Museum Novitates 2289:1–39.

Silvy, N. J., M. J. Peterson, and R. R. Lopez. 2004. The cause of the decline of pinnate grouse: the Texas example. Wildlife Society Bulletin 32:16–21.

Soulé, M. E., and L. S. Mills. 1998. No need to isolate genetics. Science 282:1658–1659.

Spaulding, A. 2007. Rapid courtship evolution in grouse (Tetraonidae): contrasting patterns of acceleration between the Eurasian and North American polygynous clades. Proceedings of the Royal Society of London B 274:1079–1086.

Spaulding, A. W., K. E. Mock, M. A. Schroeder, and K. I. Warheit. 2006. Recency, range expansion, and unsorted lineages: implications for interpreting neutral genetic variation in the Sharp-tailed Grouse (*Tympanuchus phasianellus*). Molecular Ecology 15:2317–2332.

Stiver, J. R., A. D. Apa, T. E. Remington, and R. M. Gibson. 2008. Polygyny and female breeding failure reduce effective population size in the lekking Gunnison Sage-Grouse. Biological Conservation 141:472–481.

Sugg, D. W., and R. K. Chesser. 1994. Effective population sizes with multiple paternity. Genetics 137:1147–1155.

Sutton, G. M. 1967. Oklahoma birds: their ecology and distribution, with comments on the avifauna of the Southern Great Plains University of Oklahoma Press, Norman, OK.

Taylor, M. A., and F. S. Guthery. 1980. Status, ecology, and management of the Lesser Prairie Chicken. U.S. Department of Agriculture Forest Service General Technical Report RM–77. Rocky Mountain Forest and Range Experimental Station, Fort Collins, CO.

Thompson, R. S., and K. H. Anderson. 2000. Biomes of western North America at 18,000, 6000, and 0 ^{14}C yr bp reconstructed from pollen and packrat midden data. Journal of Biogeography 27:555–584.

U. S. Fish and Wildlife Service. 2013. Endangered and threatened wildlife and plants; 6-Month Extension of Final Determination for the Proposed Listing of the Lesser Prairie-Chicken as a Threatened Species. Federal Register 78:41022.

U. S. Fish and Wildlife Service. 2014. Endangered and Threatened Wildlife and Plants; Special Rule for the Lesser Prairie-Chicken. Federal Register 79:20074–20085.

Van Den Bussche, R. A., S. R. Hoofer, D. A. Wiedenfeld, D. H. Wolfe, and S. K. Sherrod. 2003. Genetic variation within and among fragmented populations of Lesser Prairie-Chickens (*Tympanuchus pallidicinctus*). Molecular Ecology 12:675–683.

Wang, N., R. T. Kimball, E. L. Braun, B. Liang, and Z. Zhang. 2013. Assessing phylogenetic relationships among Galliformes: a multigene phylogeny with expanded taxon sampling in Phasianidae. PLoS One 8:e64312.

Waples, R. S., and O. Gaggiotti. 2006. What is a population? An empirical evaluation of some genetic methods for identifying the number of gene pools and their degree of connectivity. Molecular Ecology 15:1419–1439.

Westemeier, R. L., J. D. Brawn, S. A. Simpson, T. L. Esker, R. W. Jansen, J. W. Walk, E. L. Kershner, J. L. Bouzat, and K. N. Paige. 1998. Tracking the long-term decline and recovery of an isolated population. Science 282:1695–1698.

Wetmore, A. 1960. A classification for the birds of the world. Smithsonian Miscellaneous Collection 139:1–37.

Williams, C. K., A. R. Ives, R. D. Applegate, and J. Ripa. 2004. The collapse of cycles in the dynamics of North American grouse populations. Ecology Letters 7:1135–1142.

Williams, J. W., T. Webb III, P. H. Richard, and P. Newby. 2000. Late Quaternary biomes of Canada and the eastern United States. Journal of Biogeography 27:585–607.

Winker, K. 2010. Subspecies represent geographically partitioned variation, a gold mine of evolutionary biology, and a challenge for conservation. Ornithological Monographs 67:6–23.

Woodward, A. J. W., S. D. Fuhlendorf, D. M. Leslie, Jr., and J. Shackford. 2001. Influence of landscape composition and change in Lesser Prairie-Chicken (*Tympanuchus pallidicinctus*) populations. American Midland Naturalist 145:261–274.

Wright, S. 1978. Evolution and genetics of populations, Vol. 4. University of Chicago Press, Chicago, IL.

Zink, R. M. 2004. The role of subspecies in obscuring avian biological diversity and misleading conservation policy. Proceedings of the Royal Society of London B 271:561–564.

CHAPTER SIX

Habitat*

David A. Haukos and Jennifer C. Zavaleta

Abstract. Lesser Prairie-Chickens (*Tympanuchus palli-dicinctus*) inhabit a wide range of landscape types and environmental conditions on the Southern Great Plains. Populations experience dynamic and frequently extreme environmental conditions, and require large areas of habitat to persist, with estimates of 486 to 20,234 ha necessary for persistence. The currently occupied range has been estimated to be 15% of the perceived historical range, but extant populations of Lesser Prairie-Chickens likely represent much of the historical core areas. Despite population density estimates ranging from <1 to >50 birds/km², little is known about the carrying capacity of habitats for Lesser Prairie-Chickens. Lesser Prairie-Chickens select habitats based on vegetation structure rather than composition, requiring mid- and tall grasses and shrubs of greater stature than the vegetation of the short-grass prairie of the High Plains. In the western portion of their range, Lesser Prairie-Chickens are restricted to the sandy soils that support shrubs and other taller vegetation within the short-grass prairie matrix. Habitat quality rather than quantity is likely the driver of recent population trends of Lesser Prairie-Chickens. Nest success, annual survival, and home range size can be used to assess habitat quality across the species' range. Diet information for Lesser Prairie-Chickens is sparse and more than 30 years old, but indicates selection for invertebrates by chicks and adults throughout much of the annual cycle. Lek sites or communal display grounds, where males display for females, are a focal point for habitat use by both sexes of Lesser Prairie-Chickens, perhaps more so than for other species of prairie grouse. Nest-site selection is primarily related to visual obstruction and height of vegetation structure, rather than plant composition. Few data are available relative to habitat use by broods, but forbs are an important component. The importance of free water to Lesser Prairie-Chickens remains unknown but increasing evidence suggests that access to water is important during drought conditions. Information on habitat use during the nonbreeding season is limited, but indicates selection for residual grass cover. We hypothesize that habitat quality in the form of vegetation structure, habitat patch size and configuration, and food resources are the current dominant limiting factors interacting to exert ecological resistance on the demography and distribution of populations of Lesser Prairie-Chickens. Management practices to increase habitat quality need to be assessed at appropriate spatial and temporal scales for incorporation into an adaptive habitat management framework.

Key Words: brood, diet, habitat selection, home range, nest, survival, *Tympanuchus pallidicinctus*, water.

* Haukos, D. A. and J. C. Zavaleta. 2016. Habitat. Pp. 99–132 in D. A. Haukos and C. W. Boal (editors), Ecology and conservation of Lesser Prairie-Chickens. Studies in Avian Biology (no. 48), CRC Press, Boca Raton, FL.

opulations of Lesser Prairie-Chickens (*Tympanuchus pallidicinctus*) are restricted in time and space by available habitat (Figures 1.2 and 6.1). The species was federally listed as threatened under the Endangered Species Act in May 2014, and important contributing factors included the ongoing and probable future impacts of cumulative habitat loss and fragmentation (U.S. Fish and Wildlife Service 2014). Although the listing was vacated by judicial decision on procedural grounds in September 2015, the identified threats and impacts remain. Factors contributing to the threats include conversion of prairies to row-crop agriculture, encroachment by invasive trees and woody plants, wind energy development, petroleum production, presence of major roads, and man-made vertical structures including towers, utility lines, fences, turbines, wells, and buildings (U.S. Fish and Wildlife Service 2014). However, considerable uncertainty remains with regard to the relative influence of these factors and other possible threats to population viability of Lesser Prairie-Chickens. Knowledge of space use and habitat requirements during lekking, nesting, brood rearing, and nonbreeding periods will be necessary to frame a comprehensive conservation approach for increasing the abundance and occupied range of Lesser Prairie-Chickens. Our objective was to provide an overview of the known habitat requirements based on biologically relevant periods for Lesser Prairie-Chickens including links between habitat conditions versus measurable indices of individual or population performance. We start by framing key habitat concepts that are relevant to Lesser Prairie-Chicken. Next, we discuss the general patterns of habitat selection that dictate space use by Lesser Prairie-Chicken populations at various scales of selection (Johnson 1980). Last, specific actions for habitat management in the four major

Figure 6.1. Representative quality habitat for Lesser Prairie-Chickens in four different ecoregions: (a) Sand Shinnery Oak Prairie Ecoregion (photo by Phil Borsdorf), (b) Sand Sagebrush Prairie Ecoregion (photo by Dwayne Elmore), (c) Short-Grass Prairie/CRP Mosaic Ecoregion (photo by Dwayne Elmore), and (d) Mixed-Grass Prairie Ecoregion (photo by Jonathan Reitz). Ecoregions follow McDonald et al. (2014) and are depicted in Figure 1.2.

ecoregions of the extant range of Lesser Prairie-Chickens are addressed elsewhere in the book (Chapters 14 to 17, this volume).

HABITAT CONCEPTS

Theoretical concepts of habitat state that species coexist on a landscape by partitioning available resources such as space, food, cover, and water to maximize reproductive output and survival in the presence of competition, environmental conditions, and density-dependent effects that determine relative occurrence, abundance, and the spatial distribution of populations or species (Rotenberry 1985). Hall et al. (1997:175) defined habitat as "the resources and conditions present in an area that produce occupancy—including survival and reproduction—by a given organism." The greatest threat to bird species, including Lesser Prairie-Chickens, is the loss and degradation of habitat (Johnson 2007, U.S. Fish and Wildlife Service 2014). Habitat requirements are species specific, and thus, the concept is most meaningful when associated with a defined species, such as the habitat of Lesser Prairie-Chickens (Higgins et al. 2012). A landscape can provide habitat for many species, but patterns of use are related to individuals successfully securing and utilizing available resources. For example, the same landscape features can provide nesting habitat for one species while providing escape or wintering habitat for other species. Therefore, it is crucial that space and resource needs for each species within a community are considered in conservation planning, as management of habitat for a specific species will influence other species residing in the landscape. Due to wide-ranging movements and large space requirements, Lesser Prairie-Chickens and other prairie grouse can be considered as "keystone," "umbrella," or "surrogate" species for many species associated with native prairies (Sandercock et al. 2011). Thus, addressing habitat requirements for Lesser Prairie-Chickens will likely benefit other prairie species that are limited by similar habitat requirements.

Habitat is typically represented by correlation of measurable components of the biological and physical environment within a landscape that are perceived by humans to be important for the occurrence and abundance of a species during critical seasonal periods (MacMahon et al. 1981, Morrison et al. 2006). For example, leks are sites where male Lesser Prairie-Chickens aggregate to display and mate with visiting females. The number of attending males, number of visiting females, and duration of lek persistence may be related to elevation or substrate of lek locations or the quality of nearby nesting habitats (Haukos and Smith 1999). Thus, habitat requirements for a species are determined by a suite of limiting factors, or the biotic and abiotic factors that influence fitness of individuals, populations, or species (Block and Brennen 1993, Pulliam 2000). Generic use of "habitat" can cause considerable confusion without defining the scale of observation (Johnson 1980, Block and Brennen 1993). Factors influencing the distribution and position of individuals within a landscape usually vary across spatial and temporal scales. Therefore, when describing habitat of Lesser Prairie-Chickens or other wildlife species, scale of observation at points, patches, or landscapes is important to assess the effects on populations and relative effectiveness of conservations.

Decisions resulting in habitat use and selection are nonrandom and promote the fitness of individual(s) within a species (Pulliam and Danielson 1991, Block and Brennen 1993). Habitat use, while meaningful and necessary for conservation actions, only considers the collection of environmental components needed to meet life history requirements and is typically evaluated by seasonal periods of breeding, molting, or nonbreeding, or by feeding, predator avoidance, and other behaviors. Use is influenced by population density, competitive interactions with other species, abundance and distribution of limiting resources, and other biotic or abiotic factors (Block and Brennen 1993).

Habitat selection is a complex process whereby diverse behavioral and environment cues guide selection among available discrete landscape patches within the species range, population occurrence, and dispersal ability within populations by individuals restricted by morphological and physiological adaptations, but refined by innate and learned behavioral responses to individual fitness (Fretwell and Lucas 1970, Block and Brennen 1993). Unlike habitat use, habitat selection requires the measurement of available resources at an appropriate scale to detect disproportionate use of key resources by the species of interest (Manly et al. 2002). Jones (2001) stated that there is a hierarchical progression of behavioral and environmental processes that influence habitat selection, which inherently influences the survival and fitness of individuals. Linking habitat selection to individual reproductive output and survival and scaled to vital rates at a population level reduces uncertainty in conservation planning and implementation and allows for the identification of ecological sinks, sources, and traps that may limit populations (Schlaepfer et al. 2002, Aldridge and Boyce 2007).

Patterns of habitat selection relative to spatial and temporal variation of available resources create a selective pressure that allows researchers to measure and predict habitat quality (Cody 1985, Johnson 2007). Quality varies among available habitat patches based on accessible resources, whereby selection is an active process in which individuals can assess the quality of each patch and are free to move to the highest quality patch. "Patches" can be considered as areas with distinct boundaries, such as grasslands surrounded by cropland; grazed areas adjacent to ungrazed areas; or variation in food availability, vegetation structure, or other resources within a contiguous landscape. Theoretically, animals distribute themselves among available patches from highest to lowest quality based on energy expenditures relative to resource benefits in an ideal free distribution (Fretwell and Lucas 1970). Measures of habitat quality are most meaningful at the population scale, rather than at the level of an individual, because quality manifests itself in the persistence and density of local populations (Pidgeon et al. 2006, Doherty et al. 2010). Habitat quality at an individual level is defined as the per capita contribution to population growth expected from available resources within a given habitat (Johnson 2007). The quality of habitat patches is frequently ranked on a relative scale with observed thresholds defining the finite rate of population change (λ). Stabilizing selection rewards individuals that can distinguish between high- and low-quality habitats and make the greatest contribution to the population (Clark and Shutler 1999).

The concept of carrying capacity is implicit in the definition of habitat quality. Hobbs and Hanley (1990:515) defined carrying capacity as "the capability of land to maintain and produce animals of a given species." Carrying capacity is a measure of habitat quality and has been measured most frequently as (1) potential density of animals based on individual space needs or territory size; (2) a measure of available forage resources such as biomass, nutrition level, or potential energy that can meet minimum daily energetic or ration demands; or (3) abundance or density of unique features specific to species' needs in a defined area such as cavities or burrows, wetland type, flowering plants, or other features. In the absence of a direct measure of carrying capacity, the evaluation of habitat quality is reduced to testing for habitat selection (Fagen 1988).

A common goal in wildlife conservation is to relate features of habitat use and selection to demography and vital rates of defined populations for the purpose of developing and implementing conservation plans (Aldridge and Boyce 2007). Interpretation of habitat use and selection requires insight into those potential physical and ecological landscape features that cause a response by individuals or populations, resources necessary for survival and reproduction, and conditions that constrain use (Morrison et al. 2006). However, prior to fully addressing these ecological relationships, knowledge of habitat requirements and selection is needed to design and interpret investigations into relative influence of limiting factors and responses to conservation actions.

SCALES OF HABITAT SELECTION

First-Order Habitat Selection

Johnson (1980) provided a hierarchical framework to evaluate habitat selection based on spatial scale. First-order selection is the physical or geographic range of the species, which represents the availability of potential habitat within the abiotic range of tolerance for the species. The Lesser Prairie-Chicken is a species of prairie landscapes of the Southern Great Plains (Figure 1.2). Populations are distributed across landscapes primarily based on the relative quality of available habitat and density of individuals. In general, conversion of >40% of prairie landscapes to row-crop agriculture renders the landscape incapable of supporting self-sustaining populations of Lesser Prairie-Chickens (Crawford and Bolen 1976a). In Kansas, and likely other states, habitat quantity coarsely measured as available grassland at the landscape scale in the currently occupied range has changed little since the 1950s (Spencer 2014), and conversion of cropland to perennial grass cover via the Conservation Reserve Program (CRP) within the U.S. Department of Agriculture has offset contemporary conversion of native prairie. The only exception would be advances of irrigation technology that facilitated the conversion of prairies on deep sandy soils to row crops starting in the 1960s, thereby fragmenting previously unaltered core habitats (Sexson 1980, Spencer 2014). The principal limiting factor throughout much of the range of the Lesser Prairie-Chicken is declines in habitat quality rather than quantity.

Prairies occupied by Lesser Prairie-Chickens are characterized by mid- and tall grasses and short-statured shrubs. Much of the occupied range corresponds with sandy soil types that support this type of vegetation composition as opposed to the tighter, clay soils that typically support vegetation of smaller stature and relatively reduced vertical structure. Therefore, Lesser Prairie-Chickens are not distributed evenly across their occupied range, but rather in distinct patches related to available habitat. Lesser Prairie-Chickens currently occupy portions of the sand sagebrush (*Artemesia filifolia*) dominated native prairies of southeastern Colorado, southwestern Kansas, western and panhandle of Oklahoma, and northwest Texas; sand shinnery oak (*Quercus havardii*)-bluestem (*Andropogon* spp.) grasslands of eastern New Mexico, western Oklahoma, and northwest Texas; and mixed-grass prairie of north-central Oklahoma and south-central Kansas (Figures 1.2 and 6.1; McDonald et al. 2014). In addition, the establishment of native vegetative cover using a mix of native warm-season grasses such as little bluestem (*Schizachyrium scoparium*), sideoats grama (*Bouteloua curtipendula*), and switchgrass (*Panicum virgatum*) north of the Arkansas River in Kansas through the U.S. Department of Agriculture CRP has resulted in reoccupation of historical range by Lesser Prairie-Chickens (Rodgers and Hoffman 2005; Chapter 2, this volume).

The relative importance of grasses and shrubs in habitats of Lesser Prairie-Chickens varies from east to west, with shrubs increasing in importance as amount and reliability of precipitation decreases as one proceeds further west. In general, limiting factors, relative ranking of ecological drivers, population response to conservation, and habitat composition differ east and west of the 100° meridian of longitude, which represents a ~50 cm (20 in.) isocline in precipitation or the boundary for semiarid environmental conditions where precipitation is less than potential evaporation. Thus, proposed approaches to conservation

of Lesser Prairie-Chickens may differ between the western semiarid and eastern, more mesic, portions of their range.

The U.S. Fish and Wildlife Service (2014) estimated historical range of the Lesser Prairie-Chicken as 466,998 km² and current occupied range as 70,602 km², or an ~85% reduction in range (Figure 1.2). Boundaries of the stated historical range are likely based on anecdotal reporting of rare dispersal events or misidentification of prairie grouse species, especially in Texas and Oklahoma (Chapter 2, this volume). Scant evidence indicates that populations of Lesser Prairie-Chickens persisted long term in much of the historical range that is currently unoccupied (Chapter 2, this volume). Western regions of the depicted historical range of the Lesser Prairie-Chicken also include short-grass prairie of the High Plains, the semiarid region of the Great Plains west of the 100° meridian. True short-grass prairie dominated by blue grama (*Bouteloua gracilis*) and buffalograss (*Buchloë dactyloides*) has a structure, vegetation composition, and landscape heterogeneity that provides little suitable habitat for Lesser Prairie-Chickens (Hagen and Giesen 2004, Haukos 2011, Van Pelt et al. 2013, McDonald et al. 2014). A possible exception would be the formation of lek sites in short-grass prairie adjacent to quality nesting habitat. Much of the western region of the depicted historical range was short-grass prairie or never was a prairie such as eastern and southern portions of Texas. Thus, the current distribution of Lesser Prairie-Chickens likely represents the authentic primary range and landscapes critical for the persistence of the species. A potential exception would be the initial documentation of Lesser Prairie-Chickens in Colorado in the early 1900s following early settlement by Europeans (Chapter 2, this volume).

The current distribution of populations of Lesser Prairie-Chicken encompasses extensive landscape gradients in precipitation, temperature, and growing season. Average annual precipitation ranges from 63 cm in the east to 25 cm in the semiarid western portions of the species range. The reliability of annual precipitation increases from west to east; extreme, prolonged droughts are common, especially in the High Plains. Average growing season ranges from 220 days in the south to 160 days in the north. Average ambient low temperatures can reach -9°C in the north during January with average July highs of 33.3°C in the south, but extremes can range from -33°C to 45.5°C (Chapter 12, this volume). Freezing temperatures can occur as early as September and late as May. Potential annual evapotranspiration ranges from 53 cm in the north to 180 cm in the south. The remarkable spatial and temporal variation in precipitation and temperature creates a dynamic, unpredictable environment with habitat quality and quantity varying intra- and interannually in response to abiotic conditions, which are responsible for the rapid and extreme fluctuations in population density and habitat occupancy.

Dynamic and unpredictable environmental conditions across the occupied range of the Lesser Prairie-Chicken have resulted in selection of habitats primarily based on vegetation structure and secondarily on vegetation composition and basal coverage (Hagen et al. 2013, Larsson et al. 2013). Furthermore, the Lesser Prairie-Chicken uses a wide range of vegetation types across the species distribution. The principal habitats of the Lesser Prairie-Chicken in the High Plains are comprised of native prairies associated with sandy soils supporting mixed and tall grasses with a shrub component dominated by sand sagebrush or sand shinnery oak (Copelin 1963, Donaldson 1969, Taylor and Guthery 1980a, Hagen and Giesen 2004, Haukos 2011). The shrub component provides reliable availability of needed vertical structure should grass growth be reduced due to frequent drought, extensive grazing, or lack of sufficient response to additional disturbance factors such as fire. Presence of shrubs influences population vital rates. For example, birds using sand shinnery oak prairie with >20% shrub cover have greater survival than sites with less dense shrub cover (Patten et al. 2005a). Lesser Prairie-Chickens have greater survival in sand shinnery oak prairie compared to other habitat types (Table 6.1; Grisham 2012).

However, a fine line separates the beneficial and detrimental occurrence of woody plant species in the range of the Lesser Prairie-Chicken. Woodward et al. (2001) reported that declining Lesser Prairie-Chicken populations were closely associated with loss of shrubland cover types, including sand sagebrush and sand shinnery oak. In the eastern portion of the species range, where grass cover is reliably available and provide the primary vegetation structure, small stature shrubs, such as plum (*Prunus* spp.) and sumac (*Rhus* spp.), are found in Lesser Prairie-Chicken occupied habitat, but little evidence indicates these shrub species have similar habitat value as shrubs in western areas. Last, considerable evidence shows that Lesser Prairie-Chickens will show behavioral avoidance of trees and withdraw from habitats that include tree cover. Fuhlendorf et al. (2002) reported that the presence of invasive trees into prairies was strongly related to declining Lesser Prairie-Chickens populations. A similar finding was reported for Greater Sage-Grouse (*Centrocercus urophasianus*) in Oregon (Baruch-Mordo et al. 2013). The amount of habitat loss due to tree invasion is currently unknown, but it is pervasive throughout the range of Lesser Prairie-Chickens. Eastern red-cedar (*Juniperus virginiana*) has invaded much of the eastern range of the Lesser Prairie-Chicken (Fuhlendorf et al. 2002). However, honey mesquite (*Prosopis glandulosa*), Ashe juniper (*Juniperus ashei*), oaks (*Quercus* spp.), salt cedar (*Tamarix ramosissima*), and other tree species have invaded many other areas of potentially suitable habitat by Lesser Prairie-Chickens, significantly reducing and fragmenting available habitats (Fuhlendorf et al. 2002). Historically, natural and prescribed fire prevented trees and other tall, woody vegetation from becoming established in prairie ecosystems.

Spatial and temporal variability in habitat quality can be measured by monitoring population rates of seasonal and annual survival (Table 6.1). Survival estimates represent the interaction of occupied habitat quality and environmental conditions. As a result, long-term studies are needed to assess the impacts of fluctuating habitat quality and gradients in environmental conditions on annual and seasonal survival of Lesser Prairie-Chickens. The annual mortality varies by habitat type, year, sex, and age (Table 6.1). Survival for both males and females is lowest during the breeding season of lekking to brood rearing (Hagen et al. 2005a, 2007; Patten et al. 2005; Wolfe et al. 2007; Jones 2009; Lyons et al. 2009). Increased predation risk and energy demands for lekking likely contribute to reduced male survival during breeding (Hagen et al. 2005a, Jones 2009), whereas mating, nest initiation, incubation, and brood rearing increase female vulnerability to predation (Hagen et al. 2007).

TABLE 6.1

Estimates of annual and seasonal survival rates (S) for adult Lesser Prairie-Chickens.

State	Habitat[a]	Years	n	Sex[b]	Age[c]	Exposure period	Season[d]	S (±SE or 95% CI)	Sources
KS	SSB	1998–2002	93	M	J	12 months	BN	0.60 (±0.12)	Hagen et al. (2005a)
KS	SSB	1998–2002	82	M	A	12 months	BN	0.43 (±0.09)	
KS	SSB site 1	2000–2002	72	F	A	12 months	BN	0.30 (±0.08)	Hagen et al. (2007)
	SSB site 2	2000–2002	45	F	A	12 months	BN	0.44 (±0.08)	
	SSB site 1	2000–2002	58	F	A	12 months	BN	0.43 (±0.12)	
	SSB site 2	2000–2002	29	F	A	12 months	BN	0.59 (±0.10)	
NM	SSO	1962–1970	118	M	A	12 months	BN	0.30 (±0.07)	Campbell (1972)
	SSO	1962–1970	135	M	J	12 months	BN	0.35 (±0.04)	
KS	SSB	1997–1999	84	F	JA	6 months	B	0.74 (0.65–0.81)	Jamison (2000)
	SSB	1998–1999	36	M	JA	12 months	BN	0.57 (0.35–0.76)	
TX	SSO, SSB	2001–2003	115	MF	JA	12 months	BN	0.52 (0.32–0.71)	Jones (2009)
		2001–2003	115	MF	JA	4 months	B	0.71	
		2001–2003	115	MF	JA	8 months	N	0.72	
TX	SSB	2001–2003	115	MF	JA	1 month	B early	0.92 (±0.02)	Lyons et al. (2009)
	SSB	2001–2003	115	MF	JA	1 month	B late	0.93 (±0.02)	
	SSB	2001–2003	115	MF	JA	1 month	N	0.96 (±0.01)	
	SSO	2003–2005	72	MF	JA	1 month	B	0.85 (±0.04)	
	SSO	2003–2005	72	MF	JA	1 month	N	0.93 (±0.03)	
	SSB	2001–2003	115	MF	JA	12 months	BN	0.52 (0.32–0.71)	
	SSO	2003–2005	72	MF	JA	12 months	BN	0.31 (0.12–0.58)	
KS	SGP/CRP	2002–2003	69						Fields (2004)
		2002		F	JA	140 days	B	0.62 (0.48–0.75)	
		2003		F	JA	140 days	B	0.66 (0.53–0.77)	
TX	SSO	2008–2011	46	M	JA	6 months	B	0.57 (0.40–0.72)	Grisham (2012)
		2008	17	F	JA	5.5 months	B	0.80 (0.54–0.93)	

(Continued)

TABLE 6.1 (Continued)

Estimates of annual and seasonal survival rates (S) for adult Lesser Prairie-Chickens.

State	Habitat[a]	Years	n	Sex[b]	Age[c]	Exposure period	Season[d]	S (±SE or 95% CI)	Sources
KS	SSB	2009	16	F	JA	5.5 months	B	0.79 (0.52–0.93)	Hagen et al. (2007)
		2010	11	F	JA	5.5 months	B	0.89 (0.53–0.98)	
		2011	13	F	JA	5.5 months	B	0.71 (0.38–0.90)	
		1998–2002	58	F	J	6 months	B	0.76 (0.62–0.86)	
			72	F	A	6 months	B	0.66 (0.54–0.76)	
TX	SSO	1987–1988	55	F	JA	2 months	B	0.59 (0.41–0.78)	Haukos et al. (1989)
TX	SSO	2008–2010	42	M	A	3 months	B early	0.51 (±0.09)	Holt (2012)
						3 months	B late	0.82 (±0.08)	
			28	M	J	3 months	B early	1.0	
						3 months	B late	0.88 (±0.62)	
						3 months	B late	0.55 (±0.13)	
TX	SSO	2006–2007	23	F	JA	6 months	B	0.61 (0.41–0.80)	Leonard (2008)
		2006	37	MF	JA		B	0.71 (0.49–0.94)	
		2007	29	MF	JA		B	0.42 (0.10–0.74)	
			9	MF	JA		B		
NM	SSO	1979	16	F	JA		B	0.56	Merchant (1982)
		1980	25	F	JA		B	0.32	
TX	SSB	2001–2002	47	MF	JA	5 months	B	0.63 (±0.09)	Toole (2005)
	SSO	2001–2002	24	MF	JA	5 months	B	0.71 (±0.11)	
KS	SGP, MGP	2013–2014	201	F	JA	6 months	B	0.46 (0.38–0.53)	Plumb (2015)
	SGP	2013–2014	104	F	JA	6 months	B	0.39 (0.29–0.48)	
		2013–2014	94	F	JA	6 months	B	0.55 (0.42–0.66)	
TX	SSO, SSB	2008–2010	48	MF	JA	6 months	N	0.63 (±0.071)	Kukal (2010)
TX	SSO	2008	12	MF	JA	6 months	N	0.85 (±0.14)	Pirius (2011)
		2009	21	MF	JA	6 months	N	0.83 (±0.09)	
		2010	20	MF	JA	6 months	N	0.57 (±0.14)	
		2008–2010	53	MF	JA	6 months	N	0.72 (±0.08)	

[a] SSO, Sand Shinnery Oak Prairie; SSB, Sand Sagebrush Prairie; SGP, Short-Grass Prairie; SGP, Short-Grass Prairie/Conservation Reserve Program Mosaic; MGP, Mixed-Grass Prairie.

[b] M, male; F, female; MF, pooled sexes.

[c] J, juvenile; A, adult; JA, pooled ages.

[d] B, breeding; N, nonbreeding; BN, pooled seasons.

Second-Order Habitat Selection

Second-order habitat selection is the combined home range of all individuals or social groups within a population (Johnson 1980). Essentially, this spatial scale represents the space use and necessary resources for a population to persist on the landscape. Carrying capacity is usually measured at this scale as habitat quality is directly related to home range size of individuals comprising the population. Unfortunately, direct estimation of the carrying capacity of habitats for Lesser Prairie-Chickens has remained elusive and unreported. Empirical determination of the space needs for Lesser Prairie-Chicken populations is difficult due to the relatively large area perceived as necessary for population persistence, challenges in measuring home ranges of an adequate sample of individuals (especially during the nonbreeding season), and technological limitations for locating vagile individuals across a large landscape. The Lesser Prairie-Chicken may be limited by available area, as large areas of intact, unfragmented landscapes of suitable habitats are necessary for a population to persist (Giesen 2000, Bidwell et al. 2002, Hagen et al. 2004).

The minimum habitat patch size necessary to support populations of Lesser Prairie-Chickens is unclear, but different authors have ventured recommendations ranging from 4,900 ha to 20,236 ha of contiguous native prairie for self-sustaining leks or populations (Table 6.2). Recently, Van Pelt et al. (2013) prescribed an average focal area size of at least 20,234 ha for population management of Lesser Prairie-Chickens, with at least 70% of the area in high-quality habitat. The U.S. Fish and Wildlife Service (2014) concluded that Lesser Prairie-Chickens are limited by the lack of contiguous, large patches of remaining prairie, and reported that 98.96% and 99.97% of remaining habitat patches were <486 and 6,475 ha, respectively. Thus, few remaining prairie patches are connected without fragmentation and available to meet the suggested minimum size for population persistence. Given that restoration of sand shinnery oak and sand sagebrush prairie in semiarid regions is expensive and time consuming even under the best environmental conditions, protection and conservation of extant habitats may be the most effective and economical approach to habitat conservation. Securing existing patches of sufficient size could be a consideration in conservation planning.

Densities of Lesser Prairie-Chickens track population size, which varies temporally and spatially across the species range (Chapter 4, this volume). Most estimates of density of Lesser Prairie-Chickens are based on birds counted at leks (usually males with an assumption of 1:1 sex ratio) and counts are extrapolated to estimate population size across a surveyed area (Van Pelt et al. 2013). Delineation of the boundaries of a surveyed area can have a large effect on population estimates. Recently, Davis et al. (2008) reported average densities for Texas (2.2 birds/km^2; range of 0.84–3.32), New Mexico (1.86 birds/km^2), and Kansas (3.85 birds/km^2). In the Oklahoma Lesser Prairie-Chicken conservation plan, Haufler et al. (2012) set habitat goals based on a population goal of 1.92 birds/km^2. During the most recent peak in population numbers of Lesser Prairie-Chickens in sand shinnery oak prairie of Texas and New Mexico during the early 1980s, Olawsky and Smith (1991) used transect sampling to estimate summer densities of 20.0–25.8 and 33.8–52.7 birds/km^2 during

TABLE 6.2

Recommended total space needs for persistence of a Lesser Prairie-Chicken population.

Recommended area (ha)	Recommendation qualifiers	Sources
20,236	At least 70% in good- to high-quality habitat	Haufler et al. (2012)
20,236	Low-quality habitat	Applegate and Riley (1998)
10,118	High-quality habitat	
10,118	Lek complex	Bidwell et al. (2002)
4,900	Single lek	
6,475		Taylor and Guthery (1980a)
5,200		Mote et al. (1999)
4,900	Single lek	Davis (2005)

winter in Texas. Based on the range of potential densities, estimated current densities, and amount of currently unoccupied potential habitat, it can be concluded that factors affecting habitat quality and structure are influencing recent population trends of Lesser Prairie-Chickens. In 2013, McDonald et al. (2014) estimated the density of Lesser Prairie-Chicken populations between 0.07 and 0.27 birds/km^2 (Table 1.2).

Third-Order Habitat Selection

The use of specific sites by individuals within their home range defines third-order selection (Johnson 1980). Thus, habitat use at this scale is contingent upon habitat availability that can be measured using seasonal movements that contribute to the home range of individual Lesser Prairie-Chickens (Table 6.3). Habitat quality is frequently assessed by comparing home range size among years, populations, sex, ages, environmental conditions, and conservation practices. Until recently, available technology for tagging and tracking individual Lesser Prairie-Chickens limited detections of bird locations to small areas near capture sites and failed to document the long-distance movements needed to estimate the full extent of bird home ranges (>10 km). Recent advances in satellite transmitter technology should enable home range estimation of birds formerly censured from analyses due to movements beyond the limits of tracking range.

Characteristics of Lesser Prairie-Chicken home ranges depend on the extent and configuration of available habitat patches relative to potential movements across a landscape. Typically, home range characteristics vary with season, sex, and local landscape conditions (Table 6.3). Males usually have smaller home ranges than females, as males maintain a closer association with leks during spring (February–June) and fall (September–October) than females (Hagen and Giesen 2004, Table 6.3). Home ranges are smallest during the breeding season and largest in autumn during brood breakup, juvenile dispersal, and movements to nonbreeding habitats (Taylor and Guthery 1980b, Merchant 1982, Riley et al. 1994, Jamison 2000, Hagen 2003). Home range size increases during drought are likely due to reduced food resources and habitat quality (Merchant 1982, Hagen and Giesen 2004). Essentially, extensions of home range

size are driven by movements geared toward securing resources necessary to maximize individual fitness, with considerable variation in movements among individuals within a population (Table 6.3). Determination of timing, route, and duration of seasonal movements not only plays a critical function in landscape use and range expansion by Lesser Prairie-Chicken but is also relevant for the recovery of the species. The U.S. Fish and Wildlife Service (2014) stated that the size of management areas for Lesser Prairie-Chickens should be sufficient to incorporate the longest known movement of individual birds.

Fourth-Order Habitat Selection

Fourth-order selection describes the utilization of actual resources at the site, which is typically measured as food resources (Johnson 1980). Unfortunately, little is known about selection of food resources by Lesser Prairie-Chickens. However, daily energy demands require daily foraging as stored fat reserves comprise <4.5% of body mass (Olawsky 1987, Hagen and Giesen 2004). Food availability and quality directly impact the fitness of individual birds, especially during the lekking and nesting seasons when daily energy demand peaks for both males and females and daily survival are at the lowest level across the annual cycle.

Following European settlement and initial conversion of prairie to crops during the late 1800s, Lesser Prairie-Chicken populations greatly increased in abundance and range (Chapter 2, this volume). Population expansion was apparently due to utilization of a previously unavailable, abundant food in the form of waste grain, as settlers observed large flocks in shocked grain fields during fall and winter (Chapter 2, this volume). Much of the perceived historical range of Lesser Prairie-Chickens beyond the extant range can also be linked to observations of nomadic movements of individuals or small groups of birds responding to a lack of food resources in high-density areas (Chapter 2, this volume). The availability of invertebrates, leafy material, seeds, and other food resources for Lesser Prairie-Chickens fluctuates greatly both temporally and spatially (Jamison et al. 2002, Hagen et al. 2005b) and likely contributes to the relatively large space requirements to sustain populations.

TABLE 6.3

Estimates of movement rates and home range size for Lesser Prairie-Chickens.

State	Habitat[a]	Year	n	Sex[b]	Age[c]	Average movement or home range (±SE)	Sources
Movement							
NM	SSO	1979	16	F	JA	1.148 ± 0.11 (ha) during brooding period	Ahlborn (1980)
NM	SSO	1976–1978	23	F	JA	0.39 ± 0.23 (km) daily prenesting	Candelaria (1979)
				F	JA	0.25 ± 0.11 (km) daily nesting	
				F	JA	3.31 ± 3.21 (km) nests from trap site	
				F	JA	0.28 ± 0.55 (km) daily postnesting	
KS	SSB	1997–1999	10	MF	JA	1.62 ± 0.40 (km) lek of initial capture versus lek of recapture	Jamison (2000)
TX	SSO	2008–2009	15	M	JA	0.614 ± 0.04 (km) fall minimum daily movement	Kukal (2010)
			12	M	JA	0.483 ± 0.04 (km) winter minimum daily movement	
		2009–2010	21	M	JA	0.630 ± 0.04 (km) fall minimum daily movement	
			17	M	JA	0.482 ± 0.03 (km) winter minimum daily movement	
		2008	3	F	JA	0.593 ± 0.06 (km) fall minimum daily movement	
		2009–2010	4	F	JA	0.499 ± 0.10 (km) fall minimum daily movement	
			4	F	JA	0.391 ± 0.08 (km) winter minimum daily movement	
TX	SSO	2006–2007	18	MF	JA	2.788 ± 0.47 (km) greatest distance during breeding season	Leonard (2008)
						1.197 ± 0.15 (km) distance during breeding season	
TX	SSO	2008–2009	1	MF	JA	0.707 ± 0.12 (km) minimum weekly movement	Pirius (2011)
		2009–2010	12	MF	JA	0.676 ± 0.12 (km) minimum weekly movement	
		2010–2011	10	MF	JA	0.662 ± 0.16 (km) minimum weekly movement	
KS	SSB	1997–2003	9	M	J	1.2 ± 0.2 (km) fledging dispersal	Pitman et al. (2006a)
			3	F	J	10.49 ± 7.94 (km) fledging dispersal	
NM	SSO	1976–1978	40	F	JA	0.390 ± 0.12 (km) daily, prenesting	Riley et al. (1994)
			12	F	JA	0.250 ± 0.06 (km) daily, nesting	

(Continued)

TABLE 6.3 (Continued)

Estimates of movement rates and home range size for Lesser Prairie-Chickens.

State	Habitat[a]	Year	n	Sex[b]	Age[c]	Average movement or home range (±SE)	Sources
TX	SSO		3	F	JA	0.280 ± 0.280 (km) daily, broods	Taylor (1979)
			19	F	JA	0.220 ± 0.071 (km) daily, postnesting	
		Oct 1977	6	F	A	0.298 ± 0.106 (km) between daily locations	
		Nov 1977	6	F	A	0.454 ± 0.06 (km) between daily locations	
		Dec 1977	6	F	A	0.652 ± 0.06 (km between daily locations	
		Jan 1978	6	F	A	0.680 ± 0.07 (km) between daily locations	
		Feb 1978	6	F	A	0.383 ± 0.07 (km between daily locations	
		Oct 1977	5	M	J	0.663 ± 0.13 (km) between daily locations	
		Nov 1977	5	M	J	1.045 ± 0.11 (km) between daily locations	
		Dec 1977	5	M	J	1.069 ± 0.10 (km) between daily locations	
		Jan 1978	5	M	J	0.497 ± 0.11 (km) between daily locations	
		Oct 1977	7	M	A	0.451± 0.07 (km) between daily locations	
		Nov 1977	7	M	A	0.676 ± 0.07 (km) between daily locations	
		Dec 1977	7	M	A	0.697 ± 0.05 (km) between daily locations	
		Jan 1978	7	M	A	0.591 ± 0.06 (km) between daily locations	
		Feb 1978	7	M	A	0.390 ± 0.06 (km) between daily locations	
		Oct 1977	1	F	J	1.226 ± 0.18 (km) between daily locations	
		Nov 1977	1	F	J	0.410 ± 0.12 (km) between daily locations	
		Dec 1977	1	F	J	0.484 ± 0.12 (km) between daily locations	
		Jan 1978	1	F	J	0.267 ± 0.04 (km) between daily locations	
KS	SGP	2013	33	F	JA	1.556 ± 0.03 (km) average daily movement; breeding	Plumb (2015)
	SGP	2014	15	F	JA	1.506 ± 0.03 (km) average daily movement; breeding	
	MGP	2013	23	F	JA	1.033 ± 0.03 (km) average daily movement; breeding	
	MGP	2014	35	F	JA	1.362 ± 0.03 (km) average daily movement; breeding	

(Continued)

TABLE 6.3 (Continued)

Estimates of movement rates and home range size for Lesser Prairie-Chickens.

State	Habitat[a]	Year	n	Sex[b]	Age[c]	Average movement or home range (±SE)	Sources
Home range size							
NM	SSO	1976–1978	23	F	JA	230.75 ± 80.25 (ha) prenesting	Candelaria (1979)
				F	JA	92.02 ± 18.36 (ha) nesting period	
				F	JA	118.94 ± 58.92 (ha) postnesting	
TX	SSO	2008–2012	38	F	JA	671.4 ± 538 (sd) ha, 471.2 ± 327 ha, and 415.1 ± 306 ha with 95% LSCV, 95% Plug in, and 100% MCP, respectively; breeding	Borsdorf (2013)
			37	M	JA	306.0 ± 188 (sd) ha, 244.7 ± 131 ha, 173.2 ± 112 ha with 95% LSCV, 95% Plug in, and 100% MCP, respectively; breeding	
KS	SSB	1997–1999	10	M	JA	12–140 ha spring, 77–144 ha summer, 229–409 ha winter	Jamison (2000)
TX	SSO	2008–2009	11	M	JA	670.6 ± 98.5 (ha) fall	Kukal (2010)
			11	M	JA	514.5 ±167.3 (ha) winter	
			3	F	JA	319.5 ± 50.1 (ha) fall	
		2009–2010	18	M	JA	599.5 ± 181.1 (ha) fall	
			16	M	JA	480.8 ± 129.5 (ha) winter	
		2008	3	F	JA	319.5 ± 50.1 (ha) fall	
		2009–2010	3	F	JA	760.6 ± 452.0 (ha) fall	
			4	F	JA	282.3 ± 74.8 (ha) winter	
TX	SSO	2006–2007	18	MF	JA	265 ± 76 (ha) breeding season	Leonard (2008)
TX	SSO	2008–2011	5	F	A	503.5 ±34.9 (ha) nonbreeding	Pirius (2011)
			17	M	A	489.1 ± 34.9 (ha) nonbreeding	
		2008–2009	1	MF	JA	939.7 (ha) nonbreeding	
		2009–2010	12	MF	JA	430.3 ± 28.7 (ha) nonbreeding	
		2010–2011	10	MF	JA	506.4 ± 57.3 (ha) nonbreeding	

(Continued)

TABLE 6.3 (Continued)

Estimates of movement rates and home range size for Lesser Prairie-Chickens.

State	Habitat[a]	Year	n	Sex[b]	Age[c]	Average movement or home range (±SE)	Sources
NM	SSO	1976–1978	40	F	JA	231 ± 40.9 (ha) prenesting	Riley et al. (1994)
			12	F	JA	92 ± 2.7 (ha) nesting	
			3	F	JA	119 ± 234.1 (ha) broods	
			19	F	JA	73 ± 15.2 (ha) postnesting	
TX	SSO	1977–1978	19	MF	JA	400 ha nonbreeding	Taylor and Guthery (1980b)
TX	SSB	2001	20	MF	JA	288.5 ± 237 (ha) breeding/fall	Toole (2005)
	SSO	2001	11	MF	JA	178.4 ± 177 (ha) breeding/fall	
	SSB	2002	17	MF	JA	167.0 ± 24.8 (ha) breeding	
	SSO	2002	7	MF	JA	115.7 ± 23.9 (ha) breeding	
KS	SGP, MGP	2013–2014	56	F	JA	340 ± 53 (ha) breeding	Plumb (2015)

[a] SSO, Sand Shinnery Oak Prairie; SSB, Sand Sagebrush Prairie; SGP/CRP, Short-Grass Prairie/Conservation Reserve Program Mosaic.

[b] M, male; F, female; MF, pooled sexes.

[c] J, juvenile; A, adult; JA, pooled ages.

The available information on diets of Lesser Prairie-Chickens have been taken from field studies more than 30 years ago that examined crop contents or histological samples from fecal droppings and mainly for birds in sand shinnery oak prairie (Table 6.4). Compared to other prairie grouse, invertebrates comprise a greater component of the diet of chicks and adults in Lesser Prairie-Chickens (Table 6.4). The limited information on the diets of chicks of Lesser Prairie-Chicken suggests that young birds feed almost exclusively on invertebrates (Suminski 1977, Davis et al. 1980) and that invertebrates comprise the main part of the diet for a longer period of time up until brood breakup than seen in most other gallinaceous species (Savory 1989). Treehoppers (Membracidae), short-horned grasshoppers (Acrididae), and ants (Formicidae) were the most abundant items consumed by volume (95%).

Adults on the sand shinnery oak prairie consume mainly vegetative material, especially sand shinnery oak catkins and acorns, in the spring (Jones 1964, Suminski 1977). In the summer, Lesser Prairie-Chickens eat mostly insects, especially grasshoppers (*Acrididae*), but use seeds when plants are flowering (Jones 1964, Davis et al. 1980). During fall, the proportion of invertebrates in the diet increases (Jones 1964). During the winter, Lesser Prairie-Chickens eat mostly vegetative material, especially shinnery oak acorns (Jones 1964, Davis et al. 1980, Doerr and Guthery 1983). However, acorns of the sand shinnery oak are not available each year, because acorns are usually produced during mast years that only occur on an average of three years per decade (Pettit 1986). Temporal variation in food resources likely limits population growth in the sand shinnery oak prairies and historically contributed to dispersal of individuals into seemingly unsuitable areas. In a study that included herbicide treatment of the sand shinnery oak, Olawsky (1987) found that Lesser Prairie-Chickens ate more acorns from the sand shinnery oak in untreated areas, whereas birds in treated areas ate more foliage and plant materials. In a year with normal precipitation, birds ate more insects in untreated areas, but in a wet year, consumption of insects was similar between treatments. The study results indicate that management with herbicides and grazing can affect Lesser Prairie-Chicken food selection, especially when effects are compounded with annual variation in environmental conditions.

The dietary use of waste grain by Lesser Prairie-Chickens during the fall and winter has been reported, but the amount consumed, contribution to the diet, and seasonality are unknown (Campbell 1972, Crawford and Bolen 1976b, Ahlborn 1980, Salter et al. 2005). Lesser Prairie-Chickens tend to use cultivated grains most frequently when the availability of acorns, seeds, invertebrates, and other native foods is diminished by severe environmental conditions, such as drought and harsh winters (Copelin 1963, Ahlborn 1980). Historically, the importance of waste grain may have been overemphasized due to occasional observations of large flocks in harvested crop fields, but high densities of Lesser Prairie-Chickens can also occur in areas without available waste grain (Olawsky and Smith 1991). It is unknown what role waste grain plays in influencing population densities or viability because this food resource is not available throughout the occupied range of the Lesser Prairie-Chicken and contributes little to other stages of the annual life cycle.

Food resources can be a major limiting factor for populations of Lesser Prairie-Chickens, but limited information is available to devise conservation strategies targeting food resources. The large spatial scale required for population persistence limits the ability of managers to provide food resources at a scale that could be demographically meaningful. Given the natural spatial and temporal variation of invertebrates, acorns, and seeds, it may be difficult, if not impossible to focus habitat management efforts on food resources for Lesser Prairie-Chickens. Our inability to reliably predict available food resources restricts the understanding of the impacts of variation in food resources on the population demography of Lesser Prairie-Chickens. Given the importance of invertebrates to broods and adults, it would be useful to better understand how invertebrate abundance, especially grasshoppers (Orthoptera), is affected by environmental conditions, vegetation composition, and management practices (Fields 2004; Hagen et al. 2005b, 2009; Fields et al. 2006). Data on the composition of seasonal diets of Lesser Prairie-Chickens across their range would aid in conservation planning. Managing vegetation for invertebrate abundance is not straightforward, as environmental conditions will override designed habitat alterations (Jonas and Joern

TABLE 6.4
Diets of Lesser Prairie-Chickens.

State	Habitat type[a]	Year	Season	Method	Age	Metric	Mast and seed	Vegetative	Insect	Sources
TX	SSO	1971–1973	Fall	30 crops	Adult	% frequency		57	42.9	Crawford and Bolen (1976b)
						% mass		89.98	10	
						% volume		80.97	19	
NM	SSO	1976–1978	Spring	21 crops	Adult	% comp.	15.5 ± 5.8	78.7 ± 7.6	5.9 ± 3.8	Davis et al. (1980)
			Summer	18 crops	Adult		21.4 ± 8.2	23.3 ±7.2	55.3 ± 9.3	
				10 crops	Chicks 1–4 weeks		0	0	100	
				17 crops	Chicks 5–8 weeks		0.6 ± 0.6	0.1	99.3 ± 6.3	
TX	SSO	1978–1979	Summer	30 droppings	Adult	% frequency	0	40	59.6	Doerr and Guthery (1983)
			Fall				4.4	30.6	64.9	
			Winter				63.1	29	7.6	
			Spring				15.9	54.4	26.7	
OK	SSB	1959–1961	Summer	1,129 droppings	Adult	% frequency	17.82	6.3	46.89	Jones (1963)
			Fall				11.13	15.18	39.73	
			Winter				12.29	39.98	15.21	
			Spring				11.68	29.6	34.58	
	SSB		Summer				56.6	7	61.22	
			Fall				11.58	9.56	48.26	
			Winter				8.65	48.7	7.94	
			Spring				9.3	36.34	17.85	
OK	SSB	1959–1961		7 droppings, 1 crop, and 1 gizzard	Adult	% volume	7.7	7.1	85.1	Jones (1963)

(Continued)

TABLE 6.4 (Continued)
Diets of Lesser Prairie-Chickens.

State	Habitat type[a]	Year	Season	Method	Age	Metric	Mast and seed	Vegetative	Insect	Sources
TX, NM	SSO	1984	Summer[b]	6 crops		% volume		98.62	1.38	Olawsky (1987)
			Summer	9 crops				81.33	18.67	
		1985	Summer[b]	15 crops				65.85	34.15	
			Summer	12 crops				68.26	31.74	
NM	SSO	1976	Fall	9 crops	Adult	% comp.	75	28	7	Riley et al. (1993)
		1977	Fall	17 crops			21	49	30	
		1976–1977	Winter	6 crops			69	26	5	
NM	SSO	1976	Fall	9 crops	Adult		66	27.4	6.6	Smith (1979)
		1977	Fall	17 crops			20.5	49.9	29.6	
		1976–1977	Winter	6 crops			69.3	26	4.7	
NM	SSO	1976	Spring	9 crops	Adult		27.5	60.4	12.1	Suminski (1977)
		1976	Summer and fall	12 crops	Chicks 20 weeks		70.7	11.9	17.4	

[a] SSO, Sand Shinnery Oak Prairie; SSB, Sand Sagebrush Prairie.
[b] Sand Shinnery Oak treated with tebuthiuron.

2007). Collection and examination of crop contents is unlikely to be feasible while the species is listed as threatened, and alternative methods to determine the contemporary diet of Lesser Prairie-Chickens need to be explored. Updated techniques in the form of isotopic analysis of tissues or genetic testing of composition of fecal samples may prove useful and give more specific information on the species' diet (Pompanon et al. 2012, Blanco-Fontao et al. 2013).

LEK HABITATS

Leks are communal display grounds where Lesser Prairie-Chicken males gather, display, and establish small territories to attract and mate with females. Leks are the focal feature of all Lesser Prairie-Chicken populations; movements, home ranges, habitat use, occupied range, and conservation practices are usually measured in relation to lek locations. Male Lesser Prairie-Chickens are associated with leks for longer periods than other species of prairie grouse, with males attending leks for breeding for a 5-month period from February to June, and a shorter fall display period during late September through early October. Lesser Prairie-Chickens concentrate their daily and seasonal activities within 4.8 km of leks even during the nonbreeding season (Riley et al. 1994, Woodward et al. 2001, Hagen and Giesen 2004, Kukal 2010, Boal et al. 2014, Grisham et al. 2014). Daily activities of Lesser Prairie-Chickens are closely associated with lek locations throughout the year, and habitat use based on space requirements and vegetation type changes little among seasons. The only exception may occur during early to midwinter, when birds flock together and use the densest available grass cover. Conservation of traditional or continuously attended lek sites associated with quality nesting, brood rearing, and wintering habitats is required to support populations of Lesser Prairie-Chickens (U.S. Fish and Wildlife Service 2014).

Lek formation and density are used to index population trends and habitat quality as increasing population size is best reflected in formation of new leks, rather than increases in numbers of displaying birds at established leks (Hoffman 1963, Cannon and Knopf 1981, Merchant 1982, Haukos and Smith 1999). Traditional lek sites with continuing attendance of birds for >6–10 consecutive years are critical to population demography of Lesser Prairie-Chickens because females are more likely to visit established sites rather than temporary, annually variable satellite leks that form prior to the start of the female visitation period (Haukos and Smith 1999).

Leks are typically associated with relatively higher locations within a landscape, such as the tops of wind-swept ridges, exposed knolls, or dunes, with vegetation height <10 cm, exposed soil, and enhanced visibility of the surrounding area (Copelin 1963, Jones 1963, Taylor and Guthery 1980a). Sites of anthropogenic activities such as drilling pads, livestock watering area and confinement, fallow cropland, low traffic roads, and dune blowouts can be used as lek sites (Crawford and Bolen 1976b, Davis et al. 1979, Sell 1979, Haukos and Smith 1999). Sound transmission may also influence lek location at open sites (Butler et al. 2010). Importantly, lek formation is thought to be associated with the highest quality nesting habitat on the landscape. Based on our current knowledge of lek dynamics and formation, the availability of lek sites is unlikely to limit Lesser Prairie-Chickens if quality nesting and brood-rearing habitat are available. The ecological process underlying the formation, location, and persistence of leks remains a mystery, but developing a mechanistic understanding of lek dynamics will aid in successful conservation of Lesser Prairie-Chickens and their habitats.

NESTING HABITATS

Annual variation in precipitation, temperature, and other environmental conditions, along with landscape management practices, fragmentation, and increasing presence of anthropogenic structures, interacts to create year-to-year uncertainty in available quantity and quality of nesting habitat for Lesser Prairie-Chickens. Therefore, female Lesser Prairie-Chickens have evolved various strategies to contend with such uncertainty, the most notable being the population response of high recruitment rates of juveniles in the rare years when environmental conditions result in exceptional habitat quality (i.e., 3–4 years of 10; Tables 6.5 and 6.6; Chapter 12, this volume). A "boom or bust" reproductive response buffers the population during poor nesting years when little to no recruitment occurs (Chapter 12, this volume). However, paradoxically, annual nest

TABLE 6.5

Nest characteristics (±SE) as percent cover of grass, shrub, forb, litter, and bare ground for Lesser Prairie-Chickens.

State	Habitat[a]	Year	n	Location	Percent cover					Sources
					Grass	Shrub	Forb	Litter	Bare ground	
TX, NM	SSO	2008–2010	32	Nest	21.8	68.8	6.3		17.3	Boal et al. (2014)
NM	SSO	1976–1978	5	Subtype 1 Successful	12.3	3.0	0.2	67.2	32.9	Davis et al. (1979)
			3	Subtype 1 Unsuccessful	10.4	3.8	2.9	50.0	36.5	
			4	Subtype 2 Successful	10.0	1.5	0.1	51.9	34.9	
			17	Subtype 2 Unsuccessful	7.5	3.5	0.5	53.6	34.9	
			1	Subtype 3 Successful	5.0	15.0	2.5	41.3	36.2	
			6	Subtype 3 Unsuccessful	6.6	4.4	0.0	54.8	34.2	
CO	SSB	1986–1990	31		29.4 ± 14.9	7.2 ± 9.4		69.5 ± 14.3		Giesen (1994)
TX, NM	SSO	2001–2011	36	Nest	21.76 ± 3.6	38.61 ± 4.4	5.27 ± 2.4	26.7 ± 3.1		Grisham (2012)
				Random points	3.55 ± 0.8	4.41 ± 1.2	2.43 ± 2.9	33.6 ± 3.5		
KS	SSB	1998–2002		Site I	13.45 ± 9.7	6.86 ± 5.3	14.65 ± 10.1			Hagen et al. (2004)
				Site II	12.48 ± 16.3	11.02 ± 7.7	13.58 ± 16.4			
NM	SSO	2000–2005	45	Nest-site mean	47.31	53.82				Patten and Kelly (2010)
				Random mean	34.67	29.47				
KS	SSB	1998–2002	174	Nest	37.2 ± 2.0	15.2 ± 1.0	8.4 ± 0.6			Pitman et al. (2005)
				Paired random point	36.4 ± 2.0	8.2 ± 0.8	10.3 ± 0.7			
NM	SSO	1976–1978	37	Successful Subtype 1	64	32.5	3.5			Riley et al. (1992)

(Continued)

TABLE 6.5 (Continued)

Nest characteristics (±SE) as percent cover of grass, shrub, forb, litter, and bare ground for Lesser Prairie-Chickens.

State	Habitat[a]	Year	n	Location	Percent cover					Sources
					Grass	Shrub	Forb	Litter	Bare ground	
				Unsuccessful Subtype 1	49.6	31.4	19.1			Wisdom (1980)
				Successful Subtype 2	55.1	41.8	3.1			
				Unsuccessful Subtype 2	44.5	48.1	7.4			
				Successful Subtype 3	23.8	66.2	10			
				Unsuccessful Subtype 3	37.9	37.9	7.4			
NM	SSO	1976–1978	9	Use Subtype 1	77.7	22.2	0			
				Availability Subtype 1	39.7	30.3	11.4			
			17	Use Subtype 2	71.4	23.8	4.8			
				Availability Subtype 2	28.5	30.7	11			
			7	Use Subtype 3	28.5	57.1	14.3			
				Availability Subtype 3	24.1	45	12			

[a] SSO, Sand Shinnery Oak Prairie; SSB, Sand Sagebrush Prairie.

TABLE 6.6
Estimates of nest survival (±SE or 95% CI) for Lesser Prairie-Chickens.

State	Habitat[a]	Year	n	Apparent nest success	Daily survival rate	Derived estimate of nest survival	Sources
NM	SSO	2004–2005	23	0.76			Davis (2009)
KS	SGP	2001–2003	59	0.48		0.72 (0.60–0.84)[b]	Fields et al. (2006)
TX	SSO	2008–2011	36		0.97 (0.95–0.98)	0.43 (0.23–0.56)	Grisham et al. (2014)
TX, NM	SSO	2001–2011	229		0.98	0.57	Grisham et al. (2013)
NM	SSO	2001–2011	182		0.95 ± 0.003	0.24	Grisham (2012)
TX	SSO	1987–1988	13	0.15	0.9466 ± 0.0157	0.16	Haukos (1988)
TX	SSO	2008–2010	24	0.37	0.96 ± 0.01	0.36 ± 0.05	Holt (2012)
TX	SSO, SSB	2001–2003	21	0.67			Jones (2009)
KS	SGP, MGP	2013–2014	185	0.26		0.39 (0.34–0.43)	Lautenbach (2015)
TX	SSO	2006–2007	12	0.41			Leonard (2008)
TX	SSO, SSB	2001–2005	57	0.47			Lyons et al. (2011)
TX	SSO, SSB	2001–2003	22	0.64	0.983 ± 0.006	0.54	
TX	SSO	2003–2005	35	0.37	0.965 ± 0.007	0.29	
NM	SSO	1979	13	0.54			Merchant (1982)
		1980	11	0.0			
KS	SSB	1997–2002	196	0.26% ± 3%			Pitman et al. (2006b)
NM	SSO	1976–1978	36	0.27			Riley et al. (1992)
NM	SSO	1979	15	0.47			Wilson (1982)
NM	SSO	1980	11	0.0			
NM	SSO	1976–1978	36	0.27			Wisdom (1980)

[a] SSO, Sand Shinnery Oak Prairie; SSB, Sand Sagebrush Prairie; SGP, Short-Grass Prairie/Conservation Reserve Program Mosaic; MGP, Mixed-Grass Prairie.

[b] Based on median initiation date.

success appears to be conditionally dependent on habitat quality as years with suitable habitat for nest success do not always translate to increased recruitment into the population (D. Haukos, unpubl. data). The paradox indicates that there are processes other than habitat quantity and quality impacting reproductive success in populations of Lesser Prairie-Chickens (Hagen et al. 2009). Another strategy is the ability of the hen to alter the nest microclimate to mitigate severe environmental conditions by mediating temperature and relative humidity within the range of tolerance for egg survival (Grisham 2012). Therefore, nest-site selection is an annual decision by females using local nest sites that maximize the probability of nest success while minimizing risks of predation and environmental extremes.

Most studies of Lesser Prairie-Chicken reproductive ecology include an evaluation of nesting habitat, particularly microhabitat of the nest site, and report that nearly all nest sites are within 1.6 km of a lek (Hagen et al. 2013; Table 6.5). Thus, it has been recommended to concentrate the management of nesting habitats within a 1.6 km buffer of leks or lek clusters with ~2/3 of the area devoted to creating and maintaining nesting habitat and the remaining 1/3 of the area in early-successional vegetation states and devoted to brood habitat (Hagen et al. 2013). The behavioral process of nest-site selection remains unknown, but females likely select habitats based on their needs for concealment and thermoregulation (Hagen et al. 2007, 2013; Grisham et al. 2014). Residual cover provided by vegetation of the current and previous growing seasons is critical for quality nesting habitat (Hagen et al. 2013). Nests have been found in a variety of grass, forb, and shrub species, but nest sites are usually characterized by similar vegetation structure across the species range (Table 6.5; Hagen et al. 2013). In general, females select nest sites that have greater visual obstruction, horizontal cover, residual grass cover, and litter cover with less bare ground than associated random points (Riley 1978, Riley et al. 1992, Giesen 1994, Pitman et al. 2005, Davis 2009, Hagen et al. 2013, Grisham et al. 2014). Similarly, vegetation is taller at nest sites than adjacent random points (Davis et al. 1979, Haukos and Smith 1989, Riley et al. 1992, Giesen 1994, Hagen et al. 2004, Pitman et al. 2005). Successful nests are typically located in sites with the greatest visual obstruction and are usually located either in residual bunchgrasses or under shrubs (Fields 2004, Hagen et al. 2004, Davis 2009). Nests are commonly associated with shrubs throughout the range of Lesser Prairie-Chickens, with an increasing use of shrubs as a flexible response to the extent of grass removal caused by livestock grazing (Haukos and Smith 1989). However, the threshold of grazing intensity that results in the shift from predominant grass cover to predominant shrub cover at nests is still unknown.

Lesser Prairie-Chickens usually select for nesting habitats with similar vegetation structure but differences have been reported among ecoregions across the species range (Chapters 14 to 17, this volume). Within the Sand Sagebrush Prairie Ecoregion, nest sites had greater vertical cover compared to random sites, which is mainly due to the presence of taller sagebrush, grass, and forbs at nest sites (Table 6.3; Hagen et al. 2013; Chapter 15, this volume). Within the same ecoregion, successful nests were in patches with high shrub densities of >5,000 plants/ha (Giesen 1994, Pitman 2003). Hagen et al. (2004, 2013) recommended that quality nesting habitat in the sand sagebrush prairie should provide dense shrubs and residual bunch grasses ranging in height from >25 cm in western range to >40 cm in eastern range, with >75% vertical screening and 50% overhead cover at 30 cm above ground. At a minimum, a >60% canopy cover of shrubs, grasses, and forbs is needed for suitable nesting habitat in the Sand Sagebrush Prairie Ecoregion (Hagen et al. 2013).

In the Sand Shinnery Oak Prairie Ecoregion, females select for greater vertical structure and taller grasses at nest sites as compared to random points (Chapter 16, this volume). Grasses found in greater proportion for nesting and brooding habitats are sand bluestem (*Andropogon hallii*), little bluestem, and sand dropseed (*Sporobolus cryptandrus*) (Davis et al. 1979). Purple three-awn (*Aristida purpurea*) is used as nesting habitat in continuously grazed patches (Haukos and Smith 1989). Hagen et al. (2013) recommended management goals with vegetation height of >36 cm for shrubs and >50 cm for grasses to create and maintain nesting habitat. Female Lesser Prairie-Chickens on the Sand Shinnery Oak Prairie selected for less bare ground and increased litter relative to random points (Hagen et al. 2013). Litter is likely more important in the Sand Shinnery Oak Prairie than other ecoregions because of the need to mediate extreme temperatures and thermal stress in this ecoregion (Patten et al. 2005a, Hagen et al. 2013).

Little is known about requirements for nesting habitat in the Short-Grass Prairie/CRP Mosaic Ecoregion (Chapter 14, this volume) or Mixed-Grass Prairie Ecoregion (Chapter 16, this volume). One nesting study in the Short-Grass Prairie/CRP Mosaic Ecoregion was conducted in a combination of native prairie and CRP patches (Fields 2004) and is not directly comparable to work from other ecoregions. Fields (2004) reported that 100% visual obstruction (2.7 dm) at nest sites was about the same as sand sagebrush prairie. Nest sites of Lesser Prairie-Chickens in the mosaic ecoregion consisted primarily of grass cover (63%–88%), with forbs and shrubs as minor components. Grass cover at nest sites was greater than nests in the Sand Sagebrush and Sand Shinnery Oak Prairie, but the difference was likely due to use of CRP patches. Fields (2004) also found that females selected CRP interseeded with forbs more than expected, whereas CRP fields with grass only or a mix of forbs and grass were used in proportion to their availability. Visual obstruction readings were lower in short-grass prairie than in CRP fields, including fields interseeded with forbs or with grass only. However, both types of CRP had two to three times more grass cover than short-grass prairie. Short-grass prairie had 3–5 times as many forbs as CRP. Among all habitat types, mid- and tall-grass species such as western wheatgrass (*Pascopyrum smithii*), little bluestem, big bluestem (*Andropogon gerardii*), and switchgrass were the dominant plant types associated with the nest bowl.

In the Mixed-Grass Prairie Ecoregion, Lesser Prairie-Chicken nest sites had greater mean heights for 100% and 0% visual obstruction, less bare ground, and more forb covers within 2 m than random sites (Holt 2012). Additionally, compositional analysis indicated that Lesser Prairie-Chickens selected against shrub cover within 4 m and in favor of grass and forb cover when selecting nest sites.

Nest placement can also be determined by available habitat under different management regimes such as herbicide treatments, grazing, and CRP, but the relationships are not always clear (Olawsky 1987, Grisham 2012). Within the sand shinnery oak prairie, nest-site selection among herbicide treatment and grazing plots were used in proportion to availability, suggesting that low rates of herbicide and moderate intensity grazing treatments are not detrimental to Lesser Prairie-Chicken nesting ecology (Grisham 2012). However, Haukos and Smith (1989) reported that Lesser Prairie-Chickens selected sand shinnery oak patches not treated by herbicide in an area subjected to intensive grazing. The latter result indicates that nesting females select nest sites based on available structural characteristics of vegetation, and not the process or land use treatment that create the vegetation structure.

When evaluating available nesting habitat, consideration of the behavioral response to the presence of anthropogenic structures via avoidance buffers is necessary to determine the true distribution of potentially available habitat (Chapter 11, this volume; see Chapters 14 to 17 for specific avoidance distances). For example, anthropogenic features reduced available nesting habitat due to avoidance buffers by ~11,000 ha in a sand sagebrush prairie in Kansas (Robel et al. 2004). Females selected nesting locations that were farther from powerlines, pumpjacks, improved roads, buildings, and other structures than was expected at random (Pitman et al. 2005, Grisham et al. 2014). Within sand shinnery oak prairies, Grisham et al. (2014) found that 85% of nests were at least 0.5 km from anthropogenic features that included powerlines, buildings, pumpjacks, and developed roads, with the exceptions of undeveloped roads and

stock tanks. On the other hand, females nested closer to stock tanks than was expected at random (Grisham et al. 2014). Nests were located farthest from buildings, pumpjacks, and improved roads (Grisham et al. 2014). The behavioral mechanisms for avoidance of these anthropogenic structures are undetermined, but vertical structures may promote avoidance in a landscape that is normally open and unobstructed. The associated human activity and unnatural noise associated with presence of anthropogenic structures may contribute to avoidance as well (Robel et al. 2004, Hagen et al. 2011). Avoidance may not depend on the density of the anthropogenic features (Hagen et al. 2011). Anthropogenic features can limit the amount of available habitat, but their presence does not necessarily influence nest success (Pitman et al. 2005).

Land use and conservation practices can influence habitat quantity and quality for Lesser Prairie-Chickens. A number of management techniques can affect the nesting habitat of Lesser Prairie-Chickens, including grazing, herbicide, and prescribed fire. Unfortunately, other than herbicide treatments of shrubs, limited information is available regarding the responses of nesting Lesser Prairie-Chickens to the available management practices for habitat improvements, including grazing management, tree removal, prescribed fire, and prairie restoration. All of these actions can positively or negatively affect the structure and composition of nesting vegetation at point, patch, and landscape scales, thereby influencing habitat selection at multiple scales by Lesser Prairie-Chickens. Indeed, the response by Lesser Prairie-Chickens to applied management practices may be dependent upon the spatial scale examined. Furthermore, demographic responses of Lesser Prairie-Chickens to different applied management practices in their range are poorly understood, and research efforts to measure vital rate responses to management practices and determine thresholds for habitat use would benefit conservation efforts. Many studies have examined microhabitat characteristics at nest sites and associated random points but few studies have tested factors affecting nest-site selection by females at the patch or landscape scales. Nest sites are often reported as being within 1.5 km of a lek (Hagen et al. 2013; Table 6.3), but recent literature and data based on advanced field techniques indicate that some females will disperse farther from lek of capture than previously reported, and more than 20 km during breeding season (D. A. Haukos, unpubl. data).

Many questions remain regarding nesting habitat. One key question is as follows: at what spatial scale or patch size are females selecting nesting habitat? What patch size should be used to evaluate habitat quality or quantity and allow for comparisons among sites? Future research on breeding Lesser Prairie-Chicken response to management techniques could be used to fine-tune management strategies based on grazing, prescribed fire, and use of herbicides. Long-term studies on the effect of management, especially under multiple management strategies and environmental conditions, will be particularly useful for future planning and management prescriptions. Last, little information is available on the habitat use and selection by females that either did not initiate a nest or have lost a nest or brood. Habitat use by this subset of the population may influence female survival and future reproductive output.

BROOD HABITATS

Our understanding of chick survival, habitat use, and movements of Lesser Prairie-Chickens during the brood-rearing period is limited (Tables 6.7 and 6.8). Compared to adult-only habitat or nest sites, brood habitat is characterized by sparser and shorter vegetation, which facilitates locomotion and predator escape (Jones 1963, Riley et al. 1993). Further, habitats selected by females with broods appear to have increased food resources and better thermoregulatory capacity compared to other available habitats (Jamison et al. 2002, Hagen et al. 2005b, Bell et al. 2010). Broods occupy habitats with moderate stands of cover with 25% canopy cover of shrubs, forbs, or grasses at a height of 20–30 cm (Hagen et al. 2004). Brood habitat is further characterized by greater cover of forbs than random sites (Pitman et al. 2006a, Hagen et al. 2013).

Brood ecology has primarily been characterized for the Sand Shinnery Oak Prairie and Sand Sagebrush Prairie Ecoregions. In sand shinnery oak prairies, broods selected patches with ≥60% bare ground, a greater stem density of sand shinnery oak, lower grass and shrub height, and lower visual obstruction than random sites (Riley 1978, Riley et al. 1993, Bell et al. 2010, Hagen et al. 2013). Moreover, forb and grass composition was similar between nesting and brooding habitat, but brooding areas had more bare ground and less litter than nesting areas (Hagen et al. 2013). Broods were found in sites with taller plants and increased overhead cover when the temperature exceeded 26.4°C

TABLE 6.7

Estimates of chick and brood survival (95% CI) for Lesser Prairie-Chickens.

State	Habitat[a]	Year	n[b]	Type	Duration	Survival	Sources
KS	SGP	2001–2003	27	Brood	60 days	0.28 apparent	Fields et al. (2006)
				Brood	60 days	0.49 (0.10–0.87) raised by adults	
				Brood	60 days	0.05 (0.00–0.11) reared by yearlings	
TX	SSO	2008–2010	9	Chick	1–14 days	0.88 (±0.02) daily	Holt (2012)
				Chick	15–63 days	0.99 (±0.01) daily	
				Chick	1–14 days	0.18 (±0.01) interval	
				Chick	15–63 days	0.55 (±0.16) interval	
KS	SGP, MGP	2013–2014	43	Brood	0–56 days	0.31 (0.16–0.47)	Lautenbach (2015)
KS	SSB	1997–2003	19	Brood	60 days	0.41 apparent	Pitman et al. (2006c)
				Chick	1–14 days	0.48 (0.37–0.62) interval	
				Chick	15–60 days	0.37 (0.22–0.61) interval	
				Chick	1–60 days	0.18 (0.03–0.38) interval	

[a] SSO, Sand Shinnery Oak Prairie; SSB, Sand Sagebrush Prairie; SGP, Short-Grass Prairie/Conservation Reserve Program Mosaic; MGP, Mixed-Grass Prairie.

[b] Number of broods monitored.

TABLE 6.8
Brood movements and space use (±SE) of Lesser Prairie-Chickens.

State	Habitat[a]	Year	n	Brood movement/Space use	Sources
NM	SSO	1979	16	23.4 ± 5.9 (ha) broods <4 weeks	Ahlborn (1980)
				34.1 ± 7.9 (ha) broods >4 weeks	
				47.0 ± 2.8 (ha) entire brood period	
KS	SSB	1997–1999	14	248 ± 17 (m) daily brood movements ≤14 days	Jamison (2000)
			8	320 ± 57 (m) daily brood movements ≥14 days	
KS	SSB	1997–2003	9	273 ± 10.5 (m) average minimum daily brood movements 0–14 days	Pitman et al. (2006a)
			9	312 ± 7.9 (m) average minimum daily brood movements 15–60 days	
KS	SGP, MGP	2013–2014	19	85.3 ± 14.7 (ha)	Plumb (2015)
	SGP	2013–2014	12	858.44 ± 17.2 (m) mean daily movement	
	MGP	2013–2014	7	654.74 ± 19.3 (m) mean daily movement	

[a] SSO, Sand Shinnery Oak Prairie; SSB, Sand Sagebrush Prairie; SGP, Short-Grass Prairie/Conservation Reserve Program Mosaic; MGP, Mixed-Grass Prairie.

(Bell et al. 2010). Compared to nesting habitat, brood habitat in sagebrush prairies has less canopy cover, less grass cover, and more forb cover (Hagen et al. 2013). Additionally, broods selected patches based on vegetation composition that consisted of 15% forbs, 20% sagebrush cover, and moderate densities of shrubs at ~4,000 ha plants/ha (Hagen et al. 2004).

The availability of invertebrates, especially grasshoppers (Orthoptera), as potential food resources is important for habitat use of broods (Jamison et al. 2002, Hagen et al. 2005b). Grasshopper abundance was greater at brood sites than at random sites in sand sagebrush prairies, but can be highly patchy and dependent on herbaceous cover and structure (Hagen et al. 2005b). Forb and sagebrush cover may be predictors of invertebrate biomass, but the evidence is mixed (Jamison et al. 2002, Hagen et al. 2005b). Invertebrate biomass (primarily Orthoptera) was greater on CRP plots interseeded with forbs as compared to grass CRP plots (Fields 2004). Grasshopper abundance is difficult to predict as the proposed relationships among habitat characteristics and grasshoppers are not always straightforward (Jonas and Joern 2007).

Brood ecology, specifically habitat use and selection, is one of the greatest information gaps for understanding the ecology of Lesser Prairie-Chickens. Studies of broods are typically limited by low sample size and require pooling of data among years and sites to obtain an adequate sample of observations for analyses. Such conditions limit our understanding of variation of habitat use by broods relative to the potential range of environmental conditions, spatial scales, and available food resources. Response of broods to habitat management is primarily observational, requiring additional scientific effort and rigor to develop improved options for conservation and management (Hagen et al. 2005b, 2009).

NONBREEDING HABITATS

Few studies of Lesser Prairie-Chickens have examined habitat use during the nonbreeding season. Lesser Prairie-Chickens have more long-distance movements and larger home ranges during the nonbreeding season (September–mid-February), but most individuals occupy the same habitats as during the breeding season (Taylor and Guthery 1980a; Hagen and Giesen 2004). Pirius et al. (2013) reported that 97% of bird locations were within 1.7 km of a lek during the nonbreeding season on sand shinnery oak prairies. Kukal (2010) found that 98% of bird locations were within 2.4 km of a known lek during winter in the Texas portion of the Mixed-Grass Prairie Ecoregion, with no differences between males and females. Such findings indicate that either lek locations are important to habitat use throughout the year or that Lesser Prairie-Chicken populations are currently confined to habitat patches associated with leks due to landscape fragmentation.

If foliage of the sand shinnery oak decreases following a freezing event, Lesser Prairie-Chickens will use patches with greater grass cover and <15% of shrub cover during winter (Kukal 2010, Pirius et al. 2013). In addition, during periods of reduced food availability during the nonbreeding season, birds will congregate to forage in harvested grain fields. However, relative to the occupied range, waste grain is only available to a small percentage of extant populations of Lesser Prairie-Chickens (Jones 1964, Crawford and Bolen 1976b, Ahlborn 1980, Taylor and Guthery 1980b, Jamison 2000, Salter et al. 2005). A conventional assumption is that nonbreeding habitat is not limiting to Lesser Prairie-Chickens, but few data exist across the range of the species to support this viewpoint. With extensive overlap in habitat use between breeding and nonbreeding periods, management for quality nesting and brooding habitats should provide sufficient nonbreeding habitat.

WATER

Free water has not been considered a direct requirement of Lesser Prairie-Chickens because of the use of metabolic water from foods and the assumption that historically the species occupied prairie with little available surface water (Bidwell et al. 2002, Hagen and Giesen 2004, Elmore et al. 2009). However, the use of free water has been documented in all seasons (Copelin 1963, Jones 1964, Crawford and Bolen 1973, Sell 1979, Boal et al. 2014). If one considers the currently occupied range of Lesser Prairie-Chickens, in particular the restriction to sandy soils of the western region, the assumption that surface water was historically unavailable may not be supported. Considerable evidence indicates that numerous artesian springs in the sand country supported Lesser Prairie-Chickens prior to the advent of conversion of prairies to row crop agriculture and eventual widespread irrigation (Brune 2002). Many springs have ceased to function during the past 70 years due to ongoing declines in water levels in underlying aquifers, so it is possible that free water was important to Lesser Prairie-Chicken populations in the semiarid High Plains portion of the species range.

Recent evidence is indicating that free water may be a factor in the habitat use by Lesser Prairie-Chickens. In sand shinnery oak habitats, 99.9% and 80% of Lesser Prairie-Chicken locations were within 3.2 km and 1.7 km of an available water source, respectively (average ~1.2 km; Pirius et al. 2013). Free water may be important for egg development or other life history requirement during drought conditions (Crawford and Bolen 1973, Grisham 2012). The influence of free water on Lesser Prairie-Chickens has not been verified yet, but several lines of evidence suggest it warrants closer evaluation. If free water is shown to influence habitat use in the semiarid portion of Lesser Prairie-Chicken occupied range, then reconsideration of available habitat may be necessary.

HABITAT QUALITY

If patch size and vegetation structure are adequate, habitat quality for Lesser Prairie-Chickens is primarily determined by three factors: (1) short-term environmental conditions such as drought, precipitation, temperature, and evapotranspiration; (2) time since disturbance; and (3) local management practices including grazing, anthropogenic structures, and effects of disturbance. We hypothesize that habitat quality in the form of vegetation structure, habitat patch size and juxtaposition, and food resources are the current dominant limiting factors interacting with current and past environmental conditions and soil type to exert ecological resistance on the growth and distribution of populations of Lesser Prairie-Chickens. Habitat quality influences population growth, yet is usually assessed at fine scales by measuring the effects of individual point selection within used habitat patches. For example, much of the literature on the ecology of Lesser Prairie-Chickens focuses on the assessments of nest-site vegetation characteristics relative to random points within the same habitat patch in an attempt to judge habitat quality in relation to nest survival, which may or may not be directly or indirectly related to vegetative structure. However, habitat quality likely manifests itself at a greater spatial scale than individual points, such that increasing habitat quality across a landscape will have a greater impact on population vital rates than on local management efforts.

In general, quality habitats for Lesser Prairie-Chickens are considered to have vegetation conditions with >35% canopy cover of grasses, shrubs, and forbs, with >50% composition of preferred species of shrubs including sand shinnery oak or sand sagebrush, with an intermixing of the appropriate structure of selected nesting and brood habitat (Hagen et al. 2013). Native grasses that increase habitat quality include switchgrass, little bluestem, sideoats grama, plains bristlegrass (*Setaria vulpiseta*), big bluestem, sand bluestem, Indian grass (*Sorghastrum nutans*), or prairie sandreed (*Calamovilfa longifolia*). Van Pelt et al. (2013) outlined quality habitat as including preferred native grass species with a canopy of at least 20%–30% with variable grass heights that average >38.1 cm (Table 6.9). The canopy cover of shrubs is suggested as being 10–50%, depending on

TABLE 6.9

Recommended measures of habitat quality for Lesser Prairie-Chickens following Van Pelt et al. (2013).

Sand Shinnery Oak Prairie

  Nesting habitat

    Canopy cover of sand shinnery oak: 20%–50%

    Canopy cover of preferred grasses (native bluestems, switchgrass, Indian grass, and sideoats grama): >20%

    Canopy cover of a good mix of species of native forbs: >10%

    Variable grass heights that average: >38.1 cm

  Brood habitat

    Canopy cover of sand shinnery oak: 10%–25%

    Canopy cover of preferred native grasses: >15%

    Canopy cover of a mix of native forbs: >20%

    Variable grass heights that average: >38.1 cm

    Shrub, grass, and forb understory open enough to allow movements of chicks

Sand Sagebrush Prairie

  Nesting habitat

    Canopy cover of sand sagebrush: 15%–30%

    Canopy cover of preferred native grasses: >30%

    Canopy cover of a mix of native forbs: >10%

    Variable grass heights that average: >38.1 cm

  Brood habitat

    Canopy cover of sand sagebrush: 10%–25%

    Canopy cover of preferred native grasses: >20%

    Canopy cover of a mix of native forbs: >20%

    Variable grass heights that average: >38.1 cm

    Shrub, grass, and forb understory open enough to allow movements of chicks

Mixed-Grass Prairies and Short-Grass Prairie/CRP Mosaic without shrubs

  Nesting habitat

    Canopy cover of preferred native grasses: >50%

    Canopy cover of a mix of native forbs: >10%

    Variable grass heights that average: >38.1 cm

  Brood habitat

    Canopy cover of preferred native grasses: 30%–50%

    Canopy cover of a mix of native forbs: >20%

    Variable grass heights that average: >38.1 cm

    Shrub, grass, and forb understory open enough to allow movements of chicks whether the dominant shrub is sand shinnery oak or sand sagebrush, and whether the area is being used for nesting or brood rearing (Van Pelt et al. 2013; Table 6.9). A forb cover of >10% is preferred (Van Pelt et al. 2013; Table 6.8). Further, Van Pelt et al. (2013) recommended annual grazing utilization rates of ≤33% of available annual production under typical environmental conditions for sustaining quality Lesser Prairie-Chicken habitat.

Lesser Prairie-Chickens consistently select habitats with greater canopy cover than is available elsewhere across the range of the species. Increased cover provides concealment from predators and favorable microclimate characteristics (Patten et al. 2005a, Grisham et al. 2013, Larsson et al. 2013). The ability of Lesser Prairie-Chickens to mediate exposure to extreme environmental conditions with cold and hot temperatures by selective use of available habitat is a driving force in habitat use (Bell et al. 2010, Grisham 2012, Borsdorf 2013, Larsson et al. 2013). Habitat quality mediates the effects of drought, which is one of the key drivers of the population demography of Lesser Prairie-Chickens (Chapter 2, this volume). Prolonged drought affects composition and structure of vegetation necessary to provide suitable nesting and roosting cover, food resources, and opportunity for escape from predators (Copelin 1963, Merchant 1982, Applegate and Riley 1998, Hagen et al. 2004; Chapter 12, this volume). Home ranges expand during drought conditions in response to diminished habitat quality, thereby increasing space use requirements (Copelin 1963, Merchant 1982).

The assessment of habitat quality for Lesser Prairie-Chickens has been primarily observational and correlative. Habitat quality rather than quantity is likely the driver of recent population trends of Lesser Prairie-Chickens. Advances in our understanding of the capacity of management practices to increase habitat quality that will elicit positive population responses such as improved recruitment or survival leading to increases in density or occupied range will require controlled experiments to reduce uncertainty in conservation actions. Emphasis on quantitative data from designed experiments will expedite knowledge gain compared to the correlative results provided by observational studies. Given the dynamic and frequently harsh environmental conditions across the range of the Lesser Prairie-Chicken, the effects of management practices may not be measureable

for several years following implementation. Studies that assess the effects of conservation actions on habitat conditions for Lesser Prairie-Chickens will be most informative at a population level or second-order habitat selection. Spatial and temporal scales that measure population responses across a range of environmental conditions and management practices are needed to understand influences on habitat quality. Thus, long-term studies are necessary to evaluate the effects of conservation actions designed to increase habitat quality in an adaptive management framework (Chapter 17, this volume), because precipitation, temperature, and other environmental conditions may have a greater influence on habitat quality than applied management practices.

SUMMARY

The extant range of Lesser Prairie-Chickens is characterized by harsh, unpredictable environmental conditions. Populations are scattered throughout the southwestern Great Plains, primarily occurring in areas dominated by low-statured shrubs such as sand shinnery oak and sand sagebrush, and mid- to tall grasses that are typically associated with sandy soils. The current and historical range of Lesser Prairie-Chickens overlaps with the short-grass prairie of the semiarid High Plains, but short-grass prairie provides little habitat for the species. Habitat loss, conversion, degradation, and fragmentation are the primary cause of long-term population declines, whereas annual variation in precipitation and temperature are the main drivers of short-term population fluctuations. Core or source populations of Lesser Prairie-Chickens require large, intact patches of native prairie to persist but such landscapes are becoming increasingly isolated, resulting in a lower probability of movements of individuals among populations and a loss of connectivity. Restoration of sand shinnery oak and sand sagebrush prairie in semiarid regions is expensive and time consuming under the best possible environmental conditions, and protection and conservation of extant habitats may be the most effective and economic approach to habitat conservation. We hypothesize that habitat quality in the form of vegetation structure, patch size and configuration, and food resources are the main limiting factors that interact with current and past environmental conditions to exert ecological resistance on the recovery and distribution of populations of

Lesser Prairie-Chicken. Given the natural spatial and temporal variation of invertebrates, acorns, seeds, and other important food resources, it may be difficult, if not impossible, to focus habitat management efforts on food resources for Lesser Prairie-Chickens. Therefore, improving habitat quality of extant available habitat could be the most effective conservation action to increase abundance and occupied range of Lesser Prairie-Chickens. Controlled experiments to measure population response to applied habitat management practices will reduce the considerable uncertainty in regard to conservation of Lesser Prairie-Chickens.

ACKNOWLEDGMENTS

The Western Association of Fish and Wildlife Agencies, Kansas State University, Texas Tech University, and the U.S. Geological Survey provided support during preparation of this chapter. Reviews of initial drafts by C. Boal, C. Hagen, and V. Winder greatly improved the chapter.

LITERATURE CITED

Ahlborn, C. G. 1980. Brood-rearing habitat and fall-winter movements of Lesser Prairie Chickens in eastern New Mexico. M.S. thesis, New Mexico State University, Las Cruces, NM.

Aldridge, C. L., and M. S. Boyce. 2007. Linking occurrence and fitness to persistence: habitat-based approach for endangered Greater Sage-Grouse. Ecological Applications 17:508–526.

Applegate, R. D., and T. Z. Riley. 1998. Lesser Prairie-Chicken management. Rangelands 20:13–15.

Baruch-Mordo, S., J. S. Evans, J. P. Severson, D. E. Naugle, J. D. Maestas, J. M. Kiesecker, M. J. Falkowski, C. A. Hagen, and K. P. Reese. 2013. Saving Sage-Grouse from the trees: a proactive solution to reducing a key threat to a candidate species. Biological Conservation 167:233–241.

Bell, L. A., S. D. Fuhlendorf, M. A. Patten, D. H. Wolfe, and S. K. Sherrod. 2010. Lesser Prairie-Chicken hen and brood habitat use on sand shinnery oak. Rangeland Ecology and Management 63:478–486.

Bent, A. C. 1932. Life Histories of North American Gallinaceous Birds. United States Natural Museum Bulletin 162, Washington, DC.

Bidwell, T., S. Fuhlendorf, B. Gillen, S. Harmon, R. Horton, R. Rodgers, S. Sherrod, D. Wiedenfeld, and D. Wolfe. 2002. Ecology and management of the Lesser Prairie-Chicken. Oklahoma Cooperative Extension Service E-970, Oklahoma State University, Stillwater, OK.

Blanco-Fontao, B., B. K. Sandercock, J. R. Obeso, L. B. McNew, and M. Quevedo. 2013. Effects of sexual dimorphism and landscape composition on the trophic behavior of Greater Prairie-Chickens. PLoS One 8:e79986.

Block, W. M., and L. A. Brennan. 1993. The habitat concept in ornithology: theory and applications. Current Ornithology 11:35–91.

Boal, C. W., P. K. Borsdorf, and T. Gicklhorn. 2014. Assessment of Lesser Prairie-Chicken use of wildlife water guzzlers. Bulletin of the Texas Ornithological Society 46:10–18.

Boal, C. W., B. Grisham, D. A. Haukos, J. C. Zavaleta, and C. Dixon. 2014. Lesser Prairie-Chicken nest site selection, microclimate, and nest survival in association with vegetation response to a grassland restoration program. U.S. Geological Survey Open-File Report 2013–1235, Reston, VA.

Borsdorf, P. K. 2013. Lesser Prairie-Chicken habitat selection across varying land use practices in eastern New Mexico and west Texas. M.S. thesis, Texas Tech University, Lubbock, TX.

Brune, G. 2002. Springs of Texas, 2nd edn., Vol. 1. Texas A&M Press, College Station, TX.

Butler, M. J., W. B. Ballard, R. D. Holt, and H. A. Whitlaw. 2010. Sound intensity of booming in Lesser Prairie-Chickens. Journal of Wildlife Management 74:1160–1162.

Campbell, H. 1972. A population study of Lesser Prairie Chickens in New Mexico. Journal of Wildlife Management 36:689–699.

Candelaria, M. A. 1979. Movements and habitat-use by Lesser Prairie Chickens in eastern New Mexico. M.S. thesis, New Mexico State University, Las Cruces, NM.

Cannon, R. W., and F. L. Knopf. 1981. Lek numbers as a trend index to prairie grouse populations. Journal of Wildlife Management 45:776–778.

Clark, R. G., and D. Shutler. 1999. Avian habitat selection: pattern from process in nest-site use by ducks? Ecology 80:272–287.

Cody, M. L. 1985. Habitat selection in birds. Academic Press, Orlando, FL.

Copelin, F. F. 1963. The Lesser Prairie Chicken in Oklahoma. Technical Bulletin No. 6, Oklahoma Department of Wildlife Conservation Department Oklahoma City, OK.

Crawford, J. A., and E. G. Bolen. 1973. Spring use of stock ponds by Lesser Prairie Chickens. Wilson Bulletin 85:471–472.

Crawford, J. A., and E. G. Bolen. 1976a. Effects of land use on Lesser Prairie Chickens in Texas. Journal of Wildlife Management 40:96–104.

Crawford, J. A., and E. G. Bolen. 1976b. Fall diet of Lesser Prairie Chickens in west Texas. Condor 78:142–144.

Davis, C. A., T. Z. Riley, R. A. Smith, and M. J. Wisdom. 1980. Spring-summer foods of Lesser Prairie Chickens in New Mexico. Pp. 75–80 in P. A. Vohs and F. L. Knopf (editors), Proceedings of Prairie Grouse Symposium, Oklahoma State University, Stillwater, OK.

Davis, C. A., T. Z. Riley, H. R. Suminski, and M. J. Wisdom. 1979. Habitat evaluation of Lesser Prairie Chickens in eastern Chaves County, New Mexico. Final Report to Bureau of Land Management, Roswell, Contract YA-512-CT6–61, Department of Fishery and Wildlife Sciences, New Mexico State University, Las Cruces, NM.

Davis, D. M. [online]. 2005. Status of the Lesser Prairie-Chicken in New Mexico: recommendation to not list the species as threatened under the New Mexico Wildlife Conservation Act. Final Investigation Report, New Mexico Department of Game and Fish, Santa Fe, NM. <http://www.fws.gov/southwest/es/documents/R2ES/LitCited/LPC_2012/Davis_2005.pdf> (12 July 2014).

Davis, D. M. 2009. Nesting ecology and preproductive success of Lesser Prairie-Chickens in shinnery oak-dominated rangelands. Wilson Journal of Ornithology 121:322–327.

Davis, D. M., R. E. Horton, E. A. Odell, R. D. Rodgers, and H. A. Whitlaw. [online]. 2008. Lesser Prairie-Chicken conservation initiative. Lesser Prairie-Chicken Interstate Working Group, Colorado Division of Wildlife, Ft. Collins, CO. <http://www.wafwa.org/documents/LPCCI_FINAL.pdf> (14 July 2014).

Doerr, T. B., and F. S. Guthery. 1983. Effect of tebuthiuron on Lesser Prairie-Chicken habitat and foods. Journal of Wildlife Management 47:1138–1142.

Doherty, K. E., D. E. Naugle, and B. L. Walker. 2010. Greater Sage-Grouse nesting habitat: the importance of managing at multiple scales. Journal of Wildlife Management 74:1544–1553.

Donaldson, D. D. 1969. Effect on Lesser Prairie Chickens of brush control in western Oklahoma. Dissertation, Oklahoma State University, Stillwater, OK.

Elmore, D., T. Bidwell, R. Ranft, and D. Wolfe. 2009. Habitat evaluation guide for the Lesser Prairie-Chicken. E-1014. Oklahoma Cooperative Extension Service, Division of Agricultural Sciences and Natural Resources, Oklahoma State University, Stillwater, OK.

Fagen, R. 1988. Population effects of habitat change: a quantitative assessment. Journal of Wildlife Management 52:41–46.

Fields, T. L. 2004. Breeding season habitat use of Conservation Reserve Program land by Lesser Prairie-Chickens in west central Kansas. M.S. thesis, Colorado State University, Fort Collins, CO.

Fields, T. L., G. C. White, W. C. Gilgert, and R. D. Rodgers. 2006. Nest and brood survival of Lesser Prairie-Chickens in west central Kansas. Journal of Wildlife Management 70:931–938.

Fretwell, S. D., and H. L. Lucas Jr. 1970. On territorial behavior and other factors influencing habitat distribution in birds. I. Theoretical development. Acta Biotheoretica 19:16–36.

Fuhlendorf, S. D., A. J. W. Woodward, D. M. Leslie Jr., and J. S. Shackford. 2002. Multi-scale effects of habitat loss and fragmentation on Lesser Prairie-Chicken populations of the US Southern Great Plains. Landscape Ecology 17:617–628.

Giesen, K. M. 1994. Movements and nesting habitat of Lesser Prairie-Chicken hens in Colorado. Southwestern Naturalist 39:96–98.

Giesen, K. M. 2000. Population status and management of Lesser Prairie-Chicken in Colorado. Prairie Naturalist 32:137–148.

Grisham, B. A. 2012. The ecology of Lesser Prairie-Chickens in shinnery oak-grassland communities in New Mexico and Texas with implications toward habitat management and future climate change. Ph.D. dissertation, Texas Tech University, Lubbock, TX.

Grisham B. A., C. W. Boal, D. A. Haukos, D. M. Davis, K. K. Boydston, C. Dixon, and W. Heck. 2013. The predicted influence of climate change on Lesser Prairie-Chicken reproductive parameters. PLoS One 8:e68225.

Grisham, B. A., P. K. Borsdorf, C. W. Boal, and K. K. Boydston. 2014. Nesting ecology and nest survival of Lesser Prairie-Chickens on the Southern High Plains of Texas. Journal of Wildlife Management 78:857–866.

Hagen, C. A. 2003. A demographic analysis of Lesser Prairie-Chicken populations in southwestern Kansas: survival, population viability, and habitat use. Ph.D. dissertation, Kansas State University, Manhattan, KS.

Hagen, C. A., and K. M. Giesen. [online]. 2004. Lesser Prairie-Chicken (Tympanuchus pallidicinctus). Birds of North America 364. <http://bna.birds.cornell.edu/bna/species/364/articles/introduction> (15 November 2013).

Hagen, C. A., B. A. Grisham, C. W. Boal, and D. A. Haukos. 2013. A meta-analysis of Lesser Prairie-Chicken nesting and brood rearing habitats: implications for habitat management. Wildlife Society Bulletin 37:750–758.

Hagen, C. A., B. E. Jamison, K. M. Giesen, and T. Z. Riley. 2004. Guidelines for managing Lesser Prairie-Chicken populations and their habitats. Wildlife Society Bulletin 32:69–82.

Hagen, C. A., J. C. Pitman, T. M. Loughin, B. K. Sandercock, R. J. Robel, and R. D. Applegate. 2011. Impacts of anthropogenic features on habitat use by Lesser Prairie-Chickens. Pp. 63–75 in B. K. Sandercock, K. Martin, and G. Segelbacher (editors), Ecology, conservation, and management of grouse. Studies in Avian Biology No. 39, Cooper Ornithological Society, University of California Press, Berkeley, CA.

Hagen, C. A., J. C. Pitman, B. K. Sandercock, R. J. Robel, and R. D. Applegate. 2005a. Age-specific variation in apparent survival rates of male Lesser Prairie-Chickens. Condor 107:78–86.

Hagen, C. A., J. C. Pitman, B. K. Sandercock, R. J. Robel, and R. D. Applegate. 2007. Age-specific survival and probable causes of mortality in female Lesser Prairie-Chickens. Journal of Wildlife Management 71:518–525.

Hagen, C. A., G. C. Salter, J. C. Pitman, R. J. Robel, and R. D. Applegate. 2005b. Lesser Prairie-Chicken brood habitat in sand sagebrush: invertebrate biomass and vegetation. Wildlife Society Bulletin 33:1080–1091.

Hagen, C. A., B. K. Sandercock, J. C. Pitman, R. J. Robel, and R. D. Applegate. 2009. Spatial variation in Lesser Prairie-Chicken demography: a sensitivity analysis of population dynamics and management alternatives. Journal of Wildlife Management 73:1325–1332.

Hall, L. S., P. R. Krausman, and M. L. Morrison. 1997. The habitat concept and a plea for standard terminology. Wildlife Society Bulletin 25:171–182.

Haufler, J., D. Davis, and J. Caulfield. [online]. 2012. Oklahoma Lesser Prairie-Chicken conservation plan: a strategy for species conservation. Ecosystem Management Research Institute, Seeley Lake, MT. <http://www.emri.org/PDF%20Docs/Adobe%20files/Final_OK_LEPC_Mgmt_Plan_23Oct2012.pdf> (10 August 2014).

Haukos, D. A. 1988. Reproductive ecology of Lesser Prairie-Chickens in west Texas. M.S. thesis, Texas Tech University, Lubbock, TX.

Haukos, D. A. 2011. Use of tebuthiuron to restore sand shinnery oak grasslands of the Southern High Plains. Pp. 103–124 in M. Naguib and A. E. Hasaneen (editors), Herbicides: mechanisms and mode of action. Intech, Rijeka, Croatia.

Haukos, D. A., and L. M. Smith. 1989. Lesser Prairie-Chicken nest site selection and vegetation characteristics in tebuthiuron-treated and untreated sand shinnery oak in Texas. Great Basin Naturalist 49:624–626.

Haukos, D. A., and L. M. Smith. 1999. Effects of lek age on age structure and attendance of Lesser Prairie-Chickens (Tympanuchus pallidicinctus). American Midland Naturalist 142:415–420.

Haukos, D. A., L. M. Smith, and G. S. Broda. 1989. The use of radio-telemetry to estimate Lesser Prairie-Chicken nest success and hen survival. Pp. 238–243 in C. J. Amlaner (editor), Proceedings of the 10th International Symposium on Biotelemetry. University of Arkansas, Fayetteville, AR.

Haukos, D. A., L. M. Smith, and G. S. Broda. 1990. Spring trapping of Lesser Prairie-Chickens. Journal of Field Ornithology 61:20–25.

Higgins, K. F., K. J. Jenkins, G. K. Clambey, D. W. Uresk, D. E. Naugle, R. W. Klaver, J. E. Norland, K. C. Jensen, and W. T. Barker. 2012. Vegetation sampling and measurement. Pp. 381–409 in N. J. Silvy (editor), The wildlife techniques manual: research. Johns Hopkins Press, Baltimore, MD.

Hobbs, N. T., and T. A. Hanley. 1990. Habitat evaluation: do use/availability data reflect carrying capacity? Journal of Wildlife Management 54:515–522.

Hoffman, D. M. 1963. The Lesser Prairie Chicken in Colorado. Journal of Wildlife Management 27:726–732.

Holt, R. D. 2012. Breeding season demographics of a Lesser Prairie-Chicken (Tympanuchus pallidicinctus) population in the northeast Texas Panhandle. Ph.D. dissertation, Texas Tech University, Lubbock, TX.

Jamison, B. E. 2000. Lesser Prairie-Chicken chick survival, adult survival, and habitat selection and movements of males in fragmented rangelands of southwestern Kansas. M.S. thesis, Kansas State University, Manhattan, KS.

Jamison, B. E., R. J. Robel, J. S. Pontius, and R. D. Applegate. 2002. Invertebrate biomass: associations with Lesser Prairie-Chicken habitat use and sand sagebrush density in southwestern Kansas. Wildlife Society Bulletin 30:517–526.

Jonas, J. L., and A. Joern. 2007. Grasshopper (Orthoptera: Acrididae) communities respond to fire, bison grazing and weather in North American tallgrass prairie: a long-term study. Oecologia 153:699–711.

Jones, J. 2001. Habitat selection studies in avian ecology: a critical review. Auk 118:557–562.

Jones, R. E. 1963. Identification and analysis of Lesser and Greater Prairie Chicken habitat. Journal of Wildlife Management 27:757–778.

Jones, R. E. 1964. Habitat used by Lesser Prairie Chickens for feeding related to seasonal behavior of plants in Beaver County, Oklahoma. Southwestern Naturalist 9:111–117.

Jones, R. S. 2009. Seasonal survival, reproduction, and use of wildfire areas by Lesser Prairie-Chickens in the northeastern Texas panhandle. M.S. thesis, Texas A & M University, College Station, TX.

Johnson, D. H. 1980. The comparison of usage and availability measurements for evaluating resource preference. Ecology 61:65–71.

Johnson, M. D. 2007. Measuring habitat quality: a review. Condor 109:489–504.

Kukal, C. A. 2010. The over-winter ecology of Lesser Prairie-Chicken (*Tympanuchus pallidicinctus*) in the northeast Texas Panhandle. M.S. thesis, Texas Tech University, Lubbock, TX.

Larsson, L. C., C. L. Pruett, D. H. Wolfe, and M. A. Patten. 2013. Fine-scale selection of habitat by the Lesser Prairie-Chicken. Southwestern Naturalist 58:135–149.

Lautenbach, J. M. 2015. Lesser Prairie-Chicken reproductive success, habitat selection, and response to trees. M.S. Thesis, Kansas State University, Manhattan, KS.

Leonard, J. P. 2008. The effects of shinnery oak removal on Lesser Prairie-Chicken survival, movement, and reproduction. M.S. thesis, Texas A&M University, College Station, TX.

Lyons, E. K. 2008. Lesser Prairie-Chicken demographics in Texas: survival, reproduction, and population viability. Ph.D. dissertation, Texas A&M University, College Station, TX.

Lyons, E. K., B. A. Collier, N. J. Silvy, R. R. Lopez, B. E. Toole, R. S. Jones, and S. J. DeMaso. 2009. Breeding and non-breeding survival of Lesser Prairie-Chickens *Tympanuchus pallidicinctus* in Texas, USA. Wildlife Biology 15:89–96.

Lyons, E. K., R. S. Jones, J. P. Leonard, B. E. Toole, R. A. McCleery, R. R. Lopez, M. J. Peterson, S. J. DeMaso, and N. J. Silvy. 2011. Regional variation in nesting success of Lesser Prairie-Chickens. Pp. 223–231 in B. K. Sandercock, K. Martin, and G. Segelbacher, editors. Ecology, conservation, and management of grouse. Studies in Avian Biology (no. 39), University of California Press, Berkeley, CA.

MacMahon, J. A., D. J. Schimpf, D. C. Anderson, K. G. Smith, and R. L. Bayn, Jr. 1981. An organism-centered approach to some community and ecosystem concepts. Journal of Theoretical Biology 88:287–307.

Manly, B. F. J., L. L. McDonald, D. L. Thomas, T. L. McDonald, and W. P. Erickson. 2002. Resource selection by animals: statistical design and analysis for field studies, 2nd edn. Kluwer Academic Publishers, Dordrecht, Netherlands.

McDonald, L., G. Beauprez, G. Gardner J. Griswold, C. Hagen, D. Klute, S. Kyle, J. Pitman, T. Rintz, and B. Van Pelt. 2014. Range-wide population size of the Lesser Prairie-Chicken: 2012 and 2013. Wildlife Society Bulletin 38:536–546.

McDonald, L. L., W. P. Erickson, M. S. Boyce, and J. R. Alldredge. 2012. Modeling vertebrate use of terrestrial resources. Pp. 410–428 in N. J. Silvy (editor), The wildlife techniques manual: research. Johns Hopkins Press, Baltimore, MD.

Merchant, S. S. 1982. Habitat-use, reproductive success, and survival of female Lesser Prairie-Chickens in two years of contrasting weather. M.S. thesis, New Mexico State University, Las Cruces, NM.

Mote, K. D., R. D. Applegate, J. A. Bailey, K. E. Giesen, R. Horton, and J. L. Sheppard. [online]. 1999. Assessment and conservation strategy for the Lesser Prairie-Chicken (*Tympanuchus pallidicinctus*). Kansas Department of Wildlife and Parks. Emporia, KS. <http://www.fws.gov/southwest/es/documents/R2ES/LitCited/LPC_2012/Mote_et_al_1999.pdf> (8 August 2014).

Morrison, M. L., B. G. Marcot, and R. W. Mannan. 2006. Wildlife-habitat relationships: concepts and applications, 3rd edn. Island Press, Washington, DC.

Olawsky, C. D. 1987. Effects of shinnery oak control with tebuthiuron on Lesser Prairie-Chicken populations. M.S. thesis, Texas Tech University, Lubbock, TX.

Olawsky, C. D., and L. M. Smith. 1991. Lesser Prairie-Chicken densities on tebuthiuron-treated and untreated sand shinnery oak rangelands. Journal of Range Management 44:364–368.

Patten, M. A., and J. F. Kelly. 2010. Habitat selection and the perceptual trap. Ecological Applications 20:2148–2156.

Patten, M. A., D. H. Wolfe, E. Shochat, and S. K. Sherrod. 2005a. Effects of microhabitat and microclimate selection on adult survivorship of the Lesser Prairie-Chicken. Journal of Wildlife Management 69:1270–1278.

Patten, M. A., D. H. Wolfe, E. Shochat, and S. K. Sherrod. 2005b. Habitat fragmentation, rapid evolution, and population persistence. Evolutionary Ecology Research 7:235–249.

Pettit, R. D. 1986. Sand shinnery oak control and management. Management Note 8, Natural Resources Management, Texas Tech University, Lubbock, TX.

Pidgeon, A. M., V. C. Radeloff, and N. A. Matthews. 2006. Contrasting measures of fitness to classify habitat quality for the Black-throated Sparrow (*Amphispiza bilineata*). Biological Conservation 132:199–210.

Pirius, N. E. 2011. The non-breeding season ecology of Lesser Prairie-Chickens (*Tympanuchus pallidicinctus*) in the Southern High Plains of Texas. M.S. thesis, Texas Tech University, Lubbock, TX.

Pirius, N. E., C. W. Boal, D. A. Haukos, and M. C. Wallace. 2013. Winter habitat use and survival of Lesser Prairie-Chickens in west Texas. Wildlife Society Bulletin 37:759–765.

Pitman, J. C. 2003. Lesser Prairie-Chicken nest site selection and nest success, juvenile gender determination and growth and juvenile survival and dispersal in southwestern Kansas. M.S. thesis, Kansas State University, Manhattan, KS.

Pitman, J. C., C. A. Hagen, B. E. Jamison, R. J. Robel, T. M. Loughin, and R. D. Applegate. 2006b. Nesting ecology of Lesser Prairie-Chickens in sand sagebrush prairie of southwestern Kansas. Wilson Journal of Ornithology 118:23–35.

Pitman, J. C., C. A. Hagen, B. E. Jamison, R. J. Robel, T. M. Loughin, and R. D. Applegate. 2006c. Survival of juvenile Lesser Prairie-Chickens in Kansas. Wildlife Society Bulletin 34:675–681.

Pitman, J. C., C. A. Hagen, R. J. Robel, T. M. Loughin, and R. D. Applegate. 2005. Location and success of Lesser Prairie-Chicken nests in relation to vegetation and human disturbance. Journal of Wildlife Management 69:1259–1269.

Pitman, J. C., B. E. Jamison, C. A. Hagen, R. J. Robel, and R. D. Applegate. 2006a. Brood break-up and juvenile dispersal of Lesser Prairie-Chicken in Kansas. Prairie Naturalist 38:85–99.

Plumb, R. T. 2015. Lesser Prairie-Chicken movement, space use, survival, and response to anthropogenic structures in Kansas and Colorado. M.S. thesis, Kansas State University, Manhattan, KS.

Pompanon, F., B. E. Deagle, W. O. C. Symondson, D. S. Brown, S. N. Jarman, and P. Taberlet. 2012. Who is eating what: diet assessment using next generation sequencing. Molecular Ecology 21:1931–1950.

Pulliam, H. R. 2000. On the relationship between niche and distribution. Ecology Letters 3:349–361.

Pulliam, H. R., and B. J. Danielson. 1991. Sources, sinks, and habitat selection: a landscape perspective on population dynamics. American Naturalist 137:50–66.

Riley, T. Z. 1978. Nesting and brood-rearing habitat of Lesser Prairie Chickens in southeastern New Mexico. M.S. thesis, New Mexico State University, Las Cruces, NM.

Riley, T. Z., C. A. Davis, M. A. Candelaria, and R. Suminski. 1994. Lesser Prairie-Chicken movements and home ranges in New Mexico. Prairie Naturalist 26:183–186.

Riley, T. Z., C. A. Davis, M. Ortiz, and M. J. Wisdom. 1992. Vegetative characteristics of successful and unsuccessful nests of Lesser Prairie-Chickens. Journal of Wildlife Management 56:383–387.

Riley, T. Z., C. A. Davis, and R. A. Smith. 1993. Autumn and winter foods of the Lesser Prairie-Chicken (*Tympanuchus pallidicinctus*) (Galliformes: Tetraonidae). Great Basin Naturalist 53:186–189.

Robel, R. J., J. A. Harrington, Jr., C. A. Hagen, J. C. Pitman, and R. R. Reker. 2004. Effect of energy development and human activity on the use of sand sagebrush habitat by prairie chickens in southwestern Kansas. Transactions of the North American Wildlife and Natural Resource Conference 69:251–266.

Rodgers, R. D., and R. W. Hoffman. 2005. Prairie grouse population response to Conservation Reserve Program grasslands: an overview. Pp. 120–128 in A. W. Allen and M. W. Vandever (editors), The Conservation Reserve Program—Planting for the Future: proceedings of a National Conference, Fort Collins, CO.

Rotenberry, J. T. 1985. The role of habitat in avian community composition: physiognomy or floristics. Oecologia 67:213–217.

Salter, G. C., R. J. Robel, and K. E. Kemp. 2005. Lesser Prairie-Chicken use of harvested corn fields during fall and winter in southwestern Kansas. Prairie Naturalist 37:1–9.

Sandercock, B. K., K. Martin, and G. Segelbacher (editors). 2011. Ecology, management and conservation of grouse. Studies in Avian Biology No. 39, Cooper Ornithological Society, University of California Press, Berkeley, CA.

Savory, C. J. 1989. The importance of invertebrate food to chicks of gallinaceous species. Proceedings of the Nutrition Society 48:113–133.

Schlaepfer, M. A., M. C. Runge, and P. W. Sherman. 2002. Ecological and evolutionary traps. Trends in Ecology and Evolution 17:474–480.

Sell, D. L. 1979. Spring and summer movements and habitat use by Lesser Prairie Chicken females in Yoakum County, Texas. M.S. thesis, Texas Tech University, Lubbock, TX.

Sexson, M. L. 1980. Destruction of sandsage prairie in southwest Kansas. Pp. 113–115 in C. L. Kucera (editor), Proceedings of the Seventh North American Prairie Conference, Southwestern Missouri State University, Springfield, MO.

Sharpe, R. S. 1968. The evolutionary relationships and comparative behavior of prairie chickens. Ph.D. dissertation, University of Nebraska, Lincoln, NE.

Smith, R. A. 1979. Fall and winter habitat of Lesser Prairie Chickens in southeastern New Mexico. M.S. thesis, New Mexico State University, Las Cruces, NM.

Spencer, D. A. 2014. A historical record of land cover change of the Lesser Prairie-Chicken range in Kansas. M.S. thesis, Kansas State University, Manhattan, KS.

Suminski, H. R. 1977. Habitat evaluation for Lesser Prairie Chickens in eastern Chaves, New Mexico. M.S. thesis, New Mexico State University, Las Cruces, NM.

Taylor, M. A. 1979. Lesser Prairie Chicken use of man-made leks. Southwestern Naturalist 24: 706–707.

Taylor, M. A., and F. S. Guthery. 1980a. Status, ecology, and management of the Lesser Prairie Chicken. U. S. Department of Agriculture, Forest Service General Technical Report RM-77. Rocky Mountain Forest and Range Experiment Station, Fort Collins, CO.

Taylor, M. A., and F. S. Guthery. 1980b. Fall-winter movements, ranges, and habitat use of Lesser Prairie Chickens. Journal of Wildlife Management 44:512–524.

Toole, B. E. 2005. Survival, seasonal movements, and cover use by Lesser Prairie-Chickens in the Texas Panhandle. M.S. thesis, Texas A&M University, College Station, TX.

U.S. Fish and Wildlife Service. 2014. Endangered and Threatened Wildlife and Plants; Special Rule for the Lesser Prairie-Chicken. Federal Register 79 FR 19973:19973–20071.

Van Pelt, W. E., S. Kyle, J. Pitman, D. Klute, G. Beauprez, D. Schoeling, A. Janus, and J. Haufler. 2013. The Lesser Prairie-Chicken range-wide conservation plan. Western Association of Fish and Wildlife Agencies, Cheyenne, WY.

Wilson, D. L. 1982. Nesting habitat of Lesser Prairie Chickens in Roosevelt and Lea counties, New Mexico. M.S. thesis, New Mexico State University, Las Cruces, NM.

Wisdom, M. J. 1980. Nesting habitat of Lesser Prairie Chickens in eastern New Mexico. M.S. thesis, University of New Mexico, Las Cruces, NM.

Wolfe, D. H., M. A. Patten, E. Shochat, C. L. Pruett, and S. K. Sherrod. 2007. Causes and patterns of mortality in Lesser Prairie-Chickens *Tympanuchus pallidicinctus* and implications for management. Wildlife Biology 13:95–104.

Woodward, A. J. W., S. D. Fuhlendorf, D. M. Leslie, and J. Shackford. 2001. Influence of landscape composition and change on Lesser Prairie-Chicken (*Tympanuchus pallidicinctus*) populations. American Midland Naturalist 145:261–274.

CHAPTER SEVEN

Harvest*

David A. Haukos, James C. Pitman, Grant M. Beauprez, and Doug D. Schoeling

Abstract. Harvest management of upland game birds has evolved from restrictions on harvest seasons and release of captive-reared birds toward a scientific approach based on the demographic concepts of compensatory versus additive mortality. Modern harvest strategies are based on an adaptive approach of establishment of population goals, designing hunting seasons to meet those goals and implementing a monitoring system to reduce uncertainty. Market hunting of Lesser Prairie-Chickens (*Tympanuchus pallidicinctus*) was common in the late 1800s and contributed to population declines that resulted in halting commercial harvest of the species by the early 1900s. Sport harvest of Lesser Prairie-Chickens was characterized by wide deviations in season dates, length, and bag limits within and among states. Extreme fluctuations in population densities of Lesser Prairie-Chickens due to annual variation in environmental conditions resulted in a series of closed and open seasons that were based on anecdotal perceptions of abundance up until the mid-1960s, followed by stabilization of harvest regulations until the late 1990s, and then closure to harvest to protect remaining populations. Sport harvest is not considered to be an important impact on the population demography of Lesser Prairie-Chickens but all states closed sport harvest of the species between 1996 and 2014. Reasons for closures included concerns about low population numbers, consideration of listing the species as state threatened, and official listing of the species as threatened under the federal Endangered Species Act in 2014. Lesser Prairie-Chickens have the ability to quickly respond to improved environmental conditions if quality habitat is available, and sport harvest could be reinstated if population recovery eventually exceeds stated goals for population numbers. In the future, an adaptive approach to harvest management could be used to reduce uncertainty regarding the influence of sport harvest on population demography of Lesser Prairie-Chickens.

Key Words: adaptive management, bag limit, harvest, hunting, sport.

H unting and harvest of upland game birds are a long-standing tradition that has met the divergent purposes of human subsistence, market hunting, and a sport restricted to nobility, to a highly informed process designed to simultaneously achieve ecological goals while maximizing opportunities for recreational harvest of a diverse group of birds. In 2011, 4.5 million people participated in the hunting of upland game species in the United States, second only to big game (U.S. Department of the Interior, U.S. Fish and Wildlife Service, and

* Haukos, D. A., J. C. Pitman, G. M. Beauprez, and D. D. Schoeling. 2016. Harvest. Pp. 133–144 in D. A. Haukos and C. W. Boal (editors), Ecology and conservation of Lesser Prairie-Chickens. Studies in Avian Biology (no. 48), CRC Press, Boca Raton, FL.

U.S. Department of Commerce, and U.S. Census Bureau 2014). Participating individuals hunted a total of 51 million days and spent $2.6 billion on trips and equipment in pursuit of small game. Grouse, including prairie chickens, were sought by 812,000 hunters (U.S. Department of the Interior, U.S. Fish and Wildlife Service, and U.S. Department of Commerce, and U.S. Census Bureau 2014). Hunter expenditures made important financial contributions to local economies, landowners, and a variety of different vendors, in addition to providing license fees and excise taxes that are directly used for wildlife management by state and federal wildlife agencies.

The Lesser Prairie-Chicken (*Tympanuchus pallidicinctus*) is an iconic upland game bird of the southwestern Great Plains that has been available for sport harvest to varying degrees in five states where the species occurs: Texas, New Mexico, Oklahoma, Kansas, and Colorado. Similar to other upland game birds, variation in harvest regulations among states and over time for Lesser Prairie-Chickens has been due to tradition, societal perceptions from both hunting and nonhunting interests, biopolitics, and research findings (Hurst and Rosene 1985, Strickland et al. 1994). Kansas first regulated the harvest of Lesser Prairie-Chickens in 1861, but most other states did not regulate harvest until the early 20th century. Extensive settlement of much of the current Lesser Prairie-Chicken range did not occur until the late 1800s and presettlement populations of the species were likely isolated at low densities (Chapter 2, this volume). Initially, following settlement, populations of Lesser Prairie-Chickens increased in both density and range, with an increase in harvest opportunities prompting the states to regulate hunting (Chapter 2, this volume). However, hunting regulations for Lesser Prairie-Chickens have varied widely among years and states, perhaps more so than for any other upland game species, in response to the extreme population fluctuations characteristic of this species (Chapter 4, this volume). It was often common for hunting seasons to be closed for one or more years, followed immediately by liberal harvest regulations in response to rapid increases in population numbers. For example, the droughts of the 1930s and 1950s and associated substantial declines in population numbers resulted in closure of Lesser Prairie-Chicken hunting in all

states, with seasons remaining closed until the late 1940s and early 1960s, respectively. The petition to list the Lesser Prairie-Chicken as a threatened species in 1995 resulted in state-by-state consideration of providing harvest opportunity to the point where only Kansas offered a hunting season when the species was listed as threatened in May of 2014 (U.S. Fish and Wildlife Service 2014). Following federal listing as a threatened species, legal sport harvest of Lesser Prairie-Chickens has been discontinued range-wide. The listing rule was vacated by judicial decision in September 2015, creating considerable regulatory uncertainty at the time of this volume.

Strategies associated with the management of upland game bird harvest have varied across time (Connelly et al. 2012). Through the mid-1900s, harvest management was primarily based on manipulation of bag limits, season length, and opening/closing of seasons based on anecdotal perceptions of population status. The primary direct population management technique was stocking of birds propagated by game farms and produced by many state agencies and commercial hatcheries. The typical initial responses to reduced numbers of upland game birds from the late 1800s through the 1960s and the early 1970s were artificial propagation and release, or capture and translocation. Such management actions were advocated by many authors, including Forbush (1912:563) who stated "Those who are killing the game of this country must put back into the coverts at least two game birds for every one which they kill if we are to continue to have game, and this can be done only by means of artificial propagation on game farms and preserves." Further, Forbush (1912:595) concluded that the combination of propagation and stocking of game birds and enforcement of game laws were the principal measures of increasing game bird populations, "Some self-denial on the part of the sportsman and an aroused public interest and public sentiment, with liberality in encouraging the propagation of game birds, will bring about respect for the laws, and make the North American continent again the greatest game bird country in the world."

Unfortunately, propagation and stocking of upland game birds have not withstood scientific scrutiny as viable techniques to either increase numbers or reestablish populations in unoccupied habitat due to poor productivity, low survival, and high costs relative to increasing vital rates of extant populations by increasing habitat quantity and quality (Krauss et al. 1987, Leif 1994, Parish and Sotherton 2007, Sokos et al. 2008, Musil and Connelly 2009). Release of pen-reared upland game birds just before or during hunting season can increase local harvest, but little evidence indicates that captive rearing will increase future population size or expand the range of wild upland game birds, especially prairie grouse (Maple and Silvy 1988, Strickland et al. 1994). Toepfer et al. (1990) concluded that prairie grouse have the worst record of any group of wildlife species in regard to establishment or enhancement of populations with translocated or propagated birds. An intensive breeding program was started in 1992 for the endangered Attwater's Prairie-Chicken (*T. cupido attwateri*). The Attwater's Prairie-Chicken recovery plan has specific population goals that include the release of captive and propagated birds to supplement wild populations (U.S. Fish and Wildlife Service 2010). Unfortunately, a number of issues related to propagation remain that have limited the success of these last-ditch efforts to prevent extinction, including poor chick survival, low juvenile survival upon release, outbreaks of disease, decreased reproductive effort of released birds, altered behavior among captive birds, and genetic considerations (U.S. Fish and Wildlife Service 2010). Following the population declines in the 1930s, New Mexico (Ligon 1954) and Kansas (Coats 1955) attempted captive propagation of Lesser Prairie-Chickens. The program had limited success in raising chicks in captivity, but no information is available on the hatching or survival rates of pen-reared chicks or survival of released birds. Indeed, Ligon (1954:10) stated "No other newly hatched game bird seems more helpless in regard to its inherited instincts for survival when confined within four walls." To our knowledge, translocations of Lesser Prairie-Chickens have never resulted in a newly established population that persisted within its current range (Texas Game, Fish, and Oyster Commission 1945, Toepher et al 1990, Snyder et al. 1999). However, a translocation of Lesser Prairie-Chickens to the Hawaiian Island of Ni'ihau in the 1930s was known to persist for at least 25–30 years (Chapter 2, this volume).

As knowledge of upland game population demography and limiting factors increased during the latter part of the 20th century, along with acceptance of the demographic concepts of harvestable surplus and compensatory mortality, improved strategies for harvest management were developed. The acceptance of harvest as a compensatory form of mortality led to liberalization and stabilization of regulations to the point where seasons and bag limits for many species of upland game birds were relatively constant for decades in most states (Connelly et al. 2012). Hunting regulations for Lesser Prairie-Chickens stabilized and were relatively constant from the early 1960s through early 1990s.

In the past three decades, the assessment of the effects of exploitation on wildlife populations has benefited from the accumulation of data from long-term monitoring, advances in technical and quantitative tools for tracking animal movements (e.g., satellite telemetry, Geographic Information Systems, Global Positioning Systems), and prediction of outcomes of alternative management options through mathematical modeling. In addition, the concept of partial compensation of harvest in annual mortality began to be supported by evidence (Bergerud 1985, 1988; Caughley and Sinclair 1994; Sandercock et al. 2011). Indeed, Bergerud (1988) concluded that for many species of upland game birds, including Greater Prairie-Chicken (*T. cupido*), harvest mortality was additive to overwinter mortality. Hamerstrom and Hamerstrom (1973) concluded that mortality increased 25% above normal for Greater Prairie-Chickens when hunting was permitted. Impacts of harvest on upland game birds are now considered to be population and habitat specific with the need for informed management decisions based on empirical information in an adaptive context (Guthery 2000, Johnson and Case 2000, Sutherland 2001).

Strickland et al. (1994) indicated that successful harvest management included three basic components: (1) inventory of population numbers, (2) identification of population and harvest goals, and (3) development of regulations allowing the goals to be met. Nichols et al. (1995) further refined the idea of informed or adaptive harvest management, which included the basic requirements of (1) development of explicit goals and actions, (2) implementation of specific actions to achieve harvest management objectives, (3) knowledge of the relative population responses to alternative harvest actions, and (4) measurement of outcomes of actions in relation to management objectives with the intent to gain knowledge for future actions. Changes in population numbers of most upland game birds, including Lesser Prairie-Chickens, are typically characterized by an annual index for some portion of the species' range but harvest regulations are rarely based on known abundance or density of the species (Sutherland 2001), which is in contrast to the annual process of setting harvest regulations for migratory waterfowl (Williams and Johnson 1995, Connelly et al. 2012). Recently, Van Pelt et al. (2013) proposed a range-wide population goal of 67,000 birds for Lesser Prairie-Chickens (2014 estimate 22,415; McDonald et al. 2014a), which may be used in an adaptive manner to guide future harvest decisions if the species is eventually delisted.

Inventory efforts to determine population numbers of Lesser Prairie-Chickens have been inconsistent and variable in time, space, and field methods, which has hampered the assessments of the impacts of harvest on population demography (Chapter 4, this volume). Fortunately, since 2012, a range-wide aerial population survey has been conducted with consistent effort based on statistically valid sampling protocols resulting in the potential to estimate population size and detect annual changes in population abundance (McDonald et al. 2014b). Continuation of aerial surveys would allow the assessment of the impacts of future harvest of the Lesser Prairie-Chicken should populations increase sufficiently to merit removal from the list of threatened species.

The most common available data related to the harvest of Lesser Prairie-Chickens are estimation of total harvest, number of hunters, number of days spent afield hunting, and seasonal bags per hunter. States have emphasized the collection of harvest data using telephone and mail or online questionnaires for the assessments of harvest of upland game birds (Strickland et al. 1994). Harvest data provide insights into the magnitude and composition of harvest, but it remains difficult to assess the impacts on population demography of Lesser Prairie-Chickens or other prairie grouse (Connelly et al. 2012). For example, harvest rate is a commonly derived measure for waterfowl, usually through band-recovery analyses, but historical estimates are lacking for Lesser

Prairie-Chickens due to lack of widespread banding effort or estimates of population size subject to harvest. Therefore, should harvest become possible for populations of Lesser Prairie-Chickens, consideration of new approaches to harvest management would reduce uncertainty relative to the impacts of harvest on population growth. Our objectives for this chapter were to (1) review the published information on the potential effects of harvest on populations of Lesser Prairie-Chickens, (2) compile the known information on historical harvest regulations by state, and (3) provide temporal estimates of available harvest data for each state.

EFFECTS OF HARVEST ON POPULATION DEMOGRAPHY

Published information on harvest of Lesser Prairie-Chickens is limited but the available information supports the conclusion reached by the USFWS (2014) that harvest was not a significant contributing factor to recent population declines of the species. Harvest data are limited without an operational monitoring program for the assessment of harvest relative to population density and demography. Another potential reason for limited harvest data is that most research efforts have concentrated on marking breeding females, which comprise a subset of the fall population at risk of harvest due to low initial sample sizes and relatively low survival rates during the preceding breeding season. Hagen et al. (2007) recorded ~1% of 220 radio-tagged females as harvested, which was similar to the rate of hunter-killed birds for all banded birds in the field study (Hagen 2003). Hagen et al. (2009) concluded that the harvest of Lesser Prairie-Chickens in the Sand Sagebrush Prairie Ecoregion had little, if any, effect on the population demography or growth rate.

In a life table analysis of primarily male Lesser Prairie-Chickens captured in New Mexico from 1962 to 1970, Campbell (1972) estimated a band-recovery rate of 7%–8% with an estimated kill rate of 22.5%, after adjustments for band reporting rate and crippling rate. He concluded that the level of annual harvest was not detrimental to the population. Further, Campbell (1972) and Lee (1950) concluded that there were no differences in relative vulnerability to harvest between males and females or juveniles and adults.

Copelin (1963:50) linked habitat conditions to reproductive success in Oklahoma, and recognized the pronounced fluctuations in population density as a response to environmental conditions when he stated "Because of the population recovery of Lesser Prairie Chickens at about 10 year intervals, open seasons should be considered when population density appears to be increasing." His recommended implementation of a harvest strategy that would limit take to 25%–30% of an increasing population was primarily based on a similar recommendation for Greater Prairie-Chickens in Wisconsin (Hamerstrom et al. 1957). Last, Copelin (1963) advocated for increasing information relative to setting hunting seasons for Lesser Prairie-Chickens by endorsing annual spring surveys to monitor population trends and using summer brood and fall lek surveys to assess reproductive success.

STATE-SPECIFIC HARVEST REGULATIONS AND POPULATION TRENDS

Colorado

Lesser Prairie-Chickens are a relatively recent arrival to Colorado and were first confirmed in 1914 (Lincoln 1918). Occupied habitats in Colorado represent the western edge of the species range (Giesen 2000; Chapter 2, this volume), and the species has been state threatened since 1973 (Hagen and Giesen 2004). The first mention of a Lesser Prairie-Chicken hunting season in Colorado was in 1917, established from September 15 to October 1 with a daily bag of 10, and possession limit of 15 (E. Gorman, Colorado Wildlife and Parks, pers. comm.). Some form of harvest continued for about three decades until hunting was closed in 1937 and has remained closed thereafter.

New Mexico

During the late 1800s, Lesser Prairie-Chickens were extensively harvested in New Mexico for subsistence and markets (Bent 1932; Chapter 13, this volume). The earliest known harvest regulations were enacted in 1897 with the establishment of a September–February season for prairie chickens. It is likely, but not documented, that an open hunting season persisted for Lesser Prairie-Chickens in New Mexico until the mid-1930s.

According to Lee (1950:475), "In the early 1930s a combination of drought and resultant overgrazing the prairie chicken range had driven the birds in New Mexico to the brink of extermination," which resulted in an extended period of closed seasons. Lee (1950) reported on hunting seasons in 1948 and 1949, with the 1949 season starting at noon on November 26 and ending at 1630 on November 28. Using check stations, 167 hunters were checked in 1949 with a reported average hunter success of 1.96 birds per hunter.

Available contemporary hunt records go back to 1958 when a 2-day season with three-bird daily bag and possession limit was reported. It is unknown if a season existed between 1949 and 1958, but was likely closed for much, if not all, of this period due to drought. The season was closed in 1959 and reopened in 1960, after which a season was held annually until 1995. The season length was 2 days in 1960 and 1965–1969; 3 days in 1961–1964 and 1970–1986; and 9 days from 1987 to 1994 in response to the relative high population levels of the 1980s (Chapter 4, this volume). With the exception of 1960 and 1965–1966, the daily bag was set as three birds. During this period, the average annual harvest was 1.70 birds per hunter. Data on age and sex ratios are only available from 1958 to 1968, the bulk of which were reported by Campbell (1972). Numbers for estimated total birds harvested peaked during 1985–1988 (Figure 7.1). Harvest was relatively constant during the 1970s into the early 1980s. Starting in 1990, the estimated harvest declined annually through 1995, until the season was closed in 1996 (Figure 7.1).

Oklahoma

Prior to becoming a state, the legislature of Oklahoma Territory passed a law in 1890 which restricted the hunting of Lesser Prairie-Chickens to November 1–February 1, which was amended in 1899 to September 1–January 1. In 1909, the hunting season was established as September 1–November 1, but no hunting on Sundays. The initial daily bag and possession limits were established in 1909 as 15 and 100, respectively. In 1915, the state legislature passed a law prohibiting harvest of prairie chickens at any time following a recommendation by the State Game Warden. The Oklahoma Game and Fish Commission were

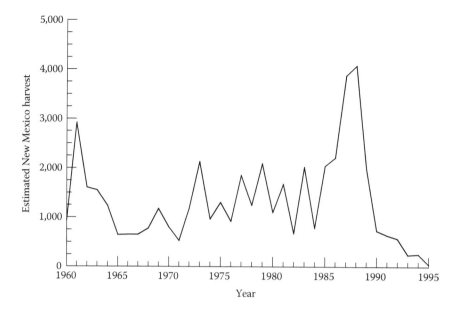

Figure 7.1. Estimated harvest of Lesser Prairie-Chickens in New Mexico, 1960–1995.

granted authority in 1929 to open a season on prairie chickens when "... the birds became so numerous as to endanger private property or farm crops" (Copelin 1963:48). Subsequently, Lesser Prairie-Chicken seasons and bag/possession limits were granted in 1929 (September 9–13, 6/12), 1931 (September 15–17, 5/10), and 1933 (October 2–4, 5/10). Hunting was then closed for more than a decade until the opening of a 1-day hunt on December 2, 1950 with a bag limit of two birds. The total harvest in 1950 was estimated at 600–1,000 Greater and Lesser Prairie-Chickens. In 1951, a 3-day hunt was conducted on December 11, 13, and 15 with a daily bag limit of two birds. The drought of the 1950s resulted in closed seasons from 1952 to 1963.

From 1964 to 1997, the harvest of Lesser Prairie-Chickens was permitted in Oklahoma with seasonal management set at a county scale. Typically, hunting was limited to Beaver, Woodward, Ellis, and Harper counties, but occasionally neighboring counties of Texas and Roger Mills were included, contingent on perceived status of Lesser Prairie-Chickens in Oklahoma. Bag and possession limits were fairly consistent with 2/4 in 1964 and 1967–1995, 3/6 in 1965 and 1966, and 1/2 in 1996–1997. The season was closed in 1998 and has not reopened.

The historical annual harvest of Lesser Prairie-Chickens in Oklahoma is not available because

harvest estimates from hunter questionnaire surveys are for the harvest of two species of prairie chickens combined (Greater and Lesser Prairie-Chicken). Check station data were not available. From 1964 to 1997, the average seasonal harvest of prairie chickens was 1.41 birds per hunter. With the exception of 1974 and 1975, the harvest of prairie chickens in Oklahoma was fairly stable from 1965 to 1990 (Figure 7.2). However, starting in 1991, the annual harvest of prairie chickens showed significant declines in Oklahoma.

Texas

The Texas Game, Fish, and Oyster Commission (1945) indicated that unregulated commercial and sport harvest prior to 1900 contributed to the significant decline in all species of prairie chickens in the state. It was stated "Many years ago prairie chickens ranked high as game birds, but now there are so few of them that an open season cannot be justified." Texas Game, Fish, and Oyster Commission (1945:75). The Texas State Legislature closed hunting for both Lesser and Attwater's Prairie-Chickens in September 1937.

The hunting season remained closed for Lesser Prairie-Chickens until 1967 when a 2-day season was allowed in the northeastern Texas Panhandle (Mixed-Grass Prairie Ecoregion), with a similar season starting in 1970 for the

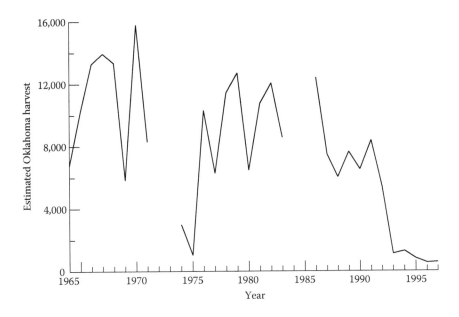

Figure 7.2. Estimated harvest of prairie chickens in Oklahoma, 1964–1997.

southwestern population (Sand Shinnery Oak Prairie Ecoregion). The typical hunting season in Texas consisted of 2 days during a weekend in October with a daily bag limit of two or three birds. The last general hunting season was held in 2004. In 2005, the harvest of Lesser Prairie-Chickens in Texas was restricted to properties involved in a state-approved wildlife management plan that included habitat enhancement, harvest recommendations/records, and population monitoring (Lionberger 2009). The number of permits issued and Lesser Prairie-Chickens harvested by year under this policy were as follows: 2005—15 permits and 12 birds, 2006—15 permits and 9 birds, 2007—15 permits and 7 birds, and no permits issued in 2008. Hunting of Lesser Prairie-Chickens has been prohibited in Texas since 2009.

Information on the composition and size of the annual harvest of Lesser Prairie-Chickens in Texas varies due to the variety of techniques used to collect such data. Hunter check stations were established from 1967 to 1974 and 1982 to 1986. Only successful hunters were required to bring harvested birds to check stations, but compliance and number of unsuccessful hunters were unknown. Therefore, for those seasons, only total number of birds checked, birds checked per hunter, and sex and age ratios of harvested birds were recorded. No reliable estimates of total number of hunters or total harvest are available. From 1975 to 1981, harvest data were collected by encountering hunters in the field, which reduced the number of hunters surveyed and total number of birds assessed for age and sex. Similar to check stations, total harvest and total number of hunters could not be estimated with this survey technique. To assist in collecting harvest data, hunters were required to obtain a no-cost special permit from 1987 to 1992 and 1997 to 2004 and questionnaires were mailed to permit holders to gather information for the estimation of total harvest and birds per hunter. The response rate to the questionnaire averaged 41.6% (range 28.0%–61.2%; Lionberger 2009). Data for 1992 are not available. From 1993 to 1996, Lesser Prairie-Chicken harvest was estimated using the Small Game Harvest Survey; thus, results for those seasons are not comparable to those generated by questionnaires of special permit holders. During the period of the required special permit and associated hunter questionnaire, the average seasonal bag was 0.92 birds per hunter (range 0.48–1.73). Birds per hunter declined steadily from 1987 to 2004 (Figure 7.3). The average annual estimated total harvest of Lesser Prairie-Chickens in Texas was 444 birds (range 59–1388) during the period. The estimated harvest declined 88% from 1987 to 2004 (Figure 7.3).

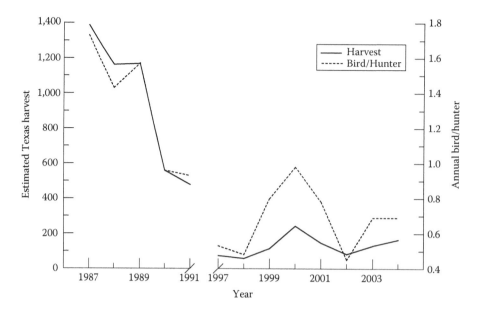

Figure 7.3. Estimated harvest and birds per hunter of Lesser Prairie-Chickens in Texas, 1987–1991 and 1998–2004.

Kansas

Kansas currently supports the greatest number of Lesser Prairie-Chickens (McDonald et al. 2014b; Chapter 4, this volume). Prairie chicken harvest in Kansas was managed as three units during 1989–2013 (Figure 7.4). Lesser Prairie-Chickens occurred in the northwest and southwest units. Complicating the issue of harvest in Kansas is the overlap of Lesser Prairie-Chickens with Greater Prairie-Chickens, a similar species subject to harvest, in the northwest unit. Following the decision with federal listing of Lesser Prairie-Chickens as threatened species in May 2014, and avoid to incidental take of Lesser Prairie-Chickens by hunters pursuing Greater Prairie-Chicken, the Kansas Department of Wildlife, Parks and Tourism closed prairie chicken harvest throughout the range of the Lesser Prairie-Chicken in Kansas, and redrew the management unit for legal harvest of Greater Prairie-Chickens in 2014 (Figure 7.4).

Kansas began regulating the harvest of prairie chickens in 1861, more than 50 years earlier than other states. Prairie chicken seasons were established by county until 1975, after which seasons were established statewide or by management unit. Separation of hunting seasons for Lesser and Greater Prairie-Chickens started in 1970. Until 1911, harvest regulations primarily were stated season dates (various season dates and lengths from August 1 to March 31) with some mention of means of take and occasional Sunday closures. In 1893, prairie chicken hunting was closed until 1897, and purchase and sale of game birds were declared illegal. In 1911, a daily bag of 12 birds was established. However, the prairie chicken season was closed from 1913 to 1916. Starting in 1917, the season length was restricted to <15 days, including 2 days from 1928 to 1935 following a closed season in 1927, typically with a daily bag of five birds. Due to the drought of the 1930s, the season was closed from 1936 to 1940 with additional closures from 1944 to 1949 and 1953 to 1956 due to low population numbers. For the 1941–1943 seasons, seasons were 1–2 days long with a bag of three birds. For 1950–1952, the season was 1 day with a bag of two birds. From 1957 to 1994, the daily bag for Lesser Prairie-Chickens was two birds. During this period, seasons were typically from early November through the end of January (1981–1982 season ended February 15) with a duration ranging from 2 to 100 days. From 1995 to 2011, the bag limit for Lesser Prairie-Chickens was decreased to one bird and then raised to two birds for 2012 and 2013 in the northwest zone. Also starting in 1995, season dates for the southwest zone differed from the northwest zones. Starting in 2012, a $2.50 permit was required to hunt prairie chickens in Kansas to identify prairie chicken hunters in an attempt to improve the accuracy of harvest estimates and species distribution.

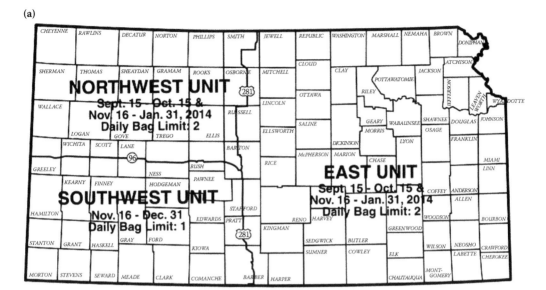

(a)

(b)

Greater Prairie-Chicken Hunting Unit

Greater Prairie-Chicken Hunting Region

Closed to Prairie-Chicken Hunting

- ☐ Greater Prairie-Chicken range
- ☐ Lesser Prairie-Chicken range
- ☐ Greater and Lesser Prairie-Chicken range

Miles
0 15 30   60

Figure 7.4. Harvest management units for prairie chickens in Kansas: (a) 1989–2013 prairie chicken harvest was management as three units with Lesser Prairie-Chickens occurring in the northwest and southwest units and (b) in 2014, prairie chicken harvest was closed throughout the range of the Lesser Prairie-Chicken in Kansas with the creation of a single management unit for Greater Prairie-Chickens.

The first statewide hunting season for prairie chickens in Kansas was established in 1975, after a 40-year period since 1935. Therefore, harvest metrics for Lesser Prairie-Chickens were comparable from 1975 to 2012 (2013 data were not available at the time of preparation). Harvest data were collected via a questionnaire of a random sample of Kansas residents and nonresidents that purchased a license type that allowed small game hunting. Separation of harvest of

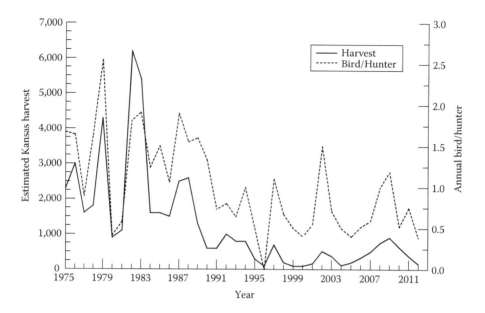

Figure 7.5. Estimated harvest and birds per hunter of Lesser Prairie-Chickens in Kansas, 1975–2012.

prairie chickens by species was delineated by county based on the dominant species in each county. Data from 25 counties were considered to be representative of the range of Lesser Prairie-Chickens in Kansas. Any harvest of Lesser Prairie-Chickens north of the Arkansas River due to the relatively recent range expansion was not accounted for in this approach (Chapter 4, this volume), because birds harvested in northern counties were considered to be Greater Prairie-Chickens (Dahlgren 2013). From 1975 to 2012, average numbers of Lesser Prairie-Chickens harvested per hunter for the entire season were 1.01 birds per hunter (range = 0–2.55, Figure 7.5). The harvest of Lesser Prairie-Chickens peaked in the early 1980s with annual estimates exceeding 5,000 birds, but by the early 1990s, the estimated harvest had declined to <1,000 birds with fluctuations between 100 and 900 birds until 2012 (Figure 7.5).

CONCLUSIONS

Sport harvest has been a relatively minor source of mortality for populations of Lesser Prairie-Chicken in the past century. Hunting season regulations for open dates, duration of season, and bag limits were quite variable among and within states until the 1960s, and were typically set in response to anecdotal perceptions of population densities, which fluctuated extensively in response to variable environmental conditions with a boom–bust reproductive success due to patterns of precipitation. Hunter success has tracked long-term population trends, with population declines since the early 1990s leading to reduced harvest by hunters. The five states that comprise the extant range of Lesser Prairie-Chickens have all closed harvest in response to low population numbers, consideration of listing the species as threatened, or to federal listing as a threatened species. The harvest of Lesser Prairie-Chickens will remain closed as long as the species is listed as threatened, but the historical record indicates that populations have often declined to a point where sport harvest was no longer considered viable, but for all states except Colorado, populations eventually recovered to sufficient levels to allow harvest. A combination of improved environmental conditions and successful conservation efforts, such as the Lesser Prairie-Chicken Range-wide Conservation Plan (Van Pelt et al. 2013), has the potential to result in future recovery of populations of the Lesser Prairie-Chicken. An adaptive approach to harvest combined with improved monitoring of the effects of harvest would enhance understanding of the impacts of harvest on population demography.

ACKNOWLEDGMENTS

We thank D. Dahlgren for compiling historic information on hunting seasons in Kansas, H. Whitlaw for information on harvest in Texas, and D. Klute and E. Gorman for assisting with information from Colorado. We are grateful for the efforts by many agency personnel in annually collecting harvest information.

LITERATURE CITED

Bent, A. C. 1932. Life histories of North American gallinaceous birds. U.S. National Museum Bulletin 162. Dover Publications, New York, NY.

Bergerud, A. T. 1985. The additive effect of hunting mortality on the natural mortality rates of grouse. Pp. 345–364 in S. L. Beason and S. F. Roberson (editors), Game harvest management. Caesar Kleberg Wildlife Research Institute, Kingsville, TX.

Bergerud, A. T. 1988. Increasing the numbers of grouse. Pp. 345–364 in A. T. Bergerud and M. W. Gratson (editors), Adaptive strategies and population ecology of Northern grouse. University of Minnesota Press, Minneapolis, MN.

Campbell, H. 1972. A population study of Lesser Prairie Chickens in New Mexico. Journal of Wildlife Management 36:689–699.

Caughley, G., and A. R. E. Sinclair. 1994. Wildlife ecology and management. Blackwell Scientific, Cambridge, MA.

Coats, J. 1955. Raising Lesser Prairie Chickens in captivity. Kansas Fish and Game 13:16–20.

Connelly, J. W., J. H. Gammonley, and T. W. Keegan. 2012. Harvest management. Pp. 202–231 in N. J. Silvy (editor), The wildlife techniques manual: management. Johns Hopkins University Press, Baltimore, MD.

Coplin, F. F. 1963. Lesser Prairie Chicken in Oklahoma. Pittman-Robertson Project Number W-62-R, Oklahoma Wildlife Conservation Department, Oklahoma City, OK.

Dahlgren, D. 2013. Small game hunter activity survey—2012. State Wildlife Research and Surveys, Grant W-39-R-19, Pittman-Robertson Federal Aid in Wildlife Restoration. Kansas Department of Wildlife, Parks, and Tourism, Pratt, KS.

Forbush, E. H. 1912. A history of the game birds, wildfowl and shore birds of Massachusetts and adjacent states. Massachusetts State Board of Agriculture, Wright & Potter Printing Company, Boston, MA.

Giesen, K. 2000. Population status and management of Lesser Prairie-Chicken in Colorado. Prairie Naturalist 32:137–148.

Guthery, F. S. 2000. On bobwhites. Texas A&M University Press, College Station, TX.

Hagen, C. A. 2003. A demographic analysis of Lesser Prairie-Chicken populations in southwestern Kansas: survival, population viability, and habitat use. Ph.D. dissertation, Kansas State University, Manhattan, KS.

Hagen, C. A., and K. M. Giesen. [online]. 2004. Lesser Prairie-Chicken (*Tympanuchus pallidicinctus*). Birds of North America 364. <http://bna.birds.cornell.edu/bna/species/364/articles/introduction> (1 October 2013).

Hagen, C. A., J. C. Pitman, B. K. Sandercock, R. J. Robel, and R. D. Applegate. 2007. Age-specific survival and probable causes of mortality in female Lesser Prairie-Chickens. Journal of Wildlife Management 71:518–525

Hagen, C. A., B. K. Sandercock, J. C. Pitman, R. J. Robel, and R. D. Applegate. 2009. Spatial variation in Lesser Prairie-Chicken demography: a sensitivity analysis of population dynamics and management alternatives. Journal of Wildlife Management 73:1325–1332.

Hamerstrom, F. N., Jr., and F. Hamerstrom. 1973. The prairie chicken in Wisconsin: highlights of a 22-year study of counts, behavior, movements, turnover, and habitat. Technical Bulletin 64, Wisconsin Department of Natural Resources, Madison, WI.

Hamerstrom, F. N., Jr., O. Mattson, and F. Hamerstrom. 1957. A guide to prairie chicken management. Technical Bulletin 15, Wisconsin Department of Natural Resources, Madison, WI.

Hurst, G. A., and W. Rosene. 1985. Regulations and restrictions pertaining to Bobwhite Quail harvests in the southeast. Pp. 301–308 in S. L. Beason and S. F. Roberson (editors), Game harvest management. Caesar Kleberg Wildlife Research Institute, Kingsville, TX.

Johnson, F. A., and D. J. Case. 2000. Adaptive regulation of waterfowl harvests: lessons learned and prospects for the future. Transactions of the North American Wildlife and Natural Resources Conference 65:94–108.

Krauss, G. D., H. B. Graves, and S. M. Zervanos. 1987. Survival of wild and game-farm cock pheasants released in Pennsylvania. Journal of Wildlife Management 51:555–559.

Lee, L. 1950. Kill analysis for the Lesser Prairie Chicken in New Mexico, 1949. Journal of Wildlife Management 14:475–477.

Leif, A. P. 1994. Survival and reproduction of wild and pen-reared Ring-necked Pheasant hens. Journal of Wildlife Management 58:501–506.

Ligon, J. S. 1954. The Lesser Prairie Chicken and its propagation. Modern Game Breeding and Hunting Club News 24:10–11.

Lincoln, F. C. 1918. Notes on some species new to the Colorado list of birds. Auk 35:236–237.

Lionberger, J. E. 2009. Lesser Prairie-Chicken monitoring and harvest recommendations. Federal Aid Grant No. W-126-R-17, Small Game Research and Surveys. Texas Parks and Wildlife Department, Austin, TX.

Maple, D. P., and N. J. Silvy. 1988. Recovery and economics of pen-reared bobwhites in north-central Texas. Proceedings of the Annual Conference of the Southeastern Association of Fish and Wildlife Agencies 42:329–332.

McDonald, L., K. Adachi, T. Rintz, G. Gardner, and F. Hornsby. [online]. 2014a. Range-wide population size of the Lesser Prairie-Chicken: 2012, 2013, and 2014. Western EcoSystems Technology, Inc., Laramie, WY. <http://www.wafwa.org/documents/LPC-aerial-survey-results-2014.pdf> (12 August 2014).

McDonald, L., G. Beauprez, G. Gardner, J. Griswold, C. Hagen, D. Klute, S. Kyle, J. Pitman, T. Rintz, and B. Van Pelt. 2014b. Range-wide population size of the Lesser Prairie-Chicken: 2012 and 2013. Wildlife Society Bulletin 38:536–546.

Musil, D. D., and J. W. Connelly. 2009. Survival and reproduction of pen-reared vs translocated wild pheasants Phasianus colchicus. Wildlife Biology 15:80–88.

Nichols, J. D., F. A. Johnson, and B. K. Williams. 1995. Managing North American waterfowl in the face of uncertainty. Annual Review of Ecology and Systematics 26:177–199.

Parish, D. M. B., and N. W. Sotherton. 2007. The fate of released captive-reared Grey Partridges Perdix perdix: implications for reintroduction programmes. Wildlife Biology 3:140–149.

Sandercock, B. K., E. B. Nilsen, H. Brøseth, and H. C. Pedersen. 2011. Is hunting mortality additive or compensatory to natural mortality? Effects of experimental harvest on the survival and cause-specific mortality of Willow Ptarmigan. Journal of Animal Ecology 80:244–258.

Snyder, J. W., E. C. Pelren, and J. A. Crawford. 1999. Translocation histories of prairie grouse in the United States. Wildlife Society Bulletin 27:428–432.

Sokos, C. K., P. K. Birtsas, and E. P. Tsachalidis. 2008. The aims of galliforms release and choice of techniques. Wildlife Biology 14:412–422.

Strickland, M. D., H. J. Harju, K. R. McCaffery, H. W. Miller, L. M. Smith, and R. J. Stoll. 1994. Harvest management. Pp. 445–473 in T. A. Bookhout (editor), Research and management techniques for wildlife and habitats. The Wildlife Society, Bethesda, MD.

Sutherland, W. J. 2001. Sustainable exploitation: a review of principals and methods. Wildlife Biology 7:131–140.

Texas Game, Fish, and Oyster Commission. 1945. Principal game birds and mammals of Texas: their distribution and management. Austin, TX.

Toepfer, J. E., R. L. Eng, and R. K. Anderson. 1990. Translocating prairie grouse: what have we learned? Transactions of the North American Wildlife and Natural Resources Conference 55:569–579.

U.S. Fish and Wildlife Service. 2010. Attwater's prairie-chicken recovery plan, Second revision. Albuquerque, NM.

U.S. Department of the Interior, U.S. Fish and Wildlife Service, and U.S. Department of Commerce, and U.S. Census Bureau. [online]. 2014. 2011 National Survey of Fishing, Hunting, and Wildlife-Associated Recreation. <https://www.census.gov/prod/2012pubs/fhw11-nat.pdf> (13 June 2014).

U.S. Fish and Wildlife Service. 2014. Endangered and threatened wildlife and plants; special rule for the Lesser Prairie-Chicken. Federal Register 79:20074–20085.

Van Pelt, W. E., S. Kyle, J. Pitman, D. Klute, G. Beauprez, D. Schoeling, A. Janus, and J. Haufler. 2013. The Lesser Prairie-Chicken range-wide conservation plan. Western Association of Fish and Wildlife Agencies, Cheyenne, WY.

Williams, B. K., and F. A. Johnson. 1995. Adaptive management and the regulation of waterfowl harvest. Wildlife Society Bulletin 23:430–436.

CHAPTER EIGHT

Predation and Lesser Prairie-Chickens*

Clint W. Boal

Abstract. Lesser Prairie-Chickens (*Tympanuchus pallidicinctus*) are subject to predation from a suite of opportunistic predators. Predation risk is associated with local environmental and landscape conditions, vegetation composition, and the predator community, all of which are influenced by human actions. Few studies have specifically assessed the relationships among Lesser Prairie-Chickens and their predators. I review the body of knowledge on predation as it relates to seasonality and life stages of Lesser Prairie-Chickens. Few quantitative data are available but some general patterns emerge. Predation is usually greater during the breeding season compared to the nonbreeding season. Seasonality in mortality risk appears to be associated with different predation pressures, with mammalian and avian predation rates similar during the breeding season, whereas raptors pose a greater threat during the nonbreeding season. All age classes of Lesser Prairie-Chickens may experience relatively constant year-round risk and risk to nests from mammals, with seasonal variation in mortality risk to nests from ravens (*Corvus* spp.), to both nests and young birds from snakes, and to juveniles and adults from raptors. A challenge for the study of predation of Lesser Prairie-Chickens is that accurate identification of the predator often is not possible, and predators are often lumped into taxonomic guilds without clarity as to which species are responsible for the impacts. Predator management has not been attempted or advocated as a conservation tool for the Lesser Prairie-Chickens but is more often considered in the context of managing for habitat quantity and quality. If predation poses population-level influences, it is likely through direct impacts on nest and brood survival, two of the most critical population parameters for the species. It remains unknown whether predation has a population-level effect on Lesser Prairie-Chickens, or what the frequency and impact of any one predator species may be. Last, increased predation rates may or may not be the proximate result of human actions. Changes in predation risk are especially relevant in the context of land cover changes due to increased livestock grazing and other anthropocentric activities or to changes in environmental conditions such as drought. Intensive population studies are still needed for a better understanding of the community–habitat influences on predator–prey relationships of prairie grouse and for the development of improved management strategies for the conservation of Lesser Prairie-Chickens.

Key Words: mammals, mortality, nest failure, predator, raptor, survival, *Tympanuchus pallidicinctus.*

* Boal, C. W. 2016. Predation and Lesser Prairie-Chickens. Pp. 145–158 in D. A. Haukos and C. W. Boal (editors), Ecology and conservation of Lesser Prairie-Chickens. Studies in Avian Biology (no. 48), CRC Press, Boca Raton, FL.

A foundational concept of wildlife management is to identify and reduce the impact of important limiting factors to increase the size of target populations. Limiting factors are density-dependent or density-independent factors that reduce reproductive output or survival, and keep a wildlife population below numbers that can be supported by the available resources. Herein lies a major challenge for the conservation of the Lesser Prairie-Chicken (*Tympanuchus pallidicinctus*); the species currently exists in a different landscape and availability of resources from the environment in which it evolved. The species' historical distribution across the Southern Great Plains was one of the last regions of North America to be settled by Europeans. Prior to settlement starting in the 1870s, Lesser Prairie-Chickens had dwelled in a region that was in a relatively natural ecological state. Therefore, the species and associated predators likely co-occurred in a community with a stable biotic structure. In such situations, prey populations are thought to ultimately be limited by food and other resources (Leopold 1933). The stability of prairie communities was disrupted by human settlement of the Southern Great Plains and the associated degradation, fragmentation, and conversion of key habitats. Coupled with landscape-level impacts were factors that can increase the mortality of Lesser Prairie-Chickens (Patten et al. 2005), including increases in predator species that could take advantage of anthropogenic features (Knight and Kawashima 1993). In a relatively short period of time, alteration of ecological drivers and landscape composition of the Southern Great Plains resulted in small, fragmented populations of Lesser Prairie-Chickens (Hagen and Giesen 2005). When a species persists in such circumstances, predation can become a limiting factor, especially when predators are generalists and predator densities are not linked to any single prey species (Newton 1993, Macdonald et al. 1999, Thirgood et al. 2000). High predation rates on ground-nesting birds are an important conservation issue in North America (Jiménez and Conover 2001), which would not be expected in a stable biotic community. Rather, contemporary concerns of population-level predation are primarily driven by (1) changes in predator community composition, (2) changes in abundance of predators, (3) changes in landscape structure

that increase susceptibility to predation, and (4) hunter concerns with low population size and dissatisfaction with regulations limiting seasons and bag limits. All of four concerns are due to impacts of human activities on biotic communities and landscapes and to management goals for sustainable harvest of the population surplus. It is important to consider these concerns in a conservation context when considering the role of predation on populations of Lesser Prairie-Chickens, especially if predation effects on wildlife species at a population level are due to landscape changes induced by human actions.

To cope with predators, Lesser Prairie-Chickens have evolved generalized antipredator strategies (Braun et al. 1994). Adaptations include cryptically colored plumage to facilitate concealment (Braun et al. 1994), freeze-or-flight behaviors (Campbell 1950, Hagen and Giesen 2005, Behney et al. 2011), and selection of vegetation cover that facilitates concealment (Jones 1963, Hagen et al. 2013) or escape from predators (Copelin 1963, Jones 1963, Borsdorf 2013). Predators can also influence the behavior of prey through disturbance or risk avoidance (Brown et al. 2001, Preisser et al. 2005). Lesser Prairie-Chicken responses to predators can be species specific (Behney et al. 2011), and birds appear to modify their movement patterns to reduce predation risk (Pruett et al. 2009). Ultimately, predator avoidance may influence behaviors such that individual health and productivity are compromised (Caro 2005).

The importance of predator effects on prairie grouse is poorly understood with relatively few field studies (Nelson 2001). Few studies have focused on large datasets to assess mortality causes of Lesser Prairie-Chickens (Patten et al. 2005, Wolfe et al. 2007). Rather, most predation information for the Lesser Prairie-Chicken is based on data collected during research on other topics, such as survival rates or habitat use. Infrequent monitoring of birds in studies of habitat use could introduce bias into mortality assessments (Murray and Fuller 2000). A challenging aspect of studying predation of Lesser Prairie-Chickens is that accurate identification of the predator is often not possible from carcass remains; rather, predators are often lumped into taxonomic guilds without the identification of which species are having the impact. Assessing predator–prey relationships and the impact of predation on Lesser Prairie-Chickens is challenging under field conditions but necessary for improved conservation actions.

COMPONENTS OF PREDATION

Predator Community

No predator specializes solely on Lesser Prairie-Chickens; rather, each predator species is a member of a biotic community that includes a variable number of opportunistic predators that include reptiles, corvids and raptors, or mammals. Probable predators of Lesser Prairie-Chickens constitute more of a list of "the usual suspects" for ground-nesting birds than species that have been confirmed to kill young or adult birds. Uncertainty arises because opportunistic predators are all likely to depredate Lesser Prairie-Chickens whenever the chance arises. The impact of any one predator species on Lesser Prairie-Chickens, however, is challenging to assess. Schroeder and Baydack (2001) and Hagen and Giesen (2005) list predators of juvenile and adult Lesser Prairie-Chickens as eight species of raptorial birds, two mammals, and a one reptile: Cooper's Hawks (*Accipiter cooperii*), Red-tailed Hawks (*Buteo jamaicensis*), Rough-legged Hawks (*B. lagopus*), Ferruginous Hawks (*B. regalis*), Prairie Falcons (*Falco mexicanus*), Great Horned Owls (*Bubo virginianus*), Golden Eagles (*Aquila chysaetos*), Northern Harriers (*Circus cyaneus*), coyotes (*Canis latrans*), badgers (*Taxidea taxus*), and gopher snakes (*Pituophis melanoleucus*). Swift foxes (*Vulpes velox*) have been confirmed as a predator of Lesser Prairie-Chickens in Colorado and western Kansas (D.A. Haukos, pers. comm.), and red foxes (*V. vulpes*) or gray foxes (*Urocyon cinereoargenteus*) could also pose a predation risk. Interestingly, bobcats (*Lynx rufus*) have not been reported as a predator of Lesser Prairie-Chickens, despite being present throughout the species' distribution and being a common predator of birds (Schmidly 2004). Nest predators have been listed as Chihuahuan Ravens (*Corvus cryptoleucus*), coyotes, badgers, striped skunks (*Mephitis mephitis*), ground squirrels (*Xerospermophilus spilosoma*), and snakes (*Pituophis* spp.) by Schroeder and Baydack (2001) and Hagen and Giesen (2005).

Mammalian carnivores are typically territorial and present year-round. If population numbers of mammalian predators are stable, opportunistic predation risk may be relatively constant. Risk may change somewhat if alternate prey populations increase or decrease and lead to numerical or functional responses in predators, during nesting when female prairie chickens may be more susceptible to predation by olfactory predators, and at independence when broods breakup and naive young of the year begin to disperse. Predation by mammals can affect nest success and survival of chicks, juveniles, and adults (VerCauteren et al. 2012). Predation risk from snakes, on the other hand, is limited to the summer months with warm temperatures when ectothermic animals are most active. Snake predation of adult Lesser Prairie-Chickens is probably rare, and risk is likely greatest for eggs and chicks. In contrast, risk from avian predators can be seasonably variable. During the breeding season, predation risk may be constrained because intraspecific and interspecific territorial behaviors can limit raptor numbers. However, large numbers and greater diversity of migrating and wintering raptors can substantively increase seasonal risk to prairie chickens (Wolfe et al. 2007, Behney et al. 2012). Thus, raptors pose a risk to juvenile and adult Lesser Prairie-Chickens, but the risk they pose to nesting success is indirect through the depredation of nesting females. Nests can still be depredated by other avian predators, and ravens (*Corvus* spp.) will eat eggs and young at nests of Lesser Prairie-Chicken (Haukos 1988, Grisham 2012).

The seasonal patterns of how predation pressure is distributed across the annual cycle are not well known (Hagen and Giesen 2005). It is possible that adult and juvenile Lesser Prairie-Chickens may experience relatively constant risk year-round from resident populations of mammalian predators. In contrast, adults and juvenile Lesser Prairie-Chickens likely experience seasonally variable risks from exposure to raptors during spring and fall migration. Risks may be lower during the raptor breeding season when raptors are territorial, and greater during migration and winter because of an influx of migrants and nonbreeding birds that increase local numbers of raptors. During the breeding season, eggs and chicks are subject to predation from snakes, mammals, and some birds such as ravens. A better understanding of predation pressures among latitudes and vegetation communities of different Lesser Prairie-Chicken populations may facilitate a better understanding of habitat management and of whether and when predator control is warranted.

Communal Display Grounds

Similar to other prairie grouse species, Lesser Prairie-Chickens congregate at traditional communal grounds or "leks" where males display to attract and mate with females (Hagen and Giesen 2005). Lek mating systems may be an effective antipredator strategy, with multiple males facilitating early detection of approaching predators and limiting predation risk (Boyko et al. 2004, Behney et al. 2011). Conversely, if males are aggregated and engaged in displaying to females, they may experience increased susceptibility to predation while lekking (Patten et al. 2005, Wolfe et al. 2007). Predation risk may increase if the energetic demands of lekking are associated with changes in body condition and a reduced ability to escape predators.

Few studies have examined predation at leks of prairie grouse (Berger et al. 1963, Hartzler 1974, Boyko et al. 2004, Behney et al. 2011), and evidence of successful predation on lekking Lesser Prairie-Chickens is rare. Campbell (1950) and Haukos and Broda (1989) observed Northern Harriers approach and flush Lesser Prairie-Chickens from leks. The authors did not observe any captures on leks, but Haukos and Broda (1989) flushed harriers from three prairie chicken kills within 100 m of leks. Grisham (2012) reported flushing a Ferruginous Hawk from a freshly killed Lesser Prairie-Chicken on a lek. Patten et al. (2005) found that 43% of male Lesser Prairie-Chicken mortalities in New Mexico and Oklahoma coincided with peak lekking activity. A follow-up study with additional data similarly revealed that male mortality was greatest coincident with the lekking period, but monthly rates of mortality were relatively low, ranging from <2% to 6% of tracked birds across the year (Wolfe et al. 2007). During the lekking periods of March–May, raptors accounted for ~50%–75% of mortalities (Wolfe et al. 2007), which coincide with peak periods of raptor migration (Behney et al. 2011). During 650 h of data collection, however, Behney et al. (2011) found that raptor encounters at leks were low (0.09/h) and only 15 of 61 (25%) encounters resulted in a capture attempt. Northern Harriers and Swainson's Hawks were the most frequently observed predators, but no successful avian or mammalian predation was recorded at leks. Raptor diversity and presence at leks may show considerable variation not only across the latitudinal gradient of Lesser Prairie-Chicken distribution, but also within localized areas. D.H. Wolfe (pers. comm.) suggests that many lek visits by avian and mammalian predators are exploratory searches for injured birds. Lekking Lesser Prairie-Chickens responded differently to different avian predators, but regardless of predator species or response type, they also quickly returned to leks and resumed lekking behaviors in an average of 4.2 ± 5.5 min (Behney et al. 2011) after a disturbance.

Lek visits by female Lesser Prairie-Chickens appear to peak when raptor presence is at its lowest at the southern extent of the species' distribution (Haukos 1988, Haukos and Smith 1999, Behney et al. 2011). Lek timing is typically following departure of harriers and other winter-resident raptors, but prior to arrival of migratory Swainson's Hawks and other breeding season residents. With latitudinal differences across the species' distribution, seasonal patterns of raptor avoidance may not hold for Lesser Prairie-Chickens at the more northern extent of their range. Seasonal timing of lekking and raptor movements may be a partial explanation for the finding that predation of females is lower during the lekking period (Patten et al. 2005, Wolfe et al. 2007).

Nesting and Brood Rearing

Nest success and brood survival are the most important demographic parameters affecting population dynamics of Lesser Prairie-Chickens (Hagen et al. 2009). Thus, predation on nests and broods can have a substantial influence on local populations. Nest predation has been primarily attributed to coyotes, mustelids, ground squirrels, and snakes, but cattle and other mammals also destroy nests (Table 8.1). Other authors have verified that ravens depredate nests of Lesser Prairie-Chickens (Haukos 1988, Grisham 2012), and may also be responsible for partial nest losses, though such events are usually attributed to snakes (Fields 2004, Grisham 2012).

Predation of nests may increase as a predator gains experience and develops a visual or olfactory search image; skunks have been experimentally shown to increase their olfactory detection of replica bird nests from an initial distance of 2.5 to as far as 25 m (Nams 1991, 1997).

TABLE 8.1

Cause-specific nest failure of Lesser Prairie-Chickens in the United States.

State	Years	Nests[a]	Nest success	Female killed	Nest abandon	Nest predation	Cause-specific mortality						Sources
							Coyote	Mustelid	Ground squirrel	Snake	Other	Unk.	
KS	97–02	196	26.0			161	73		6	31	3[b]	48	Pitman et al. (2006)
KS	02–03	60	48.3			29	P[c]	P[c]		P[c]	2[d]		Fields (2004)
NM	76–78	36	17.0	1	9	16	3	2		6	1[e]	4	Riley et al. (1992)
NM	04–05	23	69.5			7						7	Davis (2009)
TX (w)	76–77	8	37.5		2	3					3[f]		Sell (1979)
TX (w)	87–88	17	15.0		1	10	1			4	5[g]		Haukos (1988)
TX (ne)	08–10	24	37.0	2	1		2	5				5	Holt (2012)
TX (w)	08–11	35	46.0	4	6		P[c]			P[c]	1[h]	8	Grisham (2012)

[a] Total number of nests assessed.
[b] Nests trampled by cattle.
[c] Identified as probable (P) cause of nest failure by authors.
[d] Nests depredated by ants.
[e] Predation by unknown mammal.
[f] Predation by raven (1) and unknown mammals (2).
[g] Predation by ravens (2), unknown mammal (1), and cattle trampling (1).
[h] Predation by raven.

Environmental conditions also influence olfactory detection of nests. Humidity can influence nest detection (Palmer et al. 1993, Conover 2007), and olfactory detection improves when there is a breeze but degrades as wind increases because the odor plume becomes wider and concentration of odorants decreases (Conover 2007).

Little information exists about predators or predation rates of Lesser Prairie-Chicken chicks, but insights may be gained from the few studies that have used radiotelemetry to address the issue for similar species. Manzer and Hannon (2008) found that 81% of chick mortalities occurred in the first 15 days after hatching in Sharp-tailed Grouse (*Tympanuchus phasianellus*), with predation being the greatest cause of mortality (72%), and a majority of mortalities were identified as due to mammalian predation. Similarly, 98% of chick mortality occurred prior to 21-day posthatch in Greater Sage-Grouse (*Centrocercus urophasianus*), with 82% of mortalities due to mammalian (78.6%), avian (17.8%), and reptilian (3.6%) predation (Gregg et al. 2006). Schole et al. (2011) found that all radio-tagged chicks died prior to 21 days of age in their study of Greater Prairie-Chickens (*T. cupido*); they confirmed three mortalities due to exposure and two mortalities due to predation, but presumed 19 disappearances could also be due to predation. In a study of the introduced Ring-necked Pheasant (*Phasianus colchicus*), Riley et al. (1998) found that 85% of chick mortality was due to predatory mammals. No prey remains of Lesser Prairie-Chickens were identified among 266 identifiable prey delivered to five Swainson's Hawk nests in prairie chicken habitat in west Texas (Behney et al. 2010). However, any of a suite of predators occupying Lesser Prairie-Chicken habitat might capture chicks when the opportunity arises. It remains unknown whether predation is a primary source of mortality for Lesser Prairie-Chicken chicks or is secondary to other factors such as starvation or exposure.

Losses of adult females are greatest during the early nesting period (Patten et al. 2005, Wolfe et al. 2007, Grisham and Boal 2015). A majority of breeding season losses are due to predation, with roughly equal proportions attributed to mammals and raptors (Haukos et al. 1989, Holt 2012, Grisham and Boal 2015). When considering both sexes, Sell (1979) and Merchant (1982) reported that predation was primarily caused by mammals during the breeding season. Most studies report predation as the main cause of mortality but one exception is Patten et al. (2005), who found that most mortality of females during the breeding season was due to collisions with fences and power lines, especially in Oklahoma.

Predation of Adults

Exceptions occur but two general patterns emerge when examining the predation of full-grown Lesser Prairie-Chickens. First, predation is generally greatest during the lekking, nesting, and brood-rearing periods compared to the non-breeding season. In Oklahoma and New Mexico, Patten et al. (2005) found that 43% of male mortalities occurred between March and May and 52% of female mortalities occurred in May and June. With the exception of increased male mortality in September, Hagen (2003) found that monthly survival rates in Kansas were greater during the months of the nonbreeding period. In contrast, marked birds in west Texas experienced similar rates of predation mortality during the breeding season (28%; Grisham and Boal 2015) and the nonbreeding season (22%; Pirius et al. 2013).

Second, mortality risk from different predation pressures shows seasonal variation among sites. Seasonal variation in predation risk from mammalian and avian predators ranges from being roughly equal, to being highest from mammals during the breeding season (Figures 8.1 and 8.2), to raptors posing a greater threat during the nonbreeding season (Table 8.2), especially during spring and fall migration (Figure 8.2). Different risks can be associated with local environmental and landscape conditions, vegetation community, and predator community. Links between predation risk and habitat conditions occur primarily in the southern portions of the occupied range. For example, raptors posed a greater risk to Lesser Prairie-Chickens during the non-breeding season in sand shinnery oak (*Quercus havardii*) shrubland communities in the western region of Texas (Pirius et al. 2013), compared to more open prairie landscapes of the northeast panhandle of Texas (Kukal 2010; Table 8.2). If viewed on an annual basis, mammals may pose

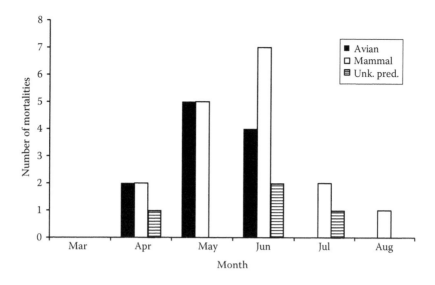

Figure 8.1. Number of avian, mammalian, and unknown predation mortalities of Lesser Prairie-Chickens in west Texas during the breeding seasons (March–August) of 2008–2011. (Modified from Grisham and Boal 2015.)

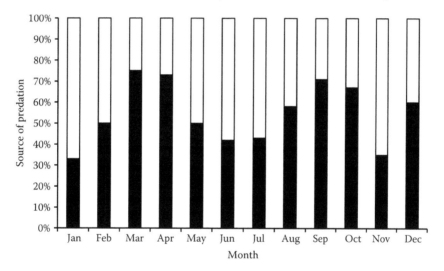

Figure 8.2. Relative monthly mammal (light) and raptor (dark) predation on the Lesser Prairie-Chicken in eastern New Mexico and western Oklahoma during 1999–2004. Spring and autumn peaks in relative predation by raptors coincided with raptor migration peaks. (Modified from Wolfe et al. 2007.)

greater relative risk to Lesser Prairie-Chickens in the northern parts of their distribution (Hagen 2003, Pitman 2003), whereas raptors pose an equal or slightly greater relative risk at the southern extent of the species' distribution (Patten et al. 2005; Table 8.2). Geographic variation in predation risk may be partially explained if latitudinal, and weather-mediated seasonal, differences in wintering distributions and compositions of raptor communities occur (Behney et al. 2012).

PROCESS AND EFFECTS OF PREDATION

Challenges of Interpreting Predation

Researchers and managers face several challenges when attempting to identify the cause-specific factors driving the mortality of Lesser Prairie-Chickens and determining population-level effects of observed rates of mortality losses. Other than rare observations of mortality events, researchers are tasked with interpreting egg or carcass remains, predator sign at a nest or kill site, and

TABLE 8.2

Cause-specific mortality of adult and juvenile Lesser Prairie-Chickens in the United States.

State	Years	Season	Sex	Mammal N	Mammal %	Raptor N	Raptor %	Snake N	Snake %	Other[a] N	Other[a] %	Total	Sources
KS	97–03	Year-round	Female	54	59.3	10	11.0	3	3.3	24	26.4	91	Hagen et al. (2007)
KS	97–03	Year-round	Male	18	40.9	9	20.5	0.0	0.0	17	38.6	44	Hagen (2003)
KS	00–03	Year-round	Both	13	61.9	2	9.5	0.0	0.0	6	28.6	21	Pitman (2003)
NM	79–80	Breeding	Female	11	91.7	1	8.3	0	0.0	0	0.0	12	Merchant (1982)
NM&OK	94–04	Year-round	Both	76	45.5	91	54.5	0.0	0.0	0.0	0.0	167	Wolfe et al. (2007)
OK	05–09, 13	Year-round	Both	27	23.1	31	26.5	1	0.8	58	49.6	117	GMSARC unpubl. data[b]
TX (w)	76–77	Year-round	Both	5	35.7	0	0.0	0	0.0	9	64.3	14	Sell (1979)
TX (w)	87–88	Lekking	Female	5	38.5	8	61.5	0	0.0	0	0.0	13	Haukos et al. (1989)
TX (ne)	08–10	Breeding	Female	3	42.9	3	42.9	0	0.0	1	14.3	7	Holt (2012)
TX (ne)	08–10	Breeding	Male	8	38.1	9	42.9	0	0.0	4	19.0	21	Holt (2012)
TX (w)	08–11	Breeding	Female	8	50.0	4	25.0	0	0.0	4	25.0	16	Grisham (2012)
TX (w)	08–11	Breeding	Male	9	39.1	7	30.4	0	0.0	7	30.4	23	Grisham (2012)
TX (ne)	08–10	Nonbreeding	Both	5	29.4	8	47.1	0	0.0	4	23.5	17	Kukal (2010)
TX (w)	08–10	Nonbreeding	Both	2	15.4	10	76.9	0	0.0	1	7.7	13	Pirius et al. (2013)

[a] "Other" includes other identified (e.g., collision, disease) and unidentified but confirmed mortalities.

[b] Unpublished data provided by D. H. Wolfe, from studies conducted through the George Miksch Sutton Avian Research Center, Bartlesville, OK.

other forensic evidence for determining the cause of mortality. Field guides (Sooter 1946, Rearden 1951) and detailed descriptions (Davis et al. 1979) are available for the identification of nest predators. However, little information is available to facilitate the identification of predators of adult Lesser Prairie-Chickens beyond simply "raptor" or "mammal." Depending on available evidence, a raptor can seldom be identified beyond being a falcon, a hawk or harrier, or an owl (Rezendes 1999, C.W. Boal, unpubl. data). Information is sometimes more available for mammals because of species-specific characteristics of physical evidence left at a kill site (Dumke and Pils 1973, Small et al. 1991, Rezendes 1999). However, most researchers conservatively, and understandably, categorize predators only as avian or mammalian.

Coupled with the lack of specificity in predator identification is the issue of proximate versus ultimate cause of death (Boal and Ballard 2013). For example, when identifying the causes of mortality, Wolfe et al. (2007) found seven Lesser Prairie-Chicken mortalities, two of which were depredated, as having preexisting collision injuries. Had a scavenging animal found the other five carcasses first, observers might have recorded the cause of mortality as a predation event. Scavenging can also result in the misidentification of the predator that actually killed a grouse (Bumann and Stauffer 2002, Boal and Ballard 2013). VerCauteren et al. (2012) provide some diagnostic characteristics of remains that were scavenged compared to those freshly killed, which may help identification in such situations. Additionally, capture myopathy can affect many wildlife species, including game birds (Nicholson et al. 2000, Hofle et al. 2004), and may result in some captured Lesser Prairie-Chickens experiencing stress-related mortality that leads to increased susceptibility to predation.

Radio transmitters and global positioning system (GPS) transmitters are important tools for research on the population biology of Lesser Prairie-Chickens, and tracking devices have provided the majority of information on survival and cause-specific mortality in prairie grouse. However, transmitter attachments can predispose some species to predation by compromising normal behavior, mobility, or concealment (Murray and Fuller 2000), and influences of transmitters on Lesser Prairie-Chickens are not well understood. Marks and Marks (1987) attached radios as necklace collars, but reported that raptors preyed selectively on radio-marked Sharp-tailed Grouse. Similarly, Haukos et al. (1989) discussed the likelihood that solar-powered backpack transmitters increased the predation of Lesser Prairie-Chickens in their study. In contrast, Hagen et al. (2006) found no evidence of a radio effect when assessing survival between radio-tagged and banded-only prairie chickens. Differences among field studies may be due to size and configuration of the transmitters. Ongoing reductions in transmitter size and improvement of attachment methods may eventually alleviate handicapping concerns, but it is an issue that warrants consideration in study design and interpretation of results.

Predator–Prey Dynamics

Examining predator–prey relationships is a more complex endeavor than simply evaluating causes of direct mortality (Lima and Steury 2005). Predation risk can be influenced by population numbers of local prey, availability of alternate prey, variation in prey vulnerability among different sexes, age classes or seasons, and environmental conditions that influence the vulnerability of prey or effectiveness of the predator. Variation in risk is further compounded by Lesser Prairie-Chickens being subject to predation from a suite of opportunistic generalist predators. Lesser Prairie-Chickens are not the primary prey for any specialist predator, but opportunistic predation may increase when populations of more common prey are low or unavailable, or when predator numbers increase in an area. The first case is a functional response in which predators switch their diet in response to prey availability, whereas the second case is a numerical response if predator populations increase through local reproduction, delayed dispersal, or immigration to an area (Boal and Ballard 2013).

Predation may also be indicative of other underlying problems, such as disease or parasitism, within populations. Hudson et al. (1992) found that Red Grouse (*Lagopus lagopus scotica*) killed by predators had greater numbers of parasitic nematodes than birds that were shot in autumn. Cecal nematode infections may result in increased scent being produced and thus facilitate detection by olfactory predators; trained dogs found fewer nests of Red Grouse that had been treated to reduce nematode infection compared to controls

(Hudson et al. 1992). There is also evidence that some raptors use cues to locate injured prey that are more vulnerable (Kenward 1978, Temple 1987, Caro 2005).

PREDATOR MANAGEMENT

Predator Control and Habitat Management

Predator control is frequently used as a management tool for many species of wildlife but has been met with mixed results and continues to be controversial (Van Ballenberghe 2006, VerCauteren et al. 2012). In a meta-analysis, Côté and Sutherland (1997) found that predator removal was effective in facilitating the production of a harvestable surplus of game birds, but less successful in increasing numbers of breeding populations. Similarly, Hewitt et al. (2001) found that predator control may increase the nest success of forest grouse but was inconsistent in increasing autumn or spring grouse numbers. When considering sage-grouse, Connelly et al. (2000) suggested that predator control programs tended to be expensive and ineffective but could provide short-term value. Hagen et al. (2007) speculated that predator control may improve female survival of Lesser Prairie-Chickens but was likely not a sustainable action. To be successful, any predator control effort should be directed at the predator species posing the greatest risk to the prey animal. Control efforts are challenging when considering Lesser Prairie-Chickens because the species is prey for a number of opportunistic predators, and impacts of predation at local scales are poorly understood.

Direct Predator Control

No predator control actions have been initiated as a conservation measure for Lesser Prairie-Chickens (Hagen and Giesen 2005). If predator control is pursued, it is important to clearly identify what the goals are and what strategy will best serve to achieve them. Predicting the results of a control program is challenging because of the dynamic nature of natural systems and changing interactions among biological, environmental, and practical factors. Additionally, predator control can have unintended consequences. For example, ground squirrels have been reported to depredate Lesser Prairie-Chicken nests (Pitman et al. 2006),

but mesopredator control can lead to competitive release of small mammals with subsequent increases in predation on bird nests (Henke and Bryant 1999, Dion et al. 2000). Increasing predator control also corresponded to an increase in parasitic infection rates among Red Grouse (Hudson et al. 1998). Packer et al. (2003) reasoned that predator control was counterproductive to conservation efforts when prey animals are infected with aggregations of macroparasites that increase morbidity but not mortality; infected animals are then more vulnerable to predation and removal from the population.

Boertje et al. (2010) provide a useful set of recommendations for the design of predator control programs. First, it should be verified that predators kill substantial numbers of the species of interest that would otherwise survive. Second, if predation is reduced, can management facilitate an increase in abundance of the target species? Third, if predation is reduced, can the available habitat sustain a larger population of the species of interest? Last, are predator populations sustainable inside treatment areas with control and adjacent areas without control? If a program meets these criteria, it will help ensure that predator control is justified and scientifically defensible. Clearly defined criteria are especially valuable given that lethal control continues to be a controversial issue (Boertje et al. 2010).

Indirect Predator Control by Habitat Management

Nelson (2001) noted that predation on prairie grouse is a function of habitat quality and distribution, suggesting that predator management should be considered in the context of habitat degradation and fragmentation. Habitat management to increase nesting and escape cover is frequently the recommended action for reducing predation of eggs and young during incubation and brood rearing (Schroeder and Baydack 2001, Fields et al. 2006, Hagen et al. 2007). Wolfe et al. (2007) further advocated that habitat management include the removal of potential raptor perches; utility poles were a significant predictor of raptor densities in Lesser Prairie-Chicken habitats in west Texas (Behney et al. 2012). Another consideration is that nest predation typically increases when nests are clumped or in close proximity (Caro 2005). Thus, an optimal strategy to avoid predation is dispersed nesting; however, adequate spacing requires large blocks of habitat for dispersed nesting, which may be the biggest conservation challenge for Lesser Prairie-Chickens.

CONSERVATION IMPLICATIONS

Consideration must be given to the fact that population-level effects of predation, and the frequency and impact of predation by any one species, remain largely unknown. The risk is that anecdotal observations become accepted dogma that lead to misunderstandings and, possibly, misguided management actions. Given this caveat, if predation does pose population-level influences, it is likely through impacts on nest and brood survival (Table 8.1), which are two of the most critical population parameters for Lesser Prairie-Chickens (Pitman et al. 2006, Hagen et al. 2009, Grisham 2012, Van Pelt et al. 2013).

Predator management has not been attempted or advocated as a conservation tool for the Lesser Prairie-Chickens (Hagen and Giesen 2005). A prevailing view among managers is that predation on prairie grouse is a function of habitat quality and distribution and that management of predator risk may be best addressed in context of habitat degradation and fragmentation (Nelson 2001). Braun et al. (1994) cautioned that fragmentation and simplification of habitat structure alter predator–prey balances by facilitating both presence and search efficiency of predators in locating nests and birds. Silvy et al. (2004) echoed this perception, attributing the decline of prairie grouse to the loss of habitat quantity; loss of habitat quantity translates to smaller, isolated populations, which may experience greater susceptibility to predation. Thus, increased predation may or may not be the proximate result of human actions. A key question regarding predator–prey interactions that applies to Lesser Prairie-Chickens is how antipredator morphology and behavior interact (Caro 2005). Interactions are especially relevant in context of land cover changes due to increased livestock grazing and other anthropocentric activities, and future habitat conditions that may include more frequent drought. Intensive population studies of the ecological links between environmental conditions and predation are needed to develop a more complete understanding of the community–habitat influences on predator–prey relationships of Lesser Prairie-Chickens, and for the development of appropriate management strategies to facilitate population recovery.

ACKNOWLEDGMENTS

I thank my mentors, colleagues, and students, with whom many discussions about predator–prey relationships have served to constantly develop my understanding and challenge my thinking on the fascinating topic. In particular, I thank W. Ballard, R.W. Mannan, B. Millsap, D. Rollins, and M. Wallace for stimulating conversations on the topic and A. Behney and R. Perkins for collaborations to address some of the questions raised. I thank D. Haukos, D. Lucia, H. Whitlaw, and B. Grisham for expanding my understanding of Lesser Prairie-Chickens and being good partners in collaborative research. I thank L. Brennan and D. Wolfe for constructive reviews of the manuscript.

LITERATURE CITED

Behney, A. C., C. W. Boal, H. A. Whitlaw, and D. R. Lucia. 2010. Prey use by Swainson's Hawks in the Lesser Prairie-Chicken range of the Southern High Plains of Texas. Journal of Raptor Research 44:317–322.

Behney, A. C., C. W. Boal, H. A. Whitlaw, and D. R. Lucia. 2011. Interactions of raptor and Lesser Prairie-Chickens at leks in the Texas Southern High Plains. Wilson Journal of Ornithology 123:332–338.

Behney, A. C., C. W. Boal, H. A. Whitlaw, and D. R. Lucia. 2012. Raptor community composition in the Texas Southern High Plains Lesser Prairie-Chicken range. Wildlife Society Bulletin 36:291–296.

Berger, D. D., F. Hamerstrom, and F. N. Hamerstrom, Jr. 1963. The effect of raptors on prairie chickens on booming grounds. Journal of Wildlife Management 27:778–791.

Boal, C. W., and W. B. Ballard. 2013. Predator–prey relationships and management. Pp. 195–213 in P. R. Krausman and J. W. Cain, III (editors), Wildlife management and conservation: contemporary principles and practices. Johns Hopkins University Press, Baltimore, MD.

Boertje, R. D., M. A. Keech, and T. F. Paragi. 2010. Science and values influencing predator control for Alaska moose management. Journal of Wildlife Management 74:917–928.

Borsdorf, P. K. 2013. Lesser Prairie-Chicken habitat selection across varying land use practices in eastern New Mexico and west Texas. M.S. thesis, Texas Tech University, Lubbock, TX.

Boyko, A. R., R. M. Gibson, and J. R. Lucas. 2004. How predation risk affects the temporal dynamics of avian leks: Greater Sage-Grouse versus Golden Eagles. American Naturalist 163:154–165.

Braun, C. E., K. Martin, T. E. Remington, and J. R. Young. 1994. North American grouse: issues and strategies for the 21st century. Transactions of the North American Wildlife and Natural Resources Conference 59:428–438.

Brown, J. S., B. P. Kotler, and A. Bouskila. 2001. Ecology of fear: foraging games between predators and prey with pulsed resources. Annales Zoologici Fennici 38:71–87.

Bumann, G. B., and D. F. Stauffer. 2002. Scavenging of Ruffed Grouse in the Appalachians: influences and implications. Wildlife Society Bulletin 30:853–860.

Campbell, H. 1950. Note on the behavior of Marsh Hawks toward Lesser Prairie Chickens. Journal of Wildlife Management 14:477–478.

Caro, T. 2005. Antipredator defenses in birds and mammals. University of Chicago Press, Chicago, IL.

Connelly, J. W., M. A. Schroeder, A. R. Sands, and C. E. Braun. 2000. Guidelines to manage Sage-Grouse populations and their habitats. Wildlife Society Bulletin 28:967–985.

Conover, M. R. 2007. Predator–prey dynamics: the role of olfaction. CRC Press, Boca Raton, FL.

Copelin, F. F. 1963. The Lesser Prairie Chicken in Oklahoma. Oklahoma Wildlife Conservation Department, Oklahoma City, OK.

Côté, I. M., and W. J. Sutherland. 1997. The effectiveness of removing predators to protect bird populations. Conservation Biology 11:395–405.

Davis, C. A., T. Z. Riley, H. R. Suminski, and M. J. Wisdom. 1979. Habitat evaluation of Lesser Prairie Chickens in eastern Chaves County, New Mexico. Final Report to Bureau of Land Management, Roswell, Contract YA-512-CT6-61. Department of Fishery and Wildlife Sciences, New Mexico State University, Las Cruces, NM.

Davis, D. M. 2009. Nesting ecology and reproductive success of Lesser Prairie-Chickens in shinnery oak–dominated rangelands. Wilson Journal of Ornithology 121:322–327.

Dion, N., K. A. Hobson, and S. Larivière. 2000. Effects of removing duck-nest predators on nesting success of grassland songbirds. Canadian Journal of Zoology 77:1801–1806.

Dumke, R. T., and C. M. Pils. 1973. Mortality of radio-tagged pheasants on the Waterloo wildlife area. Wisconsin Department of Natural Resources Technical Bulletin Number 72. Wisconsin Department of Natural Resources, Madison, WI.

Fields, T. L. 2004. Breeding season habitat use of Conservation Reserve Program (CRP) land by Lesser Prairie-Chickens in west central Kansas. M.S. thesis, Colorado State University, Fort Collins, CO.

Fields, T. L., G. C. White, W. C. Gilgert, and R. D. Rodgers. 2006. Nest and brood survival of Lesser Prairie-Chickens in west central Kansas. Journal of Wildlife Management 70:931–938.

Gregg, M. A., M. Dunbar, and J. A. Crawford. 2006. Use of implanted radio-transmitters to estimate survival of Greater Sage-Grouse chicks. Journal of Wildlife Management 71:646–651.

Grisham, B. A. 2012. The ecology of Lesser Prairie-Chickens in shinnery oak–grassland communities in New Mexico and Texas with implications toward habitat management and future climate change. Ph.D. dissertation, Texas Tech University, Lubbock, TX.

Grisham, B. A., and C. W. Boal. 2015. Causes of mortality and temporal patterns in breeding season survival of Lesser Prairie-Chickens in shinnery oak prairies. Wildlife Society Bulletin 39:536–542.

Hagen, C. A. 2003. A demographic analysis of Lesser Prairie-Chicken populations in southwestern Kansas: survival, population viability, and habitat use. Ph.D. dissertation, Kansas State University, Manhattan, KS.

Hagen, C. A., and K. M. Giesen. [online]. 2005. Lesser Prairie-Chicken (*Tympanuchus pallidicinctus*). A. Poole (ed.), The birds of North America online No. 364. Cornell Lab of Ornithology, Ithaca, NY. Doi: 10.2173/bna.364. <http://bna.birds.cornell.edu/bna/species/364/articles/introduction> (15 December 2013).

Hagen, C. A., B. Grisham, C. Boal, and D. Haukos. 2013. A meta-analysis of Lesser Prairie-Chicken nesting and brood rearing habitats: recommendations for habitat management. Wildlife Society Bulletin 37:750–758.

Hagen, C. A., J. C. Pitman, B. K. Sandercock, R. J. Robel, and R. D. Applegate. 2007. Age-specific survival and probable causes of mortality in female Lesser Prairie-Chickens. Journal of Wildlife Management 71:518–525.

Hagen, C. A., B. K. Sandercock, J. C. Pitman, R. J. Robel, and R. D. Applegate. 2006. Radiotelemetry survival estimates of Lesser Prairie-Chickens in Kansas: are there transmitter biases? Wildlife Society Bulletin 34:1064–1069.

Hagen, C. A., B. K. Sandercock, J. C. Pitman, R. J. Robel, and R. D. Applegate. 2009. Spatial variation in Lesser Prairie-Chicken demography: a sensitivity analysis of population dynamics and management alternatives. Journal of Wildlife Management 73:1325–1332.

Hartzler, J. E. 1974. Predation and the daily timing of Sage Grouse leks. Auk 91:532–536.

Haukos, D. A. 1988. Reproductive ecology of Lesser Prairie-Chickens in west Texas. M.S. thesis, Texas Tech University, Lubbock, TX.

Haukos, D. A., and G. S. Broda. 1989. Northern Harrier (*Circus cyaneus*) predation of Lesser Prairie-Chicken (*Tympanuchus pallidicinctus*). Journal of Raptor Research 23:182–183

Haukos, D. A., and L. M. Smith. 1999. Effect of lek age on age structure and attendance of Lesser Prairie-Chickens (*Tympanuchus pallidicinctus*). American Midland Naturalist 142:415–420.

Haukos, D. A., L. M. Smith, and G. S. Broda. 1989. The use of radio-telemetry to estimate Lesser Prairie-Chicken nest success and hen survival. Pp. 238–243 in C. J. Amlaner (editor), Biotelemetry X: proceedings of the 10th International Symposium on Biotelemetry, The University of Arkansas Press, Fayetteville, AR,

Henke, S. E., and F. C. Bryant. 1999. Effects of coyote removal on the faunal community in western Texas. Journal of Wildlife Management 63:1066–1081.

Hewitt, D. G., D. M. Keppie, and D. F. Stauffer. 2001. Predation effects on forest grouse recruitment. Wildlife Society Bulletin 29:16–23.

Hofle, U., J. Millan, C. Gortazar, F. J. Buenestado, I. Marco, and R. Villafuerte. 2004. Self-injury and capture myopathy in net-captured juvenile Red-legged Partridge with necklace radiotags. Wildlife Society Bulletin 32:344–350.

Holt, R. D. 2012. Breeding season demographics of a Lesser Prairie-Chicken (*Tympanuchus pallidicinctus*) population in the northeast Texas Panhandle. Ph.D. dissertation, Texas Tech University, Lubbock, TX.

Hudson, P. J., A. P. Dobson, and D. Newborn. 1992. Do parasites make prey vulnerable to predation? Red Grouse and parasites. Journal of Animal Ecology 61:681–692.

Hudson, P. J., A. P. Dobson, and D. Newborn. 1998. Prevention of population cycles by parasite removal. Science 282:2256–2258.

Jiménez, J. E., and M. R. Conover. 2001. Ecological approaches to reduce predation on ground-nesting gamebirds and their nests. Wildlife Society Bulletin 29:62–69.

Jones, R. E. 1963. Identification and analysis of Lesser and Greater Prairie Chicken habitat. Journal of Wildlife Management 27:757–778.

Kenward, R. E. 1978. Hawks and doves: factors affecting success and selection in Goshawk attacks on Woodpigeons. Journal of Animal Ecology 47:449–460.

Knight, R. L., and J. Y. Kawashima. 1993. Responses of Raven and Red-tailed Hawk populations to linear right-of-ways. Journal of Wildlife Management 57:266–271.

Kukal, C. A. 2010. The over-winter ecology of Lesser Prairie-Chicken (*Tympanuchus pallidicinctus*) in the northeast Texas Panhandle. M.S. thesis, Texas Tech University, Lubbock, TX.

Leopold, A. 1933. Game management. University of Wisconsin Press, Madison, WI.

Lima, S. L., and T. D. Steury. 2005. Perception of predation risk: the foundation of nonlethal predator–prey interactions. Pp. 166–188 in P. Barbosa and I. Castellanos (editors), Ecology of predator–prey interactions. Oxford University Press, Oxford, U.K.

Macdonald, D. W., G. M. Mace, and G. R. Barretto. 1999. The effects of predators on fragmented prey populations: a case study for the conservation of endangered prey. Journal of Zoology 247:487–506.

Manzer, D. L., and S. J. Hannon. 2008. Survival of Sharp-tailed Grouse chicks and hens in a fragmented prairie landscape. Wildlife Biology 14:16–25.

Marks, J. S., and V. S. Marks. 1987. Influence of radio-collars on survival of Sharp-tailed Grouse. Journal of Wildlife Management 51:468–471.

Merchant, S. S. 1982. Habitat-use, reproductive success, and survival of female Lesser Prairie Chickens in two years of contrasting weather. M.S. thesis, New Mexico State University, Las Cruces, NM.

Murray, D. L., and M. R. Fuller. 2000. A critical review of the effects of marking on the biology of vertebrates. Pp. 15–64 in L. Boitani and T. Fuller (editors), Research techniques in animal ecology: controversies and consequences. Columbia University Press, New York, NY.

Nams, V. O. 1991. Olfactory search images in striped skunks. Behavior 119:267–284.

Nams, V. O. 1997. Density-dependent predation by skunks using olfactory search images. Oecologia 110:440–448.

Nelson, H. K. 2001. Impact of predation on avian recruitment—An introduction. Wildlife Society Bulletin 29:2–5.

Newton, I. 1993. Predation and limitation of bird numbers. Current Ornithology 11:143–198.

Nicholson, D. S., R. L. Lochmiller, M. D. Stewart, R. E. Masters, and D. M. Leslie. 2000. Risk factors associated with capture-related death in Eastern Wild Turkey hens. Journal of Wildlife Diseases 36:308–315.

Packer, C., R. D. Holt, P. J. Hudson, K. D. Lafferty, and A. P. Dobson. 2003. Keeping the herds healthy and alert: implications of predator control for infectious disease. Ecology Letters 6:797–802.

Palmer, W. E., S. R. Priest, R. S. Seiss, P. S. Phalen, and G. A. Hurst. 1993. Reproductive effort and success in a declining Wild Turkey population. Proceedings of the Annual Conference of the Southeastern Association of Fish and Wildlife Agencies 47:138–147.

Patten, M. A., D. H. Wolfe, E. Shochat, and S. K. Sherrod. 2005. Habitat fragmentation, rapid evolution and population persistence. Evolutionary Ecology Research 7:235–249.

Pirius, N. E., C. W. Boal, D. A. Haukos, and M. C. Wallace. 2013. Winter habitat use and survival of Lesser Prairie-Chickens in west Texas. Wildlife Society Bulletin 37:759–765.

Pitman, J. C. 2003. Lesser Prairie-Chicken nest site selection and nest success, juvenile gender determination and growth, and juvenile survival and dispersal in southwestern Kansas. M.S. thesis, Kansas State University, Manhattan, KS.

Pitman, J. C., C. A. Hagen, B. E. Jamison, R. J. Robel, T. M. Loughin, and R. D. Applegate. 2006. Nesting ecology of Lesser Prairie-Chickens in sand sagebrush prairie of southwestern Kansas. Wilson Journal of Ornithology 118:23–35.

Preisser, E. L., D. I. Bolnick, and M. F. Benard. 2005. Scared to death? The effects of intimidation and consumption in predator–prey interactions. Ecology 86:501–509.

Pruett, C. L., M. A. Patten, and D. H. Wolfe. 2009. Avoidance behavior by prairie grouse: implications for development of wind energy. Conservation Biology 23:1253–1259.

Rearden, J. D. 1951. Identification of waterfowl nest predators. Journal of Wildlife Management 15:386–395.

Rezendes, P. 1999. Tracking and the art of seeing: how to read animal tracks and sign, 2nd edn. HarperCollins Publishers, Inc., New York, NY.

Riley, T. Z., W. R. Clark, E. Weing, and P. A. Vohs. 1998. Survival of Ring-necked Pheasant chicks during brood rearing. Journal of Wildlife Management 62:36–44.

Riley, T. Z., C. A. Davis, M. Ortiz, and M. J. Wisdom. 1992. Vegetative characteristics of successful and unsuccessful nests of Lesser Prairie-Chickens. Journal of Wildlife Management 56:383–387.

Schmidly, D. J. 2004. The mammals of Texas. University of Texas Press, Austin, TX.

Schole, A. C., T. W. Matthews, L. A. Powell, J. J. Lusk, and J. S. Taylor. 2011. Chick survival of Greater Prairie-Chickens. Pp. 247–254 in B. K. Sandercock, K. Martin, and G. Segelbacher (editors), Ecology, conservation, and management of grouse. Studies in Avian Biology Series (no. 39), University of California Press, Berkeley, CA.

Schroeder, M. A., and R. K. Baydack. 2001. Predation and the management of prairie grouse. Wildlife Society Bulletin 29:24–32.

Sell, D. L. 1979. Spring and summer movements and habitat use by Lesser Prairie Chicken females in Yoakum County, Texas. M.S. thesis, Texas Tech University, Lubbock, TX.

Silvy, N. J., M. J. Peterson, and R. R. Lopez. 2004. The cause of the decline of pinnated grouse: the Texas example. Wildlife Society Bulletin 32:16–21.

Small, R. J., J. C. Holzwart, and D. H. Rusch. 1991. Predation and hunting mortality of Ruffed Grouse in central Wisconsin. Journal of Wildlife Management 55:512–520.

Sooter, C. A. 1946. Habits of coyotes in destroying nests and eggs of waterfowl. Journal of Wildlife Management 10:33–38.

Temple, S. A. 1987. Do predators always capture substandard individuals disproportionately from prey populations? Ecology 68:669–674.

Thirgood, S. J., S. M. Redpath, D. T. Haydon, P. Rothery, I. Newton, and P. J. Hudson. 2000. Habitat loss and raptor predation: disentangling long- and short-term causes of Red Grouse declines. Proceedings of the Royal Society of London B 267:651–656.

Van Ballenberghe, V. 2006. Predator control, politics, and wildlife conservation in Alaska. Alces 42:1–11.

Van Pelt, W. E., S. Kyle, J. Pitman, D. Klute, G. Beauprez, D. Schoeling, A. Janus, and J. B. Haufler. 2013. The Lesser Prairie-Chicken range-wide conservation plan. Western Association of Fish and Wildlife Agencies, Cheyenne, WY.

VerCauteren, K. C., R. A. Dolbeer, and E. M. Gese. 2012. Identification and management of wildlife damage. Pp. 232–269 in N. J. Silvy (editor), The wildlife techniques manual: management, 7th edn. John Hopkins Press, Baltimore, MD.

Wolfe, D. H., M. A. Patten, E. Shochat, C. L. Pruett, and S. K. Sherrod. 2007. Causes and patterns of mortality in Lesser Prairie-Chickens Tympanuchus pallidicinctus and implications for management. Wildlife Biology 13:95–104.

CHAPTER NINE

Macroparasitic, Microparasitic, and Noninfectious Diseases of Lesser Prairie-Chickens*

Markus J. Peterson

Abstract. Abundance of Lesser Prairie-Chickens (*Tympanuchus pallidicinctus*) has been declining for several decades, and infectious and noninfectious diseases are becoming increasingly relevant to the conservation of the species. The purpose of this chapter is to summarize what is known regarding the diseases of Lesser Prairie-Chickens and then to explore the relevance of these data for wildlife policy-makers, conservationists, and researchers. A variety of ectoparasites, parasitic helminths, microparasites, and noninfectious diseases have been documented for Lesser Prairie-Chickens, but the population-level influence of only a few of these organisms is adequately understood. Based on available data, studies of related galliform species, and theoretical perspectives, the macroparasites *Oxyspirura petrowi* and *Tetrameres* sp., and the microparasites *Eimeria* spp., *Plasmodium pedioecetii*, and the infectious bronchitis virus (possibly others) have the potential to regulate populations of Lesser Prairie-Chickens. Several microparasites discussed in this chapter, such as *Pasteurella multocida*, could cause density-independent disease outbreaks that might extirpate small, isolated populations of Lesser Prairie-Chickens if they were to spillover during an epidemic in other host species. Noninfectious disease resulting from poor insect availability for chicks and collision mortalities with fence and electrical transmission wires can also have population-level consequences under certain circumstances. Mycotoxins, environmental contaminants, and the influence of female nutritional status on reproductive success could influence population dynamics of Lesser Prairie-Chickens; research projects on these topics are needed. Infectious and noninfectious diseases are most likely to affect populations of Lesser Prairie-Chickens as key components of more complex ecological interactions—including dynamics resulting from predicted changes in climate—rather than situations where a specific agent directly results in a marked decline in abundance. Surveys of infectious agents are still needed for poorly studied portions of Lesser Prairie-Chicken range, and there is critical need for research clarifying the effect of specific infectious agents and noninfectious diseases on population dynamics. In the future, conservation efforts would benefit from a better understanding of how anthropogenic habitat changes, infectious agents, noninfectious diseases, predation, and other aspects of the ecosystem interact to influence population dynamics of Lesser Prairie-Chickens.

Key Words: infectious disease, parasite, pathogen, population dynamics, population regulation, *Tympanuchus pallidicinctus*.

* Peterson, M. J. 2016. Macroparasitic, microparasitic, and noninfectious diseases of Lesser Prairie-Chickens. Pp. 159–183 in D. A. Haukos and C. W. Boal (editors), Ecology and conservation of Lesser Prairie-Chickens. Studies in Avian Biology (no. 48), CRC Press, Boca Raton, FL.

Wildlife health and disease are increasingly important aspects of wildlife conservation, particularly when human health, livestock health, or species at risk of extinction are affected (Daszak et al. 2000, Peterson and Ferro 2012). For example, Jones et al. (2008) found that 60% of 335 emerging infectious disease events occurring worldwide since 1940 were zoonoses transmitted between humans and other animals, with the majority of these diseases originating in wild animals (72%). The authors also found that the proportion of emerging infectious diseases originating in wildlife has also increased since the 1940s. Recent concerns over the H5N1 strain of avian influenza and the H1N1 swine flu pandemic—and continuing concern over Lyme disease, rabies, tuberculosis, and West Nile virus (WNV)—have sensitized the public to the role that wildlife can play in human health and personal well-being (Peterson and Ferro 2012). Wildlife biologists are now familiar with the ramifications of infectious diseases such as bovine tuberculosis, brucellosis, and chronic wasting disease in free-roaming cervid populations and, perhaps more importantly, changes in public perceptions (Peterson et al. 2006). Biologists tasked with grouse conservation now are keenly aware of the threat WNV poses for low- and mid-elevation populations of Greater Sage-Grouse (*Centrocercus urophasianus*, Walker and Naugle 2011). Conservation biologists are beginning to grapple with how infectious agents, noninfectious diseases, human activities, and other stressors interact to the detriment of species at risk of extinction (Brook et al. 2008, Fleishman et al. 2011, Taylor et al. 2013).

Infectious and noninfectious disease are directly relevant to Lesser Prairie-Chickens (*Tympanuchus pallidicinctus*) because of the conservation status of the species. Abundance of Lesser Prairie-Chickens has declined since perhaps as early as the 1880s (Crawford 1980, Taylor and Guthery 1980, Hagen and Giesen 2004). In Texas, the area inhabited by Lesser Prairie-Chickens decreased at least as rapidly since 1940 as did that of the critically endangered Attwater's Prairie-Chicken (*T. cupido attwateri*; Silvy et al. 2004). The U.S. Fish and Wildlife Service received a petition to list the Lesser Prairie-Chicken as threatened under the Endangered Species Act of 1973 during the autumn of 1995; they ruled that listing was warranted, but precluded by higher listing priorities in June 1998 (U.S. Fish and Wildlife Service 1998). Biologists from relevant state wildlife agencies and other entities often voiced concern regarding the potential role diseases may play in Lesser Prairie-Chicken populations as alternative conservation strategies were explored (Applegate and Riley 1998, Mote et al. 1999, Hagen et al. 2004). As Lesser Prairie-Chicken abundance continued to decline, the U.S. Fish and Wildlife Service (2012) published a proposed rule to list this species as threatened in December 2012, and, after a 6-month extension of the final determination to allow additional time for public comment (U.S. Fish and Wildlife Service 2013), listed this species as threatened on May 12, 2014 (U.S. Fish and Wildlife Service 2014). However, the listing rule was vacated by judicial decision in September 2015.

No empirical evidence has demonstrated that a specific infectious or noninfectious disease has contributed substantively to long-term declines in numbers of Lesser Prairie-Chickens, but lack of data does not rule out disease as a contributing factor given the paucity of research on this topic. A contact zone between the ranges of Lesser and Greater Prairie-Chicken apparently developed in western Kansas during the 1990s, with males of both species displaying on the same leks and apparent hybrids observed (Bain and Farley 2002). Mixed-species leks and hybrid mating could facilitate the transmission of infectious agents among individuals of the two congeneric species. Regardless, if Lesser Prairie-Chicken abundance continues to decline and populations become more fragmented, diseases could become critical impediments to species recovery (Mote et al. 1999, Peterson 2004, U.S. Fish and Wildlife Service 2012). For all these reasons, wildlife managers and conservation biologists will benefit from an up-to-date review of the diseases of Lesser Prairie-Chickens.

For purposes of my review, "disease" is defined as an interruption, cessation, or disorder of body functions, systems, or organs (Peterson and Ferro 2012). The definition includes toxic, genetic, metabolic, behavioral, neoplastic, and nutritional diseases, in addition to symptoms caused by infectious agents such as parasitic helminths, ectoparasites, bacteria, and viruses. It is beyond the scope of this chapter to provide a broad overview of disease processes, ecology, epidemiology, investigations, or management. Peterson and Ferro (2012) provide a primer on key topics, a glossary of disease-related terminology written for wildlife biologists, and a list resources for specific disease-related topics for practicing wildlife conservationists. Elsewhere, Begon et al. (2006) and Sinclair et al. (2006) provide excellent chapter-length primers on host–parasite ecology. Wobeser (2005) offers a useful textbook addressing the essential components of wildlife disease processes, investigations, and management. These publications are excellent primers on their respective topics but do not provide details on specific wildlife diseases. Peterson and Ferro (2012) discuss numerous sources that provide such information, and many of these publications are cited later as they apply to Lesser Prairie-Chickens, other prairie grouse (*Tympanuchus* spp.), or galliform birds in general. Some aspects of this chapter follow from my previous reviews of wildlife disease (Peterson 2003, 2004, 2007, Peterson and Ferro 2012).

My objectives for this chapter are to review what is known regarding the diseases of Lesser Prairie-Chickens and then discuss the relevance of available information for wildlife policy-makers, conservationists, and researchers. Specifically, I first briefly summarize the perspectives that wildlife scientists have held historically toward diseases of wildlife, because underlying assumptions guided what research was completed, and then how the discipline interpreted results of available studies. Second, I outline what is known about the infectious agents of Lesser Prairie-Chickens, including macroparasites such as parasitic arthropods and helminths and microparasites that include pathogenic protozoans, bacteria, and viruses. Third, I discuss what is known (and unknown) regarding selected noninfectious diseases of Lesser Prairie-Chickens. Fourth, I discuss the ecological and conservation ramifications of how infectious agents, noninfectious disease, and ecological interactions among these entities and other biotic and abiotic factors may influence population dynamics of Lesser Prairie-Chickens. Last, I briefly summarize the review and discuss the implications of these conclusions for managers tasked with developing conservation plans for Lesser Prairie-Chickens.

HISTORICAL PERSPECTIVE

Infectious agents have potential importance for the conservation of species at risk of extinction, and emerging infectious diseases of wildlife are significant threats to health of humans and livestock. Nevertheless, a long-term, sustained body of disease-related research does not exist for Lesser Prairie-Chickens or for most other wildlife species in North America. The reasons for our current gaps in knowledge are largely artifacts of history of wildlife biology (Peterson 2003, 2004, 2007, Peterson and Ferro 2012).

Early Interest

As wildlife science developed as a discipline, investigations of disease often were an integral component of wildlife-oriented research (Peterson 2004, 2007). For example, a lengthy publication titled *The Grouse in Health and Disease* (Committee of Inquiry on Grouse Disease 1911) argued that *Trichostrongylus tenuis* (= *T. pergracilis*) was the primary cause of what was known as "the Grouse Disease" occurring in populations of Red Grouse (*Lagopus lagopus scotica*) in the British Isles. The principal impetus for the study was to determine whether infectious agents controlled observed variation in grouse numbers among years. The early publication stimulated researchers in North America to attempt ecological studies that included disease investigations (Stoddard 1931, Bump et al. 1947), but also to search for their own versions of "the Grouse Disease" (Gross 1925:424, Lack 1954:164) or "the Quail Disease" (Bass 1939, 1940; Durant and Doll 1941). Aldo Leopold (1933:325), in his influential book *Game Management*, increased interest in the infectious agents of wildlife by arguing that "the role of disease in wild-life conservation has probably been radically underestimated." He also maintained "density fluctuations, such as cycles and irruptions, are almost certainly due to fluctuations in the prevalence of, virulence of, or resistance to [infectious] diseases." Thus, Leopold placed host–parasite interactions on par with other important interspecific relationships, such as predator–prey interactions. He did not offer empirical or experimental evidence to support his suppositions, however.

Parasites as By-Products of Poor Habitat

Starting around 1950, many influential wildlife scientists in North America began to assume that infectious and noninfectious diseases of free-roaming wildlife were ecologically unimportant, except as correlates of poor habitat conditions or as natural disasters (Trippensee 1948:369–384, Lack 1954:161–169, Taylor 1956:581–583). In his book *Wildlife Management*, Gabrielson (1951) did not even mention wildlife diseases or parasitism, implying that he assumed that disease was inconsequential for wild animals and their populations. Similarly, Herman (1969:325) ended his review of how diseases influenced wildlife populations by stating there was only "limited documentation that disease, as an individual factor, can drastically affect population fluctuations" and that "it is imperative that we recognize the dependency of the occurrence of disease in wildlife on habitat conditions." Herman (1963) pointed out elsewhere, however, that few studies had been conducted in such a manner that the population-level effects of infectious agents could be documented even if they were important. The criticism of the field still largely holds today (Peterson 1996, Tompkins et al. 2002). At any rate, perceiving bacterial or viral diseases, for example, as simply extensions of poor habitat conditions or as natural disasters where management could not reasonably be brought to bear—much like hurricanes or volcanic eruptions—led North American wildlife scientists to neglect these important interspecific relationships until relatively recently (Peterson 1991a). Most university programs that trained wildlife scientists in North America during this period did not include wildlife disease as part of the wildlife course curriculum.

Conversely, since the early 20th century, parasitologists interested in parasite systematics continued to study their favorite taxa in wild hosts (Peterson 2004, 2007, Peterson and Ferro 2012). Such efforts tended to emphasize host lists, parasite descriptions, and revisions of taxonomic relationships. Similarly, veterinary pathologists and microbiologists conducted numerous studies of infectious diseases in wildlife designed to determine whether wild species served as reservoir hosts for diseases occurring in livestock or humans. Most of these efforts provided a useful foundation of natural history but lacked an ecological dimension until recently and did not address many issues important to wildlife ecologists and conservationists (Peterson 1991b).

Ecological Perspective

Host–parasite interactions had been addressed earlier from an ecological perspective, but a landmark series of two-part articles by Anderson and May (1978, 1979) and May and Anderson (1978,

1979) provided the basic theoretical framework still used by ecologists for evaluating host–parasite interactions. The authors demonstrated that parasites could, under certain circumstances, not only affect the health of individual animals, but also regulate host populations, or cause marked declines in host abundance in a density-independent fashion. These models provide an ecological perspective that parasites use their hosts as habitat, depend on their host for nutrition, and cause their host "harm" during some point in their life cycle (Anderson and May 1978). Anderson and May (1979) also offered an ecologically rather than taxonomically based categorization of parasites that is directly relevant to wildlife conservationists (Peterson 2004). *Macroparasites* (parasitic arthropods and helminths) tend to have longer generation times than microparasites, direct multiplication in or on the host is absent or occurs at a low rate, and the immune response elicited by these organisms depends on the number present and typically is of short duration. For these reasons, macroparasites generally occur as endemic host infections that are more likely to cause morbidity rather than direct mortality. Conversely, *microparasites* (fungi, protozoans, bacteria, and viruses) are characterized by small size, short generation times, high rates of direct reproduction within the host, and a tendency to induce long-lasting immunity to reinfection. Microparasitic infection typically is short relative to the expected life span of the host. For this reason, microparasitic diseases often occur as epidemics where the pathogen apparently disappears as susceptible hosts die or become immune, only to reappear when sufficient densities of susceptible hosts are again available in the population. Parasites can complete their life cycles either (1) directly by contact between hosts, inhalation, ingestion, or skin penetration or (2) indirectly via biting vectors, penetration by free-living larva produced in an intermediate host, or by the host ingesting an intermediate host. One of the primary approaches to controlling infectious diseases in wildlife is to disrupt the life cycles of parasites, so understanding how infectious agents are transmitted is important to wildlife conservationists.

Since the mid-1990s, interest has been renewed in attempting to bridge the disciplinary divides that traditionally separated the fields of parasitology, microbiology, epidemiology, ecology, livestock health, human health, wildlife health, biodiversity conservation, and wildlife management with synthetic, cross-disciplinary approaches such as conservation medicine (Aguirre et al. 2002), and more recently the One Health Initiative (Rubin et al. 2013). Others maintain that ecology is the logical discipline to provide a synthesis that bridges different perspectives toward wildlife diseases (Hudson et al. 2002, Collinge and Ray 2006, Ostfeld et al. 2008). Regardless, multidisciplinary teams that include wildlife ecologists and conservationists increasingly are being employed to address emerging infectious diseases, zoonoses, and infectious and noninfectious diseases of importance to wildlife, domestic animals, or humans.

Peterson (2004) detailed how research addressing the diseases of most species of prairie grouse followed the historical pattern described earlier. Unfortunately, the Lesser Prairie-Chicken is an exception to this generalization in that no comprehensive surveys of the infectious agents associated with this species were conducted and published during the 1930s and 1940s. Below, I summarize research conducted on the infectious and noninfectious diseases of Lesser Prairie-Chickens since that period.

INFECTIOUS AGENTS

Here, I review what is known regarding the infectious agents documented for Lesser Prairie-Chickens. Where possible and appropriate, I address this goal by briefly discussing parasite prevalence, intensity, seasonality, pathogenicity, and whether previous research suggests the parasite may influence host population dynamics. I do not describe pathogenesis, clinical signs, lesions, and diagnostic techniques for infectious agents, but the cited works provide many of these specific details. Most reference books detailing the infectious diseases of wild and domestic birds address prairie grouse superficially if at all, but do include useful treatments of the major infectious agents discussed later (Friend and Franson 1999a, Davidson 2006, Atkinson et al. 2007, Thomas et al. 2007, Saif et al. 2008).

For readers more interested in focusing on the host species rather than infectious agents, I recommend three comprehensive reviews addressing infectious diseases and parasites of galliform birds that are sympatric and allopatric with the Lesser Prairie-Chicken. Peterson (2004) reviewed what was known about the infectious agents of all species of prairie grouse (*Tympanuchus*) across their respective ranges. Christiansen and Tate (2011) offer a range-wide review of the parasites and infectious diseases of Greater Sage-Grouse. Peterson (2007) reviewed the macro- and microparasites of Texas quails, including studies of Scaled Quail (*Callipepla squamata*), Gambel's Quail (*C. gambelii*), Northern Bobwhite (*Colinus virginianus*), and Montezuma Quail (*Cyrtonyx montezumae*) across their respective ranges. Many of these host species have been intensively and extensively studied compared to Lesser Prairie-Chicken with respect to macro- and microparasites, and the reviews provide a baseline for the potential importance of infectious agents for population dynamics of Lesser Prairie-Chickens.

During the past century of research that biologists have studied the macro- and microparasites discussed later, many of the specific and generic names of disease agents have changed at least once. Moreover, no single source of currently approved names is available for all parasite taxa. Thus, the binomial nomenclature associated with different taxa has become rather murky in some cases. I have used names agreed upon by current consensus where they could be verified and included the synonyms used in cited sources in parentheses. Otherwise, I followed the nomenclature used by the cited authority.

Macroparasites

Parasitic Arthropods

Chewing lice (Mallophaga) from five genera and at least seven species have been described for prairie grouse (Peterson 2004). Biologists trapping Lesser Prairie-Chickens commonly see lice on birds in hand, but few have collected these parasites, identified them, and published their results. Emerson (1951) documented *Goniodes cupido* and *Lagopoecus* sp. for Lesser Prairie-Chickens captured in Oklahoma. One should expect to find chewing lice of at least one species associated for all Lesser Prairie-Chicken populations.

Feather mites (Acari) have not been reported for Lesser Prairie-Chickens, but undoubtedly occur and have been documented for other species of prairie grouse (Peterson 2004). Additionally, because of the methods used by many ectoparasite surveys, mites probably would not have been detected even if they were present. Lesser

Prairie-Chickens sometimes host ticks, but no tick species have been reported in the literature. See Peterson (2004) for ticks and other ectoparasites reported for other species of prairie grouse.

During the first half of the 20th century, biologists studying prairie grouse often argued that ectoparasites could be detrimental to hosts—particularly chicks, incubating females, or anytime parasite intensity was high (Gross 1930:37–39; Leigh 1940). Later, the prevailing view was that parasitic arthropods were important to prairie grouse only when habitat conditions led to nutritional or other "stresses" for hosts (Hillman and Jackson 1973:28–30; Tirhi 1995:19). Dick (1981:235) analyzed ectoparasites of a population of Sharp-tailed Grouse (T. phasianellus), and argued "the role of ectoparasites on mortality and population fluctuations... is far from clear." More recently, ecologists have discovered that ectoparasites could be important mediators of host behavior, thus influencing host populations (Dobson 1988). For example, female Greater Sage-Grouse differentially selected males with fewer ectoparasites (Boyce 1990, Johnson and Boyce 1991). Conversely, male Sharp-tailed Grouse possessing central territories—and presumably doing most of the breeding—actually had more lice than birds on the periphery (Tsuji et al. 2001). Because most males were only lightly infested, however, discriminating females garnered little benefit as far as exposure to ectoparasites was concerned. At any rate, ectoparasites are commonly found on Lesser Prairie-Chickens and congeneric individuals, but their population-level significance requires further clarification.

Parasitic Helminths

Heterakis isolonche (= H. bonasae) is a cecal threadworm and is the only directly transmitted nematode documented for Lesser Prairie-Chickens. Pence et al. (1983) found no differences in H. isolonche prevalence or intensity for Lesser Prairie-Chickens between spring and autumn samples (Table 9.1). A related species, H. gallinarum, can transmit the protozoan Histomonas meleagridis, the etiologic agent of histomoniasis or blackhead disease (Davidson 2007, Yazwinski and Tucker 2008). H. meleagridis was found during most parasite surveys of Greater Prairie-Chickens (T. c. pinnatus) and Sharp-tailed Grouse, but has not been recovered from Lesser Prairie-Chickens (Peterson 2004).

Although H. isolonche causes significant disease in captive Ring-necked Pheasants (Phasianus colchicus), few pathologic changes were noted for grouse even with high parasite intensities (Fedynich 2007, Yazwinski and Tucker 2008).

The life cycle of the proventricular worm Tetrameres spp. is indirect, with insects serving as intermediate hosts (Kinsella and Forrester 2007). Robel et al. (2003) could not determine the species of Tetrameres recovered from Lesser Prairie-Chickens in southwestern Kansas because only female parasites were recovered, but they narrowed their identification to either T. americana or T. pattersoni. The authors found a higher prevalence during spring and autumn compared to winter (Table 9.1), but no statistical differences in prevalence by host age (90.9% and 90.9% of 33 juveniles and 43 adults, respectively), or by sex (100% and 69.6% of 62 males and 23 females, respectively). The mean intensity of infection for all individuals was 21 (Table 9.1), with juveniles ($\bar{x} = 24$, range = 3–66) harboring slightly more parasites than adults ($\bar{x} = 17$, range = 1–59). The pathogenicity of T. americana and T. pattersoni in prairie grouse is unknown, but Robel et al. (2003) found that Tetrameres sp. intensity was not related to movements, reproductive productivity, or survival of Lesser Prairie-Chickens. The intermediate hosts of these two Tetrameres spp. include grasshoppers (Melanoplus femurrubrum, Chortophaga iridifasciata) and cockroaches (Blattella germanica; Cram 1933). T. americana causes few pathologic changes in Northern Bobwhite, but can cause severe disease in domestic chickens under certain husbandry regimes (Cram 1931, Yazwinski and Tucker 2008). Conversely, T. pattersoni is much more pathogenic for Northern Bobwhite than is T. americana (Davidson et al. 1982, 1991, Yazwinski and Tucker 2008).

Robel et al. (2003) maintained that a Subulura sp. found in Lesser Prairie-Chickens most closely resembled S. suctoria, but could also be a new species (Table 9.1). S. suctoria uses beetles as an intermediate host (Yazwinski and Tucker 2008). Robel et al. (2003) found that prevalence was lower during spring than autumn or winter (Table 9.1), and higher for juveniles than adults (71.4% and 46.7% of 35 and 45, respectively), but did not differ by host sex (55.6% and 68.0% of 35 males and 17 females, respectively). They detected no differences in the intensity of infection by host age or sex ($\bar{x} = 30$ and 18, range = 1–319 and 1–22 for juveniles and adults,

TABLE 9.1
Prevalence and intensity of infection by parasitic helminths of Lesser Prairie-Chickens.

Parasite	State	Season	Prevalence n pos. (n ex.)[a]	%	Intensity[b] Mean	Range	Sources
Nematodes							
Heterakis isolonche	Texas	Spring	10 (15)	66.7	17.5	1–150[c]	Pence and Sell (1979), Pence et al. (1983)
		Autumn	11 (26)	42.3	66.5	1–271	
Oxyspirura petrowi	Kansas	All birds[d]	53 (56)	94.6	14	1–81	Robel et al. (2003)
		Spring	12 (13)	92.3			
		Autumn	10 (10)	100.0			
		Winter	30 (32)	93.8			
	Texas	Spring	8 (15)	53.3	3.8	1–19	Pence and Sell (1979), Pence et al. (1983)
		Autumn	17 (26)	42.3	65.4	1–12	
Physaloptera sp.	Texas	Spring	6 (15)	40.0	3.7	1–9	Pence et al. (1983)
		Autumn	10 (26)	38.5	1	1	
Subulura sp.[e]	Kansas	All birds[d]	54 (91)	59.3	28	1–319	Robel et al. (2003)
		Spring	2 (11)	18.2			
		Autumn	8 (11)	72.7			
		Winter	41 (65)	63.1			
Tetrameres sp.[f]	Kansas	All birds[d]	81 (88)	92.0	21	1–66	Robel et al. (2003)
		Spring	12 (12)	100.0			
		Autumn	11 (11)	100.0			
		Winter	54 (61)	88.5			
Cestodes							
Rhabdometra odiosa	Texas	Spring	1 (15)	6.7	1	1	Pence and Sell (1979), Pence et al. (1983)
		Autumn	14 (26)	53.8	9.4	1–29	

[a] Number of individuals positive for parasite (number of individuals examined).
[b] The number of parasitic helminths of a given species per host.
[c] The range listed in Pence et al. (1983) was mistakenly published as 1–15 rather than 1–150 (D. B. Pence, personal communication).
[d] Robel et al. (2003) listed parasite intensity for all birds examined, but not by season; all birds examined included more birds than the sum of seasonal totals.
[e] Not keyed to species. Most closely resembled S. suctoria; perhaps a new species.
[f] Because only female helminths were recovered, these parasites could not be keyed to species. Probably T. pattersoni or T. americana.

respectively; $\bar{x}$ = 26 and 34, range = 1–319 and 1–199 for males and females, respectively). Subulura spp. cause few pathologic changes in hosts (Yazwinski and Tucker 2008).

The eyeworm Oxyspirura petrowi (= O. lumsdeni) is typically found under the nictitating membrane and has commonly been reported in prairie grouse since the 1930s in studies where the eyes were examined (Peterson 2004). Addison and Anderson (1969) described this species for a number of prairie grouse species, including the Lesser Prairie-Chicken, but did not address prevalence or intensity. Pence and Sell (1979) and Pence et al. (1983) documented O. petrowi for Lesser Prairie-Chickens in the Texas Panhandle. Prevalence and intensity did not differ between spring versus autumn (Table 9.1). Robel et al. (2003) described O. petrowi in Lesser Prairie-Chickens from southwestern Kansas. No differences in prevalence by season were found (Table 9.1), but they did document higher prevalence in adult as compared to juvenile hosts (100% and 91.3% of 30 and 23, respectively), and higher intensities for juveniles than adults ($\bar{x}$ = 21 and 9, range = 1–81 and

2–32, respectively). No differences in prevalence or intensity by host sex were reported (prevalence: 93.5% and 100% of 46 and 10 males and females, respectively; intensity: $\bar{x} = 13$ and 17, range $= 1$–81 and 4–51 for males and females, respectively).

The life history of *O. petrowi* has yet to be detailed, but it is assumed to have an intermediate host that is an arthropod, and probably an insect (Pence 1972). Saunders (1935:343) observed ocular irritation in several parasitized Sharp-tailed Grouse and Greater Prairie-Chickens, leading him to conclude that "serious consequences," such as decreased foraging efficiency and increased predation, were associated with high intensities of *O. petrowi*. Similarly, Jackson (1969:70) speculated that vision of some parasitized Northern Bobwhites in his north Texas study area may have been impaired by eyeworms (Peterson 2007). Conversely, Pence (1972) found that even intensities as high as 30 worms per eye in an array of avian hosts caused no gross or histopathologic changes. Similarly, Robel et al. (2003) found that *O. petrowi* intensity was not related to movements, reproductive productivity, or survival of Lesser Prairie-Chickens. Definitive experiments designed to determine the population-level significance of *O. petrowi* have yet to be completed for any wild galliform species.

Filarial nematodes of a few species have been reported for prairie grouse (Peterson 2004). Pence et al. (1983) found *Physaloptera* sp. larvae in the crop or proventriculus of Lesser Prairie-Chickens collected in Texas (Table 9.1), with greater prevalence and intensity in spring versus autumn. Prairie-chickens presumably acquire larval *Physaloptera* sp. by ingesting infected arthropods. The population-level significance of *Physaloptera* sp. is unknown.

Cestodes of three genera and five species have been documented for prairie grouse (Peterson 2004). *Rhabdometra odiosa*, however, is the only tapeworm reported for the Lesser Prairie-Chicken (Table 9.1), and the species has not been documented for other species of prairie grouse (Peterson 2004). Avian tapeworms use arthropods or isopods as intermediate hosts (Pence et al. 1983, McDougald 2008a), so parasite prevalence and intensity should be expected to vary by season. Pence et al. (1983) found a higher prevalence and intensity of *R. odiosa* in Lesser Prairie-Chickens captured during the autumn than spring (Table 9.1). Arthropods (e.g., grasshoppers, crickets, beetles)

made up 8%, 27%, 60%, and 65% of the diet of Lesser Prairie-Chickens on their study area during winter, spring, summer, and autumn, respectively, thus offering a reasonable explanation for differential *R. odiosa* prevalence and intensity observed by season. Further, because arthropods and isopods constitute a larger proportion of the diet in young prairie grouse, young birds typically have higher cestode prevalence and intensity than adults (Peterson 2004).

Most authors reporting cestodes in prairie grouse did not observe gross pathologic changes attributable to these parasites, although some authors maintained that certain cestodes could be pathogenic for young birds (Peterson 2004). For example, Harper et al. (1967) noted inflammation where tapeworm scolices were attached in Greater Prairie-Chickens. Leigh (1940, 1941) found that sufficiently intense infections of *R. variabilis* in young Greater Prairie-Chickens occluded the intestinal lumen; he maintained that blockages could reduce host vitality and render these young birds more susceptive to predation or microparasitic infection. Regardless, possible direct and indirect effects of cestodes on Lesser Prairie-Chicken or other prairie grouse populations have yet to be explored.

Other species of prairie grouse sometimes harbor additional taxa of parasitic helminths, including thorny-headed worms (*Mediorhynchus papillosus*) and various species of trematodes (Peterson 2004). None of these parasites have been documented for Lesser Prairie-Chickens. Lack of detection could be due to incomplete sampling efforts, but the arid climate where many Lesser Prairie-Chicken populations occur also could be responsible for low parasite diversity for this host species (Peterson 1996).

Microparasites

Hematozoa

Several species of hematozoa have been documented in blood samples taken from prairie grouse throughout their ranges (Peterson 2004). Arthropod vectors are required to transmit hematozoa, so these microparasites tend to occur seasonally (Atkinson 2007a,b, Forrester and Greiner 2007). For Lesser Prairie-Chickens, Stabler (1978) identified *Plasmodium pedioecetii* on blood films from 2 of 29 (6.9%) birds from

New Mexico and 2 of 8 (25%) birds from Texas. Smith et al. (2003) also identified *P. pedioecetii* on blood films from 4 of 32 (12.5%) Lesser Prairie-Chickens captured in New Mexico. Intensity of infection was low (observed in 0.5%–0.7% of erythrocytes examined from positive birds). Culicine and anopheline mosquitoes of the genera *Culex, Aedes, Culiseta,* and *Anopheles* support development and transmission of *Plasmodium* spp. The disease caused by *Plasmodium* spp. is commonly referred to as avian malaria (Atkinson 2007a, Bermudez 2008).

Several wildlife biologists have argued that hematozoa may be a serious problem for prairie grouse, particularly young birds (Saunders 1935, Shillinger 1942, Cowan and Peterle 1957). Parasite intensities were sometimes high and associated with disease symptoms (Flakas 1952, Cowan and Peterle 1957), but studies addressing the population-level significance of these disease agents have yet to be completed for any species of prairie grouse.

Other Protozoa

It is reasonable to assume that *Eimeria* spp. occur at some level in most, if not all prairie grouse populations (Peterson 2004), but surprisingly few surveys of intestinal coccidia have been completed. Three studies published between 1936 and 1941 reported *Eimeria* spp. for Greater Prairie-Chickens or Sharp-tailed Grouse (Peterson 2004). More recently, Smith et al. (2003) described *E. tympanuchi* for 5 of 64 (7.8%) Lesser Prairie-Chickens captured in New Mexico, and Fritzler et al. (2011) described *E. attwateri* from captive Attwater's Prairie-Chickens with clinical signs of coccidiosis. The life cycle of all *Eimeria* spp. is direct, but oocysts shed in the feces of infected grouse must sporulate to become infective to a susceptive host, with a life cycle of ~7 to 14 days (Yabsley 2007, McDougald and Fitz-Coy 2008). *Eimeria* spp. can be highly pathogenic to many avian species in captive propagation programs (Shillinger and Morley 1937, McDougald and Fitz-Coy 2008), but the significance of intestinal coccidiosis in free-living prairie grouse has yet to be evaluated.

Wildlife biologists have long assumed that the flagellated protozoan *H. meleagridis* can negatively influence populations of prairie grouse (Gross 1928, 1930). The two main reasons for

this supposition are the ubiquitous nature of the cecal threadworm *H. gallinarum* as a vector, and the severity of blackhead disease or histomoniasis in prairie grouse (Peterson 2004). Conversely, *H. isolonche*—the cecal threadworm documented for Lesser Prairie-Chickens—is not a good vector for *H. meleagridis* (Davidson et al. 1978, Davidson 2007, McDougald 2008b). Gross (1928:527–528) documented *H. meleagridis*-induced mortality of Heath Hens (*T. c. cupido*), and maintained that the disease contributed to the extinction of this subspecies. Similarly, Gross (1930:40–41), Leigh (1940), and Schwartz (1945:87–88) described Greater Prairie-Chickens killed by histomoniasis in Wisconsin, Illinois, and Missouri, respectively. Although *H. meleagridis* definitely can cause significant mortality of free-living prairie grouse, its population-level influence remains uncertain. Moreover, it is unclear whether a competent vector for *H. meleagridis* exists in populations of Lesser Prairie-Chicken.

Bacteria

Peterson (2004) found few documented cases of bacterial disease in free-living prairie grouse. Sampling issues probably are largely responsible for this lack of documentation. For example, unless a major epidemic is underway, finding fresh, intact carcasses of individuals that succumbed to a bacterial infection in the wild would be rare indeed. Moreover, case reports documenting bacterial diseases in individual wild animals are not encouraged by most refereed journals, so only biologists who prepared the report would be aware of bacteria as a disease agent.

Researchers isolated *Mycoplasma* sp. from the trachea, and *Salmonella* sp. (group B) from the kidney of apparently healthy Lesser Prairie-Chickens captured in Oklahoma and Kansas during 2001–2002 (D. H. Wolfe, unpubl. data, C. A. Hagen, unpubl. data). Similarly, Hagen et al. (2007) isolated *Pasteurella multocida* from lung, liver, and spleen of a radio-marked female Lesser Prairie-Chicken that died during their study in Kansas with clinical signs of avian cholera (Samuel et al. 2007).

Serological techniques provide nonlethal methods to determine whether apparently healthy prairie grouse have developed antibodies specific to various bacteria without sacrificing the animals sampled. Peterson et al. (2002) tested blood sera collected from Lesser Prairie-Chickens captured in the northeastern Texas Panhandle (Table 9.2). All samples were negative for antibodies specific to *Chlamydophila psittaci* (= *Chlamydia psittaci*), *M. gallisepticum*, *M. synoviae*, *S. pullorum*, and *S. typhimurium*. Hagen et al. (2002), using a serum plate antigen (SPA) test, found a low seroprevalence against *M. gallisepticum*, *M. meleagridis*, and *M. synoviae* for Lesser Prairie-Chickens captured in southwest Kansas (Table 9.2). The SPA-positive samples were not confirmed by hemagglutination inhibition testing and no cultures were attempted, and the authors were uncertain which *Mycoplasma* sp. elicited the positive SPA responses

(Peterson et al. 2002). A recently completed study found specific antibody to *P. multocida* for serum from one adult male Lesser Prairie-Chicken captured in the southwestern Texas Panhandle (S.M. Presley, unpubl. data; Table 9.2).

Lesser Prairie-Chickens would likely be vulnerable to many typical diseases of galliform birds caused by pathogenic bacteria. See Peterson (2004) for a review of serological surveys conducted for bacterial agents of Greater and Attwater's Prairie-Chickens. The exact influence these agents may have on population dynamics of prairie grouse remains unknown.

TABLE 9.2

Prevalence of specific antibodies to selected pathogenic bacteria and viruses and to proviral DNA of the reticuloendotheliosis virus by polymerase chain reaction (PCR) for Lesser Prairie-Chickens.

Group		Prevalence		
Microparasite	State	n pos. (n ex.)[a]	%	Sources
Bacteria				
Chlamydophila psittaci	Texas	0 (24)	0.0	Peterson et al. (2002)
Mycoplasma gallisepticum	Kansas	8 (162)[b]	4.9	Hagen et al. (2002)
	Texas	0 (24)	0.0	Peterson et al. (2002)
M. meleagridis	Kansas	8 (162)[b]	4.9	Hagen et al. (2002)
M. synoviae	Kansas	5 (162)[b]	3.1	Hagen et al. (2002)
	Texas	0 (24)	0.0	Peterson et al. (2002)
Pasteurella multocida	Texas	1 (45)	2.2	S. M. Presley (unpubl. data)[c]
Salmonella pullorum	Texas	0 (24)	0.0	Peterson et al. (2002)
S. typhimurium	Texas	0 (24)	0.0	Peterson et al. (2002)
Viruses				
Avian influenza	Texas	0 (24)	0.0	Peterson et al. (2002)
Infectious bronchitis—AR	Texas	8 (17)	47.1	Peterson et al. (2002)
Infectious bronchitis—MA	Texas	2 (18)	11.1	Peterson et al. (2002)
Newcastle disease	Texas	0 (23)	0.0	Peterson et al. (2002)
	Texas	0 (41)[d]	0.0	S. M. Presley (unpubl. data)[c]
Reticuloendotheliosis	Kansas	0 (3)	0.0	Wiedenfeld et al. (2002)
	Oklahoma	0 (102)	0.0	Wiedenfeld et al. (2002)
	New Mexico	0 (79)	0.0	Wiedenfeld et al. (2002)
	Texas	0 (24)	0.0	Peterson et al. (2002)
West Nile	Texas	1 (98)[d]	1.0	S. M. Presley (unpubl. data)[c]

[a] Number of individuals seropositive or PCR positive for listed infectious agent (number examined).

[b] The serum plate antigen (SPA) test was used, and because SPA-positive samples were not confirmed by hemagglutination inhibition testing and no cultures were attempted, one cannot be certain which *Mycoplasma* sp. elicited the positive the SPA responses (Peterson et al. 2002).

[c] S. M. Presley, Texas Tech University, unpubl. data; sera collected 2008–2011.

[d] These results are presumptive because the researchers are still validating the enzyme-linked immunosorbent assay they developed and used in the study.

Fungi

Little has been published regarding fungal diseases of free-living prairie grouse, and I found no previous reports for Lesser Prairie-Chickens. See Peterson (2004) for a review of fungal diseases documented for other species of prairie grouse. Lesser Prairie-Chickens are almost certainly susceptible to fungal agents. Fungal diseases, such as pulmonary aspergillosis (caused by *Aspergillus* spp.), can cause serious illness but the importance of fungal infections to population dynamics of Lesser Prairie-Chickens or other prairie grouse is unknown.

Viruses

Two field studies have used methods based on polymerase chain reaction (PCR) to test blood samples for the proviral DNA associated with the reticuloendotheliosis virus (REV, Retroviridae) for Lesser Prairie-Chickens captured in the northeastern Texas Panhandle (Peterson et al. 2002), and study areas in Kansas, Oklahoma, and New Mexico (Wiedenfeld et al. 2002; Table 9.2). All samples were negative for the pathogen. Peterson et al. (2002) also failed to detect specific antibody against avian influenza (Orthomyxoviridae) and Newcastle disease (Paramyxoviridae) viruses for serum samples collected in Texas (Table 9.2). However, evidence of exposure to the Massachusetts and Arkansas serotypes of infectious bronchitis virus (Coronaviridae) was found using a microhemagglutination-inhibition test (Table 9.2). Five of eight positive individuals were juveniles, two of which were serologically positive for both serotypes. A recent survey evaluated blood sera from Lesser Prairie-Chickens from the southwestern Texas Panhandle and failed to detect specific antibody to the Newcastle disease virus, but did find presumptive evidence of exposure to WNV (Flaviviridae) for one adult male (S.M. Presley, unpubl. data; Table 9.2). See Peterson (2004) for a review of serological surveys conducted for viral agents of Greater and Attwater's Prairie-Chickens.

Newcastle disease, infectious bronchitis, reticuloendotheliosis, and highly pathogenic avian influenza viruses could cause serious disease in individual Lesser Prairie-Chickens, and may have the potential to influence host population dynamics. Some of the other viruses that cause disease in a range of galliform birds may also be pathogenic for Lesser Prairie-Chickens and negatively influence populations. Emerging viral diseases could become established in North America, disperse across the continent and negatively influence Lesser Prairie-Chicken populations. For example, mosquito-borne WNV was first recognized in North America in 1999, spread rapidly across much of the continent, and emerged as a serious threat to low- and mid-elevation populations of Greater Sage-Grouse by 2004 (Walker and Naugle 2011). The WNV example illustrates just how rapidly a microparasite–host–habitat relationship can develop or change (Peterson and Ferro 2012).

NONINFECTIOUS DISEASES

Noninfectious disease refers to interruption, cessation, or disorder of body functions, systems, or organs caused by factors such as malnutrition and starvation, environmental contaminants, physical injury, shock, biotoxins, metabolic abnormalities, and many types of neoplasia (Fairbrother et al. 1996, Peterson and Ferro 2012). For example, most wildlife biologists are familiar with capture myopathy, which is a serious noninfectious disease sometimes associated with capture and handling of wild animals (Williams and Thorne 1996). Unfortunately, little is known about many noninfectious diseases that cause illness for individual Lesser Prairie-Chickens and other prairie grouse. Here, I consider noninfectious diseases that could be important to Lesser Prairie-Chickens and their population dynamics.

Nutrition

Nutritional diseases such as starvation typically do not cause major mortality events for adult galliforms in the wild. In semiarid regions, however, nutritional status influences whether female Northern Bobwhites and Rio Grande Wild Turkeys (*Meleagris gallopavo intermedia*) enter reproductive condition during the breeding season, and the duration of the period that reproductive females remain active, which directly influences reproductive productivity (Beasom 1973, Pattee and Beasom 1979, Wood et al. 1986, Guthery et al. 1988, Davis and Gruen 1994, Guthery 2002). Some populations of Lesser Prairie-Chickens inhabit arid climatic zones with environmental conditions similar to sites where other galliform birds exhibit intermittent breeding. Insufficient data are available to determine exactly how nutrition influences the body condition of female Lesser Prairie-Chickens, but interactions between food and reproductive output are likely one of the ways that drought conditions lead to lower recruitment into the breeding population (Hagen et al. 2004, 2009, Lyons et al. 2011, Grisham et al. 2014).

Nutritional deficits caused by poor insect availability commonly occur with newly hatched chicks and young grouse of most species until birds are old enough to obtain adequate nutrition, particularly protein, from plant-based foods (Hudson 1986, Park et al. 2001, Hannon and Martin 2006). Similarly, arthropods are critically important to the diet of young Lesser Prairie-Chickens. For example, Davis et al. (1980) found that arthropods made up 99%–100% of the diet of chicks (≤4 weeks of age) and juvenile Lesser Prairie-Chickens (5–10 weeks of age). Direct mortality due to starvation may be critically important in some years, but malnutrition may also make young Lesser Prairie-Chickens more vulnerable to parasites and predation.

Biotoxins

Biotoxins are poisonous substances that are the metabolic products of a living organism, such as botulinum toxins, algal toxins such as red tide, ricin, nicotine, and various mycotoxins. Mycotoxins are toxic metabolites produced by various species of fungi (O'Hara 1996, Hoerr 2008) and are more relevant than other biotoxins for Lesser Prairie-Chickens because of the availability of waste grain in fields and the prevalence of feeding and baiting wildlife. Wildlife biologists should be familiar with the risks posed by aflatoxins, a group of highly toxic and carcinogenic mycotoxins produced by *Aspergillus flavus*, *A. parasiticus*, and *Penicillium puberulum*, and mortality caused by aflatoxins has been reported for a number of wild animal species (O'Hara 1996, Hoerr 2008).

I found no mortality records of prairie grouse caused by mycotoxins, but the species is likely to be susceptible to biotoxins (Peterson 2004). Mycotoxins can be associated with waste grains in the field, or milo and corn used as feed or bait for Northern Bobwhites and white-tailed deer (*Odocoileus virginianus*, Peterson 2007). Mycotoxins cause immunosuppression, and decreased growth, reproductive success, and survival in galliform birds that consume contaminated grains and could have a negative impact on chicks of Lesser Prairie-Chickens (Quist et al. 2000, Hoerr 2008). For these reasons, data are needed regarding how low levels (<200 ppb) of aflatoxin and other dangerous mycotoxins, such as ochratoxin and T-2 toxin, influence the demography of Lesser Prairie-Chickens. If observed levels are shown to be detrimental, state regulatory agencies could further restrict the level of mycotoxins allowed in grains used for baiting and feeding wildlife, or eliminate these practices. Similarly, because improper storage can greatly increase mycotoxin levels in these products after purchase, more effective methods for helping users learn how to safely store and use these commodities may be needed.

Environmental Contaminants

Industry and manufacturing has released a wide array of potential contaminants into the environment. Toxic materials include cyanides, heavy metals (e.g., arsenic, lead, mercury), pesticides (e.g., insecticides, herbicides, fungicides, rodenticides), polychlorinated biphenyls, petroleum, and other compounds (Fairbrother et al. 1996, Friend and Franson 1999a, Sheffield et al. 2012). Environmental contaminants can be toxic to individual animals in sufficient quantities and cause direct effects of morbidity and mortality, or indirect effects of subclinical disease, decreased reproductive success, developmental abnormalities, and bioaccumulation potentially leading to biomagnification.

Peterson (2004) found few studies of environmental contaminants associated with prairie grouse and their habitats. Watkins (1969) identified higher concentrations of dichlorodiphenyl-trichloroethane (DDT) in body tissues of ten Attwater's Prairie-Chickens that died during capture than nine surviving birds. Flickinger and Swineford (1983), however, found low levels of organochlorine pesticides, polychlorinated biphenyls, and heavy metal residues in tissues of Attwater's Prairie-Chickens regardless of whether birds originated from agricultural landscapes or grasslands. A more recent study reported low levels of organochlorine residues in livers of 13 Attwater's Prairie-Chickens, but found unusually high concentrations of cadmium and zinc in one liver sample, and a high concentration of zinc in another sample (M.A. Mora, unpubl. data).

No published studies have evaluated Lesser Prairie-Chicken tissues or habitats for environmental contaminants, but the species is likely to be susceptible to these materials. Surveys of environmental contaminants associated with Lesser Prairie-Chickens are needed for at least two reasons. First, Lesser Prairie-Chickens, in some parts of their range, inhabit areas near crop fields such as corn, wheat, and cotton, where insecticides, herbicides, and defoliants are applied, sometimes with aerial spraying. Second, petroleum extraction is expanding rapidly within the range of Lesser Prairie-Chickens. Leks sometimes form on well pads, and spills of petroleum and oil can occur near equipment. Lesser Prairie-Chickens have the potential to be exposed to different environmental contaminants, but the degree of exposure is unknown. If for no other reason, baseline data regarding environmental contaminants are needed as a point of comparison for future monitoring.

Trauma and Accidents

Physical injuries, often followed by shock, are among the most common causes of noninfectious diseases of wild animals (Cooper 1996, Mann and Helmick 1996). Sudden, violent forces that result in compression, stretching, torsion, or penetration of tissues produce traumatic injuries. Common accidental traumatic injuries of wildlife include collisions with stationary objects such as fence wires, electrical wires, guywires, wind turbines, or windows; they also include collisions with moving objects such as motor vehicles, trains, and aircraft. Accidental trauma also includes electrocution from lightning or power lines and burns from grass or forest fires. Accidental injuries that may or may not include trauma include drowning after falling through ice or into cattle tanks, and entanglements with rope or other materials. Trauma, ranging from minor injuries to death, sometimes occurs when wildlife biologists capture wild animals using traps, nets, or chemical immobilizing agents. Intra- and interspecific interactions such as territorial defense, competition for mates, injuries from predation, and crippling losses from hunting also can cause trauma.

Certain types of traumatic injuries probably are the best-understood noninfectious diseases of Lesser Prairie-Chickens. For example, predation is a common cause of traumatic injury for this species. Wolfe et al. (2007) found that predation by raptors and mammals caused the death of 35.0% and 29.2%, respectively, of 260 radio-marked Lesser Prairie-Chickens in Oklahoma and New Mexico where the cause of death could be determined (total predation mortality = 64.2%). Hagen et al. (2007) attributed the death of 59%, 11%, and 3% of 92 radio-marked female Lesser Prairie-Chickens in Kansas to mammal, raptor, and snake predation, with the remaining 18% attributed to unknown predators. Human predation of Lesser Prairie-Chickens through legal and illegal hunting undoubtedly declined as abundance of this species declined and states restricted or eliminated hunting seasons (Chapter 7, this volume). Regardless, injuries caused by hunting are greater than the number bagged due to crippling losses where injured birds are not recovered by hunters and do not survive.

Traumatic injuries of Lesser Prairie-Chickens can be caused by trapping activities conducted by wildlife biologists. For example, practitioners have reported mild to severe traumatic injuries (including death) caused by rocket and cannon nets (Taylor 1978, Davis et al. 1979, Sell 1979). Common injuries include lacerations and abrasions on the wings and head are common, and deaths sometimes occur, when walk-in drift traps are used on leks (Haukos et al. 1990, Schroeder and Braun 1991). The injuries typically occur when raptors or humans disturb trapped birds, or when multiple individuals are captured in a single trap, when males attempt to escape when females come near, or when males defend their territories while inside a trap. Risk of injury can be minimized with careful monitoring of active traps and rapid processing of birds after capture.

Lesser Prairie-Chickens are well adapted to their native range, but anthropogenic changes since the late 1800s have resulted in habitat changes that negatively influence the species. Several researchers found evidence that Lesser Prairie-Chickens avoid large physical structures such as buildings, natural gas compressor stations, major highways, wind turbines, and large electricity transmission lines even when apparently suitable potential habitat occurs near these structures (Hagen et al. 2004; Robel et al. 2004; Pitman et al. 2005; Pruett et al. 2009a,b). Avoidance behavior effectively limits usable space available through time for this species (Guthery 1997) and is likely to become increasingly important as rapid energy development continues within the range of Lesser Prairie-Chicken. Despite the avoidance of large physical structures, collisions are among the most common traumatic injuries of Lesser Prairie-Chickens. For example, Wolfe et al. (2007) found that collisions with fences, powerlines, and motor vehicles caused the death of 33.1%, 1.5%, and 1.2%, respectively, of 260 radio-marked Lesser Prairie-Chickens in Oklahoma and New Mexico where the cause of death could be determined (total collision mortality = 35.8%). Hagen et al. (2007) attributed the deaths of 4.3% and 1.1% of 92 radio-marked female Lesser Prairie-Chickens in Kansas to collisions with powerlines and motor vehicles, respectively. Not all radio-marked birds died at the point of collision (Wolfe et al. 2007).

The evolutionary history of Lesser Prairie-Chickens did not prepare the species to cope with fence or electrical transmission wires. Wolfe et al. (2007) demonstrated that fence collisions of Lesser Prairie-Chicken resulted in mortality rates (34.6%; primarily for the Oklahoma study sites) comparable to losses to raptors (35.0%) and greater than the mortality caused by mammalian predators (29.2%). The authors also reported that the mortality caused by fence collisions was greater for females than males and was related to the fence density. Studies conducted elsewhere have rarely noted high mortality caused by fence collisions, and it may be a localized hazard risk until additional studies confirm high levels of mortality elsewhere. Regardless, landscape features responsible for 35.8% of the mortality of Lesser Prairie-Chickens reported by Wolfe et al. (2007) and 5.4% of the mortality observed by Hagen et al. (2007) did not exist prior to the late 1800s. Anthropogenic features have become increasingly common since the two population studies were completed and may continue to be an issue with planned energy development within habitats used by Lesser Prairie-Chickens.

Weather-Related Mortality

Severe weather can cause disease for individual wild animals and has the potential to influence wildlife population dynamics (Leopold 1933:348–350; Cooper 1996). Endothermic animals can control their body temperature to some degree by both behavioral strategies and physiological means. If body temperature exceeds a critical level, however, the animal loses its ability to thermoregulate and becomes hyperthermic, distressed, and ill; death can occur because of irreversible pathological changes. Similarly, extreme cold can lead to frostbite, hypothermia, and death. Losses to winter weather can be accelerated by deep crusted snow or ice sheathing vegetation. Severe weather events also can cause traumatic injuries due to objects carried by the wind, flooding, large hail, and lightning.

It is well established that the combination of drought and heavy livestock grazing can be detrimental to reproductive productivity and survival of Lesser Prairie-Chickens (Lee 1950, Schwilling 1955, Copelin 1963). Recent reviews have considered the interactive effects of drought and livestock management (Massey 2001, Hagen and Giesen 2004, Hagen et al. 2009). Here, I focus on the direct effects of weather on Lesser Prairie-Chickens and possibly their populations. Thermal stress during the preincubation period may alter embryo development, disrupt hatch synchrony, reduce egg viability, and kill embryos of Lesser Prairie-Chickens because high temperatures cause similar losses in Northern Bobwhites (Reyna and Burggren 2012). Heat stress and related dehydration can increase chick mortality, particularly during drought when vegetation cover is limited (Merchant 1982:46). The importance of heat stress to adult Lesser Prairie-Chickens is unknown but could render individuals more susceptible to predation or even cause mortality in certain situations. Better data are needed regarding the influence of heat loading on eggs, chicks, juveniles, and adult Lesser Prairie-Chickens.

Hail storms are known to cause traumatic injury and death of Greater and Lesser Prairie-Chickens (Fleharty 1995:241, U.S. Fish and Wildlife Service 2012:73870), but it is unclear whether isolated weather events can influence population dynamics of prairie grouse. Similarly, tornado-force winds certainly could lead to traumatic injuries and death of Lesser Prairie-Chickens, but no data are available to evaluate the importance of such storms at the population level. Severe winter weather, such as blizzards and ice storms, may influence population dynamics of Lesser Prairie-Chickens in some portions of their range. For example, a large snowstorm and a severe blizzard occurred one week apart across southeastern Colorado during late December 2006 (Verquer 2007). Winter storms deposited >1 m of snow that persisted for >60 days. A blanket of deep snow, and resultant lack of cover and food, was likely responsible for a 47% decline in the number of leks and a 75% decline in bird numbers at leks of Lesser Prairie-Chickens in southeastern Colorado between 2006 and 2007 (34 to 18 leks and 296 to 74 individuals in 2006 and 2007, respectively). Systematic data on the influence of storms and cold weather on Lesser Prairie-Chickens and their populations are needed.

DISEASES AND POPULATIONS

Several infectious agents and noninfectious diseases have been documented for Lesser Prairie-Chickens, but little is known regarding the consequences of these sources of mortality at the population level. Disease-related research on Lesser Prairie-Chickens has been limited compared to our understanding of disease agents in other species of prairie grouse and Greater Sage-Grouse (Peterson 2004, Christiansen and Tate 2011, Walker and Naugle 2011). If predators can have a major effect on the population dynamics of Lesser Prairie-Chickens, it is possible that macro- and microparasites may have a similar potential. Here, I discuss a few specific infectious agents and noninfectious diseases that have the potential to influence Lesser Prairie-Chicken population dynamics. I then explore how infectious agents and noninfectious diseases may interact with each other, or with other biotic or abiotic factors, to influence population dynamics of Lesser Prairie-Chickens. Ecological interactions may be more important to population dynamics than the direct effects of infectious agents or noninfectious diseases alone.

Infectious Agents

Macroparasites of prairie grouse typically exhibit aggregated distributions (Dick 1981, Pence et al. 1983, Peterson et al. 1998) and thus have the potential to regulate host populations (Tompkins et al. 2002, Peterson 2004). Not enough is known regarding the ectoparasites of Lesser Prairie-Chickens to inform a meaningful evaluation of whether they could regulate host numbers. At least two parasitic helminths of Lesser Prairie-Chickens deserve further research because they could potentially regulate host abundance. If infections of O. petrowi obscure vision, one would expect higher probability of injury or predation

for Lesser Prairie-Chickens with high intensities of eyeworm infection. Various aspects of this hypothesis could be tested experimentally using captive and free-roaming Lesser Prairie-Chickens. The *Tetrameres* sp. associated with Lesser Prairie-Chickens is unknown (Table 9.1), but could be pathogenic. *T. americana* causes severe disease in domestic chickens, but few pathologic changes in Northern Bobwhite (Cram 1931, Yazwinski and Tucker 2008). Conversely, *T. pattersoni* is quite pathogenic for Northern Bobwhite (Davidson et al. 1982, 1991, Yazwinski and Tucker 2008). Field surveys are needed to identify which of these two species parasitizes Lesser Prairie-Chickens in Kansas. Studies employing captive and free-living Lesser Prairie-Chickens could determine whether the parasite is pathogenic for this host species. Both parasites are transmitted indirectly, and control measures may be possible under certain circumstances.

Endemic microparasites that reduce fecundity or recruitment of young into the breeding population in a density-dependent fashion could also regulate Lesser Prairie-Chicken populations (Tompkins et al. 2002, Peterson 2004). Two protozoans documented for Lesser Prairie-Chickens have the potential for population regulation. *Eimeria* spp. can cause decreased growth rates and increased mortality in young galliform birds (Friend and Franson 1999b, McDougald and Fitz-Coy 2008). Intestinal coccidiosis limits recruitment in free-living populations of Greater Sage-Grouse and may have similar effects in wild populations of Lesser Prairie-Chickens (Simon 1940). *P. pedioecetii* also may be able to regulate Lesser Prairie-Chicken abundance because avian malaria can cause severe anemia, weight loss, and mortality, particularly in young birds during the first few weeks after hatching (Atkinson 1999, Bermudez 2008). If the infectious bronchitis virus is as pathogenic for prairie grouse as it is for domestic chickens, it could regulate Lesser Prairie-Chicken numbers by reducing chick survival in a density-dependent fashion (Peterson et al. 2002, Cavanagh and Gelb 2008). *Mycoplasma* spp. and viruses (WNV, REV) also have the potential to regulate Lesser Prairie-Chicken numbers, but are less likely to be important than other microparasites because REV has not been documented for this host species and only low seroprevalences of *Mycoplasma* spp. and WNV-specific antibody have been demonstrated to date (Table 9.2). Further research is needed to examine the details of the host–pathogen interactions. Moreover, as fragmentation of Lesser Prairie-Chicken range and thence populations continues, host genetic drift and bottlenecking could lead to decreased immunocompetence and greater susceptibility to novel strains of microparasites.

Large-scale epidemic events with high mortality rates can occur in waterfowl (Friend et al. 2001), but have not been reported in prairie grouse (Peterson 2004). Many microparasites occur epidemically in avian populations where they cause significant mortality (Table 9.2) and could extirpate small, isolated populations of Lesser Prairie-Chickens. If a disease outbreak was to occur in a population of Lesser Prairie-Chickens, it might originate from spillover from other wild or domestic avian species. For example, epidemics of avian cholera in free-living waterfowl populations might occur in the shared range where waterfowl and Lesser Prairie-Chickens feed in crop fields in portions of the Texas Panhandle. Lesser Prairie-Chicken mortality caused by avian cholera is known to occur (Hagen et al. 2007). The threat of *P. multocida* transmission from waterfowl to pinnated grouse was sufficiently credible that staff of the Attwater's Prairie-Chicken National Wildlife Refuge collected and incinerated carcasses of waterfowl that succumbed to avian cholera during epidemics in order to reduce the risk of *P. multocida* transmission to Attwater's Prairie-Chickens (Peterson et al. 1998). Many of the other microparasites discussed in this review could spill over during an epidemic occurring primarily in other host species and cause density-independent disease in Lesser Prairie-Chickens, which could extirpate small, isolated populations. Managers tasked with conservation of Lesser Prairie-Chicken consider disease agents as a credible threat to population viability (Mote et al. 1999).

Noninfectious Diseases

Little or no data are available regarding environmental contaminants and mycotoxins within Lesser Prairie-Chicken habitats, and baseline surveys are still needed. Wildlife conservationists are aware of how critical insects are in the diet of newly hatched and young Lesser Prairie-Chickens (Davis et al. 1980). Research is needed, however, to determine how nutritional status influences reproductive condition of female Lesser Prairie-Chickens. A subset of females may not breed during drought years in semiarid portions of this species' range (Hagen et al. 2009, Grisham et al. 2014). Female nutritional status could determine the reproductive output of Lesser Prairie-Chicken, and studies designed to determine exactly how drought depresses reproductive effort and success could be critical to the conservation of this species.

Traumatic injuries are perhaps the best-understood cause of noninfectious disease for Lesser Prairie-Chickens. Predation by raptors and mammals is a major source of mortality for Lesser Prairie-Chickens (Hagen et al. 2007, Wolfe et al. 2007). Lesser Prairie-Chickens have coevolved with predator–prey interactions, but appear to be poorly adapted to anthropogenic changes in their habitats. Lesser Prairie-Chickens avoid anthropogenic structures such as buildings, natural gas compressor stations, major highways, wind turbines, and major electrical transmission lines (Hagen et al. 2004; Robel et al. 2004; Pitman et al. 2005; Pruett et al. 2009a,b). Avoidance behavior may reduce injuries, but it reduces the area of potential habitat available to the species, which is now a critically important commodity. Despite the avoidance of large structures, collisions with fence and electrical transmission wires are among the most common cause of traumatic injuries for Lesser Prairie-Chickens (Hagen et al. 2007, Wolfe et al. 2007). In modified portions of this species' range, collision mortality rates of Lesser Prairie-Chickens are comparable to losses to predation by raptors or mammals (Wolfe et al. 2007). Anthropogenic features will likely increase with continuing energy development within the remaining habitats of Lesser Prairie-Chicken. Managers tasked with the conservation of Lesser Prairie-Chickens must develop new ways to mitigate for the population-level influence of these anthropogenic landscape features.

Ecological Interactions

In my experience, many wildlife conservationists tend to look for incontrovertible evidence where wildlife diseases are concerned. If they do not find empirical evidence that demonstrates that a specific infectious or noninfectious disease directly caused a marked decline in abundance of Lesser Prairie-Chicken, they assume in and of itself that disease is unimportant. This reasoning could be faulty and may hinder Lesser Prairie-Chicken conservation.

Infectious agents and factors causing noninfectious diseases are components of ecosystems and should be studied and understood within this context. It is possible that a helminth such as O. petrowi, or a microparasite like Eimeria spp. could regulate Lesser Prairie-Chicken abundance in a density-dependent manner, or that a microparasite such as P. multocida could extirpate a small, isolated population of Lesser Prairie-Chickens during an epidemic involving other bird species. However, the greatest influence of infectious and noninfectious diseases is likely to be as critical components of more complex ecological interactions. I provide a few brief examples to illustrate my contention.

Some interactions among infectious agents, entities causing noninfectious diseases, and other aspects of the ecosystem are rather obvious. For example, Lesser Prairie-Chickens that initially survive a collision with a wire or an injury during capture almost certainly are more susceptible to predation than are uninjured individuals. Moreover, open wounds resulting from such trauma are vulnerable to bacterial infection, which could kill the bird or render it even more susceptible to predation. Similarly, chicks weakened because of lack of insects in their diet and to dehydration may be more susceptible to predation or infection by macro- and microparasites. High intensities of parasitic helminth infections or ectoparasites also could increase the risk of predation and microparasitic infection. All of these interactions could decrease the number of Lesser Prairie-Chickens recruited into the subsequent breeding population as compared to what would have occurred in their absence.

A number of less obvious, yet potentially more important, ecological interactions involve infectious diseases, noninfectious diseases, and Lesser Prairie-Chickens. First, during a severe drought, malnutrition could prevent many females from becoming reproductively active during the normal breeding season. Drought also could lead to increased predation rates for the few nests that were initiated and to decreased insect availability and increased risk of dehydration for any chicks that hatched. Severe weather events, such as blizzard or ice storms, could further reduce the numbers of juvenile and adult birds. These factors together could essentially eliminate recruitment into the subsequent breeding population (Chapter 12, this volume). Second, environmental contaminants could also be problematic (Fry 1995, Tyler et al. 1998). A large array of environmental estrogens are associated with certain herbicides, insecticides, plasticizers, resins, and pesticide manufacturing processes could be detrimental to Lesser Prairie-Chicken reproduction at concentrations far lower than the exposure required to kill individual birds. Third, although most grouse biologists are aware that West Nile virus (WNV) can influence population dynamics of Greater Sage-Grouse (Walker and Naugle 2011), but there are important interactions among human activities, the virus, and low- and mid-elevation populations of sage grouse (Taylor et al. 2013). Taylor et al. (2013) demonstrated that without a disease outbreak, high levels of oil and gas drilling in previously undeveloped landscapes led to a 61% reduction in the number of males attending leks. West Nile virus outbreaks alone led to a 55% reduction in the number of males on leks where no petroleum development occurred. However, lek inactivity among historical leks with no males quadrupled when petroleum development and WNV outbreaks co-occurred, compared to no energy development and no WNV outbreak. Additional research is required to determine the contribution of infectious and noninfectious diseases, such as these, to ecological interactions that control Lesser Prairie-Chicken population dynamics. Such efforts are needed, if for no other reason, because of the large-scale agricultural development and petroleum extraction near and within Lesser Prairie-Chicken habitats.

Climate Change

No exploration of ecological interactions influencing species at risk is complete without considering the possible effects of climate change (Chapter 12, this volume). Grisham et al. (2013) used existing models of anticipated changes in climate to predict changes in three key demographic variables for Lesser Prairie-Chicken populations. Model simulations suggested that anticipated changes in climate could result in nest survival rates too low to maintain populations by 2050. The authors did not address how anticipated changes in climate may influence host–parasite or other important population interactions. I briefly explore two such sets of interactions as illustrations.

Most species of prairie grouse harbor a greater diversity of parasitic helminths than Lesser Prairie-Chickens, including thorny-headed worms, trematodes, and other taxa (Peterson 2004). Peterson (1996) analyzed data from all available studies of the parasitic helminths of prairie grouse (*Tympanuchus* spp.) and found that long-term meteorological data collected near the study areas could account for the parasite communities documented by each study. A Lesser Prairie-Chicken population included in this analysis was at the dry extreme of the meteorological continuum. The Attwater's Prairie-Chicken, which then inhabited the coastal prairies of Texas, represented a region characterized by greater precipitation and warmer winter temperatures than all other prairie grouse populations evaluated in the study, but had never been sampled for parasitic helminths. Based on local meteorological data and his analysis of prairie grouse parasite communities, Peterson (1996) predicted that the unsampled Attwater's Prairie-Chicken populations would support parasitic helminth communities that included two parasite species associated with disease in North American grouse (i.e., *Dispharynx nasuta*, *H. gallinarum*), as well as helminth species not found by studies of other prairie grouse. Peterson et al. (1998), using opportunistically collected samples, found *D. nasuta* in one of three samples that could be evaluated for this parasite, and *H. gallinarum* and *Trichostrongylus cramae* (= *T. tenuis*)—a species not previously documented for prairie grouse—in eight of nine suitable samples.

If climate change progresses within the range of the Lesser Prairie-Chicken as predicted (Grisham et al. 2013), then on the basis of the analysis by Peterson (1996), one would expect Lesser Prairie-Chickens to become exposed to novel parasitic helminths in the future. For example, *D. nasuta* and *H. gallinarum* are associated with disease in the host species of prairie grouse with which they coevolved. *T. cramae* is quite similar (if not identical) to *T. tenuis*, which is a parasite shown through controlled manipulative experimentation to be responsible for periodic cycles in the abundance of Red Grouse (Hudson et al. 1998). The three helminths can cause population-level changes in the abundance of host species of grouse with which they coevolved, and they could prove to be important for populations of Lesser Prairie-Chickens as well. It is possible that Lesser Prairie-Chickens already are being exposed to such parasites as their range now overlaps with that of the Greater Prairie-Chicken in western Kansas, and individuals of both species now mingle on leks (Bain and Farley 2002).

One also should expect climate change to alter an array of host–parasite interactions beyond those involving parasitic helminths. For example, climate change could allow arthropods to become established within Lesser Prairie-Chicken range where they did not previously occur, or greatly increase the abundance of certain arthropods that already occur in the region. Because arboviruses such as WNV require arthropod vectors, climate change could unleash an array of arboviruses within the avian community in the Southern Great Plains in sites where they now occur rarely or not at all. It is beyond the scope of this review to explore all such eventualities, but one could evaluate other prairie grouse populations that inhabit regions with climates similar to that predicted for Lesser Prairie-Chicken range to obtain a reasonable estimate of what may be expected given predicated changes in climate.

SYNTHESIS AND IMPLICATIONS

Conservation biologists have become increasingly aware of the potential importance of wildlife health and disease, particularly where human health and species at risk of extinction are concerned (Daszak et al. 2000, Peterson and Ferro 2012). Issues of health and disease are directly relevant for managers concerned with conservation of Lesser Prairie-Chickens because of long-term declines in the abundance of this species, as well as the degree of agricultural development, and current and planned energy development within the species range. Unfortunately, unlike the situation for other species of prairie grouse (Peterson 2004), Greater Sage-Grouse (Christiansen and Tate 2011), and Northern Bobwhite (Peterson 2007), no comprehensive surveys of macro- and microparasites were completed during 1930s and 1940s for the Lesser Prairie-Chicken. For this reason, there are no baseline data prior to the late 1970s when major habitat changes already had occurred throughout much of the historical range of this species (Crawford 1980, Taylor and Guthery 1980, Hagen and Giesen 2004).

A wide array of macroparasites, microparasites, and noninfectious diseases have been documented for Lesser Prairie-Chickens, particularly since 2000. My review has identified specific parasitic helminths and microparasites that potentially could regulate host populations. If certain microparasites were to spillover to populations of Lesser Prairie-Chicken during an epidemic occurring in other host species, they could cause density-independent disease outbreaks that may extirpate small, isolated populations of Lesser Prairie-Chickens. Regarding noninfectious diseases of Lesser Prairie-Chickens, it is clear that poor insect availability for chicks and, perhaps collisions with fences and electrical transmission wires can have population-level consequences. Mycotoxins, environmental contaminants, severe weather, and the effect of female nutritional status on reproductive success also could influence Lesser Prairie-Chicken population dynamics, and initial research addressing these topics is needed.

Managers concerned with the conservation of prairie grouse must move beyond the assumption that the only diseases that matter are those where empirical evidence clearly demonstrates that a specific infectious or noninfectious disease directly caused a marked decline in abundance of Lesser Prairie-Chickens. The greatest influence of infectious and noninfectious diseases is likely to be as key components of more complex ecological interactions. Interactions may be as simple as increased predation risk for a bird with an open wound caused by an injury that became infected with opportunistic bacteria. More complex interactions are exemplified by the synergistic influence of oil and gas drilling and WNV outbreaks documented for Greater Sage-Grouse (Walker and Naugle 2011). Last, predicted changes in climate are likely to increase the prevalence of certain macro- and microparasites already known for Lesser Prairie-Chickens and allow introduction of infectious agents not yet documented for this species.

Given the paucity of data regarding most infectious agents and noninfectious diseases of Lesser Prairie-Chicken, survey data still are badly needed, particularly for populations in Colorado, Oklahoma, and New Mexico. Surveys addressing mycotoxins and environmental contaminants have yet to be completed and published. I encourage researchers to continue cataloging the prevalence of infectious agents and the concentration of toxins. More critically, studies that clarify the influence of specific macroparasites, microparasites, and noninfectious diseases on Lesser Prairie-Chicken population dynamics are badly needed throughout the species range. The most useful studies for managers tasked with the conservation of Lesser Prairie-Chickens would be new investigations that help to untangle the complex ecological interactions where infectious agents or noninfectious diseases are simply one of many players on the ecological stage. Integrative research approaches using expertise from multiple academic disciplines will be required to reach this objective.

ACKNOWLEDGMENTS

I thank C. A. Hagen, M. A. Mora, S. M. Presley, and D. H. Wolfe for allowing me access to unpublished data. The manuscript benefited from critical evaluation by R. D. Applegate, P. J. Ferro, C. W. Boal, and D. A. Haukos. The University of Texas at El Paso and the Texas A&M University System supported this effort.

LITERATURE CITED

Addison, E. M., and R. C. Anderson. 1969. *Oxyspirura lumsdeni* n. sp. (Nematoda: Thelaziidae) from Tetraonidae in North America. Canadian Journal of Zoology 47:1223–1227.

Aguirre, A. A., R. S. Ostfeld, G. M. Tabor, C. House, and M. C. Pearl (editors). 2002. Conservation medicine: ecological health in practice. Oxford University Press, New York, NY.

Anderson, R. M., and R. M. May 1978. Regulation and stability of host–parasite population interactions: I. regulatory processes. Journal of Animal Ecology 47:219–247.

Anderson, R. M., and R. M. May 1979. Population biology of infectious diseases: Part I. Nature 280:361–367.

Applegate, R. D., and T. Z. Riley. 1998. Lesser Prairie-Chicken management. Rangelands 20:13–15.

Atkinson, C. T. 1999. Hemosporidiosis. Pp. 193–199 in M. Friend and J. C. Franson (editors), Field manual of wildlife diseases: General field procedures and diseases of birds. U.S. Geological Survey Information and Technology Report 1999–001. USGS Biological Resources Division, Reston, VA.

Atkinson, C. T. 2007a. Avian malaria. Pp. 35–53 in C. T. Atkinson, N. J. Thomas, and D. B. Hunter (editors), Parasitic diseases of wild birds. Wiley-Blackwell, Ames, IA.

Atkinson, C. T. 2007b. Haemoproteus. Pp. 13–34 in C. T. Atkinson, N. J. Thomas, and D. B. Hunter (editors), Parasitic diseases of wild birds. Wiley-Blackwell, Ames, IA.

Atkinson, C. T., N. J. Thomas, and D. B. Hunter (editors). 2007. Parasitic diseases of wild birds. Wiley-Blackwell, Ames, IA.

Bain, M. R., and G. H. Farley. 2002. Display by apparent hybrid prairie chickens in a zone of geographic overlap. Condor 104:683–687.

Bass, C. C. 1939. Observations on the specific cause and the nature of "quail disease" or ulcerative enteritis in quail. Proceedings of the Society for Experimental Biology and Medicine 42:377–380.

Bass, C. C. 1940. Specific cause and nature of ulcerative enteritis of quail. Proceedings of the Society for Experimental Biology and Medicine 46:250–252.

Beasom, S. L. 1973. Ecological factors affecting Wild Turkey reproductive success in south Texas. Ph.D. dissertation, Texas A&M University, College Station, TX.

Begon, M., C. R. Townsend, and J. L. Harper. 2006. Parasitism and disease. Pp. 347–380 in Ecology: from individuals to ecosystems. Blackwell Publishers, Boston, MA.

Bermudez, A. J. 2008. Miscellaneous and sporadic protozoal infections. Pp. 1105–1017 in Y. M. Saif, A. M. Fadly, J. R. Glisson, L. R. McDougald, L. K. Nolan, and D. E. Swayne (editors), Diseases of poultry. Wiley-Blackwell, Hoboken, NJ.

Boyce, M. S. 1990. The red queen visits Sage Grouse leks. American Zoologist 30:263–270.

Brook, B. W., N. S. Sodhi, and C. J. A. Bradshaw. 2008. Synergies among extinction drivers under global change. Trends in Ecology and Evolution 23:453–460.

Bump, G., R. W. Darrow, F. C. Edminster, and W. F. Crissey. 1947. The Ruffed Grouse: life history, propagation, management. Holling Press, Buffalo, NY.

Cavanagh, D., and J. Gelb Jr. 2008. Infectious bronchitis. Pp. 117–135 in Y. M. Saif, A. M. Fadly, J. R. Glisson, L. R. McDougald, L. K. Nolan, and D. E. Swayne (editors), Diseases of poultry. Wiley-Blackwell, Hoboken, NJ.

Christiansen, T. J., and C. M. Tate. 2011. Parasites and infectious diseases of Greater Sage-Grouse. Pp. 113–126 in S. T. Knick and J. W. Connelly (editors), Greater Sage-Grouse: ecology and conservation of a landscape species and its habitats. Studies in Avian Biology (no. 38), University of California Press, Berkeley, CA.

Collinge, S. K., and C. Ray (editors). 2006. Disease ecology: community structure and pathogen dynamics. Oxford University Press, Oxford, U.K.

Committee of Inquiry on Grouse Disease. 1911. The grouse in health and disease. Smith, Elder & Company, London, U.K.

Cooper, J. E. 1996. Physical injury. Pp. 157–172 in A. Fairbrother, L. N. Locke, and G. L. Hoff (editors), Noninfectious diseases of wildlife. Iowa State University Press, Ames, IA.

Copelin, F. F. 1963. The Lesser Prairie Chicken in Oklahoma. Oklahoma Wildlife Conservation Department, Technical Bulletin No. 6. Oklahoma Wildlife Conservation Department, Oklahoma City, OK.

Cowan, A. B., and T. J. Peterle. 1957. *Leucocytozoon bonasae* Clarke in Michigan Sharp-tailed Grouse. Journal of Wildlife Management 21:469–471.

Cram, E. B. 1931. Developmental stages of some nematodes of the Spiruroidea parasitic in poultry and game birds. U.S. Department of Agriculture, Technical Bulletin No. 227. U.S. Department of Agriculture, Washington, DC.

Cram, E. B. 1933. Observations on the life history of *Tetrameres pattersoni*. Journal of Parasitology 20:97–98.

Crawford, J. A. 1980. Status, problems, and research needs of the Lesser Prairie Chicken. Pp. 1–7 in P. A. Vohs Jr. and F. L. Knopf (editors), Proceedings of the Prairie grouse symposium. Oklahoma State University, Stillwater, OK.

Daszak, P., A. A. Cunningham, and A. D. Hyatt. 2000. Emerging infectious diseases of wildlife—threats to biodiversity and human health. Science 287:443–449.

Davidson, W. R. (editor). 2006. Field manual of wildlife diseases in the Southeastern states. Southeastern Cooperative Wildlife Disease Study, Athens, GA.

Davidson, W. R. 2007. Histomonas. Pp. 154–161 in C. T. Atkinson, N. J. Thomas, and D. B. Hunter (editors), Parasitic diseases of wild birds. Wiley-Blackwell, Ames, IA.

Davidson, W. R., G. L. Doster, and M. B. McGhee. 1978. Failure of *Heterakis bonasae* to transmit *Histomonas meleagridis*. Avian Diseases 22:627–632.

Davidson, W. R., F. E. Kellogg, and G. L. Doster. 1982. An overview of disease and parasitism in southeastern Bobwhite Quail. Proceedings of the National Quail Symposium 2:57–63.

Davidson, W. R., F. E. Kellogg, G. L. Doster, and C. T. Moore. 1991. Ecology of helminth parasitism in Bobwhites from northern Florida. Journal of Wildlife Diseases 27:185–205.

Davis, B. D., and K. D. Gruen. 1994. Breeding chronology in Rio Grande Turkey hens. Federal Aid in Wildlife Restoration Project Number W-126-R, Job Number 7.07. Final Report, Texas Parks and Wildlife Department, Austin, TX.

Davis, C. A., T. Z. Riley, R. A. Smith, H. R. Suminski, and M. J. Wisdom. 1979. Habitat evaluation of Lesser Prairie-Chickens in Eastern Chaves County, New Mexico. New Mexico Agricultural Experiment Station, Las Cruces, NM.

Davis, C. A., T. Z. Riley, R. A. Smith, and M. J. Wisdom. 1980. Spring–summer foods of Lesser Prairie Chickens in New Mexico. Pp. 75–80 in P. A. Vohs Jr. and F. L. Knopf (editors), Proceeding of the Prairie Grouse symposium. Oklahoma State University, Stillwater, OK.

Dick, T. A. 1981. Ectoparasites of Sharp-tailed Grouse, *Pedioecetes phasianellus*. Journal of Wildlife Diseases 17:229–235.

Dobson, A. P. 1988. The population biology of parasite-induced changes in host behavior. Quarterly Review of Biology 63:139–165.

Durant, A. J., and E. R. Doll. 1941. Ulcerative enteritis in quail. Agricultural Experiment Station, University of Missouri, Research Bulletin 325. University of Missouri, Columbia, MO.

Emerson, K. C. 1951. A list of Mallophaga from gallinaceous birds of North America. Journal of Wildlife Management 15:193–195.

Fairbrother, A., L. N. Locke, and G. L. Hoff (editors). 1996. Noninfectious diseases of wildlife. Iowa State University Press, Ames, IA.

Fedynich, A. M. 2007. *Heterakis* and *Ascaridia*. Pp. 388–412 in C. T. Atkinson, N. J. Thomas, and D. B. Hunter (editors), Parasitic diseases of wild birds. Wiley-Blackwell, Ames, IA.

Flakas, K. G. 1952. Wildlife pathology study. Wisconsin Wildlife Research 10:69–80.

Fleharty, E. D. 1995. Wild animals and settlers on the great plains. University of Oklahoma Press, Norman, OK.

Fleishman, E., D. E. Blockstein, J. A. Hall, M. B. Mascia, M. A. Rudd, J. M. Scott, W. J. Sutherland et al. 2011. Top 40 priorities for science to inform US conservation and management policy. Bioscience 61:290–300.

Flickinger, E. L., and D. M. Swineford. 1983. Environmental contaminant hazards to Attwater's Greater Prairie-Chickens. Journal of Wildlife Management 47:1132–1137.

Forrester, D. J., and E. C. Greiner. 2007. Leucocytozoonosis. Pp. 54–107 in C. T. Atkinson, N. J. Thomas, and D. B. Hunter (editors), Parasitic diseases of wild birds. Wiley-Blackwell, Ames, IA.

Friend, M., and J. C. Franson. 1999a. Field manual of wildlife diseases: general field procedures and diseases of birds. USGS Biological Resources Division, Reston, VA.

Friend, M., and J. C. Franson. 1999b. Intestinal coccidiosis. Pp. 99–109 in M. Friend and J. C. Franson (editors), Field manual of wildlife diseases: general field procedures and diseases of birds. U.S. Geological Survey Information and Technology Report 1999–001. USGS Biological Resources Division, Reston, VA.

Friend, M., R. G. McLean, and F. J. Dein. 2001. Disease emergence in birds: challenges for the twenty-first century. Auk 118:290–303.

Fritzler, J. M., T. M. Craig, A. Elgayar, C. Plummer, R. S. Wilson, M. J. Peterson, and G. Zhu. 2011. A new eimeriid (Apicomplexa) species from endangered Attwater's Prairie-Chickens (*Tympanuchus cupido attwateri*) in Texas. Journal of Parasitology 97:671–675.

Fry, D. M. 1995. Reproductive effects in birds exposed to pesticides and industrial chemicals. Environmental Health Perspectives 103:165–171.

Gabrielson, I. N. 1951. Wildlife management. Macmillan, New York, NY.

Grisham, B. A., C. W. Boal, D. A. Haukos, D. M. Davis, K. K. Boydston, C. Dixon, and W. R. Heck. 2013. The predicted influence of climate change on Lesser Prairie-Chicken reproductive parameters. PLoS One 8:e68225.

Grisham, B. A., P. K. Borsdorf, C. W. Boal, and K. K. Boydston. 2014. Nesting ecology and nest survival of Lesser Prairie-Chickens on the Southern High Plains. Journal of Wildlife Management 78:857–866.

Gross, A. O. 1925. Diseases of the ruffed grouse. Auk 62:423–431.

Gross, A. O. 1928. The Heath Hen. Memoirs of the Boston Society of Natural History 6:490–588.

Gross, A. O. 1930. Progress report of the Wisconsin Prairie Chicken investigation. Wisconsin Conservation Commission, Madison, WI.

Guthery, F. S. 1997. A philosophy of habitat management for Northern Bobwhites. Journal of Wildlife Management 61:291–301.

Guthery, F. S. 2002. The technology of bobwhite management: the theory behind the practice. Iowa State University Press, Ames, IA.

Guthery, F. S., N. E. Koerth, and D. S. Smith. 1988. Reproduction of Northern Bobwhites in semiarid environments. Journal of Wildlife Management 52:144–149.

Hagen, C. A., S. S. Crupper, R. D. Applegate, and R. J. Robel. 2002. Prevalence of *Mycoplasma* antibodies in Lesser Prairie-Chicken sera. Avian Diseases 46:708–712.

Hagen, C. A., and K. M. Giesen. [online]. 2004. Lesser Prairie-Chicken (*Tympanuchus pallidicinctus*). Birds of North America 364. <http://bna.birds.cornell.edu/bna/species/364/articles/introduction> (15 November 2013).

Hagen, C. A., B. E. Jamison, K. M. Giesen, and T. Z. Riley. 2004. Guidelines for managing Lesser Prairie-Chicken populations and their habitats. Wildlife Society Bulletin 32:69–82.

Hagen, C. A., J. C. Pitman, B. K. Sandercock, R. J. Robel, and R. D. Applegate. 2007. Age-specific survival and probable causes of mortality in female Lesser Prairie-Chickens. Journal of Wildlife Management 71:518–525.

Hagen, C. A., B. K. Sandercock, J. C. Pitman, R. J. Robel, and R. D. Applegate. 2009. Spatial variation in Lesser Prairie-Chicken demography: a sensitivity analysis of population dynamics and management alternatives. Journal of Wildlife Management 73:1325–1332.

Hannon, S. J., and K. Martin. 2006. Ecology of juvenile grouse during the transition to adulthood. Journal of Zoology 269:422–433.

Harper, G. R., R. D. Klataske, R. J. Robel, and M. F. Hansen. 1967. Helminths of Greater Prairie Chickens in Kansas. Journal of Wildlife Management 31:265–269.

Haukos, D. A., L. M. Smith, and G. S. Broda. 1990. Spring trapping of Lesser Prairie-Chickens. Journal of Field Ornithology 61:20–25.

Herman, C. M. 1963. Disease and infection in the Tetraonidae. Journal of Wildlife Management 27:850–855.

Herman, C. M. 1969. The impact of disease on wildlife populations. Bioscience 19:321–325, 330.

Hillman, C. N., and W. W. Jackson. 1973. The Sharptailed Grouse in South Dakota. South Dakota Department of Game, Fish and Parks, Technical Bulletin No. 3. South Dakota Department of Game, Fish and Parks, Pierre, SD.

Hoerr, F. J. 2008. Mycotoxicoses. Pp. 1197–1229 in Y. M. Saif, A. M. Fadly, J. R. Glisson, L. R. McDougald, L. K. Nolan, and D. E. Swayne (editors), Diseases of poultry. Wiley-Blackwell, Hoboken, NJ.

Hudson, P. J. 1986. Red grouse, the biology and management of a wild gamebird. The Game Conservancy Trust, Fordringbridge, U.K.

Hudson, P. J., A. P. Dobson, and D. Newborn. 1998. Prevention of population cycles by parasite removal. Science 282:2256–2258.

Hudson, P. J., A. Rizzoli, B. T. Grenfell, H. Heesterbeek, and A. P. Dobson (editors). 2002. The ecology of wildlife diseases. Oxford University Press, Oxford, U.K.

Jackson, A. S. 1969. A handbook for Bobwhite Quail management in the west Texas Rolling Plains. Texas Parks and Wildlife Department, Bulletin No. 48. Texas Parks and Wildlife Department, Austin, TX.

Johnson, L. L., and M. S. Boyce. 1991. Female choice of males with low parasite loads in Sage Grouse. Pp. 377–388 in J. E. Loye and M. Zuk (editors), Bird–parasite interactions: ecology, evolution, and behaviour. Oxford University Press, Oxford, U.K.

Jones, K. E., N. G. Patel, M. A. Levy, A. Storeygard, D. Balk, J. L. Gittleman, and P. Daszak. 2008. Global trends in emerging infectious diseases. Nature 451:990–993.

Kinsella, J. M., and D. J. Forrester. 2007. Tetrameridosis. Pp. 376–383 in C. T. Atkinson, N. J. Thomas, and D. B. Hunter (editors), Parasitic diseases of wild birds. Wiley-Blackwell, Ames, IA.

Lack, D. 1954. The natural regulation of animal numbers. Oxford University Press, London, U.K.

Lee, L. 1950. Kill analysis for the Lesser Prairie Chicken in New Mexico, 1949. Journal of Wildlife Management 14:475–477.

Leigh, W. H. 1940. Preliminary studies on parasites of upland game birds and fur-bearing mammals in Illinois. Illinois Natural History Survey Bulletin 21:185–194.

Leigh, W. H. 1941. Variation in a new species of cestode, Raillietina (Skrjabinia) variabila, from the prairie chicken in Illinois. Journal of Parasitology 27:97–106.

Leopold, A. 1933. Game management. Charles Scribner's Sons, New York, NY.

Lyons, E. K., R. S. Jones, J. P. Leonard, B. E. Toole, R. A. McCleery, R. R. Lopez, M. J. Peterson, S. J. DeMaso, and N. J. Silvy. 2011. Regional variation in nest success of Lesser Prairie-Chickens in Texas. Pp. 223–231 in B. K. Sandercock, K. Martin, and G. Segelbacher (editors), Ecology, conservation, and management of grouse. Studies in Avian Biology (no. 39), University of California Press, Berkeley, CA.

Mann, F. A., and K. E. Helmick. 1996. Shock. Pp. 173–180 in A. Fairbrother, L. N. Locke, and G. L. Hoff (editors), Noninfectious diseases of wildlife. Iowa State University Press, Ames, IA.

Massey, M. 2001. Long-range plan for the management of Lesser Prairie-Chicken in New Mexico 2002–2006. Division of Wildlife, New Mexico Department of Game and Fish, Federal Aid in Wildlife Restoration Grant W-104-R41, Project 3.4. New Mexico Department of Game and Fish, Santa Fe, NM.

May, R. M., and R. M. Anderson. 1978. Regulation and stability of host–parasite population interactions: II. destabilizing processes. Journal of Animal Ecology 47:249–267.

May, R. M., and R. M. Anderson. 1979. Population biology of infectious diseases: Part II. Nature 280:455–461.

McDougald, L. R. 2008a. Cestodes and trematodes. Pp. 1056–1066 in Y. M. Saif, A. M. Fadly, J. R. Glisson, L. R. McDougald, L. K. Nolan, and D. E. Swayne (editors), Diseases of poultry. Wiley-Blackwell, Hoboken, NJ.

McDougald, L. R. 2008b. Histomoniasis (blackhead) and other protozoan diseases of the intestinal tract. Pp. 1095–1105 in Y. M. Saif, A. M. Fadly, J. R. Glisson, L. R. McDougald, L. K. Nolan, and D. E. Swayne (editors), Diseases of poultry. Wiley-Blackwell, Hoboken, NJ.

McDougald, L. R., and S. H. Fitz-Coy. 2008. Coccidiosis. Pp. 1068–1085 in Y. M. Saif, A. M. Fadly, J. R. Glisson, L. R. McDougald, L. K. Nolan, and D. E. Swayne (editors), Diseases of poultry. Wiley-Blackwell, Hoboken, NJ.

Merchant, S. S. 1982. Habitat-use, reproductive success, and survival of female Lesser Prairie Chickens in two years of contrasting weather. M.S. thesis, New Mexico State University, Las Cruces, NM.

Mote, K. D., R. D. Applegate, J. A. Bailey, K. E. Giesen, R. Horton, and J. L. Sheppard. 1999. Assessment and conservation strategy for the Lesser Prairie-Chicken (Tympanuchus pallidicinctus). Kansas Department of Wildlife and Parks, Emporia, KS.

O'Hara, T. M. 1996. Mycotoxins. Pp. 21–30 in A. Fairbrother, L. N. Locke, and G. L. Hoff (editors), Noninfectious diseases of wildlife. Iowa State University Press, Ames, IA.

Ostfeld, R. S., F. Keesing, and V. Eviner (editors). 2008. Infectious disease ecology: effects of ecosystems on disease and of disease on ecosystems. Princeton University Press, Princeton, NJ.

Park, K. J., P. A. Robertson, S. T. Campbell, R. Foster, Z. M. Russell, D. Newborn, and P. J. Hudson. 2001. The role of invertebrates in the diet, growth and survival of Red Grouse (Lagopus lagopus scoticus) chicks. Journal of Zoology 254:137–145.

Pattee, O. H., and S. L. Beasom. 1979. Supplemental feeding to increase Wild Turkey productivity. Journal of Wildlife Management 43:512–516.

Pence, D. B. 1972. Genus Oxyspirura (Nematoda: Thelaziidae) from birds in Louisiana. Proceedings of the Helminthological Society of Washington 39:23–28.

Pence, D. B., J. T. Murphy, F. S. Guthery, and T. B. Doerr. 1983. Indications of seasonal variation in the helminth fauna of the Lesser Prairie-Chicken, Tympanuchus pallidicinctus (Ridgway) (Tetraonidae) from northwestern Texas. Proceedings of the Helminthological Society of Washington 50:345–347.

Pence, D. B., and D. L. Sell. 1979. Helminths of the Lesser Prairie Chicken, Tympanuchus pallidicintus (Ridgway) (Tetraonidae), from the Texas panhandle. Proceedings of the Helminthological Society of Washington 46:146–149.

Peterson, M. J. 1991a. The Wildlife Society publications are appropriate outlets for the results of host–parasite interaction studies. Wildlife Society Bulletin 19:360–369.

Peterson, M. J. 1991b. Wildlife parasitism, science, and management policy. Journal of Wildlife Management 55:782–789.

Peterson, M. J. 1996. The endangered Attwater's Prairie-Chicken and an analysis of prairie grouse helminthic endoparasitism. Ecography 19:424–431.

Peterson, M. J. 2003. Infectious agents of concern for the Jackson Hole Elk and Bison herds: An ecological perspective. Prepared for the National Elk Refuge and Grand Teton National Park, U.S. Department of the Interior. Department of Wildlife and Fisheries Sciences, Texas A&M University, College Station, TX.

Peterson, M. J. 2004. Parasites and infectious diseases of prairie grouse: should managers be concerned? Wildlife Society Bulletin 32:35–55.

Peterson, M. J. 2007. Diseases and parasites of Texas quails. Pp. 89–114 in L. A. Brennan (editor), Texas quails: ecology and management. Texas A&M University Press, College Station, TX.

Peterson, M. J., and P. J. Ferro. 2012. Wildlife health and disease: Surveillance, investigation, and management. Pp. 181–206 in N. J. Silvy (editor), The wildlife techniques manual: research. Johns Hopkins University Press, Baltimore, MD.

Peterson, M. J., P. J. Ferro, M. N. Peterson, R. M. Sullivan, B. E. Toole, and N. J. Silvy. 2002. Infectious disease survey of Lesser Prairie-Chickens in north Texas. Journal of Wildlife Diseases 38:834–839.

Peterson, M. J., J. R. Purvis, J. R. Lichtenfels, T. M. Craig, N. O. Dronen, and N. J. Silvy. 1998. Serologic and parasitologic survey of the endangered Attwater's Prairie-Chicken. Journal of Wildlife Diseases 34:137–144.

Peterson, M. N., A. G. Mertig, and J. Liu. 2006. Effects of zoonotic disease attributes on public attitudes towards wildlife management. Journal of Wildlife Management 70:1746–1753.

Pitman, J. C., C. A. Hagen, R. J. Robel, T. M. Loughin, and R. D. Applegate. 2005. Location and success of Lesser Prairie-Chicken nests in relation to vegetation and human disturbance. Journal of Wildlife Management 69:1259–1269.

Pruett, C. L., M. A. Patten, and D. H. Wolfe. 2009a. Avoidance behavior by prairie grouse: implications for development of wind energy. Conservation Biology 23:1253–1259.

Pruett, C. L., M. A. Patten, and D. H. Wolfe. 2009b. It's not easy being green: wind energy and a declining grassland bird. Bioscience 59:257–262.

Quist, C. F., D. I. Bounous, J. V. Kilburn, V. F. Nettles, and R. D. Wyatt. 2000. The effect of dietary aflatoxin on Wild Turkey poults. Journal of Wildlife Diseases 36:436–444.

Reyna, K. S., and W. W. Burggren. 2012. Upper lethal temperatures of Northern Bobwhite embryos and the thermal properties of their eggs. Poultry Science 91:41–46.

Robel, R. J., J. A. Harrington Jr., C. A. Hagen, J. C. Pittman, and R. R. Recker. 2004. Effect of energy development and human activity on the use of sand sagebrush habitat by Lesser Prairie-Chickens in southwestern Kansas. Transactions of the North American Wildlife and Natural Resources Conference 69:251–266.

Robel, R. J., T. L. Walker Jr., C. A. Hagen, R. K. Ridley, K. E. Kemp, and R. D. Applegate. 2003. Helminth parasites of Lesser Prairie-Chicken *Tympanuchus pallidicinctus* in southwestern Kansas: incidence, burdens and effects. Wildlife Biology 9:341–349.

Rubin, C., T. Myers, W. Stokes, B. Dunham, S. Harris, B. Lautner, and J. Annelli. 2013. Review of Institute of Medicine and National Research Council recommendations for One Health Initiative. Emerging Infectious Diseases 19:1913–1917.

Saif, Y. M., A. M. Fadly, J. R. Glisson, L. R. McDougald, L. K. Nolan, and D. E. Swayne. 2008. Diseases of poultry. Wiley-Blackwell, Hoboken, NJ.

Samuel, M. D., R. G. Botzler, and G. A. Wobeser. 2007. Avian cholera. Pp. 239–269 in N. J. Thomas, D. B. Hunter, and C. T. Atkinson (editors), Infectious diseases of wild birds. Blackwell Publishing, Ames, IA.

Saunders, G. B. 1935. Michigan's studies of Sharp-tailed Grouse. Transactions of the American Game Conference 21:342–344.

Schroeder, M. A., and C. E. Braun. 1991. Walk-in traps for capturing Greater Prairie-Chickens on leks. Journal of Field Ornithology 62:378–385.

Schwartz, C. W. 1945. The ecology of the Prairie Chicken in Missouri. University of Missouri Studies 20:1–99.

Schwilling, M. D. 1955. A study of the Lesser Prairie Chicken in Kansas. Kansas Forestry, Fish and Game Commission, Job Completion Report. Kansas Forestry, Fish and Game Commission, Pratt, KS.

Sell, D. L. 1979. Spring and summer movements and habitat use by Lesser Prairie Chicken females in Yoakum County, Texas. M.S. thesis, Texas Tech University, Lubbock, TX.

Sheffield, S. R., J. P. Sullivan, and E. F. Hill. 2012. Identifying and handling contaminant-related wildlife mortality or morbidity. Pp. 154–180 in N. J. Silvy (editor), The wildlife techniques manual: research. Johns Hopkins University Press, Baltimore, MD.

Shillinger, J. E. 1942. Diseases of farm-raised game birds. Pp. 1226–1231 in G. Hambidge and M. J. Drown (editors), Keeping livestock healthy, yearbook of agriculture 1942. U.S. Department of Agriculture, Washington, DC.

Shillinger, J. E., and L. C. Morley. 1937. Diseases of upland game birds. U.S. Department of Agriculture, Farmers' Bulletin No. 1781. U.S. Department of Agriculture, Washington, DC.

Silvy, N. J., M. J. Peterson, and R. R. Lopez. 2004. The cause of the decline of pinnated grouse: the Texas example. Wildlife Society Bulletin 32:16–21.

Simon, F. 1940. The parasites of the Sage Grouse *Centrocercus urophasianus*. University of Wyoming Publications 7:77–100.

Sinclair, A. R. E., J. M. Fryxell, and G. Caughley. 2006. Parasites and pathogens. Pp. 179–195 in Wildlife ecology, conservation, and management. Blackwell Publishers, Oxford, U.K.

Smith, B. H., D. W. Duszynski, and K. Johnson. 2003. Survey for coccidia and haemosporidia in the Lesser Prairie-Chicken (*Tympanuchus pallidicinctus*) from New Mexico with description of a new *Eimeria* species. Journal of Wildlife Diseases 39:347–353.

Stabler, R. M. 1978. *Plasmodium* (Giovannolaia) *pedioecetii* from the Lesser Prairie Chicken, *Tympanuchus pallidicinctus*. Journal of Parasitology 64:1125–1126.

Stoddard, H. L. 1931. The bobwhite quail: its habits, preservation and increase. Charles Scribner's Sons, New York, NY.

Taylor, M. A. 1978. Fall and winter movements and habitat use of Lesser Prairie Chickens. M.S. thesis, Texas Tech University, Lubbock, TX.

Taylor, M. A., and F. S. Guthery. 1980. Status, ecology, and management of the Lesser Prairie Chicken. USDA Forest Service, Rocky Mountain Forest and Range Experiment Station, General Technical Report RM-77. USDA Forest Service, Rocky Mountain Forest and Range Experiment Station, Fort Collins, CO.

Taylor, R. L., J. D. Tack, D. E. Naugle, and L. S. Mills. 2013. Combined effects of energy development and disease on Greater Sage-Grouse. PloS One 8:e71256.

Taylor, W. P. 1956. The deer of North America. Stackpole, Harrisburg, PA.

Thomas, N. J., D. B. Hunter, and C. T. Atkinson (editors). 2007. Infectious diseases of wild birds. Blackwell Publishing, Ames, IA.

Tirhi, M. J. 1995. Washington state management plan for Sharp-tailed Grouse. Washington Department of Fish and Wildlife, Wildlife Management Progress Report. Washington Department of Fish and Wildlife, Olympia, WA.

Tompkins, D. M., A. P. Dobson, P. Arneberg, M. E. Begon, I. M. Cattadori, J. V. Greenman, J. A. P. Heesterbeek et al. 2002. Parasites and host population dynamics. Pp. 45–62 in P. J. Hudson, A. Rizzoli, B. T. Grenfell, H. Heesterbeek, and A. P. Dobson (editors), The ecology of wildlife diseases. Oxford University Press, Oxford, U.K.

Trippensee, R. E. 1948. Wildlife management: upland game and general principles. McGraw-Hill Cook Company, New York, NY.

Tsuji, L. J. S., J. D. Karagatzides, and G. Deiluiis. 2001. Ectoparasites in lekking Sharp-tailed Grouse, Tympanuchus phasianellus. Canadian Field-Naturalist 115:210–213.

Tyler, C. R., S. Jobling, and J. P. Sumpter. 1998. Endocrine disruption in wildlife: a critical review of the evidence. Critical Reviews in Toxicology 28:319–361.

U.S. Fish and Wildlife Service. 1998. Endangered and threatened wildlife and plants; 12-month finding for a petition to list the Lesser Prairie-Chicken as threatened and designate critical habitat. Federal Register 63:31400–31406.

U.S. Fish and Wildlife Service. 2012. Endangered and threatened wildlife and plants; listing the Lesser Prairie-Chicken as a threatened species. Federal Register 77:73827–73888.

U.S. Fish and Wildlife Service. 2013. Endangered and threatened wildlife and plants; 6-month extension of final determination for the proposed listing of the Lesser Prairie-Chicken as a threatened species. Federal Register 78:41022–41024.

U.S. Fish and Wildlife Service. 2014. Endangered and threatened wildlife and plants; special rule for the Lesser Prairie-Chicken; final rule. Federal Register 79:20073–20085.

Verquer, T. L. 2007. Colorado Lesser Prairie-Chicken breeding survey 2007. Colorado Division of Wildlife, Denver, CO.

Walker, B. L., and D. E. Naugle. 2011. West Nile virus ecology in sagebrush habitat and impacts on Greater Sage-Grouse populations. Pp. 127–142 in S. T. Knick and J. W. Connelly (editors), Greater Sage-Grouse: ecology and conservation of a landscape species and its habitats. Studies in Avian Biology (no. 38), University of California Press, Berkeley, CA.

Watkins, R. M. 1969. Nutrition and propagation of the Attwater's Prairie Chicken. III. Pesticides residues in the Attwater's Prairie Chicken and related species. Pp. 33–43 in Annual progress report, Caesar Kleberg Research Program in Wildlife Ecology. Texas Agricultural Experiment Station, College Station, TX.

Wiedenfeld, D. A., D. H. Wolfe, J. E. Toepfer, L. M. Mechlin, R. D. Applegate, and S. K. Sherrod. 2002. Survey for reticuloendotheliosis viruses in wild populations of Greater and Lesser Prairie-Chickens. Wilson Bulletin 114:142–144.

Williams, E. S., and E. T. Thorne. 1996. Exertional myopathy (capture myopathy). Pp. 181–193 in A. Fairbrother, L. N. Locke, and G. L. Hoff (editors), Noninfectious diseases of wildlife. Iowa State University Press, Ames, IA.

Wobeser, G. A. 2005. Essentials of disease in wild animals. Blackwell Publishing, Ames, IA.

Wolfe, D. H., M. A. Patten, E. Shochat, C. L. Pruett, and S. K. Sherrod. 2007. Causes and patterns of mortality in Lesser Prairie-Chickens Tympanuchus pallidicinctus and implications for management. Wildlife Biology 13:95–104.

Wood, K. N., F. S. Guthery, and N. E. Koerth. 1986. Spring–summer nutrition and condition of Northern Bobwhites in south Texas. Journal of Wildlife Management 50:84–88.

Yabsley, M. J. 2007. Tetrameridosis. Pp. 162–180 in C. T. Atkinson, N. J. Thomas, and D. B. Hunter (editors), Parasitic diseases of wild birds. Wiley-Blackwell, Ames, IA.

Yazwinski, T. A., and C. A. Tucker. 2008. Nematodes and acanthoceophalans. Pp. 1025–1056 in Y. M. Saif, A. M. Fadly, J. R. Glisson, L. R. McDougald, L. K. Nolan, and D. E. Swayne (editors), Diseases of poultry. Wiley-Blackwell, Hoboken, NJ.

PART THREE

Emerging Issues

CHAPTER TEN

Public and Private Land Conservation Dichotomy*

R. Dwayne Elmore and David K. Dahlgren

Abstract. Lesser Prairie-Chickens (*Tympanuchus pallidicinctus*) occur primarily on private lands (~94%) within portions of Colorado, Kansas, New Mexico, Oklahoma, and Texas. The current estimated occupied range of the Lesser Prairie-Chicken is 80,030 km², with ~5,000 km² in public ownership. Alternative management objectives, opportunities, and challenges are dependent on patterns of landownership. Public land management is largely driven by policy and subsequent regulations, whereas private land management is mainly influenced by incentive programs, which are impacted by dynamic changes in governmental policies. Some management actions are currently constrained on public lands because of funding priorities. Other land uses, such as energy development, are taking place range-wide irrespective of landownership. The Lesser Prairie-Chicken may be listed as threatened under the Endangered Species Act, which would likely impact management activities on private and public lands, including managed grazing, prescribed fire, energy development, and recreation. Public lands can contribute to conservation goals, but it is the influence of future policy, regulations, and conservation programs on private lands that will determine the future of Lesser Prairie-Chickens in the Southern Great Plains.

Key Words: Conservation Reserve Program, Endangered Species Act, Lesser Prairie-Chicken Conservation Initiative, National Grasslands, National Resources Conservation Service, U.S. Forest Service.

Lesser Prairie-Chickens (*Tympanuchus pallidicinctus*) have specific habitat requirements for various life history stages that are unique to the species as a sensitive grassland bird. Habitat requirements do not vary with respect to political boundaries or patterns of landownership. Lesser Prairie-Chickens currently occupy portions of Colorado, Kansas, New Mexico, Oklahoma, and Texas. However, habitat suitability does differ with respect to ownership because of varying land management objectives and constraints throughout the occupied range of the species. We address the primary differences between public and private lands in regard to impacts, habitat management, and conservation opportunities relative to Lesser Prairie-Chickens. We begin by outlining the general patterns of landownership within the five-state region where the species occurs and then summarize the main programs that are available for the conservation of Lesser Prairie-Chickens

* Elmore, R. D. and D. K. Dahlgren. 2016. Public and private land conservation dichotomy. Pp. 187–203 in D. A. Haukos and C. W. Boal (editors), Ecology and conservation of Lesser Prairie-Chickens. Studies in Avian Biology (no. 48), CRC Press, Boca Raton, FL.

within each landownership category. The Lesser Prairie-Chicken was listed as threatened under the Endangered Species Act (ESA) in May 2014, so consideration of the impact of this listing on management actions relevant to the species is warranted (U.S. Fish and Wildlife Service 2014). However, the listing rule was vacated by judicial decision in September 2015, creating considerable regulatory uncertainty at the time of this volume. Further, we discuss management actions that can both directly and indirectly affect Lesser Prairie-Chickens throughout their range. The actions include prescribed fire, grazing management, energy development, and recreation. The goal of our chapter is to develop a framework for conservation that highlights the scale of the issue, while emphasizing the concurrent, but divergent, roles that public and private land management and conservation will play for future conservation and persistence of Lesser Prairie-Chickens.

OWNERSHIP

Private Lands

The total estimated occupied range of the Lesser Prairie-Chicken is currently 80,030 km². Of that area, ~75,000 km², or an estimated 94% of the range of Lesser Prairie-Chickens are private lands. Range-wide, Lesser Prairie-Chickens occur mainly on private lands and almost exclusively in the states of Kansas, Texas, and Oklahoma. However, habitat quality and the density and distribution of Lesser Prairie-Chickens are highly variable on private lands. Private lands have undergone extensive conversion to row crops, introduced grasses, or become highly fragmented by invasive woody vegetation and anthropogenic structures (Samson et al. 2004). In addition, changes in natural disturbance factors such as fire and grazing have been altered and continue to degrade much of the remaining habitat for Lesser Prairie-Chickens. Many of these same issues, excluding conversion to row crops, threaten available habitat on public lands as well. Some of the concerns for private lands can be addressed through voluntary incentive-based programs. For instance, row crops and introduced grasses can be converted to more suitable vegetation composition and structure, and invading trees can be removed. Existing anthropogenic structures in some cases can be removed or threats mitigated, yet they are typically more permanent than other conditions. Despite a series of issues that potentially affect habitat conditions on private lands, most of the remaining large and relatively stable populations of Lesser Prairie-Chickens occur on private lands.

Federal Lands

National Grasslands are federally owned and managed by the U.S. Forest Service (USFS), an agency of the U.S. Department of Agriculture (USDA). The primary uses of National Grasslands within the current range of Lesser Prairie-Chickens are livestock grazing, energy extraction, and recreation. The Comanche National Grassland located in southeastern Colorado encompasses 1,796 km². Yet, much of this National Grassland is short-grass prairie with some canyon country, and therefore, provides marginal habitat for the Lesser Prairie-Chicken. Despite the size of the area, habitat suitability of Comanche National Grassland for Lesser Prairie-Chickens is likely limited. In 2013, only two active leks with ~11 males were found in the Comanche National Grassland (J. Reitz, Colorado Division of Wildlife, pers. comm.).

The Cimarron National Grassland consists of 438 km² in extreme southwestern Kansas. Historically, this National Grassland had many active leks, but numbers have fallen precipitously in the past 20 years. As recently as 1995, 14 active leks with an estimated total population of 284 Lesser Prairie-Chickens were present. In 2013, however, only three active leks with ~30 birds occurred at Cimarron National Grassland (A. Chappell, USFS, pers. comm.). Much of the National Grassland is dominated by sand sagebrush (Artemisia filifolia) prairie and potentially suitable habitat for Lesser Prairie-Chickens. Lack of nesting cover and brood habitat associated with long-term drought is thought to be a major factor in the decline in Lesser Prairie-Chickens at Cimarron National Grassland (A. Chappell, pers. comm.).

In western Oklahoma and northeast Texas, the McClellan Creek/Black Kettle National Grassland complex occupies 133 km². There are no active leks at these National Grasslands and only occasional Lesser Prairie-Chicken use is reported in portions of the area (C. Milner, USFS, pers. comm.). Much of the Black Kettle National Grassland is highly fragmented by tree cover and other land uses and currently offers poor habitat for the Lesser Prairie-Chicken.

Last, the Kiowa/Rita Blanca National Grassland contains 933 km² in northeastern New Mexico and the adjacent panhandles of Oklahoma and Texas. These National Grasslands are within the historical range of the Lesser Prairie-Chicken but are not currently occupied. The majority of these grasslands is short-grass prairie and highly fragmented with other landownership, so the potential to support Lesser Prairie-Chickens is limited.

Properties managed by the U.S. Department of the Interior Bureau of Land Management (BLM) are scattered throughout the range of the Lesser Prairie-Chicken, but only New Mexico contains large contiguous areas of BLM land (surface area), with ~3,341 km² within the current range of the Lesser Prairie-Chicken (Van Pelt et al. 2013). Much of the BLM lands are suitable habitat dominated by sand shinnery oak (*Quercus havardii*) prairie. Recent estimates (2013) of Lesser Prairie-Chicken leks on BLM land in New Mexico were 91 active leks (M. East, Natural Heritage New Mexico, pers. comm.). Livestock grazing, energy extraction, and recreation are primary uses of BLM land. The BLM has a Resource Management Plan Amendment and a Candidate Conservation Agreement (CCA; see later) for the Lesser Prairie-Chicken on BLM lands in New Mexico.

None of the National Wildlife Refuges (NWR) of the U.S. Fish and Wildlife Service (USFWS) are specifically designated for Lesser Prairie-Chickens, but several refuges occur within the historical range of the species and offer potential habitat. Optima NWR, located in Texas County, Oklahoma, contains 1,753 ha of which a large portion is mixed-grass and sand sagebrush prairie. Muleshoe NWR located in Bailey County, Texas, contains 2,350 ha of which a portion is upland prairie where Lesser Prairie-Chickens have occasionally been reported in the past. Buffalo Lake NWR located in Randall County, Texas, consists of 3,101 ha of which approximately half is grassland. Grulla NWR located in Roosevelt, New Mexico, contains 1,309 ha. The majority of this NWR is a saline lake bed, but Lesser Prairie-Chickens occasionally occur on the upland fringe.

State Lands

In Colorado, state-owned lands with potential to support Lesser Prairie-Chickens are extremely limited. Similarly, Texas has limited state-owned properties important to Lesser Prairie-Chicken.

Two Wildlife Management Areas (WMA) operated by Texas Parks and Wildlife Department contain potential habitat for Lesser Prairie-Chickens but are currently unoccupied. Two sites include the Gene Howe WMA (2,382 ha; Hemphill County) and the Matador WMA (11,405 ha; Cottle County). The Matador WMA is within the eastern edge of the depicted historical range of the Lesser Prairie-Chicken but has not contained any leks or birds for decades; however, portions of it do include suitable habitat in the form of sand shinnery oak. About two-thirds of the Gene Howe WMA in northeast Texas are sand sagebrush or mixed-grass prairie. No contemporary observations of Lesser Prairie-Chicken have been recorded at Gene Howe WMA but the site is closer to current occupied range than the Matador WMA. Additionally, Texas Parks and Wildlife Department owns two tracts (99 and 1,214 ha, respectively) in Yoakum County that support Lesser Prairie-Chickens (D. Lucia, USFWS, pers. comm.). The Oklahoma Department of Wildlife Conservation (ODWC) has several large WMAs with the potential to support Lesser Prairie-Chickens. Sites include Beaver River WMA (10,809 ha; Beaver County), Cimarron Hills WMA (1,526 ha; Woods County), Cimarron Bluffs WMA (1,388 ha; Harper County), Ellis County WMA (1,943 ha; Ellis County), and Packsaddle WMA (7,956 ha; Ellis County). Lesser Prairie-Chickens are occasionally seen at Cimarron Bluffs WMA with one active lek but rarely seen at the other WMAs (L. Weimers, ODWC, pers. comm.). Beaver River, Cimarron Hills, and Cimarron Bluffs WMAs are dominated by sand sagebrush prairie; Ellis and Packsaddle WMAs are dominated by sand shinnery oak prairie. Lands under the management of the Oklahoma Commissioners of the Land Office have an additional eight active Lesser Prairie-Chicken leks (A. Gregory, ODWC, pers. comm.). The Kansas Department of Wildlife, Parks, and Tourism (KDWPT) manages the Pratt Sandhills Wildlife Area (2,313 ha: Pratt County) and the Sandsage Bison Range (1,497 ha: Finney County). Limited numbers of Lesser Prairie-Chickens have been utilizing these areas with the last reports of birds during spring lek surveys in 1994 for the Sandsage Bison Range and 1999 for the Pratt Sandhills Wildlife Area (J. Pitman, KDWPT, pers. comm.). New Mexico has 10,963 ha in Lea, Roosevelt, and Chaves counties managed by the New Mexico Department of Game and Fish as

Prairie Chicken Areas (Chapter 13, this volume). Virtually, all of the properties currently have Lesser Prairie-Chickens present with 66 active leks in 2013 (M. East, pers. comm.). Further, New Mexico has ~1,000 km² within the range of the Lesser Prairie-Chicken managed by the New Mexico State Lands Office. These properties had 168 active Lesser Prairie-Chicken leks in 2013 (M. East, pers. comm.).

Ownership Summary

The vast majority of Lesser Prairie-Chicken range is in private ownership (estimated at 94%). The management on private lands is the decisive factor for the future of the species and should continue to be the significant focus of any conservation efforts for the Lesser Prairie-Chicken. The total estimated occupied range of the Lesser Prairie-Chicken as estimated from current distribution maps is 80,030 km². The total area of public lands described earlier is ~5,000 km². The public lands do not constitute all federal and state lands within the current range of the Lesser Prairie-Chicken; however, they do represent the most significant public areas that have potential to support the species. Of course, only portions of these public lands provide appropriate habitat with plant communities that support Lesser Prairie-Chicken, and much of that area would require management investment to become quality habitat. The total area of public lands is a large area, but habitat is highly fragmented in most instances and occurs sporadically across a vast spatial scale. However, some public lands may provide private land conservation opportunity by using them as focus areas to anchor regional landscape conservation programs. Further, public lands can provide demonstration sites for private landowners to observe management efforts aimed at Lesser Prairie-Chicken conservation assuming they are managed appropriately.

CONSERVATION PROGRAMS

Federal

Conservation Reserve Program

The Conservation Reserve Program (CRP) of the U.S. Department of Agriculture (USDA) was initially implemented by the Food Security Act of 1985 (or the 1985 Farm Bill) to place qualified existing cropland on highly erodible soils into perennial vegetation cover to reduce surplus commodity crops, minimize runoff, and prevent erosion for 10- or 15-year contract periods in exchange for an annual rental rate and cost share for vegetation establishment and maintenance (Farm Service Agency 2014). Currently, the CRP and associated initiatives are administered by the Farm Service Agency (FSA), with additional technical support provided by the Natural Resource Conservation Service (NRCS). Beyond the soil and water benefits, CRP lands also provide habitat for many wildlife species including the Lesser Prairie-Chicken.

The CRP is often singled out as one of the more important federal conservation programs because of the sheer scale of the program in terms of area, time on the landscape, and conservation funding. Specifically, CRP has considerable area under contract within the range of the Lesser Prairie-Chicken, with 20,379 km² under contract in 2013 (R. Wagner, USDA FSA, pers. comm.). However, the total area has decreased recently as state area capacity and reenrollment have decreased (Figure 10.1). Despite a decreased area, CRP has considerable potential to positively impact Lesser Prairie-Chicken populations by returning marginal cropland to perennial grass cover. Although the quantity of CRP is substantial, the quality of habitat it provides is variable. In some states, Kansas being notable, most CRP was planted in a diverse native mix that closely approximated native grasslands. In Oklahoma and Texas, CRP seed mixtures included exotic grasses such as old world bluestems (*Bothriochloa* spp.) and weeping lovegrass (*Eragrostis curvula*) that form monotypic stands with low species diversity. Introduced plants provide some habitat, but the quality is reduced because plant structure and composition are marginal for Lesser Prairie-Chickens. The CRP is a federal program, but both state and private partners contribute to its design and application. State agencies provide guidance on the establishment of priority areas where CRP should be focused for wildlife conservation. For example, in Kansas, CRP priority areas have been delineated to benefit the conservation of Lesser Prairie-Chickens. Further, both state biologists and private nongovernmental biologists, such as Pheasants Forever, often help deliver CRP and other Farm Bill conservation programs.

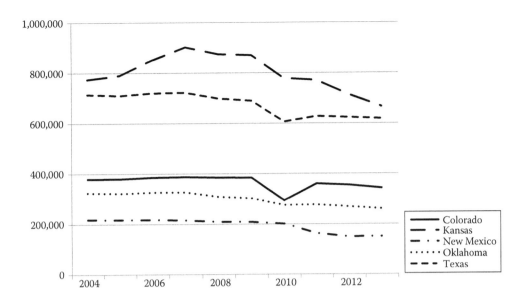

Figure 10.1. Area (ha) of lands enrolled in the Conservation Reserve Program in counties within the range of the Lesser Prairie-Chicken for the five states of Colorado, Kansas, New Mexico, Oklahoma, and Texas, 2004–2013. (Data courtesy of the U.S. Department of Agriculture 2014.)

Starting in 2008, the State Acres for Wildlife Enhancement (SAFE) initiative allows for continuous CRP enrollment as a voluntary program designed to address state and regional high priority wildlife objectives (U.S. Department of Agriculture, Farm Service Agency 2013). Each state with populations of Lesser Prairie-Chickens has opportunities for private landowners to participate in the SAFE initiative. The goal of the Colorado Lesser Prairie-Chicken SAFE project is to restore and enhance 8,705 ha of short- and midgrass sand sagebrush prairie to maintain and enhance populations of Lesser Prairie-Chickens. The goal of the Kansas Lesser Prairie-Chicken Habitat SAFE project is to enroll 21,093 ha in CRP to restore mixed-grass prairies to maintain and enhance populations of Lesser Prairie-Chickens. The goal of the New Mexico Lesser Prairie-Chicken SAFE project is to enroll 1,053 ha in CRP in the eastern part of the state to benefit the Lesser Prairie-Chicken by restoring native grasslands for breeding and brood rearing. The goal of the Oklahoma Mixed Grass Prairie SAFE project is to enroll 6,113 ha in CRP to restore mixed-grass prairie type associations in northwestern Oklahoma to benefit grassland birds. The goal of the Texas Mixed Grass SAFE project is to enroll 49,676 ha in CRP to reconnect geographically and reproductively isolated populations of LPC by creating native mixed-grass prairie and travel corridors.

NRCS Lesser Prairie-Chicken Initiative

The NRCS Lesser Prairie-Chicken Initiative is available to landowners for improving the effectiveness of voluntary conservation practices to expand habitat for Lesser Prairie-Chickens and benefit the long-term sustainability of producers' agricultural operations. The targeted program uses the Environmental Quality Incentives Program and Grassland Reserve Program to improve and protect habitat for the Lesser Prairie-Chicken. Available conservation actions under the initiative include the following: (1) supporting sustainable grazing management that results in residual nesting cover and supports native plant communities; (2) increasing connectivity of existing habitat for Lesser Prairie-Chickens; (3) improving weed and invasive species management; (4) reducing tillage on agricultural fields; (5) protecting, maintaining, and restoring large tracts of native oak/tall-grass or sand sagebrush prairie; (6) maintaining the stability of land use and conservation of shrub-dominated habitats near lek sites; and (7) promoting the use of government programs that provide incentives for the development or restoration of habitat on private lands. To date, ~4,500 km² of mostly private lands have been impacted by this program (NRCS 2013). Enrollments include 36,859 ha in Colorado, 46,082 ha in Kansas, 158,941 ha in New Mexico, 32,123 ha in Oklahoma, and 179,188 ha in Texas (C. Hagen, Oregon State University, pers. comm.). Conservation practices such as woody plant removal, prescribed fire, managed grazing, fence marking, and conservation easements can be used to achieve the program objectives.

Candidate Conservation and Safe Harbor Agreements

Candidate Conservation Agreements are a voluntary conservation agreement between the USFWS and public or private parties to identify threats to candidate species that are proposed for federal listing as threatened or endangered, and develop a plan of action to improve conditions and reduce or remove threats so that listing of the species may not be necessary (Candidate Conservation Agreements, Fact Sheet, USFWS 2011a). A Candidate Conservation Agreement with Assurances (CCAA) creates incentives for nonfederal landowners that are proactive and engage in conservation efforts for candidate species under the ESA. As a result of their efforts, landowners are given assurances from the USFWS that specifically describes what will be expected in the future. More specifically, a CCAA provides participating landowners with a permit containing assurances that if they engage in certain conservation actions for species included in the agreement, landowners will not be required to implement additional conservation measures beyond those in the CCAA even if the candidate species is listed. Each CCAA is considered separately by the USFWS and both the landowner and the USFWS come to agreement on the provisions within the CCAA. Also, additional land, water, or resource use limitations will not be imposed on them should the species become listed in the future, unless they consent to such changes (Candidate Conservation Agreements, Fact Sheet, USFWS 2011b). The primary goal of the agreements is to encourage landowner involvement in conservation activities, while reducing landowner concern about increased regulations if the species was ever to become federally listed. A variety of conservation actions

may qualify landowner protection under these agreements, including the following: (1) protecting and enhancing existing populations and habitat; (2) restoring degraded habitat; (3) creating new habitat; (4) augmenting existing populations; (5) restoring historic populations; and (5) not undertaking a specific, potentially affecting/damaging activity. Only private lands are eligible for agreements, and the three states of Texas, New Mexico, and Oklahoma have set up programs for CCAAs. When the Lesser Prairie-Chicken was listed as a threatened species in May 2014 under the ESA, agreements created when the species was still a candidate species under ESA were still operational and have the potential to contribute toward recovery.

Should federal listing of Lesser Prairie-Chickens as a threatened species reoccur, the Safe Harbor program would be the appropriate agreement structure used between the USFWS and landowners or other parties. Safe Harbor is similar to CCAA in that it is a voluntary agreement with the USFWS. Safe Harbor provides assurances from the USFWS to the private party that if the private party fulfills the conditions specified in the agreement, no additional activities will be required (USFWS 2013). When the Lesser Prairie-Chicken was listed no Safe Harbor agreements were developed.

USFWS Partners for Fish and Wildlife

Partners for Fish and Wildlife is a federal aid program administered by the U.S. Fish and Wildlife Service (USFWS). The program provides both technical and financial assistance for qualified practices for wildlife enhancement on private lands. Often, the resources of this program are combined with other federal or state resources to increase conservation benefits for specific projects. The Partners program primarily targets federal trust resources and wildlife species that are in peril such as the Lesser Prairie-Chicken. Practices such as tree removal, prescribed fire, grazing management, and vegetation establishment are the primary practices that have potential to benefit the Lesser Prairie-Chicken under the Partners program. Most often the Partners program is implemented in concert with other state or federal programs. For instance, in Kansas the removal of eastern red-cedar (*Juniperus virginiana*) projects has been expanded to multiple landownerships and become larger landscape projects by combining the Partners program with other state and NRCS projects.

State

In addition, in the initial listing decision, the USFWS concurrently published a special rule under section 4(d) of the ESA that exempts certain activities from the prohibitions of the Act. One of the provisions in the special 4(d) allows for activities being implemented under the Western Association of Fish and Wildlife Agencies' Lesser Prairie-Chicken Range-wide Conservation Plan (RWP) to be exempt from the ESA. The rule essentially enabled states to provide viable options for industry and landowners for the management of Lesser Prairie-Chickens. The conservation plan has multifaceted management purposes: it identifies range-wide and subpopulation goals, identifies habitat amounts and conditions to achieve goals, defines focal areas and connectivity zones, relies on voluntary landowner incentive programs for habitat management, promotes minimization of impacts or mitigation when necessary, establishes a mitigation framework and funding source, identifies needed research and monitoring, develops an adaptive management framework for new research or monitoring to change management options if necessary, and addresses stakeholder input (Van Pelt et al. 2013). Overall, the plan allows for continued economic development within a voluntary framework and compliance opportunities within the federal listing under the Endangered Species Act. The plan prioritizes funding into 25% permanent conservation and 75% for short-term management contracts. Landowners may receive up to 125% of actual cost for implementing conservation actions (Van Pelt et al. 2013). The approach to a federal listing is an unprecedented process with cooperation from federal agencies, state wildlife agencies, private industry, and private landowners.

State wildlife agencies have historically led conservation efforts on both public and private lands for the Lesser Prairie-Chicken. The initial listing of the Lesser Prairie-Chicken under the ESA with a 4(d) provision allowed the states to continue to manage conservation efforts (USFWS 2014). Some examples of state programs available to private landowners include the following: Habitat Improvement Program (Colorado), Wildlife Habitat Protection Program (Colorado), Wildlife Habitat Improvement Program (Oklahoma), Landowner Incentive Program (Texas), and Wildlife Habitat Improvement Program (Kansas). States also have State Wildlife Grants (SWG), which is funding granted to the state from the USFWS that may be targeted for conservation on private lands. For example, Kansas has historically had a priority area within their SWG grant proposals to provide additional assistance to private landowners within the range of Lesser Prairie-Chickens. In addition to providing unique state level programs, state wildlife agencies are integral in administering many of the federal programs. For example, most states have Farm Bill Coordinators who provide technical assistance to the implementation of USDA conservation programs, especially for wildlife resources. Efforts by these coordinators in recent years to increase priority of Lesser Prairie-Chickens have been integral to targeting species conservation efforts. Another example is in Kansas where biological technical assistance has been provided to NRCS programs by state-agency area biologists since 1994 through a contractual agreement between the NRCS and KDWPT. In addition to technical assistance, most states provide financial resources for partnership positions for biologists from Pheasants Forever as a nongovernmental organization that assists with the implementation of USDA conservation programs. Often it is by the unsung "behind the scenes" efforts and local contacts of state biologists that state and federal conservation programs are applied to private land one parcel at a time.

Private

Conservation Banking

Conservation banks can serve as powerful tool for conserving imperiled species on private lands. Banks are permanently protected lands that provide habitat for a target species or resource. The purpose is to offset actions carried out at one site, by protecting and enhancing another site. Conservation banks for threatened and endangered species typically involve a landowner, a developer, a third party to manage the bank, and the USFWS. When a developer must carry out some action that is expected to have negative effects on a sensitive species, they may purchase "credits" from a conservation bank (Conservation Banking, Fact Sheet, USFWS 2012). The USFWS approves what a credit consists of, such as number of acres, but the market determines the price of an individual credit.

The owner of the conservation bank has an incentive to manage for the imperiled species because of the value that the market has placed on that species through the sale of these credits. The USFWS requires that conservation banks have an agreement with the USFWS, grant a conservation easement that restricts certain land uses and development that could be detrimental to the target species, develop a management plan for the conservation bank, and provide funding for the management (Conservation Banking, Fact Sheet, USFWS 2012). A third party, such as a land trust or nonprofit organization, will hold the easement for the conservation bank and ensures that the easement and management actions are carried out. Once these steps are in place, the landowner with the conservation bank can sell a certain number of credits to a developer that should offset impacts. Conservation banks can be an effective method to create incentives for private landowners to conserve imperiled species while allowing development in an area.

Multiple Use vs. Single Purpose Management

Approaches and development of conservation programs differ between private and public lands because of economics, legal authorities, and existing policies. Further, the strength and the direction of the relationship between demands and realized management may differ markedly between public and private lands. However, because the Lesser Prairie-Chicken was a valued game species prior to being listed as a threatened species, both private landowners and state and federal agencies have an additional incentive to prioritize management for the species. Management for Lesser Prairie-Chickens can be compatible with other land uses, but populations cannot be maximized without affecting other land outputs. Therefore, it becomes an issue of trade-offs that managers must decide within the bounds of regulatory policy what uses, products, and services a tract of land will produce. Prioritizing among multiple uses is a paradoxical challenge for managers of public and private lands when the population status of many species, including Lesser Prairie-Chickens,

are influenced by external factors beyond their control such as management practices on adjacent lands, effective scale of management, and climatic influences.

At the outset, most private landowners might appear to manage for a single focus—maximization of profit. Yet, this is seldom the case, as many private landowners have multiple objectives and are not driven solely by economic demands (Torell and Bailey 2000). Private landowners have many of the same demands and objectives as public land managers: economics, laws and regulations, stakeholder demands, and personal objectives.

Most public lands are managed for multiple objectives of which conservation of Lesser Prairie-Chicken is a single goal. For example, on National Grasslands, the USFS must comply with the Multiple-Use and Sustained Yield Act of 1960. Many of the competing objectives are compatible, but they are not likely mutually inclusive. For example, species occurring together have various, although sometimes similar habitat requirements. Therefore, it is impossible to simultaneously maximize all potential biotic populations on a given area. Hence, management on public land often attempts to optimize, rather than maximize, multiple species or multiple objectives. The approach may work well when all species have similar value to society, but when one species has more societal value than another, single focus management may be favored.

Summary of Conservation Programs

The available programs that we have described for the conservation of Lesser Prairie-Chickens is not exhaustive, but we have summarized the most relevant programs due to their scale of implementation. What should be evident is that private landowners have an array of available programs to assist with the conservation of Lesser Prairie-Chickens. The availability of alternative conservation programs is not a trivial issue because ~94% of the range of Lesser Prairie-Chickens is in private ownership. It is clear that future conservation will require landowner participation, and continued funding and potentially expansion of incentive-based programs such as those listed here will be critical. It cannot be overemphasized that conservation efforts on private lands often hinge on incentive programs, and this paradigm distinguishes private land conservation from public land conservation. On public land, policy directly guides conservation actions. On private lands, policy indirectly guides conservation action through incentive programs. The contrast between conservation plans is particularly evident with cost-share programs that target certain management practices and actions.

ENDANGERED SPECIES ACT

Should the federal listing of the Lesser Prairie-Chicken as a threatened species under the Endangered Species Act in May 2014 be reinstated, it will have direct and indirect effects on both private and public land management. The purpose of the ESA is to protect and recover imperiled species and the ecosystems upon which they depend. Take is defined as "to harass, harm, pursue, hunt, shoot, wound, kill, trap, capture, or collect or attempt to engage in any such conduct." Within take, "harm" is a broad term that includes habitat modification and degradation if it "kills or injures wildlife by significantly impairing essential behavioral patterns, including breeding, feeding, or sheltering." As defined under the federal legislation, harm can include habitat modification demonstrated to cause loss of a listed species.

Limited data indicate that anthropogenic structures, such as the infrastructure associated with energy development, can negatively impact Lesser Prairie-Chickens (Hagen et al. 2011). Thus, the ESA listing may influence energy production. Under certain circumstances, the conversion of native grassland or prairie may also constitute take. Take of Lesser Prairie-Chickens as a threatened species could require mitigation for actions on private lands where native grassland is proposed for conversion to cropland. On private lands, livestock grazing and other normal agricultural operations such as existing croplands will not be impacted by the listing decision. In addition, the development of compatible grazing practices may change within USDA or other federal programs to further benefit the Lesser Prairie-Chicken.

Federal protection of the Lesser Prairie-Chicken under the ESA will potentially provide opportunities for conservation by more landowners on private lands than any previous species listing within the Southern Great Plains. Opportunities arise because Lesser Prairie-Chickens are considered a resident game bird, occurs across five states, and its habitat overwhelmingly occurs on private lands. Species persistence will depend on management at the landowner level across landscape scales. Conservation practices are primarily undertaken by private landowners within incentive-based programs, but the ESA listing may create misunderstanding and fear, which may initially limit landowners from taking advantage of the variety of opportunities to work toward conserving the listed species (Brook et al. 2003). A listing decision can direct federal resources toward areas where a listed species occurs and raise awareness of species status (Morrow et al. 2004), but landowner acceptance may not be adequate to make significant recovery because of fears of regulation in the short term. In the case of the federally endangered Attwater's Prairie-Chicken (*Tympanuchus cupido attwateri*), concerns about increased regulation and oversight of federal actions proved to be short-lived and conflicts were local (Morrow et al. 2004). Engagement of landowners in advance of a listing decision through the conservation agreement process and working within approved federal programs to provide predictability inside NRCS and other conservation plans may help to reduce negative reactions by landowners. Further, financial incentives to private landowners through mitigation or conservation banking may also alleviate negative perceptions. Recognizing cultural and social issues that are obstacles for conservation progress will be just as important as sound habitat management, while accepting that federal listing is another tool available for species recovery.

MANAGEMENT AND USES

We have categorized the distribution of Lesser Prairie-Chickens into public and private lands with different approaches to conservation, but several management issues apply to all lands without regard to landowner category. The three most common conservation practices shared among private and public lands management are applications of prescribed fire, control of trees and shrubs, and livestock grazing. In addition, anthropogenic impacts such as infrastructure and energy development are shared among all lands.

Prescribed Fire and Brush Management

Large-scale changes in tree cover are a primary factor associated with the loss of Lesser Prairie-Chickens in the eastern portion of the species range (Fuhlendorf et al. 2002). At local scales of home ranges or nest sites, no direct evidence has indicated how fire impacts Lesser Prairie-Chickens. However, because fire both directly and indirectly changes the structure and composition of vegetation, it is plausible that prescribed fire could be used to improve vegetation conditions for different life stages of Lesser Prairie-Chickens. Changes might be short-lived because the plant communities are fire adapted and rapidly return to preburn conditions with the exception of tree cover encroachment (Boyd and Bidwell 2001). Many land managers are reluctant or unable to use prescribed fire to manipulate woody plants. Often mechanical removal or chemical treatments are utilized as surrogates for prescribed fire. Removal and herbicides do not have the same ecological benefits as fire but can be used to reduce woody plant structure and composition and change the herbaceous composition. At a broad spatial scale, mechanical or chemical tree removal should have similar impacts as fire for Lesser Prairie-Chickens if treatments prevent conversion of grassland and shrubland plant communities to woodlands. Shrub management through various forms of disturbance may have both positive and negative impacts on Lesser Prairie-Chickens depending on the scale, but eradication of shrub communities

for Lesser Prairie-Chicken management is not supported in the literature, especially in western portions of their range (Bell et al. 2010, Thacker et al. 2012, Pirius et al. 2013). However, eradication of tree and invasive woody plant cover does have support in the literature (Fuhlendorf et al. 2002).

Woody plant management is often implemented on private lands through various state and federal cost-share programs as a method of ecological restoration for the Lesser Prairie-Chicken (Figure 10.2). The programs are largely based on anecdotal and empirical evidence that Lesser Prairie-Chickens avoid areas with tree cover (Fuhlendorf et al. 2002). Landowners are often willing to remove woody cover to benefit cattle production and meet other land objectives; therefore, these cost-share programs are generally popular and will likely continue into the future.

However, management for natural disturbance on public lands is less likely, particularly regarding applications of prescribed fire on federal lands. Challenges arise due to the difficulty in federal agencies acquiring adequate resources and training to conduct prescribed fires. For example, most fire resources within the U.S. Forest Service are spent on wildfire suppression

Figure 10.2. Encroachment of trees and woody plants is a major factor in loss of habitat in some areas of the range of Lesser Prairie-Chickens. Prescribed fire is often used on private and state-owned lands to control woody plants such as the eastern red-cedar. (Photo courtesy of R. Dwayne Elmore, Oklahoma State University.)

and preparedness, leaving few resources for pre-scribed fire. For instance from 2002 to 2012, only 17% of wildlife protection funding on federal lands was allocated to fuels reduction versus 32% on suppression and another 32% on prepared-ness (Gorte 2011). Fuels reduction includes many practices of which prescribed fire is one option. Models point to future climatic conditions that favor increased wild fire frequency and intensity in the southwestern United States (The Nature Conservancy 2013), and this scenario is unlikely to change in the immediate future. Therefore, most fire on public lands will likely be in the form of unplanned wildfires that are typically large in area and intense in nature. Unplanned fires may help to minimize encroachment by invasive woody plants or trees (at least non-sprouting species such as eastern red-cedar), but may not meet annual management needs for brood habitat in some of the more productive rangelands. State agencies can implement pre-scribed fire at a lower cost in terms of manpower than federal agencies and may be more likely to conduct prescribed fires. For instance, the Oklahoma Department of Wildlife Conservation actively carries out prescribed fires on Wildlife Management Areas that support Lesser Prairie-Chickens or have the potential to support the spe-cies. Due to policy restrictions, mechanical and chemical control will likely be used rather than prescribed fire on public lands for the foreseeable future. Treatment can be effective at prevent-ing woodland conversion, but costs are gener-ally higher which limits the scale of application. Despite constraints for federal agencies, there are guidelines in place regarding fire and the Lesser Prairie-Chicken. The USFS Comanche and Cimarron National Grasslands use prescribed fire to "treat" Lesser Prairie-Chicken habitat in a way that provides a mosaic of vegetation types. The goal is not to "blacken" an area; rather, the goal will be to treat an area by allowing for a mosaic burn pattern. Within 3–5 years, the National Grasslands will develop a Lesser Prairie-Chicken habitat assessment that utilizes prescribed fire to improve habitat. A burn plan, describing the prescribed use of fire within a specific and well-defined area, should be completed by an inter-disciplinary team prior to implementation. The plan should include the current status of habitat and how the burn will move vegetation toward quality habitat as described under the Lesser Prairie-Chicken Habitat Requirements. Annual assessments will also be conducted on previ-ously conducted prescribed burns to determine how effective they were in providing appropri-ate vegetation as described under the Habitat Requirements (USFS 2014).

In summary, fire or other significant ecological disturbance are necessary at periodic intervals to maintain productivity and vegetation composition of grasslands and shrublands as well as to prevent tree encroachment over much of the extant range of Lesser Prairie-Chickens. Disturbance can also be used to temporally modify vegetation compo-sition and cover to meet requirements of various life stages of Lesser Prairie-Chickens. Tree removal is a desirable practice for many landowners, and various state and federal programs can provide cost sharing for implementation, and manage-ment of woody plants will likely continue and potentially expand on private lands. State-owned lands likewise have the potential to continue using disturbance to manage plant communities. Due to restrictive administrative requirements and a lack of resources, federal lands are less likely to use prescribed fire at any relevant scale and more likely to use mechanical or chemical methods at limited spatial scales due to costs.

Livestock Grazing

On native grasslands throughout the range of the Lesser Prairie-Chicken, grazing for livestock production is a primary land use. Livestock production in native rangelands is one reason that existing habitat still remains for the Lesser Prairie-Chicken, albeit the quality of this habi-tat is not optimal for much of this area. Patterns of livestock grazing show considerable regional variation throughout the range of the Lesser Prairie-Chicken. Many public lands are grazed to provide multiple uses or achieve various man-agement objectives. State wildlife agencies man-age many Wildlife Management Areas by setting relatively conservative stocking rates of livestock to provide residual cover for wildlife including Lesser Prairie-Chickens. During droughts, public and private land managers will reduce stocking rates or may not stock livestock at all. Stocking rates are highly variable across each property. BLM land managers implement grazing in a man-ner to meet the standards of Public Land Health (BLM 2008). Further, grazing management is indicated as a management tool when vegetation becomes "decadent" (BLM 2008). Each allotment has a grazing plan that details stocking rate, season of use, grazing system, and any management needed. Changes are put into place under new grazing plans if monitoring data on plants and soil conditions indicate that changes are needed (BLM 2008). The Cimarron and Comanche National Grasslands grazing management plans consider stocking rates, rotation patterns, grazing intensity and duration, and contingency plans for prolonged drought for allotments where Lesser Prairie-Chickens are one management goal (USFS 2014). Escape ramps are also used to limit drowning of birds in stock water tanks. Additionally, the Cimarron and Comanche National Grasslands follow the grazing guidelines of Van Pelt et al. (2013) (33% utilization of annual production) for grazing management for the Lesser Prairie-Chicken.

One might assume that public lands would be more lightly stocked than private lands, but this is not necessarily true. The large variation in annual precipitation within the region makes overstocking likely unless producers conservatively stock grasslands each year. Recommendations for highly arid lands are not >35% utilization of annual forage production (Holecheck et al. 1999). During drought years, many grasslands are overstocked except ungrazed sites without livestock such as CRP lands. Areas that have drought management plans requiring destocking may avoid this issue if the plan is followed. Limited areas that are overstocked for short periods are not likely to be of negative consequence to populations of Lesser Prairie-Chicken. Creation of brood habitat by heavy grazing and subsequent annual forb production is a notable example. However, as the spatial scale of land management becomes larger, landscape changes will impact Lesser Prairie-Chickens (Fuhlendorf et al. 2002). The majority of federal, state, and private grasslands are grazed annually, and stocking rate and duration of grazing have the potential to impact Lesser Prairie-Chickens range-wide regardless of landownership. Lands enrolled in the Conservation Reserve Program (CRP) are a special case because they are only managed periodically with mid-contract management under the terms of the CRP contract. During severe drought, emergency haying and grazing has been allowed in CRP lands across much of the range of Lesser Prairie-Chickens in recent years. U.S. Department of Agriculture policy for emergency drought conditions does require half of a CRP field remain unhayed, or, if grazed, stocking rates are reduced to 75% of NRCS rates. The land use policies ensure that at least some vegetative cover remains from year to year. However, declarations of emergency drought that open CRP lands to emergency haying and grazing in consecutive years may prove to be problematic for maintaining suitable habitat conditions.

Similar to fire, direct impacts of livestock grazing on the Lesser Prairie-Chicken have not been well quantified. Yet, because we have ample information regarding how grazing impacts vegetation characteristics and the habitat requirements of Lesser Prairie-Chickens for different life stages, we can link existing data to form predictions regarding grazing impacts specific to Lesser Prairie-Chickens. Specifically, Lesser Prairie-Chickens require some level of residual cover for nesting, with forb composition and an open understory structure necessary for brooding (Hagen et al. 2004). Thus, stocking rates of livestock grazing that lead to optimal habitat conditions would be favorable to Lesser Prairie-Chickens at the scale of management. Deviation from those requirements would be unfavorable for Lesser Prairie-Chicken management. If pastures are heavily stocked with livestock in relation to vegetation production, residual cover will be low across the landscape. Conversely, when landscapes are lightly stocked with livestock relative to vegetation production during wet periods, residual cover will increase. Therefore, stocking rates should be closely monitored and adjusted as needed, regardless of the grazing system utilized, if maintaining or increasing Lesser Prairie-Chickens is a management goal.

Recreation

The Lesser Prairie-Chicken is a highly desirable species for purposes of outdoor recreation. Bird-watchers, photographers, and hunters seek opportunities to view displaying birds at leks in the spring and harvest prairie grouse in the fall hunting season. As the species has become more range restricted and less abundant, demand has concentrated and perhaps increased for some recreational uses.

The Lesser Prairie-Chicken is listed as a game species in Kansas, New Mexico, Oklahoma, and Texas. However, until federal listing as a threatened species, the species was hunted only in Kansas because all other states had previously closed their hunting seasons (Chapter 7, this volume). Harvest for Lesser Prairie-Chickens will remain closed as long as the species is federally listed as a threatened species. Despite a small number of hunters who pursued the Lesser Prairie-Chicken in Kansas (200–1,200 annually) and a limited harvest, the species did offer a specialty form of recreation for upland bird hunters (Dahlgren 2012). The vast majority of all Lesser Prairie-Chickens harvested in Kansas were on private lands or private lands open to the public through lease agreements under a state program for Walk-in Hunting Access (WIHA). Kansas has an extensive walk-in hunting program and hunting was not highly concentrated, reducing the likelihood of high levels of take from individual populations of the Lesser Prairie-Chicken. Similar to Kansas, virtually all historical harvest of Lesser Prairie-Chickens in Oklahoma and Texas was on private lands because limited public lands that support Lesser Prairie-Chicken. Only in New Mexico do substantial numbers of Lesser Prairie-Chickens occur on public lands, which could allow for Lesser Prairie-Chicken public access if hunting was again allowed within that state.

Wildlife viewing and photography of Lesser Prairie-Chickens is typically focused during the breeding season when males display at leks (Chapter 13, this volume). The predictability of traditional lek sites, charismatic display behaviors, and behavioral tolerance of the birds to disturbance from viewing blinds make prairie-chicken viewing a desirable form of recreation. Some level of Lesser Prairie-Chicken viewing and photography takes places in all five states where the species occurs. The Cimarron National Grassland in Kansas previously provided two blinds located on Lesser Prairie-Chicken leks that are available on a first-come, first-served basis. The Comanche National Grassland in Colorado traditionally had a blind located on a lek of Lesser Prairie-Chickens, but the facility was recently closed to the public. New Mexico, Oklahoma, and Texas currently have no public viewing opportunities for Lesser Prairie-Chickens. However, private opportunities exist in all states.

Specifically, Oklahoma and New Mexico each hold a Lesser Prairie-Chicken festival in April. Private landowners in Colorado, Oklahoma, and Texas additionally offer fee-based opportunities to view Lesser Prairie-Chickens.

Listing of the Lesser Prairie-Chicken as a threatened species may reduce recreational activities. Harvest opportunity has ceased following the federal listing in 2014. Listing could potentially increase public demand for viewing opportunities, while decreasing access on public lands because of potential for negative impacts. Recreational use could provide additional income opportunities for private landowners who allow viewing of Lesser Prairie-Chickens. Alternatively, some landowners fearful of the perceived or real implications of federal listing may limit public access for viewing opportunities.

Energy Development

Industrial development of grasslands has the potential to impact Lesser Prairie-Chickens because of installation of vertical structures, noise, vehicle traffic, and loss of vegetation cover (Figure 10.3). Limited data indicate that energy development can cause change in behavior of Lesser Prairie-Chickens (Pitman et al. 2005, Pruett et al. 2009, Hagen et al. 2011). If energy development is determined to result in take under the ESA, then it could be restricted or modified in certain areas where Lesser Prairie-Chickens occur. Many areas of the federal lands are already either developed for energy production or have lease agreements in place. Additionally, federal lands such as BLM and USFS already have restrictions in place to limit energy extraction impacts to the Lesser Prairie-Chicken (BLM 2008, USFS 2014). The Cimarron and Comanche National Grasslands have specific guidelines relative to energy development including the following: turbines and power lines will be located outside Lesser Prairie-Chicken habitat, transmission lines will be buried, existing corridors will be used, development will be focused in areas already developed that are outside Lesser Prairie-Chicken habitat, no oil and gas surface occupancy within 3.2 km of any leks of Lesser Prairie-Chickens that have been active since 2003, and must minimize new surface disturbance (USFS 2014). Further, the BLM has the following guidelines in place for the 23,472 ha Lesser Prairie-Chicken Habitat

Figure 10.3. Industrial development of native grasslands can cause habitat fragmentation that may impact Lesser Prairie-Chickens. Public and private lands are under increasing pressure for energy development of oil, gas, and renewable resources. (Photo courtesy of R. Dwayne Elmore, Oklahoma State University.)

Preservation Area of Critical Environmental Concern (ACEC): The ACEC will be closed to future oil and gas leasing; the ACEC will be closed to locatable, leaseable, and salable mineral entry; existing oil and gas leases will be developed in accordance with those prescriptions applicable in the Core Management Area and sand dune lizard habitat; and vegetation will be managed to meet the goals of the ACEC (BLM 2008).

The ownership of mineral rights by state agencies varies, but many state-owned lands have severed mineral rights from surface ownership. Thus, the agency itself has little control over energy development. However, should the Lesser Prairie-Chicken be listed under the ESA, energy companies would be required to comply with any ESA restrictions on state-owned lands. Currently, multiple energy companies are in the process of drafting Habitat Conservation Plans under Section 10 of the ESA with the USFWS that will guide development to minimize impacts to Lesser Prairie-Chickens.

Energy development pressure is similar on both public and private lands, but private landowners may be disproportionately affected by development restrictions. Impacts could be either positive or negative depending on their land management objectives and whether they hold mineral and wind rights. For instance, if a landowner wishes to capitalize on mineral rights, restrictions may be viewed as an impediment to such development in areas where Lesser Prairie-Chickens occur. However, this may be more of a perception than reality because steps could likely be taken to mitigate potential take. However, if a landowner does not maintain mineral rights, restrictions could be viewed positively because they might minimize surface disturbance. Wind rights are typically held by the landowner. If restrictions are put in place with regard to fossil fuel or renewable energy development, it will encourage state and federal agencies to create incentive programs or mitigation strategies such that landowners restricted from development are compensated for the ecological values their land provides to society while offsetting any loss of potential economic gains from development. Mitigation banking and alternative methods can provide the necessary tools to create incentive programs.

Management and Uses Summary

Fire, mechanical, and herbicide means are used to manage woody vegetation for the Lesser Prairie-Chicken. However, some level of shrub cover appears to be important for Lesser Prairie-Chickens, especially in western portions of their range where shrub eradication may be detrimental to Lesser Prairie-Chickens. Managers often attempt to balance woody plant and herbaceous cover for both livestock and Lesser Prairie-Chickens simultaneously. On private lands, many cost-share programs are available to assist landowners in managing desirable and undesirable species of woody plants relative to the Lesser Prairie-Chicken. Livestock grazing is practiced over most of the range of the species, but appropriate stocking rates are necessary to ensure adequate residual nesting and winter cover for Lesser Prairie-Chickens. In a region so characterized by annual variation in grass production, maintaining appropriate livestock stocking rates is challenging and grasslands are often improperly stocked on both private and public lands. Hunting is currently unavailable for the Lesser Prairie-Chicken, but wildlife viewing opportunities are in moderate demand. Recreational use creates some incentives for private landowners to manage for Lesser Prairie-Chickens.

SUMMARY

Most of the area within the current range of the Lesser Prairie-Chicken is in private ownership (~94%) that varies in habitat suitability. Further, the majority of Lesser Prairie-Chickens occur on these private lands. Incentive-based programs will therefore be critical for the conservation of this species and several state and federal programs are currently available to private landowners. The Conservation Reserve Program is arguably one of the more important federal programs in regard to conservation of Lesser Prairie-Chickens, but has been greatly reduced in recent years because of federal policy changes. Increasing area caps for counties that support populations of Lesser Prairie-Chickens is needed to ensure adequate habitat for this species. Directed conservation policy for Lesser Prairie-Chickens from state and federal stakeholders will be necessary to continue and expand incentive programs on private lands, while ensuring that public lands are managed for the benefit of Lesser Prairie-Chickens. Despite the fact that most Lesser Prairie-Chickens occur on private lands, public lands could serve as core areas to ensure stability among regional populations of Lesser Prairie-Chicken and provide a reference point for best practices for conservation efforts on adjacent private lands. Public lands provide higher levels of assurance of long-term Lesser Prairie-Chicken habitat suitability and are important for ESA decisions and recovery planning. Livestock grazing is an important activity within the region on public and private lands where Lesser Prairie-Chickens occur and will likely continue into the foreseeable future. Grazing can be a compatible practice with population viability of Lesser Prairie-Chickens, but consideration should be given to habitat heterogeneity, residual cover, and fence construction. For example, stocking rates should be at a level so that each year, a portion of the area within each Lesser Prairie-Chicken population has sufficient grass height at the beginning of the nesting season to accommodate nesting females. Fire is a critical process in grassland ecology. The suppression of fire has led to woody plant encroachment in many areas, which is a primary driver of regional Lesser Prairie-Chicken declines in some areas. Incentives must be increased to encourage fire on private lands and policy must be streamlined on public lands to enable managers to carry out needed practices. Last, energy extraction is important to local economies within the Southern Great Plains and to the national energy portfolio of the United States. Careful planning is needed to ensure that continuing industrial development of native grasslands does not impede the recovery of Lesser Prairie-Chickens.

ACKNOWLEDGMENTS

The authors thank J. Pitman (Kansas Department of Wildlife, Parks, and Tourism), A. Janus (Oklahoma Department of Wildlife Conservation), A. Chappell (U.S. Forest Service), R. Wanger (USDA-Farm Service Agency), J. Reitz (Colorado Division of Parks and Wildlife), D. Lucia (U.S. Fish and Wildlife Service), A. Gregory (Oklahoma Department of Wildlife Conservation), L. Weimers (Oklahoma Department of Wildlife Conservation), M. East (Natural Heritage New Mexico), and S. A. Carlton (U.S. Geological Survey-New Mexico Cooperative Fish and Wildlife Research Unit) for assistance with acquiring information on landownership. The authors also acknowledge E. Tanner (Oklahoma State University) for assistance with summarizing landownership data.

LITERATURE CITED

Bell, L. A., S. D. Fuhlendorf, M. A. Patten, D. H. Wolfe, and S. K. Sherrod. 2010. Lesser Prairie-Chicken hen and brood habitat use on sand shinnery oak rangeland. Rangeland Ecology and Management 63:478–486.

Boyd, C. S., and T. G. Bidwell. 2001. Influence of pre-scribed fire on Lesser Prairie-Chicken habitat in shinnery oak communities in western Oklahoma. Wildlife Society Bulletin 29:938–947.

Brook, A., M. Zint, and R. DeYoung. 2003. Landowners' response to an Endangered Species Act listing and implications for encouraging conservation. Conservation Biology 17:1638–1649.

Bureau of Land Management (BLM). 2008. Special status species: Record of decision and approved resource management plan amendment. Bureau of Land Management BLM NM/PL-08-05-1610. Bureau of Land Management, Pecos District Office, Roswell, NM.

Dahlgren, D. 2012. Small game hunter activity sur-vey. Performance Report, Federal Aid in Wildlife Restoration Grant W-39-R-19. Kansas Department of Wildlife, Parks, and Tourism, Hays, KS.

Fuhlendorf, S. D., A. J. W. Woodward, D. M. Leslie, Jr., and J. S. Shackford. 2002. Multi-scale effects of habitat loss and fragmentation on Lesser Prairie-Chicken populations of the US Southern Great Plains. Landscape Ecology 17:617–628.

Gorte, R. W. 2011. Federal funding for wildfire control and management. CRS Report for Congress, RL33990. Congressional Research Service, Washington, DC.

Hagen, C. A., B. E. Jamison, K. M. Giesen, and T. Z. Riley. 2004. Guidelines for managing Lesser Prairie-Chicken populations and their habitats. Wildlife Society Bulletin 32:69–82.

Hagen, C. A., J. C. Pitman, T. M. Loughin, B. K. Sandercock, R. J. Robel, and R. D. Applegate. 2011. Impacts of anthropogenic features on habitat use by Lesser Prairie-Chickens. Pp. 63–75 in B. K. Sandercock, K. Martin, and G. Segelbacher (eds.), Ecology, conservation, and management of grouse, Studies in Avian Biology (no. 39), University of California Press, Berkeley, CA.

Holecheck, J. L., M. Thomas, F. Molinar, and D. Galt. 1999. Stocking desert rangelands: what we've learned. Rangelands 21:8–12.

Morrow, M. E., T. A. Rossignol, and N. J. Silvy. 2004. Federal listing of prairie grouse: lessons from the Attwater's Prairie-Chicken. Wildlife Society Bulletin 32:112–118.

Pirius, N. E., C. W. Boal, D. A. Haukos, and M. C. Wallace. 2013. Winter habitat use and survival of Lesser Prairie-Chickens in West Texas. Wildlife Society Bulletin 37:759–765.

Pitman, J. C., C. A. Hagen, R. J. Robel, T. M. Loughlin, and R. D. Applegate. 2005. Location and success of Lesser Prairie-Chicken nests in relation to vegeta-tion and human disturbance. Journal of Wildlife Management 69:1259–1269.

Pruett, C. L., M. A. Patten, and D. H. Wolfe. 2009. Avoidance behavior by prairie grouse: implications for development of wind energy. Conservation Biology 23:1253–1259.

Samson, F. B., F. L. Knopf, and W. Ostlie. 2004. Great Plains ecosystems: past, present and future. Wildlife Society Bulletin 32:6–15.

Thacker, E. T., R. L. Gillen, S. A. Gunter, and T. L. Springer. 2012. Chemical control of sand sage-brush: implications for Lesser Prairie-Chicken habitat. Rangeland Ecology and Management 65:516–522.

The Nature Conservancy. [online]. 2013. Climate Wizard. <http://www.climatewizard.org/> (26 November 2013).

Torell, L. A., and S. A. Bailey. 2000. Is the profit motive an important determinant of grazing land use and rancher motives? Journal of Agricultural and Resource Economics 25:725.

U.S. Department of Agriculture, Farm Service Agency. [online]. 2013. Conservation Reserve Program, State Acres for Wildlife Enhancement (SAFE) approved projects. <http://www.fsa.usda.gov/Internet/FSA_File/safe2013.pdf> (1 April 2014).

U.S. Department of Agriculture, Farm Service Agency. [online]. 2014. Conservation Reserve Program. <http://www.fsa.usda.gov/FSA/webapp?area=home&subject=copr&topic=crp> (1 April 2014).

U.S. Department of Agriculture Natural Resource Conservation Service (NRCS). [online]. 2013. Lesser Prairie-Chicken initiative. <http://www.nrcs.usda.gov/wps/portal/nrcs/detail/national/programs/farmbill/initiatives> (11 November 2013).

U.S. Fish and Wildlife Service (USFWS). [online]. 2011a. Fish and Wildlife Service Conference Report for the Natural Resource Conservation Service's Lesser Prairie Chicken Initiative, FWS/R2/RD/048810. U.S. Fish and Wildlife Service, Albuquerque, NM. <http://www.nrcs.usda.gov/Internet/FSE_DOCUMENTS/stelprdb1044884.pdf> (5 April 2014).

U.S. Fish and Wildlife Service (USFWS). [online]. 2011b. Candidate conservation agreements, fact sheet. <http://www.fws.gov/endangered/esa-library/pdf/CCAs.pdf> (4 April 2014).

U.S. Fish and Wildlife Service (USFWS). [online]. 2012. Conservation banking, fact sheet. <https://www.fws.gov/endangered/esa-library/pdf/conservation_banking.pdf> (2 April 2014).

U.S. Fish and Wildlife Service (USFWS). [online]. 2013. Safe harbor agreements, fact sheet. <http://www.fws.gov/endangered/landowners/safe-harbor-agreements.html> (2 April 2014).

U.S. Fish and Wildlife Service. (USFWS). 2014. Endangered and threatened wildlife and plants; special rule for the Lesser Prairie-Chicken. Federal Register 79:20074–20085.

U.S. Forest Service (USFS). [online]. 2014. Lesser Prairie-Chicken management plan: Cimarron and Comanche National Grasslands. Cimarron and Comanche National Grasslands, Pueblo, CO. <http://www.fs.usda.gov/Internet/FSE_DOCUMENTS/stelprd3804315.pdf> (14 April 2014).

Van Pelt, W. E., S. Kyle, J. Pitman, D. Klute, G. Beauprez, D. Schoeling, A. Janus, and J. Haufler. [online]. 2013. The Lesser Prairie-Chicken range-wide conservation plan. Western Association of Fish and Wildlife Agencies. Cheyenne, WY. <http://www.wafwa.org/Documents%20and%20Settings/37/Site%20Documents/Initiatives/Lesser%20Prairie%20Chicken/2013LPCRWPfinalfor4drule12092013.pdf> (12 November 2013)

CHAPTER ELEVEN

Impacts of Energy Development, Anthropogenic Structures, and Land Use Change on Lesser Prairie-Chickens*

Anne M. Bartuszevige and Alex Daniels

Abstract. Lesser Prairie-Chickens (*Tympanuchus pallidicinctus*) are an iconic bird of the Southern Great Plains of the United States. Since the bird was first recognized as a separate species, habitat loss and fragmentation, first through agricultural expansion and then through energy development, have threatened the existence of this prairie grouse. Historically, grazing, fire, and drought created fragmentation and patches of grassland or native shrubland habitats with suitable structure for Lesser Prairie-Chickens, and the species was adapted to natural disturbance in prairie landscapes. However, suppression of natural disturbances and addition of novel anthropogenic disturbances have had negative impacts on their populations. In this chapter, we reviewed the literature on the impact of anthropogenic landscape change on populations of Lesser Prairie-Chickens. We then used a Geographic Information System to complete a preliminary analysis to evaluate the hypothesis that Lesser Prairie-Chicken lek locations are negatively associated with anthropogenic structures. Results from the preliminary analysis indicate that lek presence is related to the percent of the landscape that is unimpacted by anthropogenic features at multiple spatial scales. Largest Patch Index was the only variable that measured landscape structure that was related to lek presence. More information is needed on the impact of anthropogenic features at the landscape scale on populations of Lesser Prairie-Chicken. In the meantime, conservation of the species will require collaborations at multiple scales and among different government agencies and conservation groups.

Key Words: agriculture, anthropogenic impacts, Conservation Reserve Program, energy development, mixed-grass prairie, short-grass prairie, Southern Great Plains.

The Lesser Prairie-Chicken (*Tympanuchus pallidicinctus*) was officially recognized as a separate species in 1885 (Ridgeway 1885) and has been regarded as a bird of the western plains, wide-open spaces, and grasslands. However, shortly after the first description as a new species, initial reports of declining populations were recorded. Colvin (1914) observed that Lesser Prairie-Chickens would be lucky to still be around 50 years hence due to conversion of native

* Bartuszevige, A. M. and A. Daniels. 2016. Impacts of energy development, anthropogenic structures, and land use change on Lesser Prairie-Chickens. Pp. 205–220 in D. A. Haukos and C. W. Boal (editors), Ecology and conservation of Lesser Prairie-Chickens. Studies in Avian Biology (no. 48), CRC Press, Boca Raton, FL.

prairie to row crop agriculture. Henika (1940) subsequently observed that the Lesser Prairie-Chicken had been largely relegated to remnant patches of sand sagebrush (*Artemisia filifolia*) and sand shinnery oak (*Quercus havardii*) prairie that were associated with sandy soils not generally suitable for row crop farming. Despite little information on the ecology of a relatively "new" species, early field biologists immediately made a connection between population declines and conversion of grassland to row crop agriculture as well as degradation of grasslands by unmanaged continuous livestock grazing (Colvin 1914, Henika 1940).

Lesser Prairie-Chickens were once widely distributed across portions of five U.S. states but are now patchily distributed in the five states of Colorado, Kansas, New Mexico, Oklahoma, and Texas. The species range has shrunk to ~10% of its original extent (Taylor and Guthrey 1980), and total population numbers have shrunk to ~10% of the original estimated size (Crawford 1980). However, the methods used to estimate historical population sizes and range of Lesser Prairie-Chickens are not well documented (Hagen and Giesen 2005; Chapter 2, this volume). Threats to the species include habitat conversion and subsequent fragmentation, energy development and, to a lesser degree, disease, and climate change (Chapters 9 and 12, this volume). Ongoing population declines and known threats resulted in a federal listing of the Lesser Prairie-Chicken as a threatened species by the U.S. Fish and Wildlife Service (USFWS 2014). However, the listing rule was vacated by judicial decision in September 2015. Regardless, to understand the impacts of large-scale habitat changes on Lesser Prairie-Chickens, it is important to understand the ecological context for which the bird is adapted (Samson et al. 2004). Species within the Great Plains ecosystem evolved in a dynamic system of drought and rainfall, with large-scale biotic and abiotic disturbances. In fact, due to a high frequency of occurrence, Samson et al. (2004) described the lack of natural disturbances as true disturbance to dry prairies because Lesser Prairie-Chickens and other prairie organisms are adapted to an unpredictable environment. Thus, prairie wildlife can have boom and bust fluctuations in population numbers or have evolved to disperse long distances to find suitable habitat.

In the Southern Great Plains, spatial and temporal variability in rainfall patterns are pronounced.

Drought is common and the historical record contains several periods with severe conditions (e.g., 1890s, 1930s, 1950s, 1970s, and 1990s), including droughts in 2011 and 2012 that were among the worst conditions ever recorded (Karl et al. 2012). Augustine (2010) found that spatial variability in precipitation across eastern Colorado ranged from 40 to 120 km—such that the variability at that scale was similar to the temporal variability in precipitation. He concluded that the within-season and among-season variation in precipitation had important consequences for vegetation dynamics and migration of important herbivores such as American bison (*Bison bison*). Population numbers of Lesser Prairie-Chickens are known to fluctuate with drought cycles, with population numbers decreasing in drought years and recovering when precipitation increases (Henika 1940, Giesen 2000, Hagen and Giesen 2005). Moreover, Hagen and Giesen (2005) hypothesized that current land use patterns may limit population recovery following drought.

Prior to European settlement, bison and black-tailed prairie dogs (*Cynomys ludovicianus*) were the dominant grazers in the short- and mixed-grass prairie. About 30–60 million bison once roamed the central grasslands (McHugh 1972, Flores and Evans 1991). Described as a keystone species in the prairies, bison would heavily graze the prairie, concentrating foraging activity at recently burned areas and near prairie-dog towns (Krueger 1986). However, bison were largely extirpated as European-American settlers moved west and were replaced with domestic cattle (*Bos taurus*). Along with the near elimination of bison, prairie dogs have been considered "varmints" and were routinely removed from the landscape with lethal control. The degree to which cattle ranching can or does mimic the native grazing pattern by bison has been debated, but current evidence suggests differences in the types of plant material consumed and loafing behavior (Allred et al. 2011). Allred et al. (2011) hypothesized that many of the differences documented between bison and cattle behavior can be traced to management. Bison are generally allowed to roam over broad pastures in family groups, which allow some "natural" behavior and resultant pattern on the landscape. In addition, bison can be thought of as a tool to restore prairie and prairie ecosystem functions (Allred et al. 2011). Careful grazing management may facilitate the use of cattle as a conservation tool, but cattle are primarily raised as a commodity product and confined to precise grazing regimes in fenced pastures and not as family groups (Allred et al. 2011). As a result of the intense fencing of the landscape to manage cattle in a uniform fashion, grazing patterns on the prairie have changed from a heterogeneous landscape of highly grazed and lightly grazed areas to a situation in which homogeneous grazing management is the norm across the region (Samson et al. 2004).

Fire was also an important disturbance in the short- and mixed-grass prairies (Wright and Bailey 1982, Brockway et al. 2002, Samson et al. 2004). Often started though natural processes, such as lightning strikes, fire was also used as a tool to manage vegetation and grazing animals by Native Americans (Anderson 2006). Fires would have moved freely across the prairie, stopping only at barriers where fuel from residual vegetation was too sparse to carry flames such as the short grazing lawns of bison, sandy areas dominated by shrubs, or riparian areas with open water or wet vegetation.

The spatial and temporal scale of disturbance is important to consider when considering population dynamics of Lesser Prairie-Chickens. A herd of 4–6 million bison roaming freely would have removed nearly all the vegetation in an area of several thousand hectares before moving on. Return intervals for bison ranged from 1 to 8 years, allowing for the recovery of vegetation structure dependent on precipitation (Samson et al. 2004). Historical estimates of fire return intervals in the mixed-grass prairie are 5–10 years (Wright and Bailey 1982; Brockway et al. 2002; Samson et al. 2004). Thus, one possible adaptation of Lesser Prairie-Chickens to large-scale spatial and temporal disturbances might be the ability of birds to disperse large distances to find suitable habitat for nesting, brood rearing, or feeding. Prior to European-American settlement, large patches of suitable habitat were available. Following settlement, however, the scale and type of disturbance began to change, making suitable habitat and necessary resources progressively more difficult to find. In this chapter, we review the literature on the impact of anthropogenic landscape changes on populations of Lesser Prairie-Chickens. We then use a Geographic Information System to complete a preliminary analysis to test the hypothesis that lek locations of Lesser Prairie-Chickens are negatively associated with anthropogenic structures.

ANTHROPOGENIC IMPACTS ON LESSER PRAIRIE-CHICKENS

In the late 1800s, the Southern Great Plains were opened to homesteading through the Homestead Act of 1862. Through this Act, settlers could claim 160 acres (64.7 ha) of land and "prove up" the land—build a home, plant crops, and raise livestock to support one's family. After five years, the land was theirs if they were still living on it in a permanent dwelling and growing crops. The policy led to a slow conversion of prairie to row crop farming. The semiarid nature of the High Plains prevented large-scale conversion of the prairie until the early 1900s, when a series of high rainfall years allowed for extensive conversion of the Southern Great Plains to row crop agriculture (Egan 2006). Land use conversion, and the subsequent drought of the 1930s, nearly doomed the Lesser Prairie-Chicken. After the drought of the 1930s, some land reverted back to prairie after row crop agriculture was abandoned. Subsequently, the development of technology to exploit the High Plains (or Ogallala) Aquifer after World War II made row crop farming a more economically stable endeavor. Currently, <50% of the historic range of short- and mixed-grass prairie remains as native grasslands (Samson et al. 2004).

Lesser Prairie-Chickens will use crop fields in the winter while foraging for waste grain, but large conversions of native grassland to cropland have had a negative impact on the species (Crawford and Bolen 1976). Early studies of population responses of Lesser Prairie-Chickens to habitat conversion in Texas indicated that high population size and densities of the species were supported in areas in which greater than 63% of a 2,331 ha landscape remained in native prairie (Crawford and Bolen 1976). However, finding locations on the landscape where habitat is available in large enough patches to support the nesting and brooding activities of the species is challenging. In one study in New Mexico, only 16% of Lesser Prairie-Chicken habitat was in patches >3,200 ha, the minimum size deemed necessary to support a population of the species (Johnson et al. 2006). Only 11% of habitat was in patches of a more conservative 7,238 ha patch size (Johnson et al. 2006).

Woodward et al. (2001) documented that landscapes with declining lek attendance by Lesser Prairie-Chickens had less native shrubland and greater rates of landscape change compared to sites where lek attendance was stable. It appears that land cover type was less important than the rate of change of the land cover. In a follow-up study, Fuhlendorf et al. (2002) investigated how landscape structure at different spatial scales was related to the lek status of Lesser Prairie-Chicken as declining or stable. Declining lek sites had higher edge density at smaller spatial scales (452, 905 and 1,810 ha) and greater landscape change at large spatial scales (3,619 and 7,238 ha). The authors concluded that fragmentation of continuous grasslands and native shrublands was a critical reason for declining Lesser Prairie-Chicken populations.

A. Bartuszevige and A. Daniels (unpubl. data) evaluated the relationship between lek presence and density of Lesser Prairie-Chickens to the amount of grassland and U.S. Department of Agriculture Conservation Reserve Program (CRP) grassland on the landscape at multiple spatial scales. The authors found that the amount of grassland and CRP was greater at lek sites than at available random sites but, counterintuitively, was greater at sites with a low lek density of 1 lek per 4.8 km radius area versus sites with a high lek density of >1 lek per 4.8 km radius buffer, or the available random sites. The results might be explained by the connectivity of landscape structure or by grassland condition, neither of which was measured.

One result of the conversion of the prairie to agricultural crop production and use for cattle grazing is the number of fences erected. Patten et al. (2005) documented greater fence line density in Oklahoma compared to New Mexico; the differences in fence density were linked to different patterns of historical land settlement between the two states. Differences in fencing appear to be linked to state differences in rates of collision mortality of Lesser Prairie-Chickens, with losses being more than twice as high in Oklahoma (32.3%) compared with New Mexico (13.3%, Patten et al. 2005). Grisham (2012) found no evidence of fence collision mortalities for Lesser Prairie-Chickens in west Texas, but noted a low density of fences in his study area. Mortality rates of Lesser Prairie-Chicken collisions with fences have not been studied throughout the range of the species, and the population level effects due to collisions with anthropogenic structures and horizontal wires are unknown.

Most conversion of prairie to row crop agriculture in the Southern Great Plains happened before 1959 (Licht 1997), although one recent study found that 6% of a study area in the range of Lesser Prairie-Chickens in New Mexico had been converted from native habitat types in the 22-year period between 1981 and 2002 (Johnson et al. 2006). However, loss and fragmentation of native prairies have not stopped; increased commodity prices, energy development, roads, and transmission lines have been documented to have negative impacts on Lesser Prairie-Chicken habitats (Pitman et al. 2005, Pruett et al. 2009a). Once again, early warnings were given by wildlife managers. Based on observations at telegraph lines, Ligon (1951) predicted that roads, powerlines, and oil and gas development would all have significant impacts on prairie chickens.

Hagen et al. (2011) found that areas used by Lesser Prairie-Chickens were less likely to contain roads, oil and gas wells, and transmission towers than unused areas. More specifically, Pitman et al. (2005) documented that Lesser Prairie-Chickens nests were located further from anthropogenic features at two field sites including transmission lines, oil and gas wellheads, buildings, improved roads, and center-pivot irrigation pumps. Pruett et al. (2009a) found that mortality of Lesser Prairie-Chickens occurred due to collisions with powerlines, but that mortality rates were low. In addition, radio-marked Lesser Prairie-Chickens avoided habitats that were close to roads and transmission lines (Pruett et al. 2009a), and the small number of birds that were found near these features rarely crossed beneath or over the feature. No research to date has specifically addressed wind energy development but inferences have been made from impacts of other features and investigations of wind energy on Greater Prairie-Chickens (*Tympanuchus cupido*; Winder et al. 2014).

Apparent avoidance of vertical structures has been demonstrated in Lesser Prairie-Chickens (Pitman et al. 2005, Pruett et al. 2009a, Hagen et al. 2011), but true avoidance based on behavioral observations has not been documented in the species, except in congeneric populations of Great Prairie-Chickens (Winder et al. 2014). In addition, the mechanisms responsible for the apparent avoidance have not yet been determined. Two hypotheses are often posited. First, vertical structures are thought to create perches for avian predators that may increase predation risk for prairie grouse (Walters et al. 2014). Second, other studies have predicted that Lesser Prairie-Chickens are open country birds that may demonstrate neophobia or a fear of new vertical structures in their environment. To date, no study has documented increased predation rates near vertical structures and few have documented increased use by avian predators at vertical structures (Behney et al. 2012). In addition, no studies have directly tested the neophobia hypothesis (Walters et al. 2014).

Regardless, energy development has proceeded throughout the range of Lesser Prairie-Chickens, and wind energy development in particular has caused increased concerns (Pruett et al. 2009b, Jarnevich and Laubhan 2011). Various studies have estimated the amount of usable Lesser Prairie-Chicken habitat available when anthropogenic structures are considered in habitat calculations. Pitman et al. (2005) estimated that the presence of anthropogenic features made 7,114 ha (53%) of a 13,380 ha field site in Kansas unusable by Lesser Prairie-Chickens. Dusang (2011) estimated a potential 1,200 km² loss of habitat (7%) due to anthropogenic features on 18,000 km² available for the species in Oklahoma.

The studies described here evaluate apparent avoidance by individuals to anthropogenic features. Threshold distances were then used to extrapolate to a potential loss of habitat, which infers a population-level response. However, no range-wide studies have evaluated potential loss of accessible habitat due to the presence of anthropogenic structures (Eigenbrod et al. 2008). In addition, the potential loss of habitat from anthropogenic structures should be evaluated in the context of current habitat conditions following historic conversions of grassland to cropland. We tested the hypothesis that leks were present and at greater density in landscapes with more usable habitat not affected by apparent avoidance of anthropogenic features, which was arranged in larger patches and with less edge density.

METHODS

The basis of our analysis was the locations of leks of Lesser Prairie-Chickens, with leks serving as a surrogate for population presence in the prairie landscape. Leks are focal breeding areas for this species where males aggregate to display and mate with females during the breeding season. Leks are often traditional sites used in multiple years, and due to high site fidelity of males, leks are often used as an index of population distribution and status (Fuhlendorf et al. 2002, Hagen and Giesen 2005).

We obtained lek locations from all five states within the distribution of Lesser Prairie-Chickens. In 2012, the Western Association of Fish and Wildlife Agencies (WAFWA) funded a helicopter survey throughout the range of Lesser Prairie-Chickens. The pilot survey used a random sampling approach to select blocks and transects to survey, thus, providing publicly available, randomly sampled, lek data using a consistent protocol (McRoberts et al. 2011a,b). The WAFWA helicopter surveys were based on randomly selected 256, 15 km × 15 km blocks throughout the Lesser Prairie-Chicken range. Two transects spaced 750 m apart were flown within each block, with three observers noting lek location and number of birds per lek. Additional details can be found in McDonald et al. (2014). Using leks located in this survey for our landscape-scale analysis allowed us to infer results across the entire range of the species. Previous analyses have been limited to specific study site locations due to the ground-based methods used to locate leks.

We used 40 leks in our analysis (Figure 11.1), plus 30 sample points where Lesser Prairie-Chicken leks were not detected. Unused points were randomly selected from among the surveyed blocks of the WAFWA group helicopter survey in which a Lesser Prairie-Chicken lek was not observed. Unused points were selected from these blocks to increase probability that absence points were measured. In addition, unused points were >20 km from the next nearest known lek point or other unused point to reduce overlap among buffers.

The Playa Lakes Joint Venture (PLJV) maintains a six-state land cover dataset with 30 m spatial resolution. The land cover was created to assess the types and distribution of habitats available to birds in the Short- and Mixed-Grass Prairie Ecoregions. PLJV developed the land cover by integrating the most current and detailed data into a seamless thematic map such that land cover classes were consistent across state boundaries (McLachlan 2012). The base spatial layer in the land cover is the U.S. Geological Survey (USGS) Gap Analysis Program (GAP), the

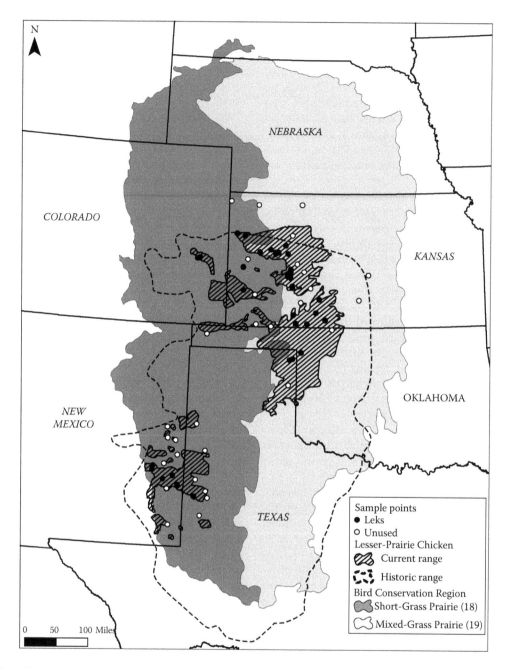

Figure 11.1. Lek and unused sample sites in the current and historic range of Lesser Prairie-Chickens in the Southern Great Plains of the United States. Four unused sites were outside of the historic range of the Lesser Prairie-Chicken but were sampled in helicopter surveys of the Western Association of Fish and Wildlife Agencies (McDonald et al. 2014), and included as unused sites in our sampling design.

USGS Southwest Regional GAP (ReGAP), and the NatureServe Ecosystems dataset. The accuracy of the land cover was assessed on a state-by-state basis because each state had different component layers. Accuracy ranged from 44% to 74%; grassland types and native shrubland types used by Lesser

Prairie-Chickens were most often confused with each other (McLachlan 2012); therefore, we considered the Grassland category sufficiently accurate for our purposes. One exception was Texas (44% accuracy); thus, the land cover was updated with recent GIS data from the Texas Ecological Systems Mapping

TABLE 11.1
Sources for spatial data for locations of anthropogenic features in the range of Lesser Prairie-Chickens.

Feature	Source
Oil and gas wells	Colorado Oil and Gas Conservation Commission: http://cogcc.state.co.us
	Kansas Geological Survey: http://www.kgs.ku.edu
	Oklahoma Corporation Commission—Oil & Gas Division: http://www.occeweb.com/
	Texas Rail Road Commission: http://www.rrc.state.tx.us
	New Mexico Oil & Natural Gas Administration & Revenue Database (ONGARD): http://gotech.nmt.edu/gotech
Transmission lines	Kansas Corporation Commission
	Oklahoma Wind Power Initiative
	Energy Information Administration States 3.0 Project
	Ventyx Velocity Suite
Roads	ESRI ArcGIS Data & Maps v. 10 North America Detailed Streets layer
Wind turbines	Federal Aeronautics Administration Digital Obstacle File: https://nfdc.faa.gov/

NOTE: Anthropogenic features were buffered by distances of 10–1.4 km (Table 11.2), and areas outside of the buffered areas were classified as unimpacted.

Project, which has an accuracy of 75% (Missouri Resource Assessment Partnership 2015). The accuracy of land enrolled in the U.S. Department of Agriculture Farm Services Agency CRP layer (2009) ranged from 87% to 100% (McLachlan 2012). The land cover was updated with 2011 CRP data (U.S. Department of Agriculture Farm Services Agency 2011). In the data layer, there were 22 land cover types designated. We aggregated land cover types into six major categories (Cropland, CRP, Developed, Grassland, Wetland, and Woodland). However, only CRP and Grassland were used in statistical analyses because cover in Cropland and Grassland was inversely related and we were most interested in describing grassland dynamics. Thus, Cropland was excluded from analyses. The Grassland category included short- and mixed-grass prairie, sand sagebrush prairie, and sand shinnery oak prairie. The Grassland category does not imply suitable condition of the vegetation communities; rather, it is a broad category that includes vegetation community types that may be used by Lesser Prairie-Chicken if local habitats are in suitable condition.

We modified the land cover dataset to identify areas impacted by anthropogenic infrastructure (Table 11.1). Vertical structures and roads are features thought to influence habitat use by Lesser Prairie-Chickens; we investigated impacts of those features commonly found in the Southern Great Plains where management recommendations are made (Table 11.2). We used recommended buffer distances around anthropogenic infrastructure from Hagen et al. (2011; Table 11.2) and categorized any habitat within those buffer distances as "impacted." For unimproved roads and power lines (<69 kV), we used recommended buffer distances taken from the Lesser Prairie-Chicken Range-wide Conservation Plan (Van Pelt et al. 2013; Table 11.2).

There is no clear landscape scale defined in the literature as being important to maintain populations of Lesser Prairie-Chickens.

TABLE 11.2
Distances used to buffer anthropogenic features in the range of Lesser Prairie-Chickens.

Feature	Distance (m)	Source
Oil and gas wells	300	Hagen et al. (2011)
Transmission lines (≥69 kV)	700	Hagen et al. (2011)
Distribution lines (<69 kV)	10	Van Pelt et al. (2013)
Roads (improved)	850	Hagen et al. (2011)
Roads (unimproved)	67	Van Pelt et al. (2013)
Buildings	1400	Hagen et al. (2011)
Wind turbines	1400	Hagen et al. (2011)

Woodward et al. (2001) found differences in lek response at 4.8 km radius buffer (7,238 ha) scale. Fuhlendorf et al. (2002) also showed that landscape-scale changes at 4.8 km best explained the difference between leks classified as declining versus stable. However, A. Bartuszevige and A. Daniels (unpubl. data) detected differences in landscape composition at lek sites out to 10-km buffer radius (31,416 ha). Thus, we chose to evaluate potential differences in landscape structure and composition at three buffer distances corresponding to landscape scales at which differences have been detected among leks of Lesser Prairie-Chickens (Woodward et al. 2001, Fuhlendorf et al. 2002, Bartuszevige and Daniels, in review). Using Playa Lakes Joint Venture's land cover data, we derived the area of CRP and Grassland within circular 3,000, 5,000, and 10,000 m radius buffers around Lek and unused points. The buffers correspond to areas of 2,827 ha, 7,885 ha, and 31,416 ha. We calculated the area of the landscape that was assumed to be unimpacted as the amount of area outside the impact buffer of an anthropogenic feature regardless of habitat type or suitability, and a combined Grassland+CRP category. We calculated four metrics of landscape structure using the tools of FRAGSTATS (McGarigal et al. 2012): Percent Land Cover, Largest Patch Index, Edge Density, and Coefficient of Variation of Patch size.

Statistical Analysis

When any two points in space are closer to each other than they are to another point, the two close points may have more in common with each other than with the other points in space due to positive spatial autocorrelation. A Moran's I analysis confirmed positive spatial autocorrelation in the data for all but three variables ($\alpha = 0.05$). Thus, we used a Generalized Linear Mixed Model logistic regression with a spatial autocorrelation correction (Dormann et al. 2007). Generalized Linear Mixed Models describe a class of models that account for both fixed effects and random effects; in this case, those effects associated with data points that are clustered in space. We tested models with multiple fixed effects (Unimpacted and Grassland+CRP landscape parameters), and models with single fixed effects within each buffer class (Unimpacted or Grassland+CRP landscape parameters). We used the MASS package

in program R to complete statistical analyses. All the buffers and landscape variables were selected *a priori*, and we did not correct for multiple comparisons, but we considered tests significant at $\alpha < 0.10$. However, data among buffers were autocorrelated and our results should be considered within that context (i.e., the Largest Patch Index in the 3,000-m buffer affects the result for the Largest Patch Index in the 5,000-m and 10,000-m buffers).

RESULTS

3,000-m Buffer

At a 3,000-m buffer distance, the Percent of Landscape that was unimpacted by anthropogenic features was significantly related to lek presence (Table 11.3). However, the Unimpacted Percent of Landscape and Grassland+CRP Percent of Landscape had a high correlation value ($r = -0.52$) when both factors were included in the model together. When evaluated separately, Unimpacted Percent of Landscape and Grassland+CRP Percent of Landscape both appear related to lek presence (Table 11.4, Figures 11.2 and 11.3).

Unimpacted Largest Patch Index and Grassland+CRP Largest Patch Index explained lek presence but both variables were correlated when they were included in the model together ($r = -0.78$). When evaluated separately, both Unimpacted Largest Patch Index and Grassland+CRP Largest Patch Index appear related to lek presence (Table 11.4, Figures 11.4 and 11.5).

Unimpacted Edge Density and Grassland+CRP Edge Density did not explain lek presence when tested in the same model or individually (Tables 11.3 and 11.4), nor were they correlated when included in the model together ($r = -0.26$). Unimpacted Coefficient of Variation of Patch Area and Grassland+CRP Coefficient of Patch Area did not explain lek presence when tested in the same model or individually (Tables 11.3 and 11.4) and they were not correlated when they were included in the model together ($r = -0.38$).

5,000-m Buffer

At the 5,000-m buffer distance, Grassland+CRP Percent of Landscape was related to lek presence (Table 11.3). Unimpacted Percent of Landscape and Grassland+CRP Percent of Landscape were

TABLE 11.3

Logistic parameter estimates from a Generalized Linear Mixed Model of Lek Sites of Lesser Prairie-Chickens versus unused sites at 3,000-m, 5,000-m, and 10,000-m buffer distances.

Scale	Estimate	SE	df	t	P ≤
3,000 m					
Grassland+CRP Percent of Landscape[a]	0.014	0.010	67	1.37	0.18
Unimpacted Percent of Landscape[a]	0.030	0.017	67	1.75	0.08
Grassland+CRP Largest Patch Index[a]	5.4e13	2.1e13	67	2.52	0.01
Unimpacted Largest Patch Index[a]	-6.7e13	1.8e13	67	-3.64	0.001
Grassland+CRP Edge Density	0.020	0.021	67	0.94	0.35
Unimpacted Edge Density	-0.028	0.031	67	-0.92	0.36
Grassland+CRP CV Patch Area	-0.002	0.003	67	-0.60	0.55
Unimpacted CV Patch Area	0.001	0.004	67	0.17	0.87
5,000 m					
Grassland+CRP Percent of Landscape	0.026	0.014	67	1.82	0.07
Unimpacted Percent of Landscape	0.027	0.019	67	1.45	0.15
Grassland+CRP Largest Patch Index[a]	-1.1e11	2.6e13	67	-0.004	0.99
Unimpacted Largest Patch Index[a]	2.3e13	2.3e13	67	0.98	0.33
Grassland+CRP Edge Density	-5.0e13	2.7e13	67	-1.82	0.07
Unimpacted Edge Density	2.0e13	3.3e13	67	0.62	0.54
Grassland+CRP CV Patch Area	0.0002	0.002	67	0.09	0.93
Unimpacted CV Patch Area	0.001	0.004	67	0.26	0.79
10,000 m					
Grassland+CRP Percent of Landscape[a]	0.032	0.021	67	1.51	0.13
Unimpacted Percent of Landscape[a]	0.031	0.023	67	1.21	0.23
Grassland+CRP Largest Patch Index	0.032	0.031	67	1.02	0.31
Unimpacted Largest Patch Index	0.033	0.033	67	1.02	0.31
Grassland+CRP Edge Density	0.046	0.031	67	1.49	0.14
Unimpacted Edge Density	-0.003	0.015	67	-0.19	0.85
Grassland+CRP CV Patch Area	0.001	0.001	67	0.59	0.56
Unimpacted CV Patch Area	0.0003	0.004	67	0.07	0.95

NOTE: Models tested the contributions of Unimpacted area and Grassland or CRP areas to explain differences between Lek and unused sites. Unimpacted Area refers to areas not impacted by anthropogenic features and outside of buffer distances (Table 11.2). Grassland+CRP refers to area that has been classified as either Grassland or Conservation Reserve Program (CRP) grassland.

[a] Indicates that the variables were correlated with each other (r > |0.5|) in the models.

not correlated when they were included in the model together (r = -0.50). When evaluated separately, both Unimpacted Percent of Landscape and Grassland+CRP Percent of Landscape appear to be related to lek presence (Table 11.4, Figures 11.2 and 11.3).

Unimpacted Largest Patch Index and Grassland+CRP Largest Patch Index did not explain lek presence when modeled together, but the variables were negatively correlated when included in the model together (r = -0.83). When evaluated separately, both Unimpacted Largest Patch Index and Grassland+CRP Largest Patch Index appear to be related to lek presence (Table 11.4, Figures 11.4 and 11.5).

Grassland+CRP Edge Density was related to lek presence when tested in the same model as Unimpacted Edge Density (Table 11.3). Unimpacted Edge Density and Grassland+CRP Edge Density were not correlated (r = -0.11). However, when

TABLE 11.4
Logistic parameter estimates from a Generalized Linear Mixed Model of Lek Sites of Lesser Prairie-Chickens versus unused sites at 3,000-m, 5,000-m, and 10,000-m buffer distances.

Scale	Unimpacted area					Grassland+CRP area				
	Estimate	SE	df	t	P	Estimate	SE	df	t	P
3,000 m										
Percent of Landscape	0.041	0.016	68	2.58	0.01†	0.022	0.008	68	2.69	0.009†
Largest Patch Index	0.015	0.008	68	1.91	0.06†	0.021	0.010	68	2.16	0.03†
Edge Density	-0.021	0.030	68	-0.70	0.48	0.015	0.020	68	0.74	0.46
CV Patch Area	-0.0003	0.004	68	-0.07	0.95	-0.001	0.002	68	-0.58	0.56
5,000 m										
Percent of Landscape	0.043	0.017	68	2.48	0.02†	0.035	0.011	68	3.00	0.004†
Largest Patch Index	4.5e13	1.1e13	68	3.97	0.0002†	0.033	0.015	68	2.24	0.03†
Edge Density	-0.019	0.024	68	-0.79	0.43	0.025	0.024	68	1.06	0.29
CV Patch Area	0.001	0.004	68	0.34	0.74	0.0004	0.002	68	0.22	0.83
10,000 m										
Percent of Landscape	0.053	0.021	68	2.49	0.02†	0.047	0.017	68	2.77	0.007†
Largest Patch Index	0.048	0.028	68	1.70	0.09†	0.054	0.031	68	1.72	0.09†
Edge Density	-0.004	0.015	68	-0.25	0.80	0.046	0.030	68	1.51	0.13
CV Patch Area	0.001	0.004	68	0.21	0.84	0.001	0.001	68	0.63	0.53

NOTE: Models estimated the individual contributions of landscape parameters to explaining differences between Lek and unused sites. Unimpacted Area refers to areas not impacted by anthropogenic features and outside of buffer distances (Table 11.2). Grassland+CRP refers to area that has been classified as either Grassland or Conservation Reserve Program (CRP) grassland. An † indicates the variable was significant at $\alpha \leq 0.10$.

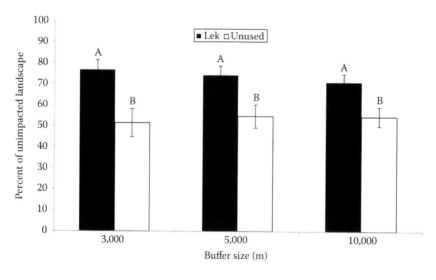

Figure 11.2. Percent of the landscape ($\bar{x} \pm 90\%$ CI) not impacted by anthropogenic features (roads, transmission and powerlines, buildings, oil and gas wells, and wind turbines) at Leks of Lesser Prairie-Chicken and unused sites at 3,000-m, 5,000-m, and 10,000-m buffer distances. Different letters within buffer distances indicate significant differences at $\alpha = 0.10$.

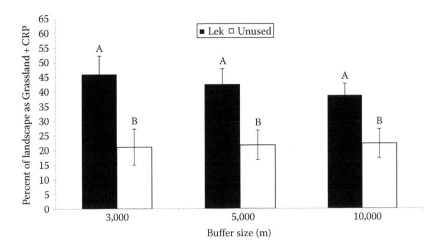

Figure 11.3. Percent of Landscape ($\bar{x} \pm 90\%\,CI$) categorized as Grassland or Conservation Reserve Program (CRP) grasslands at Leks of Lesser Prairie-Chicken and unused sites at 3,000-m, 5,000-m, and 10,000-m buffer distances. Different letters within buffer distances indicate significant differences at $\alpha = 0.10$.

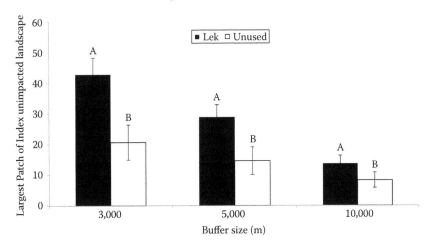

Figure 11.4. Largest Patch Index ($\bar{x} \pm 90\%\,CI$) for areas not impacted by anthropogenic features (roads, transmission and powerlines, buildings, oil and gas wells, and wind turbines) at Leks of Lesser Prairie-Chicken and unused sites at 3,000-m, 5,000-m, and 10,000-m buffer distances. Different letters within buffer distances indicate significant differences at $\alpha = 0.10$.

tested individually, neither Unimpacted Edge Density nor Grassland+CRP Edge Density appeared to be related to lek presence (Table 11.4).

Unimpacted Coefficient of Variation of Patch Area and Grassland+CRP Coefficient of Patch Area did not explain lek presence when tested in the same model or individually (Tables 11.3 and 11.4), and were not correlated ($r = -0.41$).

10,000-m Buffer

At the 10,000-m buffer distance, neither Unimpacted Percent of Landscape nor Grassland+CRP Percent of Landscape appeared to be related to lek presence

(Table 11.3). However, the Unimpacted Percent of Landscape and Grassland+CRP Percent of Landscape were negatively correlated when they were included in the model together ($r = -0.57$). When evaluated separately, both Unimpacted Percent of Landscape and Grassland+CRP Percent of Landscape appeared to be related to lek presence (Table 11.4, Figures 11.2 and 11.3).

Unimpacted Largest Patch Index and Grassland+CRP Largest Patch Index did not explain lek presence when tested in a model together (Table 11.3) and were not correlated when they were included in the model together ($r = -0.43$). When evaluated separately, both Unimpacted

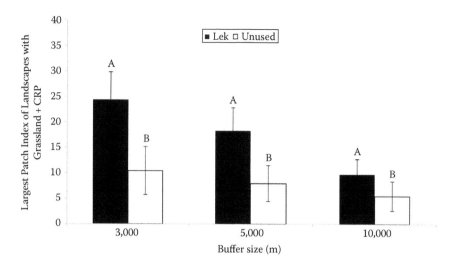

Figure 11.5. Largest Patch Index ($\bar{x} \pm 90\%\,\text{CI}$) for areas categorized as either Grassland or Conservation Reserve Program (CRP) grasslands at Leks of Lesser Prairie-Chicken and unused sites at 3,000-m, 5,000-m, and 10,000-m buffer distances. Different letters within buffer distances indicate significant differences at $\alpha = 0.10$.

Largest Patch Index and Grassland+CRP Largest Patch Index appeared to be related to lek presence (Table 11.4, Figures 11.4 and 11.5).

Unimpacted Edge Density and Grassland+CRP Edge Density did not explain lek presence when tested in the same model or individually (Tables 11.3 and 11.4), nor were they correlated when included in the model together ($r = 0.06$). Similarly, Unimpacted Coefficient of Variation of Patch Area and Grassland+CRP Coefficient of Patch Area did not explain lek presence when tested in the same model or individually (Tables 11.3 and 11.4), nor were they correlated when included in the model together ($r = -0.24$).

DISCUSSION

The Great Plains of the United States is commonly referred to as the "bread basket" of the United States, and the region has one of the highest densities of agricultural production in the western hemisphere. In addition, the Southern Great Plains is increasingly relied on to produce energy for the country. Oil and gas development is ongoing and expanding, with four of the five states with Lesser Prairie-Chickens ranked in the top 10 for wind energy development potential (National Renewable Energy Lab 2011). Several studies have documented that radio-marked Lesser Prairie-Chickens use habitat near anthropogenic features less than expected (Pitman et al. 2005,

Pruett et al. 2009a, Hagen et al. 2011). Continued anthropogenic encroachment will be combined with projected effects of climate change due to increasing temperature, less precipitation, and a higher frequency of extreme events, such as drought (Kunkel et al. 2013a,b; Shafer et al. 2014). An understanding of how populations of Lesser Prairie-Chickens respond to anthropogenic features, and the scale at which grassland amount and configuration is important to supporting populations will be essential for creating viable conservation plans for the species. The development of such conservation plans for populations of Lesser Prairie-Chickens has reached a critical juncture because population numbers declined to a level that led to federal listing as a threatened species under the Endangered Species Act in May 2014 (U.S. Fish and Wildlife Service 2014). However, the listing was vacated by judicial decision in September 2015.

Studies that investigate apparent avoidance distances of anthropogenic features by Lesser Prairie-Chicken have documented distances of up to 1 km from a feature (Pitman et al. 2005, Hagen et al. 2011). Models developed to estimate potential habitat availability have then used those distances to buffer features and determine the amount of otherwise suitable habitat avoided or lost to the species (Dusang 2011). Because no relationship of anthropogenic infrastructure to population presence or status had been investigated, we conducted preliminary analyses to investigate the relationship between the area impacted by anthropogenic features defined from the literature and an index of population distribution, here measured as lek presence.

The percent of the landscape that was defined as unimpacted by anthropogenic features was greater in areas with leks of Lesser Prairie-Chicken. The relationship extended to our largest buffered area (10,000 m, 31,416 ha) and demonstrated that presence of leks of Lesser Prairie-Chicken is negatively related to anthropogenic features, but mechanisms for this pattern remain unknown. Many studies have hypothesized causes for the apparent avoidance, but none have tested any hypotheses explicitly (Walters et al. 2014). Regardless, understanding the mechanism by which Lesser Prairie-Chickens avoid vertical structures and other anthropogenic features is important for devising effective methods of mitigation such as mufflers on oil and gas pumps or wind turbines to reduce noise pollution.

Percent of Landscape in Grassland+CRP was also an important factor for predicting lek presence. We classified all native prairie types as Grassland and included CRP in this category because previous work has shown that CRP was an important predictor of Lesser Prairie-Chicken presence (A. Bartuszevige and A. Daniels, unpubl. data). Therefore, the amount of native prairie and CRP in the landscape at broad spatial scales may be an important consideration for the conservation of Lesser Prairie-Chickens (but see Woodward et al. 2001, Fuhlendorf et al. 2002). However, the relative importance of amount of native prairie and CRP versus extent of impact by anthropogenic features is still unknown. We found many of the landscape variables had high values with each other in the correlation matrix for models where both variables were included (e.g., Unimpacted Percent of Landscape and Grassland+CRP Percent of Landscape at 3,000-m buffer distance). In addition, our preliminary results did not consider population persistence or stability. Male Lesser Prairie-Chickens are known to exhibit site fidelity to leks, and it may take multiple years before population loss or other potential impacts of anthropogenic features are detected.

Female Lesser Prairie-Chickens usually nest within 3 km of a lek, but not necessarily the lek at which they were captured or observed mating (Hagen and Giesen 2005, Pitman et al. 2005). Home range size for breeding birds is about 211–6,190 ha (Hagen and Giesen 2005). However, this area is the amount of habitat potentially used by a single bird. Using leks as a surrogate for population presence and persistence, researchers have found that much larger areas are associated with lek presence and stability. Fuhlendorf et al. (2002) documented that the lek stability of Lesser Prairie-Chickens was negatively related to landscape change at large spatial scales (7,238 ha). A. Bartuszevige and A. Daniels (unpubl. data) found that lek presence was related to the amount of grassland at all spatial scales tested, including the largest spatial scale of 31,416 ha.

However, the area available on the landscape is only part of the equation; our preliminary work demonstrates and agrees with previous work that landscape structure is also important (Fuhlendorf et al. 2002). One implicit assumption of the past studies reviewed in this chapter is that it is not just habitat loss that is detrimental to the Lesser Prairie-Chicken but also patterns of habitat fragmentation. We measured several landscape structure variables and found Largest Patch Index was the only factor related to lek presence of Lesser Prairie-Chickens. This result held for both the Unimpacted area analysis and the Grassland+CRP analysis.

Habitat loss due to conversion of grassland to agriculture has largely stabilized in the Southern Great Plains, but habitat loss through encroachment by invasive woody shrubs in the eastern portion of the species range is ongoing (Fuhlendorf et al. 2002, Johnson et al. 2006). Conversion of CRP fields back to agriculture may increase due to decreases in acreage allotments to CRP in the U.S. Farm Bill (U.S. Department of Agriculture, Farm Services Agency 2014). In addition, our compilation of results provides further evidence that habitat loss is occurring through increases in amount of human infrastructure. A further concern is that projected climate change has the potential to negatively impact the demography of Lesser Prairie-Chickens and may cause changes in structure of current habitat types (McLachlan et al. 2011, Grisham et al. 2013; Chapter 12, this volume). Conservation programs are capable of reversing or mitigating these losses, in particular U.S. Department of Agriculture Farm Bill Programs, but activities need to be focused in areas that will provide enough habitat in a large enough landscape to support populations of these birds (e.g., >30,000 ha).

Like most issues in conservation, there is not one clear answer that is responsible for declines of the species. The decline of the Lesser Prairie-Chicken is the cumulative result of multiple factors and long-term changes to their habitat. Thus, we must take action to create and conserve large, intact landscapes before the bird becomes extinct due to synergistic effects of multiple sources of anthropogenic change. Species recovery will not be easy and will require novel ideas and unconventional collaborations to achieve.

ACKNOWLEDGMENTS

Thanks to C. Boal and D. Haukos for inviting us to contribute to this volume of *Studies of Avian Biology* and for providing comments to improve our chapter. The Playa Lakes Joint Venture supported the work described in this chapter. Thanks to two anonymous reviewers for providing helpful comments that improved previous versions of our manuscript.

LITERATURE CITED

Allred, B. W., S. D. Fuhlendorf, and R. G. Hamilton. 2011. The role of herbivores in Great Plains conservation: comparative ecology of bison and cattle. Ecosphere 2:art26.

Anderson, R. C. 2006. Evolution and origin of the central grassland for North America: climate, fire, and mammalian grazers. Journal of the Torrey Botanical Society 133:626–647.

Augustine, D. J. 2010. Spatial versus temporal variation in precipitation in a semiarid ecosystem. Landscape Ecology 25:913–925.

Behney, A. C., C. W. Boal, H. A. Whitlaw, and D. R. Lucia. 2012. Raptor community composition in the Texas Southern High Plains Lesser Prairie-Chicken range. Wildlife Society Bulletin 36:291–296.

Brockway, D. G., R. G. Gatewood, and R. B. Paris. 2002. Restoring fire as an ecological process in shortgrass prairie ecosystems: initial effects of prescribed burning during the dormant and growing seasons. Journal of Environmental Management 65:135–152.

Colvin, W. S. 1914. The Lesser Prairie Hen. Outing: Sport, Adventure, Travel, Fiction 63:608–614.

Crawford, J. A. 1980. Status, problem, and research needs of the Lesser Prairie Chicken. Pp. 1–7 *in* P. A. Vohs, and F. L. Knopf (editors), Proceedings Prairie Grouse Symposium, Oklahoma State University, Stillwater, OK.

Crawford, J. A., and E. G. Bolen. 1976. Effects of land use on Lesser Prairie-Chickens in Texas. Journal of Wildlife Management 40:96–104.

Dormann, C. F., J. M. McPherson, M. B., Araújo, R. Bivand, J. Bolliger, G. Carl, R. G. Davies, A. Hirzel, W. Jetz, W. D. Kissling, I. Kuhn, R. Ohlemuller, P. R. Peres-Neto, B. Reineking, B. Schroder, F. M. Schurr, and R. Wilson. 2007. Methods to account for spatial autocorrelation in the analysis of species distributional data: a review. Ecography 30:609–628.

Dusang, D. 2011. Impacts of energy development on the Lesser Prairie-Chicken ecology and management. M.S. thesis, University of Oklahoma, Norman, OK.

Egan, T. 2006. The worst hard time. Houghton Mifflin, New York, NY.

Eigenbrod, F., S. J. Hecnar, and L. Fahrig. 2008. Accessible habitat: an improved measure of the effects of habitat loss and roads on wildlife populations. Landscape Ecology 23:159–168.

Flores, D. A., and R. D. Evans. 1991. Bison ecology and bison diplomacy: the southern plains from 1800–1850. Journal of American History 78:465–485.

Fuhlendorf, S. D., A. J. W. Woodward, D. M. Leslie, Jr., and J. S. Shackford. 2002. Multi-scale effects of habitat loss and fragmentation on Lesser Prairie-Chicken populations in the U.S. Southern Great Plains. Landscape Ecology 17:617–628.

Giesen, K. M. 2000. Population status and management of Lesser Prairie-Chicken in Colorado. Prairie Naturalist 32:137–148.

Grisham, B. A. 2012. The ecology of Lesser Prairie-Chickens in shinnery oak-grassland communities in New Mexico and Texas with implications toward habitat management and future climate change. Ph.D. dissertation, Texas Tech University, Lubbock, TX.

Grisham, B. A., C. W. Boal, D. A. Haukos, D. M. Davis, K. K. Boydston, C. Dixon, and W. R. Heck. 2013. The predicted influence of climate change on Lesser Prairie-Chicken reproductive parameters. PLoS One 8:e68225.

Hagen, C. A., and K. M. Giesen. [online]. 2005. Lesser Prairie-Chicken (*Tympanuchus pallidicinctus*). Birds of North America 364. <http://bna.birds.cornell.edu/bna/species/364/articles/introduction> (1 November 2013).

Hagen, C. A., J. C. Pitman, T. M. Loughin, B. K. Sandercock, R. J. Robel, and R. D. Applegate. 2011. Impacts of anthropogenic features on habitat use by Lesser Prairie-Chickens. Pp. 63–75 *in* B. K. Sandercock, K. Martin, and G. Segelbacher (editors), Ecology, conservation, and management of grouse, Studies in Avian Biology (no. 39), University of California Press, Berkeley, CA.

Henika, F. S. 1940. Present status and future management of the prairie chicken in region 5. Texas Fish, Game and Oyster Commission, Division of Wildlife Restoration. Project 1-R. Special Report. Lubbock, TX.

Jarnevich, C. S., and M. K. Laubhan. 2011. Balancing energy development and conservation: a method utilizing species distribution models. Environmental Management 47:926–936.

Johnson, K., T. B. Neville, and P. Neville. 2006. GIS habitat analysis for Lesser Prairie-Chickens in southeastern New Mexico. BMC Ecology 6:art18.

Karl, T. R., B. E. Gleason, M. J. Menne, J. R. McMahon, R. R. Heim, Jr., M. J. Brewer, K. E. Kunkel, D. S. Arndt, J. L. Privette, J. J. Bates, P. Y. Groisman, and D. R. Easterling. 2012. U.S. temperature and drought: recent anomalies and trends. Eos, Transactions, American Geophysical Union 93:473–474.

Krueger, K. 1986. Feeding relationships among bison, pronghorn, and prairie dogs: an experimental analysis. Ecology 67:760–770.

Kunkel, K. E., L. E. Stevens, S. E. Stevens, L. Sun, E. Janssen, D. Wuebbles, M. C. Kruk, D. P. Thomas, M. D. Shulski, N. A. Umphlett, K. G. Hubbard, K. Robbins, L. Romolo, and A. Akyuz. [online]. 2013a. Regional climate trends and scenarios for the U.S. national climate assessment. Part 4: climate of the U.S. Great Plains. NOAA Technical Report NESDIS 142-4, Washington, DC. <http://www.nesdis.noaa.gov/technical_reports/NOAA_NESDIS_Tech_Report_142-4-Climate_of_the_U.S.%20Great_Plains.pdf> (31 July 2014).

Kunkel, K. E., L. E. Stevens, S. E. Stevens, L. Sun, E. Janssen, D. Wuebbles, K. T. Redmond, J. G. Dobson. [online]. 2013b. Regional climate trends and scenarios for the U.S. national climate assessment. Part 5: climate of the Southwest U.S. NOAA Technical Report NESDIS 142-5, Washington, DC. <http://www.nesdis.noaa.gov/technical_reports/NOAA_NESDIS_Tech_Report_142-5-Climate_of_the_Southwest_U.S.pdf> (31 July 2014).

Licht, D. S. 1997. Ecology and economics of the Great Plains. University of Nebraska Press, Lincoln, NE.

Ligon, J. S. 1951. Prairie chickens, highways and power lines. New Mexico Magazine 29:29.

McDonald, L., G. Beauprez, G. Gardner, J. Griswold, C. Hagen, F. Hornsby, D. Klute, S. Kyle, J. Pitman, T. Rintz, D. Schoeling, and W. Van Pelt. 2014. Range-wide population size of the Lesser Prairie-Chicken: 2012 and 2013. Wildlife Society Bulletin 38:536–546.

McGarigal, K., S. A. Cushman, and E. Ene. [online]. 2012. FRAGSTATS v4: spatial Pattern analysis program for categorical and continuous maps. University of Massachusetts, Amherst, MA. <http://www.umass.edu/landeco/research/fragstats/fragstats.html> (31 July 2014).

McHugh, T. 1972. The time of the Buffalo. Knopf, New York, NY.

McLachlan, M., A. Bartuszevige, and D. Pool. [online]. 2011. Evaluating the potential of Conservation Reserve Program to offset projected impacts of climate change on the Lesser Prairie-Chicken (*Tympanuchus pallidicinctus*). Report submitted to Natural Resources Conservation Service, Playa Lakes Joint Venture, Lafayette, CO. <http://www.nrcs.usda.gov/Internet/FSE_DOCUMENTS/stelprdb1041603.pdf> (31 July 2014).

McLachlan, M. M. 2012. Playa Lakes Joint Venture Landcover Accuracy Assessment Report. Playa Lakes Joint Venture, Lafayette, CO. <www.pljv.org>.

McRoberts, J. T., M. J. Butler, W. B. Ballard, M. C. Wallace, H. A. Whitlaw, and D. A. Haukos. 2011a. Response of Lesser Prairie-Chickens on leks to aerial surveys. Wildlife Society Bulletin 35:27–31.

McRoberts, J. T., M. J. Butler, W. B. Ballard, H. A. Whitlaw, D. A. Haukos, and M. C. Wallace. 2011b. Detectability of Lesser Prairie-Chicken leks: a comparison of surveys from aircraft. Journal of Wildlife Management 75:771–778.

Missouri Resource Assessment Partnership. [online]. 2015. Texas ecological systems classification. <http://morap.missouri.edu/index.php/texas-ecological-systems-classification/> (20 October 2015).

National Renewables Energy Laboratory. [online]. 2011. NREL triples previous estimates of U.S. wind power potential. NREL/FS-6A42-51555, Golden, CO. <http://www.nrel.gov/docs/fy11osti/51555.pdf> (31 July 2014).

Patten, M. A., D. H. Wolfe, E. Shochat, and S. K. Sherrod. 2005. Habitat fragmentation, rapid evolution and population persistence. Evolutionary Ecology Research 7:235–249.

Pitman, J. C., C. A. Hagen, R. J. Robel, T. M. Loughin, and R. D. Applegate. 2005. Location and success of Lesser Prairie-Chicken nests in relation to vegetation and human disturbance. Journal of Wildlife Management 69:1259–1269.

Pruett, C. L., M. A. Patten, and D. A. Wolfe. 2009a. Avoidance behavior by prairie grouse: implications for development of wind energy. Conservation Biology 23:1253–1259.

Pruett, C. L., M. A. Patten, and D. A. Wolfe. 2009b. It's not easy being green: wind energy and a declining grassland bird. BioScience 59:257–262.

Ridgeway, R. 1885. Some amended names of North American birds. Proceedings of the U.S. National Museum 8:354.

Samson, F. B., F. L. Knopf, and W. Ostlie. 2004. Great Plains ecosystems: past, present and future. Wildlife Society Bulletin 32:6–15.

Shafer, M., D. Ojima, J. M. Antle, D. Kluck, R. A. McPherson, S. Petersen, B. Scanlon, and K. Sherman. 2014. Great Plains. Pp. 441–461 in J. M. Melillo, T. C. Richmond, and G. W. Yohe (editors), Climate change impacts in the United States: the third national climate assessment, U.S. Global Change Research Program, Washington, DC.

Taylor, M. A., and F. S. Guthrey. 1980. Status, ecology and management of the Lesser Prairie Chicken. U.S. Department of Agriculture Forest Service General Technical Report RM-77, Rocky Mountain Forest and Regional Experiment Station, Fort Collins, CO.

U.S. Department of Agriculture Farm Services Agency. 2009. Common Land Unit Data Layer. Washington, D.C.

U.S. Department of Agriculture Farm Service Agency. [online]. 2011. Conservation Reserve Program fiscal year summary 2011. U.S. Department of Agriculture Farm Service Agency. <http://www.fsa.usda.gov/Internet/FSA_File/annualsummary2011.pdf> (31 July 2014).

U.S. Department of Agriculture Farm Service Agency. [online] 2014. 2014 Farm Bill fact sheet. U.S. Department of Agriculture Farm Service Agency. <http://www.fsa.usda.gov/Internet/FSA_File/2014_farm_bill_customers.pdf> (31 July 2014).

U.S. Fish and Wildlife Service. 2014. Endangered and threatened wildlife and plants; special rule for the Lesser Prairie-Chicken. Federal Register 79:20074–20085.

Van Pelt, W. E., S. Kyle, J. Pitman, D. Klute, G. Beauprez, D. Schoeling, A. Janus, and J. Haufler. [online]. 2013. The Lesser Prairie-Chicken range-wide conservation plan. Western Association of Fish and Wildlife Agencies, Cheyenne, WY. <http://www.wafwa.org/documents/2013LPCRWPfinalfor4drule12092013.pdf> (31 July 2014).

Walters, K., K. Kosciuch, and J. Jones. 2014. Can the effect of tall structures on birds be isolated from other aspects of development? Wildlife Society Bulletin 38:250–256.

Winder, V. L., L. B. McNew, A. J. Gregory, L. M. Hunt, S. M. Wisely, and B. K. Sandercock. 2014. Space use by female Greater Prairie-Chickens in response to wind energy development. Ecosphere 5:art3.

Woodward, A. J. W., S. D. Fuhlendorf, D. M. Leslie, Jr., and J. Shackford. 2001. Influence of landscape composition and change on Lesser Prairie-Chicken (*Tympanuchus pallidicinctus*) populations. American Midland Naturalist 145:261–274.

Wright, H. A., and A. W. Bailey. 1982. Fire ecology: United States and southern Canada. John Wiley & Sons, Inc., Hoboken, NJ.

CHAPTER TWELVE

Climate Change*

Blake A. Grisham, Alixandra J. Godar, and Cody P. Griffin

Abstract. The Great Plains are anticipated to experience significant changes in temperature and precipitation due to climate change. Changes in environmental conditions are likely to influence the population viability of Lesser Prairie-Chickens (*Tympanuchus pallidicinctus*) through changes in habitat availability, suitability, and connectivity, with direct effects on nesting and brood-rearing activities. Conservation measures are needed to buffer against potential changes to land use patterns or to severe drought conditions due to climate change, but the necessary information for their development has not been synthesized. Here, we provide the information needed to develop improved conservation strategies for Lesser Prairie-Chickens in the context of climate change. Future management over the next half century will have to meet four challenges: (1) recording and projecting changes to temperature and precipitation for each ecoregion by 2050 and 2080; (2) understanding the potential impacts of climate change on habitat availability on the Great Plains; (3) developing definitions for annual, seasonal, and extreme weather conditions and determining the effects of climate on the ecology of Lesser Prairie-Chickens; and (4) ranking of priorities for management and research. We place emphasis on the importance of current and future habitat management because our synthesis suggests that localized extirpations are likely due to reoccurring drought conditions. Range-wide extinctions can be avoided if habitat conservation and management is made a priority within the next 15 years. Engaging policy-makers in climate change discussions to set and achieve a goal of decreasing atmospheric concentrations of carbon dioxide below 350 parts per million may serve as the most beneficial conservation action.

Key Words: ecoregion, Lesser Prairie-Chicken, precipitation, temperature, *Tympanuchus pallidicinctus,* weather.

According to the Intergovernmental Panel on Climate Change (IPCC), 1995–2012 ranked among the warmest years on record since 1850 (IPCC 2013). Temperatures in the northern hemisphere alone during the second half of the 20th century were greater than any other 50-year period in the past 500 years (IPCC 2013). Global temperatures continue to warm and influence terrestrial systems at various spatial scales and magnitudes. Biologists are faced with the task of assessing the impacts of climate change on ecosystems and wildlife populations. Reliable impact assessments are becoming increasingly important, as state and federal agencies lack

* B. A. Grisham, A. J. Godar, and C. P. Griffin. 2016. Climate change. Pp. 221–242 in D. A. Haukos and C. W. Boal (editors), Ecology and conservation of Lesser Prairie-Chickens. Studies in Avian Biology (no. 48), CRC Press, Boca Raton, FL.

information for long-term conservation of imperiled species. For example, the Lesser Prairie-Chicken (*Tympanuchus pallidicinctus*) was listed as a threatened species under the Endangered Species Act in May 2014 (USFWS 2014), although the listing rule has been subsequently vacated by judicial decision in September 2015. Regardless, this species of conservation concern is known to be sensitive to variation in environmental conditions (Grisham et al. 2013) and landscape change (Woodward et al. 2001, Fuhlendorf et al. 2002).

The current distribution of Lesser Prairie-Chickens is a five-state area of the Southern Great Plains and the bird is considered a principal indicator species for unique portions of an arid-zone ecosystem (Hagen and Giesen 2005). Negative relationships between drought and the production and survival of prairie chickens are well established (Crawford 1980, Merchant 1982, Fields et al. 2006, Grisham et al. 2013). Negative impacts of drought are disconcerting because climate change forecasts indicate that the southwestern Great Plains will become hotter and drier with reoccurring severe drought conditions (Karl et al. 2009). Here, we review and evaluate existing literature to determine if potential impacts to Great Plains ecosystems and the ecology of Lesser Prairie-Chickens will require development of new conservation strategies.

Our three objectives are as follows: (1) to conduct a literature review on the potential impacts of climate change to the Great Plains ecosystems, with emphasis on habitats of the four main ecoregions of Lesser Prairie-Chickens: Mixed-Grass Prairie, Sand Sagebrush (*Artemisia filifolia*) Prairie, Sand Shinnery Oak (*Quercus havardii*) Prairie, and Short-Grass Prairie/Conservation Reserve Program Mosaic; (2) to assess the relationships between climatic variables and Lesser Prairie-Chickens; and (3) to use existing relationships and current available information to highlight the potential influence of climate change on the species to rank management and research priorities for the conservation of Lesser Prairie-Chickens.

EFFECTS OF CLIMATE CHANGE ON THE GREAT PLAINS

The landscape of the Great Plains is expected to be transformed due to the effects of climate change on ecosystem processes and human activity. Drought conditions and heat waves are expected to become more intense as temperatures increase and precipitation decreases because of increasing greenhouse gas concentrations (Karl et al. 2009, IPCC 2013). The 20th century experienced the highest warming trend of this millennium with significant temperature increases toward the end of the century (Crowley 2000, Jones et al. 2001). According to an ensemble average of climate models where half of the models project a greater amount of change and half of the models project a lesser amount of change, the average temperature of the Great Plains by 2050 is projected to be 1.4°C–2.5°C warmer under the low (B1) carbon emission scenario and 2.8°C–3.3°C warmer under the high (A2) carbon emission scenario (Girvetz et al. 2009). The average precipitation is expected to decrease by 8% and 16% under low and high carbon emission scenarios, respectively. Multiscenario climate models show that the southwestern United States is identified as a hotspot where climate change could have extreme impacts (Diffenbaugh et al. 2008). Models from the Coupled Model Intercomparison Project 5 (CMIP5) suggest that the southern portions of the Great Plains will have the largest decreases in annual mean precipitation by mid-21st century. The North American Regional Climate Change Assessment Program model simulates an increase of ≥20 annual mean days with temperatures ≥35°C by midcentury throughout the range of Lesser Prairie-Chickens (Kunkel et al. 2013). Temperature and precipitation projections are similar for the four ecoregions that comprise the range of Lesser Prairie-Chickens (Figures 12.1 to 12.8).

Precipitation is the primary ecological driver of grassland productivity because of its limited availability in these ecoregions. Currently, the average annual precipitation within the distribution of Lesser Prairie-Chickens ranges from ~20 cm per year in the southwestern Sand Shinnery Oak Prairie Ecoregion to ~60 cm in the eastern Mixed-Grass Prairie Ecoregion (Figure 12.9). The amount of precipitation in the Great Plains is projected to increase in the north, but decrease in the south (High Plains Regional Climate Center 2013). Different projections vary in regard to the intensity, timing, and amount of percent change, but a consistent trend among all projections is decreased precipitation during the growing season (Kunkel et al. 1999). Projected changes in precipitation will result in decreased soil moisture (Gregory et al. 1997). Reduced moisture will influence vegetative cover and food for grassland species, including Lesser Prairie-Chickens. Changes in rainfall phenology outside of the growing season and increases in

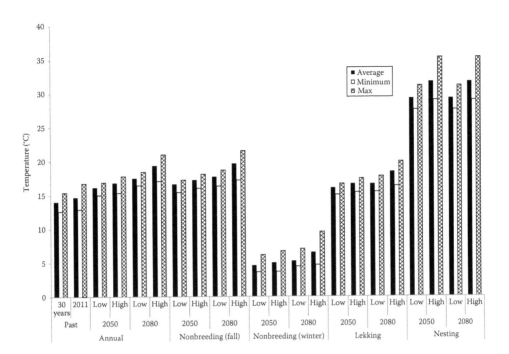

Figure 12.1. Projected changes in temperature for the Mixed-Grass Prairie Ecoregion during seasonal periods of the life cycle of Lesser Prairie-Chickens. Long-term averages are based on the 30-year period, 1982–2012. (Data from Girvetz et al. 2009.)

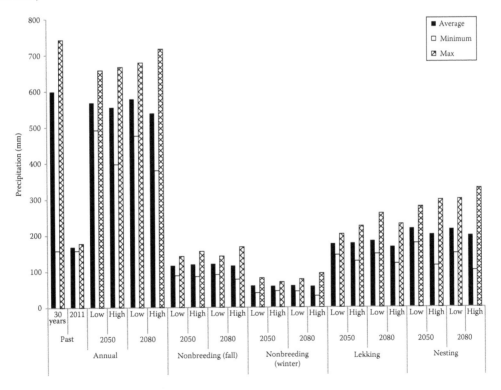

Figure 12.2. Projected changes in precipitation for the Mixed-Grass Prairie Ecoregion during seasonal periods of the life cycle of Lesser Prairie-Chickens. Long-term averages are based on the 30-year period, 1982–2012. (Data from Girvetz et al. 2009.)

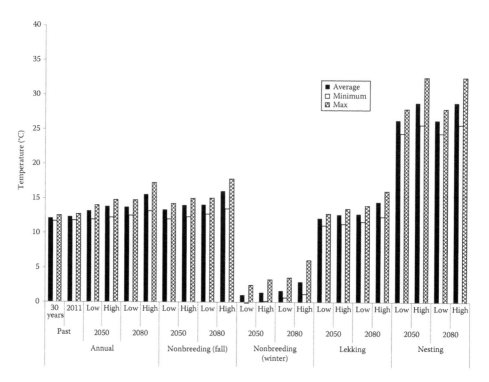

Figure 12.3. Projected changes in temperature for the Sand Sagebrush Prairie Ecoregion during seasonal periods of the life cycle of Lesser Prairie-Chickens. Long-term averages are based on the 30-year period, 1982–2012. (Data from Girvetz et al. 2009.)

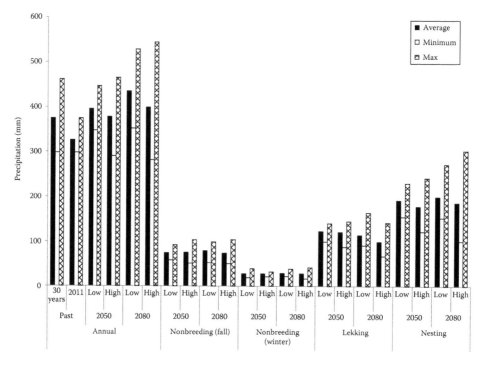

Figure 12.4. Projected changes in precipitation for the Sand Sagebrush Prairie Ecoregion during seasonal periods of the life cycle of Lesser Prairie-Chickens. Long-term averages are based on the 30-year period, 1982–2012. (Data from Girvetz et al. 2009.)

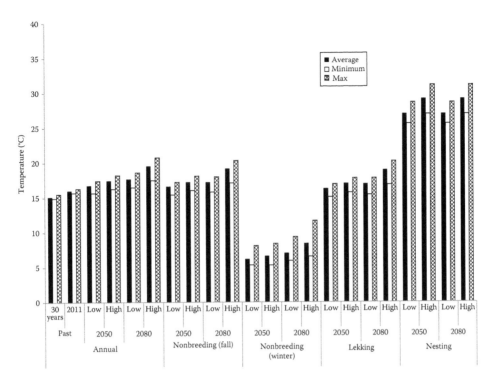

Figure 12.5. Projected changes in temperature for the Sand Shinnery Oak Prairie Ecoregion during seasonal periods of the life cycle of Lesser Prairie-Chickens. Long-term averages are based on the 30-year period, 1982–2012. (Data from Girvetz et al. 2009.)

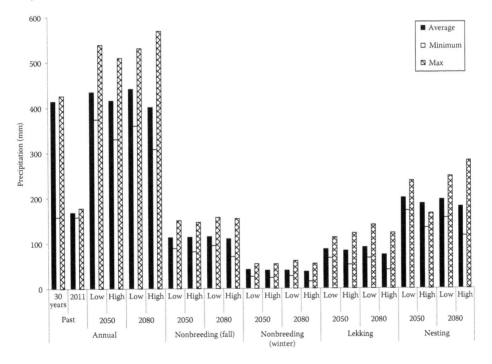

Figure 12.6. Projected changes in precipitation for the Sand Shinnery Oak Prairie Ecoregion during seasonal periods of the life cycle of Lesser Prairie-Chickens. Long-term averages are based on the 30-year period, 1982–2012. (Data from Girvetz et al. 2009.)

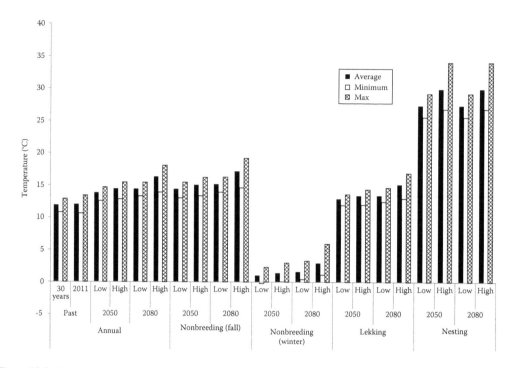

Figure 12.7. Projected changes in temperature for the Short-Grass Prairie/CRP Mosaic Ecoregion during seasonal periods of the life cycle of Lesser Prairie-Chickens. Long-term averages are based on the 30-year period, 1982–2012. (Data from Girvetz et al. 2009.)

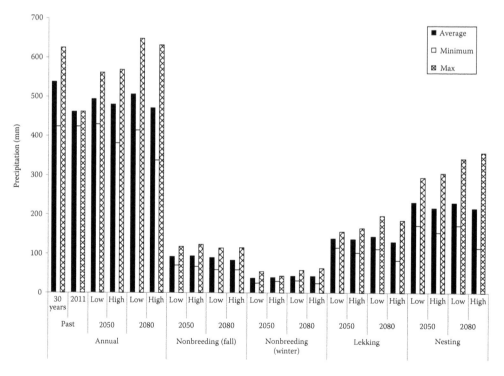

Figure 12.8. Projected changes in precipitation for the Short-Grass Prairie/CRP Mosaic Ecoregion during seasonal periods of the life cycle of Lesser Prairie-Chickens. Long-term averages are based on the 30-year period, 1982–2012. (Data from Girvetz et al. 2009.)

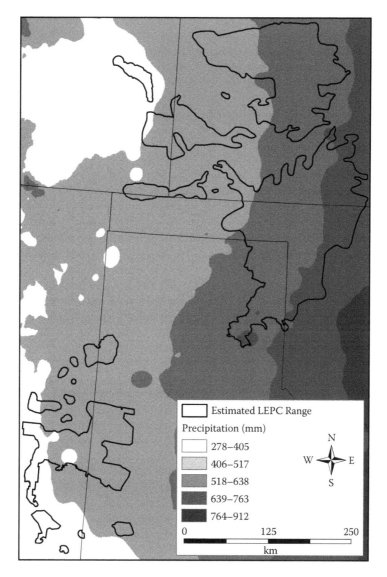

Figure 12.9. Precipitation gradient across the occupied range of Lesser Prairie-Chickens (LEPC) in the central United States, 1982–2012. PRISM Climate Group, Oregon State University, http://prism.oregonstate.edu, created May 22, 2014. Southern Great Plains Crucial Habitat Assessment Tool. August 2013. http://kars.ku.edu/maps/sgpchat, created May 22, 2014.

intensity with greater runoff could have a stronger effect on annual productivity and vegetation composition compared to changes in total rainfall (Zavaleta 2012, High Plains Regional Climate Center 2013). Combined effects of changes in mean and variance of rainfall events will cause longer and more frequent drought conditions, which will alter annual productivity and influence landscape use (Veldkamp and Lambin 2001, Zondag and Borsboom 2009). Changes in landscape use could influence the amount of suitable habitat for Lesser Prairie-Chickens (Hagen et al. 2011).

Land use practices are the main cause of changes in plant composition and vegetative structure, with increasing concentrations of atmospheric carbon dioxide exacerbating community shifts (Archer et al. 1995, Walther et al. 2002). Energy development and agricultural practices have the greatest potential to impact Lesser Prairie-Chickens because types of land use tend to occur in rural areas and influence large expanses of habitat. Energy development will increase in the future because of increases in population and per capita energy demands. According to forecasts of the U.S. Energy Information Administration (2013), crude oil production is expected to increase over the next decade before declining gradually in 2020 until 2040, with considerable production occurring near the Sand Shinnery Oak Prairies of New Mexico and Texas. Natural gas production is expected to increase over the next decade with net exports of liquid natural gas by 2016 and dry natural gas by 2019 projected to grow by 102 and 113 billion m^3, respectively. Renewable fuel use is expected to increase faster than fossil fuel use, with generated renewable energy projected to grow from 13% in 2011 to 16% in 2040. Increased demand for energy sources will result in landscape impacts as more land on the Great Plains is converted to unsuitable habitat via energy production. The Great Plains have high potential for wind energy production, particularly in areas within the distribution of the Lesser Prairie-Chicken (Greene et al. 2012; Figure 12.10).

In addition to energy development, the habitat availability of Lesser Prairie-Chickens will be influenced by changes in row crop agriculture. Crop yield is determined by suitable climates, so changes in seasonal temperatures could diminish productivity potential for current crops, resulting in outright loss of farmland or conversion to crops that are better suited to hotter, drier climates (Mendelsohn et al. 1994). Studies will differ on the projected change in amount and juxtaposition of cropland on the Great Plains by 2050 (Wheeler et al. 2000, Battisti and Naylor 2009), but most suggest that climate change could render row crops less profitable because of increasing costs of production. Land values are currently declining in counties on the western periphery of the Great Plains, and changes in irrigation regulations, projected droughts, and heat waves associated with climate change could reduce land values in the future (Polsky 2004, Karl et al. 2009). Karl et al. (2009) suggest that diversification of livestock and crops, as well as a transition from irrigated crop production to dryland farming or ranching, may be a necessary response to climate change. Currently, row crop production within the range of Lesser Prairie-Chickens depends heavily on ground water pumped from the Ogallala Aquifer in the High Plains. Current anthropogenic demands on the Ogallala Aquifer within the range of Lesser Prairie-Chickens exceed the replenishing rate, and water levels are decreasing at an alarming rate in the Southern Great Plains (Sophocleous 2012,

Sophocleous and Merriam 2012). The Ogallala Aquifer sustains ca. U.S.$20 billion of agricultural production on an annual basis, and stakeholders are concerned about the fate of agriculture should the aquifer no longer be able to support irrigated agriculture (Hernandez et al. 2013).

Ranching profitability, production, and operations will most likely decrease with climate change (Coppock 2011, Polley et al. 2013). Ranching becomes economically impractical in hot, dry conditions because ranchers struggle to regulate cattle temperature, hydration, and forage availability (Coppock 2011, Polley et al. 2013). A survey in Utah showed that ranchers responded to drought by enrolling in government relief, relying on emergency feed purchases, and selling off livestock herds (Coppock 2011). Coppock (2011) indicated that ranchers felt more prepared for future droughts after one extreme event. Therefore, implementing emergency drought strategies such as improving water for livestock, diversifying family income, improving irrigation efficiency for hay production, improving land management, and reducing stocking rates may allow for sustainable ranching operations in the future, which would protect native grasslands for Lesser Prairie-Chickens (Coppock 2011).

In addition to changing anthropogenic landscapes, climate change may alter natural landscapes. Climate change causes expansions and contractions of plant distributions (Walther et al. 2002) and could alter vegetation community structure and composition to habitats unable to support Lesser Prairie-Chickens (Jump and Penuelas 2005, Hagen et al. 2013). Changes in vegetation structure threaten to further degrade remaining fragmented Lesser Prairie-Chicken habitat. One potential implication of climate change for habitats of Lesser Prairie-Chickens in the Great Plains is encroachment of woody vegetation. Honey mesquite (*Prosopis glandulosa*) and eastern red-cedar (*Juniperus virginiana*) exhibit more plasticity compared to grasses in response to increased carbon dioxide levels, higher temperatures, and less precipitation (Brown and Archer 1999, Wilson et al. 2001, Jurena and Archer 2003). Mesquite primarily occurs on the nonsandy soils within the Sand Shinnery Oak Prairie Ecoregion of the southern portion of the range of Lesser Prairie-Chickens in the Southern High Plains of New Mexico and Texas, whereas eastern red-cedar mostly occurs in the Mixed-Grass Prairie Ecoregion in the eastern

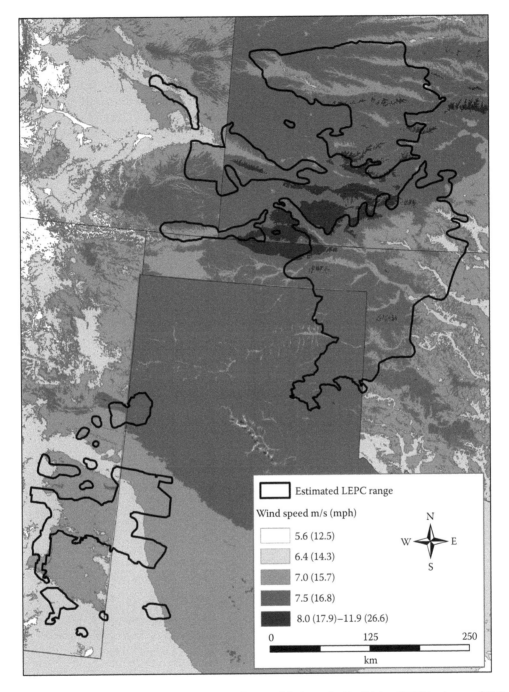

Figure 12.10. Wind energy potential within the occupied range of the Lesser Prairie-Chicken (LEPC) in the central United States. Southern Great Plains Crucial Habitat Assessment Tool. August 2013. http://kars.ku.edu/maps/sgpchat, created May 22, 2014. National Renewable Energy Laboratory Wind Data, http://nrel.gov/gis, created May 22, 2014.

portion of the range of Lesser Prairie-Chicken in Oklahoma and Kansas (USDA NRCS 2013). Mesquite and eastern red-cedar have experienced considerable range expansion since human settlement in the Southern Great Plains (Woodward et al. 2001, Fuhlendorf et al. 2002). Increases in both species of woody plants have been attributed to fire suppression and unmanaged grazing. Bachelet et al. (2001) created predictive models that simulate vegetation distributions in the United

States based on seven different climate scenarios. The authors found that all but two models indicated shrub-dominated areas will likely replace grasslands in the Great Plains. A climate model of the Hadley Climate Center (HADCM2SUL) suggested that the majority of the current remaining grasslands in the Texas Panhandle, east-central New Mexico, the eastern Oklahoma Panhandle, southeast Colorado, and southwest Kansas will be dominated by shrubs or woodland by 2050, although the specific species of shrubs is not clear. If the projected changes result in areas dominated by mesquite or cedar, then the projection is not favorable for Lesser Prairie-Chickens because unsuitable woody vegetation already occurs in the majority of their distribution. Fuhlendorf et al. (2002) found that the cover of eastern red-cedar was the strongest predictor of ongoing declines of Lesser Prairie-Chickens in Oklahoma. Two other models, the MAPSS (Mapped Atmospheric–Plant–Soil Systems) and MC1 model, also predicted decreases in grasslands compared to increases in shrub and woodlands (Bachelet et al. 2001).

The distribution and frequency of common grassland plant species are expected to change in addition to woody shrub encroachment. Grassland plants are divided into two functional groups based on the main photosynthetic pathway: C3 or cool-season plants and C4 or warm-season plants (Table 12.1). Early studies suggested

TABLE 12.1

Photosynthetic pathways for important plant species found throughout the distribution of Lesser Prairie-Chickens.

Scientific name	Common name	Citation
C3 plants		
Pascopyrum smithii/Agropyron smithii	Western wheatgrass	Volin et al. (1997), Morgan et al. (1998)
Xanthocephalum sarothrae/Gutierrezia sarothrae	Broom snakeweed	Wan et al. (1993)
Helianthus annuus	Common sunflower	Wang et al. (2006)
Vulpia octoflora	Six weeks fescue	Kemp (1983)
Hesperostipa comata	Needle-and-thread	Symstad (2000)
Ambrosia psilostachya	Western ragweed	Barnes et al. (1983)
Quercus havardii	Sand Shinnery Oak	McKinley and Blair (2008)
Prosopis glandulosa	Honey mesquite	Biggs et al. (2002)
Artemisia filifolia	Sand sagebrush	Wang (2002)
C4 plants		
Andropogon gerardii	Big bluestem	Sage and Monson (1999)
Andropogon hallii	Sand bluestem	Sage and Monson (1999)
Aristida purpurea	Purple three-awn	Sage and Monson (1999)
Bothriochloa saccharoides	Silver bluestem	Sage and Monson (1999)
Bouteloua curtipendula	Sideoats grama	Volin et al. (1997), Sage and Monson (1999)
Bouteloua gracilis	Blue grama	Morgan et al. (1998), Sage and Monson (1999)
Buchloë dactyloides	Buffalograss	Sage and Monson (1999)
Schizachyrium scoparium	Little bluestem	Volin et al. (1997), Sage and Monson (1999), Symstad (2000)
Sorghastrum nutans	Indiangrass	Sage and Monson (1999), Symstad (2000)
Sporobolus cryptandrus	Sand dropseed	Wan et al. (1993), Sage and Monson (1999)
Eragrostis trichodes	Sand lovegrass	Sage and Monson (1999)
Salsola kali	Russian thistle	Wang et al. (2006)
Panicum virgatum	Switchgrass	Symstad (2000)
Eragrostis curvula	Weeping lovegrass	Sinha and Kellogg (1996)

that increased carbon dioxide concentrations would give C3 plants a competitive advantage over C4 plants (Wand et al. 1999, Lecain et al. 2003), because C4 plants dedicate extra energy to artificially increasing the internal concentration of carbon dioxide to aid with photosynthesis (Pearcy and Ehleringer 1984). However, the future distribution of these plants will be determined by factors in addition to carbon dioxide concentrations because competitive interactions among multiple species in various conditions are poorly understood (Read and Morgan 1996, Lecain et al. 2003). Most importantly, soil moisture is the determining factor for predicting the dominant plant species; plants that are more efficient with water use are more likely to persist in the face of reoccurring drought.

The long-term effects of drought conditions exacerbated by climate change on plant community composition and structure will be the main factor driving the population viability of Lesser Prairie-Chickens at local and range-wide scales. However, data on community responses to annual variation in precipitation are lacking range-wide and should be considered a research priority (Table 12.2). Data on community responses are needed to assess the population response of Lesser Prairie-Chickens to landscape changes in plant composition and structure. Most importantly, it is fundamental to understand how Lesser Prairie-Chickens respond to changes in plant communities because the species evolved in open prairie landscapes that are exposed to considerable intra-annual and interannual variation in weather conditions.

SEASONAL WEATHER AND EXTREME EVENTS

Population studies of Lesser Prairie-Chickens under natural conditions have shown that the species is well adapted to cope with variation in environmental conditions (Crawford 1980, Merchant 1982, Fields et al. 2006, Grisham et al. 2013, 2014). The underlying relationship between weather and Lesser Prairie-Chicken demographic parameters has been evaluated in several studies (Crawford 1980, Merchant 1982, Fields et al. 2006, Grisham et al. 2013). Fields et al. (2006) found that daily nest survival rates declined with increasing temperatures during incubation ($\hat{\beta}_{temperature} = -0.05$, 95% CI = -0.13 to $+0.02$, $\Sigma AIC_{wi} = 0.32$) and brood survival decreased

as precipitation increased during brood rearing ($\hat{\beta}_{precipitation} = -2.02$, 95% CI = -3.54 to -0.50, $\Sigma AIC_{wi} = 0.39$). Grisham (2012) reported lower survival for males late in the breeding season (June–August: 0.98, SE = 0.01, 95% CI = 0.93–0.99) compared to the early breeding season March–May: 0.82, SE = 0.04, 95% CI = 0.74–0.89), and the seasonal difference in survival was more pronounced in drought years ($\hat{\beta}_{drought} = 17.01$; $\hat{\beta}_{nondrought} = 4.64$). Grisham et al. (2013) found that winter temperature prior to incubation had the greatest effect on nest survival ($\hat{\beta}_{wintertemperature} = -0.37$, $\Sigma AIC_{wi} = 0.70$). Winter temperatures were correlated to La Niña events, which are a good predictor of drought conditions throughout the distribution of Lesser Prairie-Chickens. La Niña events are ultimately a function of yearly variation in the Southern Oscillation in the Pacific Ocean, but regional and local droughts are hard to predict beyond 6 months because the annual variation in the Oscillation is also difficult to assess (Cook et al. 2004). Comparatively, long-term drought occurs with a ~20-year periodicity on the Great Plains, with historical multiyear droughts occurring in the 1930s, 1950s, 1970s, 1990s, and 2010s (Trenberth et al. 1988, Chen and Newman 1998). Hence, an understanding of long-term drought conditions on the ecology of Lesser Prairie-Chickens is available through lek monitoring, hunter harvest, and radio-tagging data at the seasonal (1–4 months) and yearly scale (Crawford 1980, Merchant 1982, Hagen and Giesen 2005, Fields et al. 2006, Grisham et al. 2013). Moreover, observed fluctuations in the population densities and range of Lesser Prairie-Chickens are directly related to the 20-year drought periodicity (Crawford and Bolen 1976, Merchant 1982, Grisham et al. 2013, McDonald et al. 2014).

Drought is a seasonal phenomenon and the relationship between seasonal weather conditions and vital rates of Lesser Prairie-Chickens has been evaluated for the southern and northern extent of the distribution (Fields et al. 2006, Grisham et al. 2013). Latitudinal variation in temperature and longitudinal variation in precipitation are most pronounced during breeding and dispersal activities (Figures 12.9 and 12.11), and evidence suggests that seasonal weather conditions negatively affect Lesser Prairie-Chickens (Fields et al. 2006, Grisham et al. 2013). However, seasonal weather is likely to disproportionately affect the vital rates of

TABLE 12.2

Research needs for conservation of Lesser Prairie-Chickens in the context of climate change.

Research needs (ranked high to low)	Questions	Ecoregions	Implications for climate change mitigation
Habitat availability and connectivity			
	How do Lesser Prairie-Chickens respond to landscape change (energy production, crop production)?	Range-wide	Identify juxtaposition amount and connectivity of habitat needed for Lesser Prairie-Chicken conservation.
	What will be the impacts to vegetation communities due to precipitation variability and increased temperatures?	Range-wide	Identify potential habitats for conservation.
	What is the role of Conservation Reserve Program lands in Lesser Prairie-Chicken ecology/ population persistence?	Range-wide	Identify potential for private land conservation easements to increase and connect Lesser Prairie-Chicken habitat.
Habitat suitability			
	What is the necessary vegetation composition and structure needed to maximize reproductive output and survival?	Range-wide	Increase current populations to buffer against increased likelihood of drought.
	What are the best methods to reduce woody encroachment in native grasslands?	Mixed-grass and Sand Shinnery Oak Prairies	Reduce encroachment of honey mesquite and eastern red-cedar to buffer against projected distribution shifts into grasslands.
Surface water (stock tanks/wildlife guzzlers)			
	Is surface water used?	Range-wide	Identify whether surface water is cost-efficient/viable management strategy.
	Improve survival and reproductive output?	Range-wide	Identify whether surface water improves demographic parameters during drought years.
	Influence of free water availability on landscape to nest site location?	Range-wide	Identify whether juxtaposition/ number of stock tanks/ guzzlers on landscape influences nest site location.
Disease prevalence/transmission			
	What is the influence of West Nile virus on demographic parameters?	Range-wide	Identify whether West Nile virus has potential to decrease demographic parameters.
	What is the influence of climate change on West Nile virus prevalence/transmission?	Range-wide	Assess whether West Nile virus has potential to have range-wide population-level effects given environmental change.
	What is the influence of free water availability on disease prevalence and transmission?	Range-wide	Identify whether free water has potential for disease potential.

(Continued)

TABLE 12.2 (*Continued*)

Research needs for conservation of Lesser Prairie-Chickens in the context of climate change.

Research needs (ranked high to low)	Questions	Ecoregions	Implications for climate change mitigation
Genetic intermixing with Greater Prairie-Chickens			
	What is the ratio of hybrid offspring produced in areas where genetic intermixing occurs and do these offspring have the ability to reproduce?	Short-grass and mixed-grass prairies in Kansas	Identify the amount of hybrid offspring produced and their ability to pass along Lesser Prairie-Chicken genetic material.
	What is the potential for existing Greater Prairie-Chicken habitat to support Lesser Prairie-Chickens in the context of precipitation and temperature change?	Short-grass, mixed-grass, and tall-grass prairies in Kansas	Identify areas for Lesser and Greater Prairie-Chicken conservation and implications of hybridization on both species.

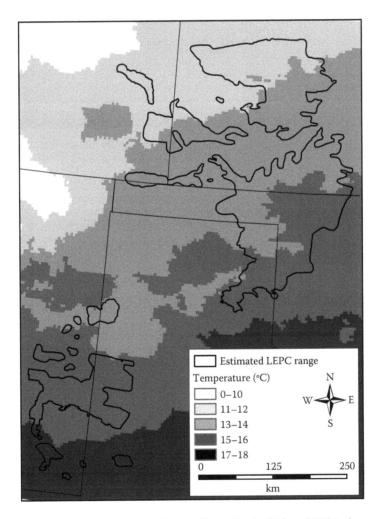

Figure 12.11. Temperature gradient across the occupied range of Lesser Prairie-Chickens (LEPC) in the central United States, 1982–2012. PRISM Climate Group, Oregon State University, http://prism.oregonstate.edu, created May 22, 2014. Southern Great Plains Crucial Habitat Assessment Tool. August 2013. http://kars.ku.edu/maps/sgpchat, created May 22, 2014.

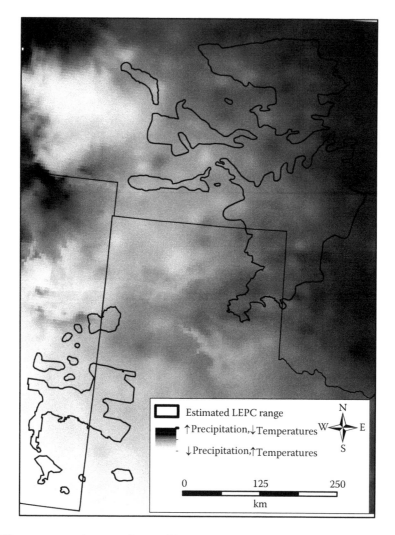

Figure 12.12. Climate regions in the occupied range of the Lesser Prairie-Chicken (LEPC) in the central United States based on a 30-year record of climate data, 1982–2012. The estimated Lesser Prairie-Chicken range layer was taken from the Southern Great Plains Critical Habitat Assessment Tool (kars.ku.edu/geodata/maps/sgpchat/) and the state boundary layer from the National Atlas of the United States of America (www.nationalatlas.gov). We created an overlay of the two raster datasets with a designated value for each pixel so that the final overlay would project regions with similar annual 30-year precipitation and similar annual 30-year temperature values. We reclassified both rasters with nine classes each to classify pixel values. We used the weighted overlay tool in ArcGIS 10.2 to overlay reclassified rasters and specified 50% influence for each with an evaluation scale of -5 to 5 by 1. The final output is the estimated range and state boundaries for the distribution of Lesser Prairie-Chickens with similar climatic regions based on designated values from the evaluation scale of the overlay. PRISM Climate Group, Oregon State University, http://prism.oregonstate.edu, created May 22, 2014. Southern Great Plains Crucial Habitat Assessment Tool. August 2013. http://kars.ku.edu/maps/sgpchat, created May 22, 2014.

Lesser Prairie-Chickens at the southern and western edges of the species' distribution because individuals at range boundaries are often living at the edge of their physiological tolerances (Parmesan et al. 2000; Figure 12.12).

An important aspect of weather that is often overlooked in impact assessments is the influence of floods, tornados, heat or cold waves, and other short-term extreme events on wildlife populations. The consequences associated with temperature and precipitation gradients across the distribution of Lesser Prairie-Chickens are underappreciated and may influence the four ecoregions differentially through variation in climate that drives extreme weather events. The variation experienced with a 15°C–19°C temperature gradient is an uncommon eco-logical phenomenon (Figure 12.11), because the phenology of breeding activities of Lesser Prairie-Chickens is consistent across latitudinal and longitudinal gradients (Boal et al. 2010). A predicted mean change in global temperatures of 2°C–4°C within the next 80 years is expected to have devastating effects on landscape suit-ability for wildlife populations, agriculture, and human society (IPCC 2013), due to long-term changes in seasonal weather conditions (mean) and associated extreme events (devia-tions from the mean). Thus, variability around the current mean temperature alone can result in daytime temperatures exceeding 40°C on the Southern High Plains or dropping to <0°C in Kansas during nest incubation (Grisham et al. 2013). The effects of extreme events are often difficult to model at a population scale because they typically occur at low frequencies within small spatial and temporal scales (Lubchenco and Karl 2012). However, most weather-related phenomena that biologists label as extreme events are misnomers (Lubchenco and Karl 2012). Avalanches, landslides, floods, tornados, hail storms, and other intense, localized events are not necessarily stochastic. Rather, extreme events are often a function of the seasonal condi-tions that define climate. Extreme events occur within the expected range of their probability of occurrence, allowing them to be modeled at localized scales. Relative extreme events can be defined as events that occur within 2–3 standard deviations from the 30-year mean and occur at random based on their expected probability of occurrence over the same interval (Lubchenco and Karl 2012). The effects of relative extreme events on Lesser Prairie-Chicken ecology are not well understood, but evidence from other assessments on prairie grouse suggests that most extreme events are negative. Extreme events may cause direct mortality or indirect mortality from starvation, displacement, or increased energetic and thermoregulation demands that negatively influences vital rates and population persistence (Smyth and Boag 1984, Horak and Applegate 1998, Grisham 2012, Grisham et al. 2013; Chapter 9, this volume). Absolute events can be incorpo-rated into ecoregion-level assessments by using the frequency of occurrence of select events over a 30-year period and then inserted into model-ing procedures based on their mean frequency of occurrence, variation around the mean, and probability distribution. Relative extreme events can be incorporated into similar modeling pro-cedures as stochastic events.

EFFECTS OF WEATHER ON LESSER PRAIRIE-CHICKEN ECOLOGY

The breeding activities of Lesser Prairie-Chickens are driven by photoperiod (Hagen and Giesen 2005), and begin in late February to early March with the congregation of males at leks (Haukos and Smith 1999, Behney et al. 2012). Peak female attendance occurs during the last week of March and the first week of April, and peak copulation attempts occur within the second to third weeks in April (Haukos and Smith 1999, Behney et al. 2012). Male Lesser Prairie-Chickens will display at leks in a variety of environmental conditions to maintain and defend established territories (Copelin 1963, Behney et al. 2012).

Lekking is energetically demanding because of increased activity and decreased foraging time, and males lose mass during the breed-ing season (Wolfe et al. 2007). The cumulative effects of display and environmental conditions appear to affect the survival of males later in the breeding season (May–mid-June; Grisham 2012). Males tend to spend less time displaying during extreme drought. In Texas, males tended to stop display activities 45 min earlier in the day during a year of extreme environmental conditions with high temperatures and low rela-tive humidity (2011: $\bar{x} = 07{:}45$ CST; n = 35 lek observations), compared to 3 other years when temperatures and relative humidity were consis-tent with 30-year means (2009, 2010, and 2012: $\bar{x} = 08{:}30$ CST, n = 62; B.A. Grisham and C.W. Boal, unpubl. data). The phenology of female lek attendance is less likely to be influenced by weather because females are not associated with a defendable territory and ovulation is stimu-lated by photoperiod (Hagen and Giesen 2005). However, weather influences a variety of nesting activities, including nest vegetation availability (Crawford 1980, Merchant 1982, Grisham 2012, Grisham et al. 2013), thermal stress on eggs and incubating females (Grisham 2012), and nest survival (Fields et al. 2006, Davis 2009, Grisham et al. 2014).

Timing of incubation and nest survival appear to be correlated, and nests of Lesser Prairie-Chicken may have a greater probability of survival if they are initiated earlier in the year (Fields et al. 2006, Pitman et al. 2006, Grisham 2012). The above-average precipitation in the spring and winter facilitates vegetation growth earlier in the growing season (Zavaleta 2012). Early growth results in more incubation attempts earlier in the breeding season (Boal et al. 2010). However, the effect of spring vegetation on timing of nesting appears to be most pronounced on the arid Southern High Plains (Grisham 2012, Grisham et al. 2013), likely resulting from less precipitation and increased potential evaporation on the southwestern edge of the species' distribution. Extreme temperatures during incubation influence nest attentiveness, egg survival, and number of nesting attempts (Smyth and Boag 1984, Grisham 2012). Data from nest thermal profiles during extreme drought conditions in Texas indicate that temperatures on the ground during daylight exceeded 54.4°C for three nests between May 18 and 22, 2011, and extreme upper limits may explain why 12 of 15 radio-tagged females did not attempt to incubate eggs in 2011 (Grisham et al. 2014). Our data indicate that there are potential upper-level temperature thresholds for incubating females (Grisham et al. 2013). Pitman et al. (2006) and Holt (2012) found no interannual variation in nesting parameters in Sand Sagebrush Prairie and Mixed-Grass Prairie Ecoregions, respectively. Conversely, Grisham et al. (2014) found that nest initiation was delayed and nest abandonment was common during drought conditions, resulting in interannual variation in nesting parameters for the Sand Shinnery Oak Prairie Ecoregion.

Poor recruitment into the next breeding season has been identified as the key demographic parameter for population declines of Lesser Prairie-Chickens (Hagen 2003, Pitman et al. 2006, Hagen et al. 2009). Chick survival can be a function of variation in daily precipitation and temperature (Fields et al. 2006). Precipitation events early in the brood-rearing period result in increased mortality because chicks have poor thermoregulation abilities up to 14 days post-hatch (Ahlborn 1980). Chick survival has been positively correlated with invertebrate availability and biomass (Hagen et al. 2005), with chick survival until independence at brood breakup is considered a critical factor associated with juvenile recruitment into the breeding population (Hagen 2003). Davis et al. (1979) found

that diets of juvenile Lesser Prairie-Chickens in New Mexico consisted primarily of short-horned grasshoppers (Acrididae) and beetles (Coleoptera). Similar invertebrate diversity was found in Kansas and Texas (Hagen et al. 2005, Grisham 2012). Grisham (2012) suggested that below-average summer temperatures coupled with above-average rainfall amounts in the spring and summer maximize invertebrate availability, and Zavaleta (2012) found that winter precipitation was the highest ranking precipitation index for invertebrate abundance and diversity during the following brood-rearing season. Fields et al. (2006) suggested that reduced precipitation during the brood-rearing season reduced the abundance of grasshoppers in Kansas. Chick mortality can occur via heat stress through direct mortality or lack of food (Merchant 1982). Flanders-Wanner et al. (2004) found that the number of days with temperatures >35°C was negatively associated with annual production of Sharp-tailed Grouse (*T. phasianellus*). If Lesser Prairie-Chickens have a similar response, production in the northern extent of their distribution could also be impacted. Temperatures exceeded 35°C for 23–32 days during two brood-rearing seasons in Kansas, and maximum temperatures increased as the season progressed, resulting in lower brood survival compared to years with cooler conditions (Fields et al. 2006).

INFLUENCE OF CLIMATE CHANGE ON LESSER PRAIRIE-CHICKENS AND RANKING OF RESEARCH AND MANAGEMENT PRIORITIES

Lesser Prairie-Chickens evolved with recurring drought in the western Great Plains, and populations have the ability to recover from long-term, extreme drought events. Historically, inclement environmental conditions associated with drought led to boom–bust fluctuations in the population numbers of Lesser Prairie-Chickens (Hagen et al. 2009). Populations had greater ability to recover because of high reproductive potential in years following periods of increased precipitation and because of immigration from other populations due to increased connectivity and long-distance dispersal across relatively large spatial scales (Grisham et al. 2014). Currently, the future influence of drought is a conservation concern for Lesser Prairie-Chickens because climate forecasts have identified drastic changes in future temperature and precipitation on the Great Plains (Grisham et al. 2013), and strong connections have been established between climatic conditions and the ecology of Lesser Prairie-Chickens (Crawford 1980, Merchant 1982, Fields et al. 2006, Grisham et al. 2013). The current distribution of Lesser Prairie-Chickens includes a diversity of environmental conditions and climate conditions are expected to change by 2050 (Figure 12.12), but extirpation or extinction of Lesser Prairie-Chickens is not an inevitable event. Lesser Prairie-Chickens are highly plastic in behavior, as demonstrated by their ability to occupy areas with various climatic conditions, vegetation communities, and to colonize vegetation communities that were not originally considered suitable, including lands enrolled in the Conservation Reserve Program (Chapter 10, this volume). Ultimately, the impacts of climate change on population persistence of Lesser Prairie-Chickens until 2050 will be influenced more by habitat availability, quality, and connectivity as opposed to changes in environmental conditions.

Local extirpations of Lesser Prairie-Chickens may occur by 2050 because of increased frequency and intensity of drought conditions that cause changes in community structure and composition (Grisham et al. 2013). This scenario is perhaps more likely in the southern and western parts of the species' current distribution (Grisham et al. 2013). Extirpations could be avoided rangewide if conservation measures to protect and increase Lesser Prairie-Chicken habitat in and around current occupied areas are established within the next 15 years or ~5 generations of the species. In addition to increasing the amount of Lesser Prairie-Chicken habitat, emphasis should be placed on protecting and restoring existing habitats to maintain and increase current populations. Restoration might be achieved through multiple mechanisms, including removal of invasive woody vegetation, restoration of prescribed fire and other ecological drivers (Chapter 6, this volume), protecting and increasing lands enrolled in the Conservation Reserve Program within all ecoregions (Chapter 10, this volume), developing incentives for consideration of Lesser Prairie-Chicken ecological requirements during anthropogenic development in occupied areas (Chapter 11, this volume), increasing the number

of available water sources and improved restoration of habitats in Sand Shinnery Oak Prairie (Chapter 17, this volume), and specific recommendations for habitat management in each of the four main ecoregions (Chapters 14 to 17, this volume).

Adaptive habitat conservation and management is vital to population persistence of Lesser Prairie-Chickens by 2050. Other potential impacts, including increased disease potential and genetic intermixing with Greater Prairie-Chickens (*T. cupido*), are likely to be secondary to potential impacts of changes in habitat availability, quality, and connectivity (Chapters 5 and 9, this volume). In the long term, landscape- and fine-scale-level effects of climate change on Lesser Prairie-Chicken demography and available habitats should be considered a research need. The development of conservation actions that mitigate changing habitat conditions by 2050 is a research priority, with emphasis on changes in vegetation for each ecoregion under different scenarios for changes in precipitation. Conservation measures would need to provide quality habitat in the short term by facilitating dispersal during drought years, and, in the long term by identifying habitats within and outside the existing range that may be considered suitable in the future given precipitation variability (Table 12.2).

Considerable uncertainty surrounds the influence of climate change on anthropogenic activities on the Great Plains. Farm and ranching production are projected to decrease, but current projections do not account for potential improvements to farm technology or adaptation by local producers. Fuhlendorf et al. (2002) found that areas with declining Lesser Prairie-Chicken populations had greater percentages of cropland. Indeed, native prairie conversion to cropland is the current leading cause of population declines for Lesser Prairie-Chickens (Hagen and Giesen 2005). Future prairie conversion might be curtailed by management agencies and nongovernmental organizations acting to engage private landowners in conservation measures for Lesser Prairie-Chickens to maintain and increase the amount of habitat through conservation easements and other incentives (NRCS 2012). Most importantly, private lands management should be done within the framework of adaptive resource management to implement and monitor new strategies as the influence of climate change becomes more prevalent on the Great Plains (Chapter 13, this volume).

Energy production is expected to increase by 2050, and future projections are fairly reliable based on human population growth and energy demands (IPCC 2013). The projections are disconcerting because energy production affects the habitat availability of Lesser Prairie-Chickens (Pitman et al. 2005, Pruett et al. 2009, Hagen et al. 2011, Grisham et al. 2014). Home ranges of Lesser Prairie-Chickens tend to exclude powerlines and other tall anthropogenic structures (Pruett et al. 2009, Hagen et al. 2011), and nesting females appear to avoid anthropogenic structures (Pitman et al. 2005). Therefore, energy production will likely be detrimental to population persistence of Lesser Prairie-Chickens if energy producers are not engaged in conservation strategies. As with private lands conservation, management agencies and nongovernmental organizations should engage energy producers and other stakeholders in context of adaptive resource management to identify areas that should be avoided for development by 2020 and 2050.

The outlook for range-wide population persistence of Lesser Prairie-Chickens until 2080 is not positive. The single, long-term solution to protect the Great Plains and Lesser Prairie-Chicken populations is to reduce atmospheric concentrations of carbon dioxide below 350 parts per million (ppm) to avoid extreme increases to global and regional temperatures (Rockström et al. 2011). The Great Plains are expected to see temperature increases up to 10°C in some seasons by 2080 (IPCC 2013). Climate change has the potential to influence human society to such a great extent that concerns for ecosystem and species persistence may become a low priority. Unfortunately, the current national and international political atmosphere is not conducive for implementation of new policy that can buffer against the effects of climate change. Therefore, all stakeholders, including scientists, energy and crop producers, state and federal agencies, and nongovernmental organizations should engage policy-makers in climate change discussions to highlight the potential devastating effects of climate change by 2080. The potential to offset long-term effects, including negative effects to Lesser Prairie-Chicken populations, exists if national and international policies are enacted within the next 15 years.

LITERATURE CITED

Ahlborn, C. G. 1980. Brood-rearing habitat and fall–winter movements of Lesser Prairie Chickens in eastern New Mexico. M.S. thesis, New Mexico State University, Las Cruces, NM.

Archer, S., D. S. Schimel, and E. A. Holland. 1995. Mechanisms of shrubland expansion—Land-use, climate or CO_2? Climatic Change 29:91–99.

Bachelet, D., R. P. Neilson, J. M. Lenihan, and R. J. Drapek. 2001. Climate change effects on vegetation distribution and carbon budget in the United States. Ecosystems 4:164–185.

Barnes, P. W., L. L. Tieszen, and D. J. Ode. 1983. Distribution, production, and diversity of C3-and C4-dominated communities in a mixed prairie. Canadian Journal of Botany 61:741–751.

Battisti, D. S., and R. L. Naylor. 2009. Historical warnings of future food insecurity with unprecedented seasonal heat. Science 323:240–244.

Behney, A. C., B. A. Grisham, C. W. Boal, H. A. Whitlaw, and D. A. Haukos. 2012. Sexual selection and mating chronology of Lesser Prairie-Chickens. Wilson Journal of Ornithology 124:96–105.

Biggs, T. H., J. Quade, and R. H. Webb. 2002. δ13C values of soil organic matter in semiarid grassland with mesquite (Prosopis) encroachment in southeastern Arizona. Geoderma 110:109–130.

Boal, C. W., D. A. Haukos, and B. A. Grisham. [online]. 2010. Final report—phase 1: understanding ecology, habitat use, phenology and thermal tolerance of nesting Lesser Prairie-Chickens to predict population level influences of climate change. <http://www.greatplainslcc.org/PDFs/2010reports/Boal_GPLCC_phase1_final_report.pdf> (15 November 2013).

Brown, J. R., and S. Archer. 1999. Shrub invasion of grassland: recruitment is continuous and not regulated by herbaceous biomass or density. Ecology 80:2385–2396.

Chen, P., and M. Newman. 1998. Rossby wave propagation and the rapid development of upper-level anomalous anticyclones during the 1988 US drought. Journal of Climate 11:2491–2504.

Cook, E. R., C. A. Woodhouse, C. M. Eakin, D. M. Meko, and D. W. Stahle. 2004. Long-term aridity changes in the Western United States. Science 306:1015–1018.

Copelin, F. F. 1963. The Lesser Prairie Chicken in Oklahoma. Oklahoma Wildlife Conservation Department, Oklahoma City, OK.

Coppock, D. L. 2011. Ranching and multiyear droughts in Utah: production impacts, risk perceptions, and changes in preparedness. Rangeland Ecology and Management 64:607–618.

Crawford, J. A. 1980. Status, problem, and research needs of the Lesser Prairie-Chicken. Pp. 1–7 in P. A. Vohs, and F. L. Knopf (editors), Proceedings of the Prairie Grouse Symposium, Oklahoma State University, Stillwater, OK.

Crawford, J. A., and E. G. Bolen. 1976. Effects of land use on Lesser Prairie Chickens in Texas. Journal of Wildlife Management 40:96–104.

Crowley, T. J. 2000. Causes of climate change over the past 1000 years. Science 289:270–277.

Davis, D. M. 2009. Nesting ecology and reproductive success of Lesser Prairie-Chickens in shinnery oak–dominated rangelands. Wilson Journal of Ornithology 121:322–327.

Davis, C. A., T. Z. Riley, H. R. Suminski, and M. J. Wisdom. 1979. Habitat evaluation of Lesser Prairie Chickens in eastern Chaves County, New Mexico. Final Report to Bureau of Land Management, Roswell, Contract YA-512-CT6-61. Department of Fishery and Wildlife Sciences, New Mexico State University, Las Cruces, NM.

Diffenbaugh, N. S., F. Giorgi, and J. S. Pal. 2008. Climate change hotspots in the United States. Geophysical Research Letters 35:L16709.

Fields, T. L., G. C. White, W. C. Gilgert, and R. D. Rodgers. 2006. Nest and brood survival of Lesser Prairie-Chickens in west central Kansas. Journal of Wildlife Management 70:931–938.

Flanders-Wanner, B. L., G. C. White, and L. L. McDaniel. 2004. Weather and prairie grouse: dealing with effects beyond our control. Wildlife Society Bulletin 32:22–34.

Fuhlendorf, S. D., A. J. W. Woodward, D. M. Leslie, Jr., and J. S. Shackford. 2002. Multi-scale effects of habitat loss and fragmentation on Lesser Prairie-Chicken populations of the US Southern Great Plains. Landscape Ecology 17:617–628.

Girvetz, E. H., C. Zganjar, G. T. Raber, E. P. Maurer, P. Kareiva, and J. J. Lawler. 2009. Applied climate-change analysis: the climate wizard tool. PLoS One 4:e8320.

Greene, J. S., M. Chatelain, M. Morrissey, and S. Stadler. 2012. Projected future wind speed and wind power density trends over the western US high plains. Atmospheric and Climate Sciences 2:32–40.

Gregory, J. M., J. F. B. Mitchell, and A. J. Brady. 1997. Summer drought in northern midlatitudes in a time-dependent CO_2 climate experiment. Journal of Climate 10:662–686.

Grisham, B. A. 2012. The ecology of Lesser Prairie-Chickens in shinnery oak–grassland communities in New Mexico and Texas with implications toward habitat management and future climate change. Ph.D. dissertation, Texas Tech University, Lubbock, TX.

Grisham, B. A., C. W. Boal, D. A. Haukos, D. M. Davis, K. K. Boydston, C. Dixon, and W. R. Heck. 2013. The predicted influence of climate change on Lesser Prairie-Chicken reproductive parameters. PLoS One 8:e68225.

Grisham, B. A., P. K. Borsdorf, C. W. Boal, and K. K. Boydston. 2014. Nesting ecology and nest survival of Lesser Prairie-Chickens on the Southern High Plains of Texas. Journal of Wildlife Management 78:857–866.

Hagen, C. A. 2003. A demographic analysis of Lesser Prairie-Chicken populations in southwestern Kansas: survival, population viability, and habitat use. Ph.D. dissertation, Kansas State University, Manhattan, KS.

Hagen, C. A. 2010. Impacts of energy development on prairie grouse ecology: a research synthesis. Transactions of the North American Wildlife and Natural Resources Conference 75:96–103.

Hagen, C. A. and K. M. Giesen. [online]. 2005. Lesser Prairie-Chicken (*Tympanuchus pallidicinctus*). Birds of North America 364. <http://bna.birds.cornell.edu/bna/species/364/articles/introduction> (15 November 2013).

Hagen, C., B. Grisham, C. Boal, and D. Haukos. 2013. A meta-analysis of Lesser Prairie-Chicken nesting and brood rearing habitats: recommendations for habitat management. Wildlife Society Bulletin 37:750–758.

Hagen, C. A., G. C. Salter, J. C. Pitman, R. J. Robel, and R. D. Applegate. 2005. Lesser Prairie-Chicken brood habitat in sand sagebrush: invertebrate biomass and vegetation. Wildlife Society Bulletin 33:1080–1091.

Hagen, C. A., B. K. Sandercock, J. C. Pitman, R. J. Robel, and R. D. Applegate. 2009. Spatial variation in Lesser Prairie-Chicken demography: a sensitivity analysis of population dynamics and management alternatives. Journal of Wildlife Management 73:1325–1332.

Hagen, C. A., J. C. Pitman, T. M. Loughin, B. K. Sandercock, R. J. Robel, and R. D. Applegate. 2011. Impacts of anthropogenic features on habitat use by Lesser Prairie-Chickens. Pp. 63–75 in B. K. Sandercock, K. Martin, and G. Segelbacher (editors), Ecology, conservation, and management of grouse. Studies in Avian Biology (no. 39), University of California Press, Berkeley, CA.

Haukos, D. A., and L. M. Smith. 1999. Effects of lek age on age structure and attendance of Lesser Prairie-Chickens (*Tympanuchus pallidicinctus*). American Midland Naturalist 142:415–420.

Hernandez, J. E., P. H. Gowda, T. H. Marek, T. A. Howell, and W. Ha. 2013. Groundwater levels in northern Texas High Plains: baseline for existing agricultural management practices. Texas Water Journal 4:22–34.

High Plains Regional Climate Center. [online]. 2013. Climate change on the prairie: a basic guide to climate change in the High Plains region. <http://www.hprcc.unl.edu/publications/files/HighPlainsClimateChangeGuide-2013.pdf> (1 November 2013).

Holt, R. D. 2012. Breeding season demographics of a Lesser Prairie-Chicken (*Tympanuchus pallidicinctus*) population in the northeast Texas panhandle. Ph.D. dissertation, Texas Tech University, Lubbock, TX.

Horak, G. J., and R. D. Applegate. 1998. Greater Prairie-Chicken management. Kansas School Naturalist 45:3–15.

Intergovernmental Panel on Climate Change (IPCC). [online]. 2013. Working group 1 contributions to the IPCC fifth assessment report (AR5), climate change 2013: the physical science basis. <http://www.climatechange2013.org/images/uploads/WGIAR5_WGI-12Doc2b_FinalDraft_All.pdf> (15 October 2013).

Jones, P. D., T. J. Osborn, and K. R. Briffa. 2001. The evolution of climate over the last millennium. Science 292:662–667.

Jump, A. S., and J. Penuelas. 2005. Running to stand still: adaptation and the response of plants to rapid climate change. Ecology Letters 8:1010–1020.

Jurena, P. N., and S. Archer. 2003. Woody plant establishment and spatial heterogeneity in grasslands. Ecology 84:907–919.

Karl, T. R., J. M. Melillo, and T. C. Peterson. 2009. Global climate change impacts in the United States. Cambridge University Press, New York, NY.

Kemp, P. R. 1983. Phenological patterns of Chihuahuan Desert plants in relation to the timing of water availability. Journal of Ecology 71:427–436.

Kunkel, K. E., K. Andsager, and D. R. Easterling. 1999. Long-term trends in extreme precipitation events over the conterminous United States and Canada. Journal of Climate 12:2515–2527.

Kunkel, K. E., L. E. Stevens, S. E. Stevens, L. Sun, E. Janssen, D. Wuebbles, M. C. Kruk, D. P. Thomas, M. D. Shulski, N. A. Umphlett, K. G. Hubbard, K. Robbins, L. Romolo, and A. Akyuz. 2013. Regional climate trends and scenarios for the U.S. national climate assessment. Part 4. Climate of the U.S. Great Plains. NOAA Technical Report NESDIS 142-4. USDC NOAA, National Environmental Satellite, Data, and Information Service, Washington, DC.

Lecain, D. R., J. A. Morgan, A. R. Mosier, and J. A. Nelson. 2003. Soil and plant water relations determine photosynthetic responses of C3 and C4 grasses in a semi-arid ecosystem under elevated CO_2. Annals of Botany 92:41–52.

Lubchenco, J. and T. R. Karl. 2012. Predicting and managing extreme weather events. Physics Today 65:31–37.

McDonald, L., G. Beauprez, G. Gardner, J. Griswold, C. Hagen, F. Hornsby, D. Klute, S. Kyle, J. Pitman, T. Rintz, D. Schoeling, and W. Van Pelt. 2014. Range-wide population size of the Lesser Prairie-Chicken: 2012 and 2013. Wildlife Society Bulletin 38:536–546.

McKinley, D. C. and J. M. Blair. 2008. Woody plant encroachment by *Juniperus virginiana* in a mesic native grassland promotes rapid carbon and nitrogen accrual. Ecosystems 11:454–468.

Mendelsohn, R., W. Nordhaus, and D. Shaw. 1994. The impact of global warming on agriculture: a Ricardian analysis. American Economic Review 84:753–771.

Merchant, S. S. 1982. Habitat-use, reproductive success, and survival of female Lesser Prairie-Chickens in two years of contrasting weather. M.S. thesis, New Mexico State University, Las Cruces, NM.

Morgan, J. A., D. R. LeCain, J. J. Read, H. W. Hunt, and W. G. Knight. 1998. Photosynthetic pathway and ontogeny affect water relations and the impact of CO_2 on *Bouteloua gracilis* (C4) and *Pascopyrum smithii* (C3). Oecologia 114:483–493.

Parmesan, C., T. L. Root, and M. R. Willig. 2000. Impacts of extreme weather and climate on terrestrial biota. Bulletin of the American Meteorological Society 81:443–450.

Pearcy, R. W. and J. Ehleringer. 1984. Comparative ecophysiology of C3 and C4 plants. Plant, Cell and Environment 7:1–13.

Pitman, J. C., C. A. Hagen, R. J. Robel, T. M. Loughin, and R. D. Applegate. 2005. Location and success of Lesser Prairie-Chicken nests in relation to vegetation and human disturbance. Journal of Wildlife Management 69:1259–1269.

Pitman, J. C., C. A. Hagen, B. E. Jamison, R. J. Robel, T. M. Loughin, and R. D. Applegate. 2006. Nesting ecology of Lesser Prairie-Chickens in sand sagebrush prairie of southwestern Kansas. Wilson Journal of Ornithology 118:23–35.

Polley, H. W., D. D. Briske, J. A. Morgan, K. Wolter, D. W. Bailey, and J. R. Brown. 2013. Climate change and North American rangelands: trends, projections, and implications. Rangeland Ecology and Management 66:493–511.

Polsky, C. 2004. Putting space and time in Ricardian climate change impact studies: agriculture in the U.S. Great Plains, 1969–1992. Annals of the Association of American Geographers 94:549–564.

Pruett, C. L., M. A. Patten, and D. H. Wolfe. 2009. Avoidance behavior by prairie grouse: implications for development of wind energy. Conservation Biology 23:1253–1259.

Read, J. J. and J. A. Morgan. 1996. Growth and partitioning in *Pascopyrum smithii* (C3) and *Bouteloua gracilis* (C4) as influenced by carbon dioxide and temperature. Annals of Botany 77:487–496.

Rockström, J., W. Steffen, K. Noone, Å. Persson, F. S. Chapin, III, E. F. Lambin, T. M. Lenton et al. 2011. A safe operating space for humanity. Nature 476:472–475.

Sage, R. F. and R. K. Monson. 1999. C4 plant biology. Academic Press, San Diego, CA.

Sinha, N. R. and E. A. Kellogg. 1996. Parallelism and diversity in multiple origins of C4 photosynthesis in the grass family. American Journal of Botany 83:1458–1470.

Smyth, K. E. and D. A. Boag. 1984. Production in Spruce Grouse and its relationship to environmental factors and population parameters. Canadian Journal of Zoology 62:2250–2257.

Sophocleous, M. 2012. Conserving and extending the useful life of the largest aquifer in North America: the future of the High Plains/Ogallala aquifer. Ground Water 50:831–839.

Sophocleous, M. and D. Merriam. 2012. The Ogallala formation of the Great Plains in central US and its containment of life-giving water. Natural Resources Research 21:415–426.

Symstad, A. J. 2000. A test of the effects of functional group richness and composition on grassland invasibility. Ecology 81:99–109.

Trenberth, K. E., G. W. Branstator, and P. A. Arkin. 1988. Origins of the 1988 North American drought. Science 242:1640–1645.

U.S. Department of Agriculture Natural Resources Conservation Service. (USDA NRCS). [online]. 2012. USDA conservation program contributions to Lesser Prairie-Chicken conservation in the context of project climate change. http://www.nrcs.usda.gov/Internet/FSE_DOCUMENTS/stelprdb1080286.pdf (November 14, 2013).

U.S. Department of Agriculture Natural Resources Conservation Service. (USDA NRCS). [online]. 2013. The PLANTS database. National Plant Data Team, Greensboro, NC. <http://plants.usda.gov> (26 November 2013).

U.S. Energy Information Administration. [online]. 2013. Annual energy outlook. <http://www.eia.gov/forecasts/aeo/pdf/0383(2013).pdf> (15 November 2013).

U.S. Fish and Wildlife Service. (USFWS). 2014. Endangered and threatened wildlife and plants; special rule for the Lesser Prairie-Chicken; final rule. Federal Register 79:20073–20085.

Veldkamp, A., and E. F. Lambin. 2001. Predicting land-use change. Agriculture Ecosystems and Environment 85:1–6.

Volin, J. C., P. B. Reich, and T. J. Givnish. 1997. Elevated carbon dioxide ameliorates the effects of ozone on photosynthesis and growth: species respond similarly regardless of photosynthetic pathway or plant functional group. New Phytologist 138:315–325.

Walther, G. R., E. Post, P. Convey, A. Menzel, C. Parmesan, T. J. C. Beebee, J. M. Fromentin, O. Hoegh-Guldberg, and F. Bairlein. 2002. Ecological responses to recent climate change. Nature 416:389–395.

Wan, C., R. E. Sosebee, and B. L. McMichael. 1993. Soil water extraction and photosynthesis in *Gutierrezia sarothrae* and *Sporobolus cryptandrus*. Journal of Range Management 46:425–430.

Wand, S. J. E., F. F. Midgely, M. H. Jones, and P. S. Curtis. 1999. Responses of wild C4 and C3 grass (Poaceae) species to elevated atmospheric CO_2 concentration: a meta-analytic test of current theories and perceptions. Global Change Biology 5:723–741.

Wang, R. Z. 2002. Photosynthetic pathways, life forms, and reproductive types for forage species along the desertification gradient on Hunshandake Desert, North China. Photosynthetica 40:321–329.

Wang, R. Z., X. Q. Liu, and Y. Bai. 2006. Photosynthetic and morphological functional types for native species from mixed prairie in Southern Saskatchewan, Canada. Photosynthetica 44:17–25.

Wheeler, T. R., P. Q. Craufurd, R. H. Ellis, J. R. Porter, and P. V. Vara Prasad. 2000. Temperature variability and the yield of annual crops. Agriculture, Ecosystems and Environment 82:159–167.

Wilson, T. B., R. H. Webb, and T. L. Thompson. 2001. Mechanisms of range expansion and removal of mesquite in desert grasslands of the southwestern United States. USDA Forest Service, Rocky Mountain Research Station General Technical Report RMRS-GTR-81. USDA Forest Service, Rocky Mountain Research Station, Fort Collins, CO.

Wolfe, D. H., M. A. Patten, E. Shochat, C. L. Pruett, and S. K. Sherrod. 2007. Causes and patterns of mortality in Lesser Prairie-Chickens *Tympanuchus pallidicinctus* and implications for management. Wildlife Biology 13:95–104.

Woodward, A. J. W., S. D. Fuhlendorf, D. M. Leslie, Jr., and J. Shackford. 2001. Influence of landscape composition and change on Lesser Prairie-Chicken (*Tympanuchus pallidicinctus*) populations. American Midland Naturalist 145:261–274.

Zavaleta, J. C. 2012. Community response to use of prescribed grazing and herbicide for restoration of sand shinnery oak grasslands. M.S. thesis, Texas Tech University, Lubbock, TX.

Zondag, B. and J. Borsboom. 2009. Driving forces of land-use change. Presented at the 49th European Regional Science Association, Regional Agency for Rural Development (Agenzia regionale per los viluppo rurale) Conference, August 2009, Londz, Poland (in Polish).

CHAPTER THIRTEEN

Conservation of Lesser Prairie-Chickens*

A CASE STUDY

Patricia McDaniel and Betty Williamson

Abstract. Conservation of Lesser Prairie-Chickens (*Tympanuchus pallidicinctus*) as a potential threatened species is complicated, with considerable uncertainty regarding the ecology, different landscapes, relative effects of impacts to populations and habitats, and a diversity of stakeholders. Conservation goals can be developed for the species range-wide, but the objectives and strategies to achieve these goals are best tailored by ecoregion or at finer spatial scales. We illustrate the successful historical and contemporary conservation actions for the Lesser Prairie-Chicken in the Sand Shinnery Oak Ecoregion. The need for conservation actions within the agricultural community was recognized initially by long-time ranchers and farmers in eastern New Mexico. Early conservation actions were in response to severe population declines during the 1930s and principally involved state purchase of land in core habitats of Lesser Prairie-Chickens. However, land purchases were insufficient to curtail further population declines, which require public–private partnerships to increase the quantity and quality of Lesser Prairie-Chicken habitat across the ecoregion. Interest in conservation of Lesser Prairie-Chickens grew among landowners in eastern New Mexico with the availability of governmental funding and other incentives. Development of partnerships and collaborations for conservation were facilitated by the locally created and administered Grasslans Charitable Foundation [*sic*]. Multiple conservation strategies have made an investment of US$29 million in eastern New Mexico. More than 50,520 ha (124,835 acres) of potential and suitable habitat has been purchased and provided permanent protection for Lesser Prairie-Chickens. By 2013, beneficial conservation practices had been applied to >199,050 ha (491,863 acres) of private land. At the end of 2014, more than 1,171,940 ha (2,894,695 acres) had been enrolled in Candidate Conservation Agreements with Assurances and 780,348 ha (1,927,461 acres) have been enrolled in Candidate Conservation Agreements in eastern NM. Education and outreach efforts were essential to generate awareness of the status of Lesser Prairie-Chickens and develop and maintain societal support for conservation actions. It is critical to monitor population and habitat responses to conservation efforts to determine whether adaptations need to be made to management actions.

Key Words: conservation agreements, education, Natural Resources Conservation Service, New Mexico, outreach, ranching, sand shinnery oak, *Tympanuchus pallidicinctus*.

---

* McDaniel, P. and B. Williamson. 2016. Conservation of Lesser Prairie-Chickens: a case study. Pp. 243–256 in D. A. Haukos and C. W. Boal (editors), Ecology and conservation of Lesser Prairie-Chickens. Studies in Avian Biology (no. 48), CRC Press, Boca Raton, FL.

Successful conservation of Lesser Prairie-Chickens (*Tympanuchus pallidicinctus*) and other wildlife species requires several hierarchical stages, including the following: (1) recognizing that a stated conservation issue has merit; (2) obtaining information on the underlying conditions creating the conservation issue; (3) establishing societal consensus that the conservation issues should be addressed; (4) establishing definitive and measurable goals, objectives, and strategies to frame, monitor, and assess the progress of a conservation response to an issue; (5) obtaining scientific knowledge of the species' response to conservation measures; (6) obtaining funding to build capacity and implement priority conservation measures; and (7) establishing commitment by stakeholders and partnerships to engage and cooperatively complete the delivery of conservation goals. Strategic Habitat Conservation is a framework used by the U.S. Fish and Wildlife Service to integrate scientific information with management decisions (U.S. Fish and Wildlife Service 2006). The Strategic Habitat Conservation response to conservation issues incorporates a structured cycle with four components: (1) biological planning, (2) conservation design, (3) conservation delivery, and (4) monitoring and research. Components are defined for desired biological outcomes and to predict consequences of specific actions on landscapes, habitats, and species. Adaptive management is a framework that seeks to reduce uncertainty about the impacts of conservation and management approaches through a learning-based or adaptive decision process, which is a vital component of successful Strategic Habitat Conservation (Aldridge et al. 2004, Williams and Brown 2012).

Lesser Prairie-Chickens were listed as a federally threatened species in May 2014, due to ongoing population declines and continuing threats to populations and key habitats (USFWS 2014). However, the listing rule was vacated by judicial decision in September 2015, creating considerable regulatory uncertainty at the time of this volume. A lack of clarity remains regarding appropriate conservation goals, objectives, and strategies, primarily because of knowledge gaps regarding the ecological needs, economic impacts, and response to different management strategies by the species. However, years of consideration and potential listing as threatened have triggered numerous conservation efforts including the development of a Lesser Prairie-Chicken Range-wide Conservation Plan (Van Pelt et al. 2013), the Lesser Prairie-Chicken Initiative of the Natural Resource Conservation Service under the U.S. Department of Agriculture (NRCS 2014), the Candidate Conservation Agreements and Candidate Conservation Agreements with Assurances of the U.S. Fish and Wildlife Service (USFWS 2014), and creation of nongovernmental approaches to habitat conservation with mitigation banks. However, considerable uncertainty remains regarding Lesser Prairie-Chicken response to conservation strategies to achieve a variety of stated goals that principally revolve around increasing population numbers and improving habitat quality.

The extant range of Lesser Prairie-Chickens includes populations in four ecoregions across five states, and most habitat occurs on private land. Therefore, targeted strategies to achieve conservation goals should be implemented at four ecoregions or at finer spatial scales given spatial variation in population trends, habitat types, and relative impacts (Chapters 2 and 10, this volume). The development of successful local conservation strategies that can be expanded to meet ecoregional or species habitat and population goals requires organization and community support at many levels. The objective of our chapter is to provide a case study of successful conservation activities for Lesser Prairie-Chickens in eastern New Mexico. The local history provides insights into the foundation for conservation actions. Last, the multitude of conservation activities initiated before the species was listed as threatened can serve as a useful model for the development of similar strategies for other ecoregions and local sites.

HISTORY

In the southeast corner of New Mexico—the Land of Enchantment—a black wrought-iron sign welcomes drivers on State Road 206 to "Milnesand, Prairie Chicken Capital of New Mexico." The small community huddles around the intersection of NM 206 and NM 262. It can be challenging to find residents within the city limits, but the post office serves 15 families in the surrounding area, many of whom live on ranches in the short- and midgrass prairies and knee-high forest of sand shinnery oak (*Quercus havardii*) that have provided quality habitat for Lesser Prairie-Chickens for the past few centuries. The tiny, close-knit assemblage of farmers, ranchers, and other citizens is at the heart of the Sand Shinnery Oak Prairie Ecoregion of eastern New Mexico (McDonald et al. 2014).

Milnesand anchors the southern end of Roosevelt County, tucked against the Texas state line. The unusual name comes from a creative spelling of Mill-in-the-Sand, a windmill and important watering hole in the late 1800s on the legendary DZ Ranch. Thanks to its geographic location in the middle of one of the best remaining areas for Lesser Prairie-Chicken in New Mexico, and a cast of characters who became part of this story almost by accident, Milnesand has become an unlikely model for conservation success when agencies and other groups focus efforts on a threatened species.

Milnesand gets its warm days and cool nights from a deceivingly high elevation of ca. 1,220 m (4,000 ft), atop the "greatest of mesas" known as the Llano Estacado or Southern High Plains of New Mexico and Texas. In his book, *El Llano Estacado: Exploration and Imagination on the High Plains of Texas and New Mexico, 1536–1860*, author John Miller Morris (1997:1) describes this land mass as follows: "50,000 square miles of the American Southwest, from the Canadian River in the north to the Edwards Plateau in the south, from the Pecos River in the west to the fantastic canyonland tributaries of the Red, Pease, Brazos, and Colorado Rivers in the east." Morris continues, "It is perhaps the largest isolated, non-mountainous area in North America, with a single mesa at the center exceeding the combined size of seven of the original thirteen states of the United States. The tableland surface alone is some 30,000 square miles of pure, featureless plain. The semi-arid climate restricts the vertical architecture common elsewhere of trees, vines, and shrubs, leaving merely shortgrass plains."

Archeologist and historian Don Clifton noted that the area was home to the Clovis Culture some 11,000 years ago when the conditions were much moister than today, but a lack of reliable surface water kept much of Llano Estacado free of any sort of permanent settlements over the centuries. Francisco Vásquez de Coronado's expedition in the area from 1540 to 1542 likely brought him into contact with Apaches, who were regular visitors to hunt the herds of American bison (*Bison bison*) that crossed the Great Plains. By the mid-1700s, D. Clifton (pers. comm.) said: "The Comanche Indians had displaced the Apaches on the plains and had a thriving trade developed between the Spanish settlements along the northern Rio Grande and the Comanches. This trade continued until the defeat of the Comanches by the U.S. Army in 1874 in Palo Duro Canyon."

Elimination of Native Americans from the plains opened the door for cattlemen, and by the 1880s, large ranches—including the DZ that straddled Milnesand—were utilizing the vast tracts of short-grass prairie for free grazing. Land use on the Southern High Plains of New Mexico was forever altered with the appearance of the namesake Milnesand windmill in 1882 and the arrival of the Pecos Valley and Northeastern Railroad in southeastern New Mexico in the 1890s. The advent of the railroad provided access for families and speculators to take advantage of the Homestead Act, legislation first passed in 1862 that offered 65 ha (160 acres) of land for individuals willing to commit to the residence requirements to scratch out an existence in the 5-year required commitment. On July 3, 1916, Congress passed the Enlarged Homestead Act, which doubled available land to a more realistic 130 ha (320-acre) plots. By the end of 1916 in Roosevelt County, D. Clifton (pers. comm.) stated that "Nearly every 320 acres were homesteaded, and there was a population of almost 16,000 people in the county."

A century later, aerial images still show scars where the sod was broken in some scattered fields, but most of Roosevelt County surrounding Milnesand was too sandy for even the most adventurous settlers to consider plowing. Early settlers quickly discovered the difficulty of surviving in country where the average annual rainfall is ~43 cm (17 in.) and highly variable from year to year. Ninety years of climate records at the New Mexico Climate Center (weather.nmsu.edu) show this variability that includes years such as 2011 that received low single digit rainfall and the legendary year of 1941 when eastern New Mexico was drenched with >100 cm (>40 in.) of rain.

EARLY OBSERVATIONS OF LESSER PRAIRIE-CHICKENS IN EASTERN NEW MEXICO

From the beginning of permanent habitation in eastern New Mexico, Lesser Prairie-Chickens played a role as a source of easily harvested protein, as documented by more than one turn-of-the-century inhabitant of the area. In the book, *Six Miles to the Windmill*, early settler John Gordon Greaves (Greaves et al. 1976:77) noted that "prairie chickens were plentiful that fall and the settlers had never heard of such a thing as the game law. These chickens would come into the fields by the thousands just before sundown." His wife, and coauthor of the memoir, Annie King Greaves wrote, "[The prairie chickens] came in droves. They covered the feed shocks until it looked as if there would be no grain left for the horses. [Gordon] killed them by the dozens. We ate chickens until I never wanted to see another. I dressed and salted them down and then it was cold enough to freeze and keep them until we wanted some again. I had fried the young ones, made dressing or dumplings or stew out of the older ones" (Greaves et al. 1976:27).

It is likely that the early settlers were more concerned with warding off starvation than abiding by any sort of regulations concerning game management. The territorial government of New Mexico concluded shortly after the turn of the century that prairie chickens and other wildlife were a resource to be monitored and potentially marketed. The first Territorial Game Warden, Page Otero, was appointed in 1903, but the territorial government had previously established a set of "wildlife rules" in 1895. By 1897, a "more comprehensible, and more sensible" act was passed for the protection of wildlife (M. Frentzel, New Mexico Department of Game and Fish [NMDGF], pers. comm.). The 1897 regulations included a September through February season for quail, grouse, prairie chicken, pheasant, partridge, and Wild Turkey (*Meleagris gallopavo*).

Jim Williamson arrived in eastern New Mexico as a toddler in 1915. For decades he was fascinated with the birds, and he often recalled riding across southern Roosevelt County by horseback on early spring mornings to a chorus of "booming" from leks in every direction. For more than half a century, he enjoyed loading visitors into a pickup truck and bouncing them through the sandhills on his ranch near Pep, New Mexico, to witness the magic of the spring ritual on a lek near the family home. Interviewed in the Lesser Prairie-Chicken Interstate Working Group video "The Lesser Prairie-Chicken: Echoes of the Past?" in 2005 (Chapter 2, this volume), Williamson said, "Prairie chickens have actually been a kind of thing with me for, I guess, over 50 years. I've come to love the prairie chickens. There's a lot of people that ranch in the area that didn't even get up early enough to know about them booming."

Although not every rancher may have been aware of the annual early morning courtship displays, these birds were such an integral part of life

on the prairie that plenty of people noticed when the numbers plummeted along with the economy during the Dust Bowl years that ravaged the nation's midsection during the 1930s. Jim Hirsch, previously from the NMDGF, wrote "in the 1930s much of the landscape in southeast New Mexico was in severe drought. Bare ground farming also occurred in the region, and pastures were heavily grazed by domestic livestock. These compounding factors contributed to population declines of Lesser Prairie-Chicken" (Massey 2001:18).

By many accounts, the birds "were near extermination" from 1933 to 1935 (M. Frentzel, NMDGF, pers. comm.). In an unpublished manuscript, former New Mexico State Game Warden Elliott Barker wrote "1934 was the driest season the State has ever known and in the northeast part of the State the blowing sand and dust was intolerable. Livestock and game were hit hard. Quail suffered tragically—virtually no nesting for two years. In desperation, cattlemen grazed the shinnery in the sandhills of the prairie chicken country until there was no cover or food at all. The chickens were almost wiped out. We resolved then to buy up and fence sections of land through the area to prevent a recurrence in the future."

The Pittman–Robertson Federal Aid in Wildlife Restoration Act, signed into law in 1937 by President Franklin Roosevelt, provided funds at an opportune time for New Mexico's Lesser Prairie-Chicken population. In 1939 and 1940, the first two of what would become 29 total Prairie Chicken Areas (PCAs) were purchased by the State Game Commission using Federal Aid funds. "These acquired properties were often farms and ranches that had failed during the Dust Bowl and Great Depression and were scattered throughout De Baca, Lea, and Roosevelt Counties," Hirsch wrote in an unpublished document regarding the PCAs. "The basis for this purchase strategy was that wide distribution of protected areas would be more beneficial to Lesser Prairie-Chicken conservation than conserving a large area in only one part of this species' range" (Massey 2001:18). The 29 properties currently encompass >10,900 ha (27,000 acres), with properties ranging in size from 11 to 2,910 ha (28–7,189 acres), according to Hirsch, and they "are managed primarily to provide habitat for Lesser Prairie-Chickens, but also to provide benefits to other wildlife species" (Massey 2001:18).

New Mexico Magazine included an optimistic article in February 1944 titled, "Prairie Chickens

Increasing in New Mexico." Written by Paul Russell, the article stated "Recent inspections have revealed that prairie chickens are rapidly increasing and spreading over their natural range in eastern New Mexico. Large numbers of the birds have been seen many miles from the only areas where chickens were left a few years ago when the program for their restoration was started." Russell specifically mentioned the Game Department's purchase of "two restoration units near Milnesand in the heart of prairie chicken country," and said that the plan "involves the purchase and development of restoration units to provide an abundance of food, cover, and water, thereby creating natural game production areas where birds increase unmolested under healthy conditions" (Russell 1944:23). The purchase and conservation of core Lesser Prairie-Chicken habitat as state PCAs with Federal Aid funding provided the source population for expanding populations of Lesser Prairie-Chickens in eastern New Mexico, following the near extirpation of the population in the 1930s, as well as more contemporary periods of population declines.

By January 1949, conditions had improved enough that an editorial written by State Game Warden Barker in *New Mexico Magazine*'s standing section, The Conservationist: News and Views of the State Department of Game and Fish, was headlined, "An Effective Restoration Job." Noting that the department had been working with Lesser Prairie-Chickens for the previous 10 years, Barker wrote that the shortage was largely blamed on "an unprecedented and disastrous drouth [sic] of the dark dust bowl years," as well as "too much illegal and legal shooting up until the early 1930s." He told the story of a 3-day trip he had taken through southern Roosevelt and northern Lea counties in 1937, where NMDGF officials were only able to count 27 birds. "Of course, there were more than that," he wrote, "but that is all we could positively account for. The birds were almost gone." Thanks to the PCAs, as well as "better seasons and lessons learned by land users during the Dust Bowl era," Barker wrote, "the chickens responded and staged a rather remarkable comeback. Soon we had enough birds to justify trapping and transplanting to other areas still in need of seed stock" (Barker 1949:23).

The recovery due to conservation of habitats in core population areas and improved environmental conditions was strong enough that Barker said the department considered a hunting season in 1947, but decided to wait another year. In 1948, a 2.5-day season with a four-bird season bag limit was held, opening on November 27 on 1,500 miles^2 (~388,500 ha) of land in eastern New Mexico. Barker said ≥1,500 hunters participated and that "all got good shooting" and "many got the bag limit." "It was truly one of the nicest bird seasons we have ever had," Barker further stated, "An effective restoration job had been done. The results were most gratifying to those who have worked at it so long and hard, as well as to the spokesman." Barker closed his editorial by saying that while there was still much work to do on habitat restoration, "We can definitely say…we are over the hump" (Barker 1949:25).

Information on hunting regulations and harvest in subsequent years is limited (Chapter 7, this volume). Grant Beauprez, a biologist with NMDGF, found records from a 2-day hunt in 1958 with a bag limit of three birds per day (three in possession). A flyer released by NMDGF in 1969 shows that residents could purchase a $5 bird fee (all dollar values in U.S. currency) and hunt a rectangle of southeastern New Mexico that included portions of Curry, De Baca, Chaves, Lea, Roosevelt, and Eddy counties in search of their three-bird bag limit (six in possession) during December 6–8. The last recorded hunt was in 1995, a 9-day season with a bag limit of three birds per day (six in possession). A 2001 NMDGF management plan addressed the economic impact that hunting of Lesser Prairie-Chickens could theoretically have as follows:

> Hunters contribute to New Mexico's economy through the purchase of guns, ammunition, clothing, food, lodging, and fuel. Small game hunters spent $21.8 million in New Mexico in 1996 and the average expenditure each trip per small game hunter was $245 (USFWS 1996). This would result in $490,000 being contributed to local economies if the estimated demand of 2,000 Lesser Prairie-Chicken hunters is fulfilled. Outfitters and guides generally have not relied upon Lesser Prairie-Chickens or other grouse species as their main income because availability of birds is unpredictable. However, when numbers of birds have been adequate and the season was open, outfitters and local residents have supplemented their income by guiding hunters and charging trespass fees on private lands.
>
> MASSEY (2001:25)

ORIGINS OF CONTEMPORARY PARTNERSHIPS

It was hunting, albeit via a more nontraditional method, that brought Jim Weaver to New Mexico for the first time in 1976. The visit was the start of what would eventually become an important element in the conservation story of Lesser Prairie-Chickens at Milnesand. Weaver's name is familiar to many in the birding world, thanks to his ornithology work at Cornell University, his involvement with the North American Falconers Association, and his role as one of the founders of the Peregrine Fund. In 1976, it was the prospect of hunting Lesser Prairie-Chickens and doing it in a mild winter climate that brought Weaver and his falcon to New Mexico. He had hunted that winter across Canada and into the bitter cold of Montana. "New Mexico seemed like a nice place after 60-something below in Montana," he said (J. Weaver, pers. comm.).

In addition to finding abundant opportunities to fly his falcon after Lesser Prairie-Chickens, Weaver fell in love with the land in eastern New Mexico. Three years later he bought the first of his holdings northeast of Milnesand and moved to the area to stay. Almost immediately, he began looking for ways to improve the land he owned and provide more favorable habitats for local wildlife. The land he bought had included failed crop fields that had been converted from sand shinnery oak prairie by early homesteaders. Weaver's first order of business was restoration of those fields by reseeding native grasses. With the help of a neighbor, he laid miles of underground water pipelines, making water tanks available in areas they had never accessed to improve livestock grazing management.

U.S. DEPARTMENT OF AGRICULTURE–LANDOWNER PARTNERSHIP

While Weaver worked on improving habitat and on acquiring additional land, he watched population numbers of Lesser Prairie-Chickens in the Sand Shinnery Oak Prairie Ecoregion start to plummet again in the late 1980s (Chapter 4, this volume). Droughts are regular occurrences in this part of the Southwest and were a major factor, but Lesser Prairie-Chickens can readily recover from natural drought (Chapters 2 and 12, this volume). It became apparent that extensive landscape changes during the previous 50 years had

reduced the capacity of the Prairie Chicken Areas as a source population for Lesser Prairie-Chickens and that conservation actions would be needed on private lands to recover local populations. In 1995, the year that USFWS was petitioned to list the Lesser Prairie-Chicken as threatened, Weaver was invited to his first meeting of the Western Governors' Association. Participants discussed the potential listing of the bird with representatives from the governors' offices, officials from wildlife agencies of the five affected states, and a host of other agency people. A product of that meeting was the formation of the Lesser Prairie-Chicken Interstate Working Group.

The 1996 U.S. Department of Agriculture Federal Agriculture Improvement and Reform Act (Farm Bill) created the Environmental Quality Incentives Program, which would become a major component of funding for Milnesand area habitat, especially as awareness grew that good management of remaining Lesser Prairie-Chicken habitat was in the hands of private landowners and that financial incentives were available. Kenny Walker, retired area conservationist for the East Area, Natural Resources Conservation Service (NRCS) in New Mexico, credits Weaver with raising awareness of the situation with Lesser Prairie-Chickens in Roosevelt County when Weaver visited the office in 1999 to talk about the possibility of working with NRCS on habitat improvements on his lands near Causey, New Mexico. Other existing incentive programs had provided funds for sand shinnery oak management and various conservation efforts, but none had previously targeted habitat for a particular species. With a possible listing decision for Lesser Prairie-Chickens in the future, the viewpoint soon changed because of the focused involvement of Jim Weaver.

From the first Western Governors' Association meeting he attended in 1995, Weaver argued that a Grazing Resource Program (GRP) was a potential answer. "I said, you have CRP, you should have GRP. You pay farmers to not farm bad land. You have ranchers who are struggling—they would be more than happy to raise prairie chickens. You have to provide incentives to ranchers who have been struggling for years" (J. Weaver, pers. comm.). Weaver was also interviewed in the Kansas Parks and Wildlife video, "The Lesser Prairie-Chicken: Echoes of the Past?" where he said, "It's important that grazers and people engaged in all forms of agriculture have a sustainable option to deal with if they do want to stay on the land. That's one of the major impacts of the research we've started here. It's a way to show that you can have both good diversity and still a meaningful way of life for the people that have been here all these years." Kenny Walker and the NRCS agreed, and by 1999, the Local Working Group of the Lesser Prairie-Chicken Interstate Working Group was discussing the issues of Lesser Prairie-Chicken habitat and management. It was determined that the only way to get Environmental Quality Incentives Program money targeted in large amounts to help the Lesser Prairie-Chicken was to establish a Geographical Priority Area with the U.S. Department of Agriculture as an area identified as a high resource or high value land that can be the focus of special funds.

All five of the Lesser Prairie-Chicken states had been in a five-state Geographical Priority Area, but that initiative did not directly focus on Lesser Prairie-Chickens. The El Llano Resource Conservation District decided to seek its own Geographical Priority Area to focus specifically on Lesser Prairie-Chickens. The district also secured a grant from National Fish and Wildlife Foundation to do outreach and disseminate information. The Lesser Prairie-Chicken specific Geographical Priority Area for eastern New Mexico was officially approved in 1999, providing additional options and incentives for private landowners interested in managing habitats for Lesser Prairie-Chickens.

Roy and Shirley Creamer moved to eastern New Mexico from Lamar, Colorado, in early 1999 after buying a ranch near Milnesand, known historically as the Jody Ranch. Their ranch manager, Kyle Dillard, said they had ranched in a number of locations, often juggling U.S. Forest Service property and the realities of owning land inhabited with popular large game animals. Dillard had looked forward to the relocation to eastern New Mexico "because we thought we were moving to a place that wasn't quite so desirable" to outsiders. Within months, a neighboring rancher who was aware of the growing interest in the local Lesser Prairie-Chicken population pointed out that they had unusually high numbers of the prairie grouse. "We didn't even know what they were," Dillard recalls. Before long, the Jody Ranch became a hotspot of interest to biologists and wildlife agencies. "We had 15 different people here from three or four agencies," Dillard said. "I don't know if it

is true or not, but the way I remember it is when they came out and started counting, they said they counted more chickens on us than they had known were in all of southeastern New Mexico" (K. Dillard, pers. comm.).

One of those interested people was Chuck Mullins, who was the state coordinator at the time for the Partners for Fish and Wildlife program (USFWS) in New Mexico. He arranged a formal agreement between the Creamers as private landowners and the USFWS. The project outlined in the cooperative agreement was developed with the goal to maintain and improve habitat for Lesser Prairie-Chickens and other wildlife in the sand shinnery oak prairie on the Jody Ranch, which was by then known as the Creamer Ranch. The funding was used to add 24 km (15 miles) of pipeline to improve the rotational management of cattle. Tish McDaniel, who would later lead the Short-Grass Prairie Project for The Nature Conservancy (TNC), was hired to monitor the Lesser Prairie-Chicken population and habitat response to the conservation practices. "We only knew of eight leks on that land prior to that," she said. "After that first spring, I had 30 leks. It turned into a full-time job for me doing monitoring, documenting bugs, conducting vegetation surveys" (P. McDaniel, pers. comm.). The abundance of Lesser Prairie-Chickens on the Creamer Ranch came as no surprise to Jim Williamson—he had crossed that property often on horseback as a young man and listened to the springtime performance. Nor was this McDaniel's first encounter with the birds. She had her own introduction to the Lesser Prairie-Chicken as an infant riding along with her parents on one of Williamson's pickup tours of lek sites.

CREATION OF ADDITIONAL PARTNERSHIPS

None of the conservation planning happened overnight, but by the year 2000, the Milnesand area was recognized by many as the core Lesser Prairie-Chicken habitat in eastern New Mexico, and a number of organizations, agencies, and individuals were determined to keep it that way. However, the lack of local organization and leadership precluded implementation of several noteworthy conservation efforts.

After being told that the NMDGF could not accept a $250,000 grant available from an outside source for Lesser Prairie-Chicken work because the agency lacked a person on staff to administer the grant, Jim Weaver created the Grasslans Charitable Foundation [sic] in 2000. The Grasslans Charitable Foundation has served as the focal point for the development of numerous management and research partnerships that have been keys to successful conservation efforts for Lesser Prairie-Chickens in eastern New Mexico. Collaborations and funding agreements with universities, governmental agencies, and nongovernmental organizations are initiated, facilitated, and administered by the foundation for the benefit of Lesser Prairie-Chickens and local landowners. Virtually, all funding spent by the Grasslans Charitable Foundation since inception has been directed toward the conservation of Lesser Prairie-Chicken either directly or indirectly, with total funding exceeding $3 million. One of the first projects was a 10-year research study assessing the responses of Lesser Prairie-Chickens, grassland birds, small mammals, herptiles, and invertebrates in the Sand Shinnery Oak Prairie Ecoregion to landscape management practices including herbicide and grazing treatments. The research portion of this investment has been used in an adaptive fashion for the conservation of Lesser Prairie-Chickens because of a number of scientific publications that have increased our knowledge of the ecology Lesser Prairie-Chickens (Patten et al. 2005a,b; Bell et al. 2007, 2010; Wolfe et al. 2007; Patten and Kelly 2010; Grisham 2012; Grisham et al. 2013), as well as the vegetation, mammalian, and invertebrate communities (Zavaleta 2012), and songbird populations in the Sand Shinnery Oak Prairie Ecoregion (Smythe and Haukos 2009, 2010).

CONSERVATION ACCOMPLISHMENTS FOR LESSER PRAIRIE-CHICKENS IN NEW MEXICO

With the support and trust of local landowners and other stakeholders, the amount of conservation funding and projects to benefit Lesser Prairie-Chickens in this remote corner of the Llano Estacado is impressive. In addition to the multitude of landowners involved in Lesser Prairie-Chicken conversation, support through technical assistance and funding has been provided by a diversity of sponsors including the following: New Mexico Department of Game and Fish, U.S. Fish and Wildlife Service, Natural Resources Conservation Service (U.S. Department

of Agriculture), The Nature Conservancy, U.S. Bureau of Land Management, local Resource Conservation Districts, Sutton Avian Research Center, Playa Lakes Joint Venture, New Mexico State Land Office, Center of Excellence, county governments, Texas Tech University, New Mexico State University, Wildlife Plus Consulting, and other groups. Multiple strategies have been implemented to benefit the Lesser Prairie-Chickens, including direct purchase of habitat, conservation agreements, site-specific application of conservation practices, and educational outreach. More than $2 million has been spent to directly benefit habitat for Lesser Prairie-Chickens in the Sand Shinnery Oak Prairie Ecoregion.

Potential Lesser Prairie-Chicken habitat that was permanently protected in the Sand Shinnery Oak Prairie Ecoregion in 2000 totaled 22,523 ha (55,657 acres), which included the New Mexico PCAs; Muleshoe National Wildlife Refuge, Texas; Grulla National Wildlife Refuge, New Mexico; and Weaver Ranch Conservation Easement. By 2009, the area of permanent protection had nearly doubled to 40,101 ha (99,093 acres) with the purchases of Milnesand Prairie Preserve (including the former Creamer and Johnson Ranches; ~$4 million) and Yoakum Dunes Preserve in Texas (~$4.8 million) by TNC and Sandhills Prairie Conservation Area (~$1.2 million) by NMDGF. Further, the Weaver Ranch added additional area to their existing conservation easement. By 2013, the amount of permanently protected potential Lesser Prairie-Chicken habitat had increased to 50,519 ha (124,835 acres) with purchases of the Sand Ranch by the BLM (~$2.2 million) and additional area added to the TNC Yoakum Dunes Preserve (Figure 13.1).

The Nature Conservancy has arguably made one of the largest and most permanent commitments to the birds and the region with the purchase of 12,140 ha (30,000 acres) of core Lesser Prairie-Chicken habitat in the Milnesand area from 2005 to 2009. The first part of the land purchase—the chicken-rich Creamer Ranch—is now called Milnesand Prairie Preserve North, and it is joined on the south by equally promising habitat from the former Johnson Ranch, known as Milnesand Prairie Preserve South. Tish McDaniel, who has been with TNC since 2004, manages the land in collaboration with the former Creamer Ranch manager, Kyle Dillard, who stayed to lease the properties and run his own cow–calf operation.

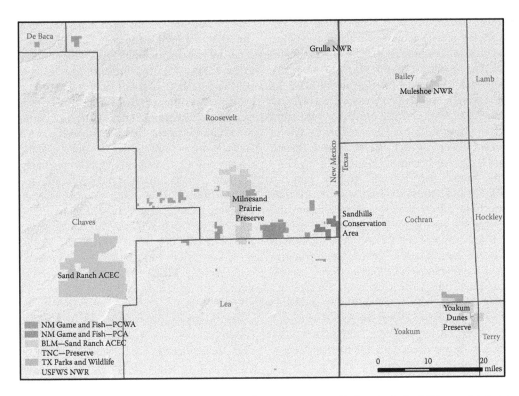

Figure 13.1. Permanently protected areas potentially benefiting Lesser Prairie-Chickens in the Sand Shinnery Oak Prairie Ecoregion.

"It's been a good relationship," Dillard says, "We've had differences of opinions, but we've always been able to work through it. As a whole, we both have the same goals in mind." Dillard explained that the improvements done on the ranches, funded by various agencies, make it easier for him to utilize the prairie in a way that benefits both livestock and wildlife. "When I look at the deal, I never focused on the chicken," he said. "That's not my job. The rangeland as a whole is my job. What's good for the cattle is good for the chicken is good for the lizard. Our goal is to make the ranch and the whole rangeland the very best that it can be. If it's good for that, it's good for everything" (K. Dillard, pers. comm.).

A number of conservation activities have initially occurred on the Milnesand Prairie Preserve. Stewardship start-up costs by TNC totaled $84,000. The State of New Mexico expended $100,000 to restore abandoned oil pads and roads. The National Fish and Wildlife Foundation contributed $50,000 for stewardship, education/restoration, and promotion of a grazing cooperative.

Conservation practices applied on private lands to benefit Lesser Prairie-Chickens since 1998 have been diverse and extensive in eastern New Mexico. The projects are site specific and have included such diverse actions as reseeding areas with native grasses, restoration of sand shinnery oak prairie, reclamation of abandoned oil field pads and roads, removal of invasive trees, dismantling of vertical structures, marking fences for greater visibility, surveying for new leks, water and fencing to improve livestock management, and enrollment in conservation programs. The Natural Resources Conservation Service (USDA) has used the Lesser Prairie-Chicken Initiative (33 contracts, 148,106 ha [365,978 acres]), Watershed Initiative, Environmental Quality Incentives Program (combined 52 contracts, 105,145 ha [259,818 acres], Wildlife Habitat Incentive Program (5 contracts, 26,630 ha [65,803 acres], and Conservation Reserve Program (~120,192 ha [297,000 acres]) to target improvements of Lesser Prairie-Chicken habitat on 283,538 ha (700,638 acres) from 1999 to 2013 (J. Miller, NRCS, pers. comm.). The benefits totaled a minimum investment of $6.9 million. Specific practices to improve nesting cover include 10-month and 1-year deferments (72,729 ha [179,717 acres]), brush management (59,954 ha [148,150 acres]), fence construction (~243 km [146 miles]) to improve grazing management, prescribed grazing (22,670 ha [56,019 acres]), upland wildlife habitat management (4,517 ha [11,162 acres]), restoration to native grasses (2,215 ha [5,473 acres]), and oil pad/road restoration (13.4 ha [33 acres]). It is estimated that ~35% of occupied Lesser Prairie-Chicken range in eastern New Mexico has benefited from these practices.

The Partners for Fish and Wildlife program (USFWS) and other funding sources (Private Stewardship Grant, American Reinvestment and Recovery Act) have focused on additional conservation activities to benefit Lesser Prairie-Chickens in eastern New Mexico. Conservation practices such as native grass restoration, improved grazing management through fencing, water distribution, grazing deferral, fence marking, and invasive tree control have impacted 28,491 ha (70,404 acres) at a cost of $655,868. Industry contributions include $2 million to clean and restore 3,440 ha (8,500 acres) impacted by oil and gas exploration. The National Fish and Wildlife Foundation through the Playa Lakes Joint Venture provided $699,790 to reclaim 44.5 ha (110 acres), mark fences, and remove invasive trees on 140 ha (345 acres). The BLM restored or enhanced 18,886 ha (46,668 acres) by fence removal, invasive tree control, fence marking, and restoration of oil pads and roads at a cost of $1.8 million.

Provisions in the Endangered Species Act provide for the proactive conservation of candidate species prior to a decision to list the species as threatened or endangered through the application of Candidate Conservation Agreements. The Center of Excellence, located in Carlsbad, New Mexico, is a nonprofit organization dedicated to improving wildlife habitat with holistic management approaches that benefit species of concern and also provides environmental enhancements that are beneficial to all organisms in those habitats. The group worked for 2.5 years and published the "Collaborative Conservation Strategies for the Lesser Prairie-Chicken and Sand Dune Lizard in New Mexico" in August 2005. The strategy provided guidance in the development of the BLM's 2008 Special Status Species Resource Management Plan Amendment, which also addresses the concerns and future management of habitats for the Lesser Prairie-Chicken and dunes sagebrush

lizard (*Sceloporus arenicolus*). In consultation with the USFWS, BLM, landowners, and industry, land use prescriptions were developed to serve as baseline mitigation for both species as part of Candidate Conservation Agreements. The Center of Excellence has been proactive in educating and enrolling ranches and industries in Roosevelt County, New Mexico, with Candidate Conservation Agreements and Candidate Conservation Agreements with Assurances to provide programmatic safeguards for states, landowners, and lessees if the Lesser Prairie-Chicken became listed as threatened. By the end of 2013, >56,656 ha (>140,000 acres) in Roosevelt County, with >8,300 ha (>20,500 acres) enrolled by industry, were included in these voluntary conservation agreements, which included $644,529 to restore habitat (Figure 13.2).

Last, strategies to increase local support for Lesser Prairie-Chicken conservation efforts have been implemented as part of an extensive education and outreach program, which included frequent informational workshops, landowner field days, community discussions, newsletters, and regular face-to-face discussions among stakeholders that included conservationists, producers, industry, and governmental agencies. The activities generated awareness of the status of Lesser Prairie-Chickens, including the development and maintenance of necessary support for successful conservation actions. The event that perhaps has brought the most attention to Milnesand is the community-initiated and community-organized annual High Plains Prairie-Chicken Festival, which started in 2002 and created the opening for the most visible part of the history of Lesser Prairie-Chickens in Milnesand (Figure 13.3). From a vantage point of a dozen years later, it is hard to remember who had the idea first for this community-changing event, or how exactly it came to be. What happened is that in a rather short period of time that first year, Tish McDaniel and Willard Heck (manager of Weaver Ranch) with ample assistance from the NMDGF, the New Mexico Wildlife Federation, TNC, the Milnesand Community, and Grasslans Charitable Foundation, pulled together a weekend event to celebrate the springtime mating ritual of the Lesser Prairie-Chicken. For $35 each, the first 100 participants received five meals, two early morning opportunities for lek viewing, and an opportunity to join in a number of field

Ranches and industry enrolled in the candidate conservation agreement with assurances

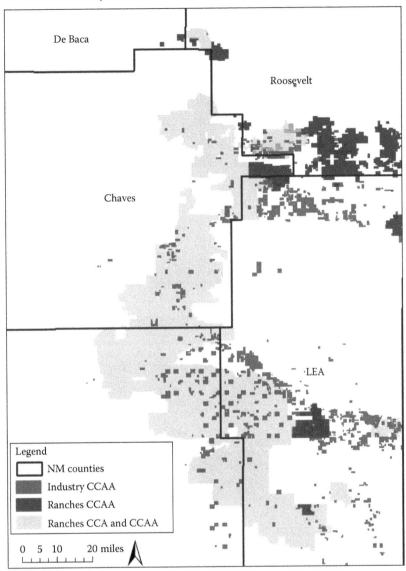

Figure 13.2. Areas of the Sand Shinnery Oak Prairie Ecoregion enrolled in conservation agreements prior to listing of the Lesser Prairie-Chicken as a threatened species in May 2014. Agreements included Candidate Conservation Agreements (CCA) and Candidate Conservation Agreement with Assurances (CCAA).

trips designed to showcase the history, flora, and fauna of the area. Organizers tried the second year with 125 participants and quickly realized it was too many. The maximum number of spaces has remained at 100 ever since, with reservations filling all available spaces several months in advance.

The High Plains Prairie Chicken Festival has developed community pride, stewardship, and determination for the conservation of Lesser Prairie-Chickens and their habitats. The New Mexico Wildlife Federation handled funds for the first two years, but Grasslans Charitable Foundation took over in the third year. Since 2004, all profits from the annual event were earmarked for the city of Milnesand and used for the maintenance of the community building and support for the volunteer fire department, which is the heart and soul of the community. From 2004 to 2012, the festival has brought $19,364 to the community and $35,622 to the volunteer fire department.

(a)

(b)

Figure 13.3. Engaging local landowners and other stakeholders through (a) field days and (b) the High Plains Prairie Chicken Festival are critical components for successful conservation of Lesser Prairie-Chickens in the Sand Shinnery Oak Prairie Ecoregion.

From the beginning, the festival relied heavily on volunteer participation from local ranch families. Locals pitched in to cook all the meals and were available all weekend to meet visitors and share information about the area. A number of families opened their homes to festival-goers for overnight accommodations and as a way to earn extra income. While the majority of festival-goers hailed from the Santa Fe and Albuquerque, New Mexico, areas, the annual event has brought visitors from across the United States. Even a handful of international visitors have made the trek to witness one of nature's most engaging rituals. Many visitors come back for repeat experiences, and frequently follow up with remarks such as, "I really admire how you have broadened the focus to the Llano Estacado ecosystem, including how people (past and present) live with the land. And I really enjoyed spending time with folks who appreciate the austere beauty of that landscape."

Unfortunately, the extreme drought since 2011 that gripped much of the Llano Estacado caused the festival to be cancelled in 2013 and 2014. Organizers hope conditions will improve and the event will once again serve as a way for landowners and visitors to share the beauty of the Lesser Prairie-Chickens. As one participant wrote after the last festival, "I will never forget it, and I hope to be one of those folks who comes back again and again."

With the exception of festival weekend, most days are quiet in Milnesand. A passing motorist is unlikely to encounter a Lesser Prairie-Chicken, or even guess at the millions of dollars that have been spent in this region in an effort to help. Past Roosevelt County economic director Doug Redmond said that as in most regions "every dollar spent here circulates about seven times" in the community.

Combined, the efforts to conserve the Lesser Prairie-Chicken in eastern New Mexico have created greater awareness of the status of the species, generated stakeholder interest and support for conservation actions, and increased available funding in support of these efforts. The trust and interpersonal relationships generated during the past decade will assist in meeting the goals of the Lesser Prairie-Chicken Range-wide Conservation Plan (Van Pelt et al. 2013) and the Lesser Prairie-Chicken Initiative (NRCS 2014). By monitoring the effects of implemented management actions and practices, an adaptive framework founding is being built to guide future actions. Likely due to these conservation efforts, the Lesser Prairie-Chicken population in eastern New Mexico has not experienced recent declines similar to other populations (Van Pelt et al. 2013). More importantly for the people of eastern New Mexico, greater than any economic impact measured in dollars will be the successful conservation of a bird loved by many in the sandhills surrounding Milnesand.

ACKNOWLEDGMENTS

We greatly appreciate the insights and assistance from G. Beauprez and M. Frentzel of New Mexico Department of Game and Fish; D. Lynn and E. Wirth of the Center of Excellence; J. Swafford of Pheasants Forever; J. Miller and K. Walker of U.S. Department of Agriculture Natural Resources Conservation Service; J. Weaver and W. Heck of Weaver Ranch and Grasslans Charitable Foundation; N. Baczek of U.S. Fish and Wildlife Service—Partners for Wildlife; and K. Dillard, rancher.

LITERATURE CITED

Aldridge, C. L., M. S. Boyce, and R. K. Baydack. 2004. Adaptive management of prairie grouse: how do we get there? Wildlife Society Bulletin 32:92–103.

Barker, E. S. 1949. An effective restoration job. New Mexico Magazine 27:23–25.

Bell, L. A., S. D. Fuhlendorf, M. A. Patten, D. H. Wolfe, and S. K. Sherrod. 2010. Lesser Prairie-Chicken hen and brood habitat use on sand shinnery oak. Rangeland Ecology and Management 63:478–486.

Bell, L. A., J. C. Pitman, M. A. Patten, D. H. Wolfe, S. K. Sherrod, and S. D. Fuhlendorf. 2007. Juvenile Lesser Prairie-Chicken growth and development in southeastern New Mexico. Wilson Journal of Ornithology 119:386–391.

Greaves, A. K., J. G. Greaves, and V. Acker. 1976. Six miles to the windmill: the personal recollections of Annie King Greaves and John Gordon Greaves for the period 1908–1913. Eastern New Mexico University, Portales, NM.

Grisham, B. A. 2012. The ecology of Lesser Prairie-Chickens in shinnery oak–grassland communities in New Mexico and Texas with implications toward habitat management and future climate change. Ph.D. dissertation, Texas Tech University, Lubbock, TX.

Grisham, B. A., C. W. Boal, D. A. Haukos, D. M. Davis, K. K. Boydston, C. Dixon, and W. R. Heck. 2013. The predicted influence of climate change on Lesser Prairie-Chicken reproductive parameters. PLoS One 8:e68225.

Massey, M. [online]. 2001. Long-range plan for the management of Lesser Prairie-Chickens in New Mexico 2002–2006. Federal Aid in Wildlife Restoration Grant W-104-R41, Project 3.4. New Mexico Department of Game and Fish, Santa Fe, NM. <http://www.wildlife.state.nm.us/conservation/documents/PCLongRange.pdf> (September 2013).

McDonald, L., G. Beauprez, G. Gardner, J. Griswold, C. Hagen, D. Klute, S. Kyle, J. Pitman, T. Rintz, and B. Van Pelt. 2014. Range-wide population size of the Lesser Prairie-Chicken: 2012 and 2013. Wildlife Society Bulletin 38:536–546.

Morris, J. M. 1997. El Llano Estacado: exploration and imagination on the high plains of Texas and New Mexico, 1536–1860. Texas State Historical Association, Austin, TX.

Natural Resources Conservation Service (NRCS). [online]. 2014. Lesser Prairie-Chicken initiative. U.S. Department of Agriculture, Washington, DC. <http://www.nrcs.usda.gov/wps/portal/nrcs/detailfull/null/?cid=nrcsdev11_023912> (14 November 2013).

Patten, M. A., and J. F. Kelly. 2010. Habitat selection and the perceptual trap. Ecological Applications 20:2148–2156.

Patten, M. A., D. H. Wolfe, E. Shochat, and S. K. Sherrod. 2005a. Habitat fragmentation, rapid evolution and population persistence. Evolutionary Ecology Research 7:235–249.

Patten, M. A., D. H. Wolfe, E. Shochat, and S. K. Sherrod. 2005b. Effects of microhabitat and microclimate selection on adult survivorship of the Lesser Prairie-Chicken. Journal of Wildlife Management 69:1270–1278.

Russell, P. 1944. Prairie chickens increasing in New Mexico. New Mexico Magazine 22:23.

Smythe, L., and D. A. Haukos. 2009. Nesting success of grassland birds in shinnery oak communities treated with tebuthiuron and grazing in eastern New Mexico. Southwestern Naturalist 54:136–145.

Smythe, L., and D. A. Haukos. 2010. Response of grassland birds in sand shinnery oak communities restored using tebuthiuron and grazing in eastern New Mexico. Restoration Ecology 18:215–223.

U.S. Fish and Wildlife Service. (UFWS). [online]. 2006. Strategic habitat conservation: a report from the National Ecological Assessment Team, Washington, DC. <http://www.fws.gov/landscape-conservation/pdf/SHCReport.pdf> (15 November 2013).

U.S. Fish and Wildlife Service. (UFWS). 2014. Endangered and threatened wildlife and plants; special rule for the Lesser Prairie-Chicken; final rule. Federal Register 79:20073–20085.

Van Pelt, W., S. Kyle, J. Pitman, D. Klute, G. Beauprez, D. Schoeling, A. Janus, and J. Haufler. [online]. 2013. The Lesser Prairie-Chicken range-wide conservation plan. Western Association of Fish and Wildlife Agencies, Cheyenne, WY. <http://www.wafwa.org/documents/2013LPCRWPfinalfor4drule12092013.pdf> (October 2013).

Williams, B. K., and E. D. Brown. 2012. Adaptive management: the U.S. Department of the Interior applications guide. Adaptive Management Working Group, U.S. Department of the Interior, Washington, DC.

Wolfe, D. H., M. A. Patten, E. Shochat, C. L. Pruett, and S. K. Sherrod. 2007. Causes and patterns of mortality in Lesser Prairie-Chickens *Tympanuchus pallidicinctus* and implications for management. Wildlife Biology 13:95–104.

Zavaleta, J. C. 2012. Community response to use of prescribed grazing and herbicide for restoration of sand shinnery oak grasslands. M.S. thesis, Texas Tech University, Lubbock, TX.

PART FOUR

Conservation and Management

CHAPTER FOURTEEN

Grasslands of Western Kansas, North of the Arkansas River*

David K. Dahlgren, Randy D. Rodgers, R. Dwayne Elmore, and Matthew R. Bain

Abstract. Lesser Prairie-Chickens (*Tympanuchus pallidicinctus*) occur in short-grass and mixed-grass prairies and associated grasslands restored through the Conservation Reserve Program (CRP) north of the Arkansas River in Kansas. The Short-Grass Prairie/CRP Mosaic Ecoregion currently supports ~65% of the range-wide population of Lesser Prairie-Chickens. CRP lands provide important grassland habitats, especially nesting and brood-rearing areas for breeding Lesser Prairie-Chickens. A combination of implementation of CRP grasslands at a landscape scale and favorable environmental conditions is thought to have led to a significant increase in the occupied range and population density of Lesser Prairie-Chickens in the ecoregion. Spring lek surveys since 1999 have documented the northern expansion of the observed distribution of Lesser Prairie-Chickens in 2008, 2011, and 2012. An expanding distribution has led to the development of a contact zone of sympatry between Lesser Prairie-Chickens and Greater Prairie-Chickens (*T. cupido*). Hybridization between the two congeneric species has been documented and is currently estimated to occur at a rate of ~5%. The potential effects of hybridization on the genetic structure of Lesser Prairie-Chickens are poorly understood. Conservation of Lesser Prairie-Chickens in the region will be best accomplished by maintaining current habitat and provide management tools, guidelines, etc. and implement, recommend management practices, such as grazing, prescribed fire, herbicide application, and prairie restoration, to improve habitat quality at smaller spatial scales. The ecoregion has been only recently occupied by substantial numbers of Lesser Prairie-Chickens, and new data are needed to develop conservation and management plans. Current knowledge gaps include information on population demographics, limiting factors, habitat use and seasonal movements at various scales, habitat management techniques, energy development impacts, and climate change. Additionally, improved land use policies are needed for long-term protection of habitat within the region, beyond the typical duration of 10–15 years for CRP contracts. If conservation goals are met, the ecoregion north of the Arkansas River in Kansas could continue to remain a stronghold for the Lesser Prairie-Chicken in the future.

Key Words: Conservation Reserve Program, habitat management, hybrid, Kansas, Short-Grass Prairie/CRP Mosaic Ecoregion, sympatric range, *Tympanuchus pallidicinctus*.

---

* Dahlgren, D. K., R. D. Rodgers, R. D. Elmore, and M. R. Bain. 2016. Grasslands of Western Kansas, North of the Arkansas River. Pp. 259–279 in D. A. Haukos and C. W. Boal (editors), Ecology and conservation of Lesser Prairie-Chickens. Studies in Avian Biology (no. 48), CRC Press, Boca Raton, FL.

The Short-Grass Prairie/Conservation Reserve Program (CRP) Mosaic Ecoregion lies on the eastern extent of the short-grass prairie and the transition to mixed-grass prairie in western Kansas, north of the Arkansas River (Figure 14.1). The short-grass prairie is unique to the western edge of the Great Plains abutting the eastern front range of the Rocky Mountains (Samson and Knopf 1996). Short-grass prairie extends into the western quarter of Kansas, and the area is often referred to as the northern High Plains (Shiflet 1994). Mixed-grass prairie also occurs in the region, especially along the eastern edge of the distribution of Lesser Prairie-Chickens (*Tympanuchus pallidicinctus*) in Kansas, but also as inclusions further west (Kansas Native Plant Society 2014). McDonald et al. (2014) and Van Pelt et al. (2013) have described the area as the Short-Grass Prairie/CRP Mosaic Ecoregion for Lesser Prairie-Chickens, and the name describes the gradient of grassland types found in the ecoregion.

Before European settlement, the ecoregion was a landscape of generally flat short-grass prairie interspersed with mixed-grass prairie and small tracts of sand sagebrush (*Artemisia filifolia*) prairie along some drainages and unique soil types. Playas or small ephemeral wetlands were historically a common feature across this area, especially within the large expanse of flat "table" lands (Haukos and Smith 1994). Following European settlement, much of the extant prairies were cultivated and playas were incorporated into larger crop fields. Center-pivot irrigation systems became widespread in the 1960s and 1970s, allowing farmers to tap into the Ogallala Aquifer to provide season-long water supply for increased crop production. The use of groundwater revolutionized agriculture in the area, even during periods of drought. During the 1960s to the 1980s, large areas of prairies were again broken and plowed as a result of this advancement in agricultural technology (Waddell and Hanzlick 1978, Sexson 1980).

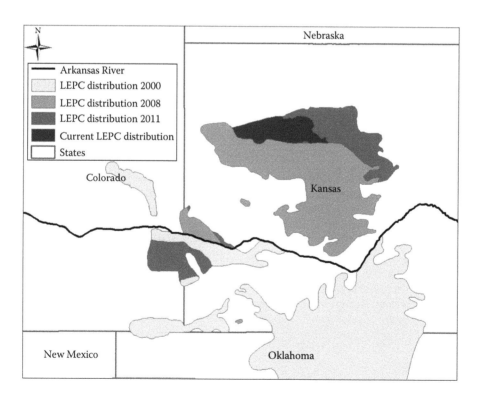

Figure 14.1. Changes in the distribution of Lesser Prairie-Chickens (LEPC) in Kansas and Colorado since the early 2000s. Spring lek surveys were conducted by the Colorado Parks and Wildlife, Kansas Department of Wildlife, Parks, and Tourism, and partners and resulted in multiple northward changes in distribution boundaries during the past decade (also see Figure 14.3). Distributional limits and dates represent where and when biologists officially moved the known boundaries for the species distribution, but not necessarily how Lesser Prairie-Chicken were using the landscape in real time.

Plant cover and vegetative structure of native short-grass prairie are not generally considered suitable habitat for the life cycle of Lesser Prairie-Chickens (Hagen et al. 2004). For example, Lesser Prairie-Chickens tend to select plant heights and visual obstruction for nesting cover much greater than the habitats provided by typical short-grass communities (Hagen et al. 2004). In other parts of their range, Lesser Prairie-Chickens often use shrub-dominated landscapes that provide needed vegetation structure in semiarid environments (Chapters 15 and 17, this volume). Historically, the sand sagebrush and mixed-grass prairies along the larger drainages in this region may have provided habitat for Lesser Prairie-Chickens. It is unknown what proportion of the eastern short-grass and western mixed-grass regions in Kansas were historically occupied by Lesser Prairie-Chickens or how densities varied among vegetation types (Hagen 2003). However, voluntary conversion of cropland into perennial grass cover through the Conservation Reserve Program of the U.S. Department of Agriculture has improved habitat conditions. The state of Kansas differed from other states by requiring seed mixes to resemble native mixed-grass and tall-grass communities, which increased potential habitat, reduced landscape fragmentation, and resulted in increased population abundance and occupancy of Lesser Prairie-Chickens (Rodgers 1999, Fields 2004, Rodgers and Hoffman 2005, Fields et al. 2006; Figure 14.2).

The objectives of our chapter are to (1) provide a synthesis of known ecological information regarding Lesser Prairie-Chickens and their habitat requirements in the short-grass, mixed-grass, and CRP prairie complex of the ecoregion; (2) consider distribution changes and northern expansion in the ecoregion since the early 2000s; (3) provide insights regarding the sympatric distributions of Lesser and Greater Prairie-Chickens (*Tympanuchus cupido*) and evidence for hybridization; (4) provide management recommendations specific to the ecoregion; and (5) describe the most important research and information needs that are still needed for Lesser Prairie-Chickens within the Short-Grass Prairie/CRP Mosaic Ecoregion.

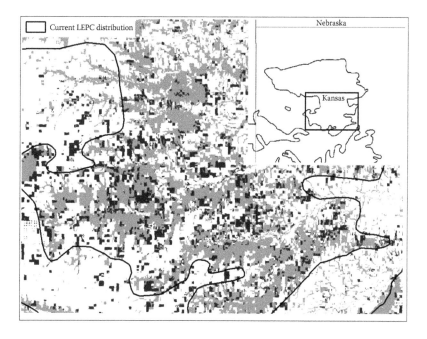

Figure 14.2. Vegetative land cover at the northern extent of the range of Lesser Prairie-Chickens (LEPC) in Kansas. Light gray represents grasslands, black represents Conservation Reserve Program (CRP) cover in 2005, and white represents cropland. The interspersion of CRP and grassland seems to be important to prairie chickens in this ecoregion. The area includes a contact zone where Lesser and Greater Prairie-Chickens are sympatric (see Figure 14.3).

CHARACTERISTICS OF THE GRASSLAND–CRP MOSAIC

The Short-Grass Prairie/CRP Mosaic Ecoregion is in a rain shadow of the Rocky Mountains and far removed from the moist influence of the Gulf of Mexico. The environment is typically semiarid with most precipitation falling as rain during the warm growing season. The average annual precipitation ranges from 28 to 51 cm, with amounts increasing west to east. Short-grass prairie is dominated by grasses such as buffalograss (*Buchloë dactyloides*) and blue grama (*Bouteloua gracilis*). Midgrasses usually present in mixed-grass prairies include sideoats grama (*B. curtipendula*), little bluestem (*Schizachyrium scoparium*), sand dropseed (*Sporobolus cryptandrus*), and western wheatgrass (*Pascopyrum smithii*). Woody species are generally not abundant, although some small inclusions of sand sagebrush occur along a few drainage corridors (Küchler 1974).

Estimates can vary but ~60% of short-grass and mixed-grass prairies remain in North America (Bragg and Steuter 1996, Samson and Knopf 1996, Weaver et al. 1996). However, within the Short-Grass Prairie/CRP Mosaic Ecoregion, at least 73% of the landscape has been converted to cropland, with ~7% of the area currently in CRP (M. Houts, Kansas Biological Survey, unpubl. data). Areas with less productive soils, steeper slopes, or insufficient precipitation or groundwater resources have remained grasslands. Livestock grazing is the predominant land use for the remaining grasslands. Much of the High Plains that is not dominated by large expanses of cropland is currently a complex of grazed short- and mixed-grass prairies, seeded native grasses in CRP fields, and intermixed cropland (Figure 14.2).

The Conservation Reserve Program is currently administered by the Farm Service Agency (FSA) of the U.S. Department of Agriculture (USDA). The program was created in 1986 and provided funds for the conversion of marginal croplands to perennial grasslands (Figure 14.2). In this particular region, nearly all of the seed mixes for establishing CRP consisted of native grasses. However, the species of grass seeded were rarely the historically dominant grass species for a given ecological site. In Kansas, the USDA, in cooperation with the state wildlife agency, provided technical guidance for grass seed mixes in CRP plantings primarily consisting of native

tall- and midgrass species, including the following: big bluestem (*Andropogon gerardii*), little bluestem, Indian grass (*Sorghastrum nutans*), switchgrass (*Panicum virgatum*), and sideoats grama, with occasional additions of western wheatgrass, blue grama, and buffalograss. Forbs were included in CRP seeding mixes for new contracts beginning with the 1996 Federal Agriculture Improvement and Reform Act (Farm Bill), and most mixes commonly included a variety of native forbs: Maximillian sunflower (*Helianthus maximiliani*), purple prairie clover (*Petalostemon pupureum*), prairie coneflower (*Ratibida columnifera*), and Illinois bundleflower (*Desmanthus illinoensis*). Introduced forbs such as alfalfa (*Medicago sativa*), white sweet clover (*Melilotus alba*), and yellow sweet clover (*M. officinalis*) were also permitted in the ecoregion (Fields 2004).

The CRP seed mixes produced grasslands, which provided similar structure to mixed- and tall-grass prairies. The resultant stands were interspersed among native prairies and cropland, providing nesting cover for Lesser Prairie-Chickens that was adjacent to shorter, more open grassland habitats preferred for brood rearing (Fields 2004, Hagen et al. 2004; Figure 14.2). Both nest success and chick survival are critical factors in the population growth of Lesser Prairie-Chickens (Hagen et al. 2009). Existing evidence suggests the interactive effects of "newly" available habitat in CRP cover and favorable environmental conditions likely contributed to significant expansion and growth of Lesser Prairie-Chicken populations in the ecoregion in the 1990s and early 2000s (Channell 2010).

CHANGES IN THE DISTRIBUTION OF LESSER PRAIRIE-CHICKENS

As recently as 2000, published distributions generally delineated the Arkansas River as the northern extent for the contemporary range of Lesser Prairie-Chickens, and associated with sand sagebrush vegetation in Colorado and Kansas (Bailey and Niedrach 1965, Andrews and Righter 1992, Jensen et al. 2000; Figure 14.1). However, there were numerous reports of Lesser Prairie-Chicken occurring in areas north of the Arkansas River prior to 2000 (Jensen et al. 2000, Hagen 2003). Lek surveys for Lesser Prairie-Chickens in the ecoregion began in earnest from 1999 to 2004

(R. Rodgers, unpubl. data). The estimated distribution of Lesser Prairie-Chickens was expanded to incorporate more northern locations in 2008 (Figure 14.1). Some detected leks were only attended by Lesser Prairie-Chickens, but mixed leks of Lesser Prairie-Chickens and Greater Prairie-Chickens were also located in an area of sympatric distribution (Figure 14.3). From 2010 to 2013, intensive lek searches along the northern Lesser Prairie-Chicken distribution boundary were completed (M. Bain and D. Dahlgren, unpubl. data). Based on information gathered, the known boundary of Lesser Prairie-Chicken distribution was again moved north in 2011 and 2012 (Figures 14.1 and 14.3).

In recent years, the highest densities of Lesser Prairie-Chickens within Kansas and range-wide were in the Short-Grass Prairie/CRP Mosaic Ecoregion north of the Arkansas River in Kansas (Pitman 2013, McDonald et al. 2014). In fact, during the recent range-wide survey and subsequent population analysis for Lesser Prairie-Chickens, an estimated 65% of the remaining range-wide populations occurred in the ecoregion (McDonald et al. 2014). Based on the best available data, it appears that populations of Lesser Prairie-Chickens have grown substantially in the grasslands north of the Arkansas River since the mid-1980s and were associated with introduction of CRP plantings (Rodgers 1999, Rodgers and Hoffman 2005).

SYMPATRIC RANGE AND HYBRIDIZATION

The grasslands north of the Arkansas River represent the only portion of the range of Lesser Prairie-Chickens where the species is sympatric with Greater Prairie-Chickens (Figure 14.3). Most of the known leks in the ecoregion are either Lesser Prairie-Chickens only or mixed-species leks (Figure 14.3). Hybridization has been documented between Lesser Prairie-Chickens and Greater Prairie-Chickens, and hybrid vocalizations have been found to be intermediate between the two species (Bain and Farley 2002). Further, some vocalizations in the area of sympatry are

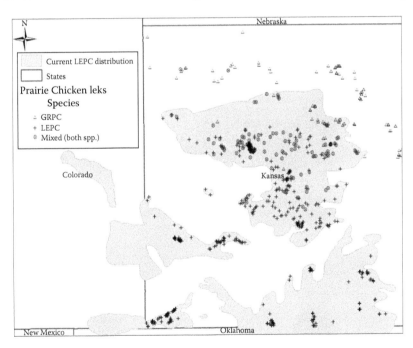

Figure 14.3. Lek locations of Greater Prairie-Chickens (GRPC), Lesser Prairie-Chickens (LEPC), and mixed-species leks in a contact zone where both species are sympatric in west-central and northwest Kansas. Data on lek locations were provided by the Kansas Department of Wildlife, Parks, and Tourism. The map includes all historic lek locations, but some leks may not be currently active. These data represent only known locations, but do not account for variation in abundance or areas without surveys. Two mixed-species leks located north of the current range of Lesser Prairie-Chicken were occasional occurrences of a single Lesser Prairie-Chicken at leks of Greater Prairie-Chicken and were omitted from delineation of the current distributions.

intermediate between hybrid and parental forms, suggesting possible backcrossing (i.e., F2 or F3 offspring, M. Bain and D. Dahlgren, pers. obs.). Intermediate vocalizations suggest that hybrids can produce viable offspring, but this observation has not been confirmed with molecular methods and the extent of genetic introgression remains unknown. Distinguishing a first-generation hybrid (F1) or hybrid offspring (F2, F3) from either parent species due to morphology or plumage characteristic is difficult. However, like vocalizations, some plumage characteristics seem to be intermediate between the two species (Figure 14.4).

If hybridization reduces fitness, isolating mechanisms are likely to evolve. However, postzygotic isolation evolves slowly and low levels of hybridization may occur with no loss of fitness (Grant and Grant 1996). Reproductive isolation between Lesser Prairie-Chicken and Greater Prairie-Chicken appears to be weak in this region, particularly gametic and postzygotic isolating mechanisms. However, prezygotic behavioral isolating mechanisms could include female choice, male competition, lek segregation and other lek attendance attributes, or display behaviors that could minimize the occurrence of hybridization. The effects of genetic introgression on recruitment of Lesser Prairie-Chickens are unknown, as are the possible negative effects of deleterious alleles, or the positive consequences of hybrid vigor and adaptation to a dynamic landscape.

Evidence at different spatial scales suggests that some degree of lek segregation occurs and the hybridization rate is <5%, which has not changed along lek survey routes in this ecoregion during the last decade (Bain 2002, Pitman 2013). Bain (2002) found that hybrid males had high attendance rates, agonistic behavioral traits, and lek territories similar to males of the parent species. However, copulation by hybrid males was never detected despite a large number of observation periods. Female choice might prevent further genetic introgression by avoiding hybrid males as mates. Therefore, it is likely that hybrid genetics would introgress at a greater rate through hybrid females. Characteristics involved in attracting females are likely the first to diverge (Ellsworth et al. 1994). Acoustic masking of display vocalizations might discourage

Figure 14.4. Hybrid Lesser-Greater Prairie-Chicken on a lek in western Kansas. The eye combs are characteristic of Lesser Prairie-Chicken and the air sacs are intermediate in color between the two parental species. Lower pictures are rump feathers of Lesser, hybrid, and Greater Prairie-Chickens (left to right). Note the contrast in the amount of dark and light brown for both upper and lower barring, with the hybrid intermediate between the two parental species.

mixed-species leks (Gibson et al. 1991), because optimization of sound windows might have contributed to the divergence of *Tympanuchus* (Sparling 1983). Relatively, minor differences in display behaviors such as boom duration can affect antiphonal booming, other display characteristics, and lek segregation (Bain 2002). Therefore, it is likely that even the most different traits, such as behaviors associated with breeding displays, could have diverged very recently.

The issues of hybridization and the possibility of viable offspring and potential backcrossing to parent species warrant consideration, as does the degree of speciation between *Tympanuchus* species. The Lesser Prairie-Chicken was initially classified as a unique species in 1885 (Ridgway 1885). Due to the fact that genetic information was not available at the time, the distinction was clearly based on morphological, behavioral, and distributional differentiation. Moreover, there are differing species definitions in the scientific literature for the two species of prairie chickens. Jones (1964) and Crawford (1978) considered Lesser and Greater Prairie-Chickens to be separate species, but suggested that reproductive isolation and species distinction could be tested in a zone of sympatry. Aldrich and Duvall (1955) and Johnsgard (1983) considered Lesser Prairie-Chicken to be a subspecies of Greater Prairie-Chicken, and the American Ornithologists' Union (1998) considered them to be a single superspecies. More recent analysis using modern genetic methodology amplifies the lack of evidence for definitive genetic speciation (Chapter 5, this volume). Luchinni et al. (2001) considered the Lesser Prairie-Chicken and Greater Prairie-Chicken "nominal" species; meaning phenotypically, but not necessarily genetically different. The authors have stated that "Nominal species of *Tympanuchus* hybridize extensively where they are in contact…; their mtDNA haplotypes are not fixed among species… and show shallow genetic distances, suggesting that speciation has been recent and perhaps incomplete" (Luchinni et al. 2001:159). Additionally, Oyler-McCance et al. (2010) provide little evidence for genetic divergence among *Tympanuchus* spp. If molecular data indicate that the Lesser Prairie-Chicken and Greater Prairie-Chicken are not yet fully genetically divergent, current distribution boundaries might be considered less taxonomically important. For practical purposes, it is likely that the application of conservation practices will benefit both species in their sympatric range. Because the ecoregion currently supports the highest densities of birds reported for Lesser Prairie-Chickens, >65% of the extant range-wide population may be exposed to potential hybridization with Greater Prairie-Chickens (McDonald et al. 2014). Hybridization has critical implications associated with systematics, and the potential impacts of genetic introgression or "dilution" of the genes of Lesser Prairie-Chickens. However, little is known about the consequences of hybridization between these two species or implications for management or conservation measures in the future. Understanding hybridization between the two species remains a future research need.

The systematics literature suggests that Lesser Prairie-Chickens and Greater Prairie-Chickens are genetically similar and are species that have only recently diverged. However, we are not suggesting that Lesser Prairie-Chickens are not unique or on significantly divergent evolutionary trajectories, nor that any area that supports *Tympanuchus* spp. is not critical to their conservation. Rather, we argue that clear articulation is needed regarding genetic and functional population goals for both Lesser Prairie-Chickens and Greater Prairie-Chickens. We also suggest that sympatric zones, distributional shifts, and patterns of speciation and introgression need to be better understood and considered as conservation efforts move forward and objectives for recovery must be clearly stated (Chapter 5, this volume).

ECOLOGICAL DRIVERS

Drought

The northern High Plains is subject to periodic drought (Samson and Knopf 1996). In fact, it may be more appropriate to describe climate patterns in the region as regularly in drought conditions, but periodically interrupted by wetter periods. Similar to most of the Great Plains, climatic conditions are highly variable. Timing, frequency, and amount of precipitation, in relation to soil type, growing season, and temperature, are major drivers of plant species composition, annual growth, and production (Holecheck et al. 2000), as well as associated invertebrate and vertebrate animal communities. Most precipitation comes during the growing season, generally favoring warm-season grasses in both short-grass and mixed-grass communities (Samson and Knopf 1996). Recent drought years of 2010–2013 have been moderate to severe in the ecoregion. Residual grass and shrub cover associated with CRP and lightly stocked grasslands are of particular importance for Lesser Prairie-Chickens during drought, as annual biomass production is minimal and grazing pressure removes much of the cover outside of CRP fields.

Grazing

Short-grass and mixed-grass prairies coevolved with periodic seasonal grazing by large ungulates such as American bison (*Bison bison*), pronghorn (*Antilocapra americana*), elk (*Cervus canadensis*), mule deer (*Odocoileus hemionus*), and white-tailed deer (*O. virginianus*). Other important herbivores have included smaller organisms such as insects, songbirds, and black-tailed prairie dogs (*Cynomys ludovicianus*). Prairie dogs, in particular, played key roles in soil disturbance, affecting the distribution of large ungulates (Coppock et al. 1983, Krueger 1986, Whicker and Detling 1988). Short-grass prairie can be particularly resilient to grazing pressure (Shiflet 1994). Current grazing practices in the ecoregion are primarily cow-calf and stocker cattle operations. High grazing intensity has generally reduced or completely removed midgrass species such as little bluestem and sideoats grama from many ecological sites where these plants were once common. However, in recent years, some producers are utilizing more moderate stocking rates and intensities. Changes in grazing management are especially true for landowners involved in contracts with the Lesser Prairie-Chicken Initiative (LPCI) administered by the Natural Resources Conservation Service (USDA NRCS 2013). Data are currently lacking on the potential responses of Lesser Prairie-Chickens to different grazing practices in the ecoregion.

The CRP fields, based on policy of the FSA, are generally not subject to grazing during the contract period. However, FSA can permit haying and grazing of CRP stands during emergency drought conditions within a county during a given growing season, which allows producers to hay or graze their own CRP contracted fields. According to Kansas FSA policy, if landowners choose to hay, they must leave at least 50% of the field in standing cover and a stubble height on average of 25.4 cm (10 in.) within the distribution of Lesser Prairie-Chickens (Shaughnessy 2014). If landowners choose to graze CRP fields with domestic livestock, they are required to have a stocking rate of no greater than 75% of NRCS established rates. Again, landowners must retain a stubble height that averages 25.4 cm (10 in.) within the known distribution of Lesser Prairie-Chickens (Shaughnessy 2014). In recent drought years, most counties in the ecoregion have authorized emergency haying and grazing in multiple consecutive years (Shaughnessy 2014). In 2013, the Kansas FSA required that any portion of CRP fields within the range of Lesser Prairie-Chickens can only be emergency hayed or grazed in 1 of 3 consecutive years. In 2014, the policy was adopted as part of the biological opinion in the federal listing of the Lesser Prairie-Chicken under the Endangered Species Act (Shaughnessy 2014).

Fire

Fire is an important ecological process influencing grassland systems in the Great Plains. Prior to European settlement, Native Americans used fire to influence grassland and animal communities (Moore 1972, Frost 1998). Fire-return intervals have been largely determined by climate, physiographic, edaphic, and vegetation conditions and resiliency (Daubenmire 1968, Wright and Bailey 1982). Since European settlement, natural fire has largely been suppressed across most of the Great Plains, including the Short-Grass Prairie/CRP Mosaic Ecoregion. Both short-grass and mixed-grass prairies respond well to fire if moisture is available following a burning event (Frost 1998, Brockway et al. 2002). In many native grasslands in this ecoregion, sideoats grama may be the only species capable of producing adequate structure for nesting. If fire is followed by intensive grazing or drought, sideoats grama can take on a lower growing, sod-forming structure and competition appears to favor buffalograss and blue grama; thus, fire can diminish nesting habitat in some cases (M. Bain, pers. obs.; Archer and Smeins 1991, McPherson 1995). However, in stands of grass that have not received disturbance, such as some CRP fields, or where woody plant invasion has occurred, fire can be a cost-efficient tool for improving nesting and brood-rearing habitats for Lesser Prairie-Chickens. When grasslands in the ecoregion were less fragmented, intact, contiguous, and comprised of a greater proportion of midgrass species, fire likely played an important role in maintaining habitat for Lesser Prairie-Chickens. In recent years, there has been an increased effort by multiple agencies and nongovernmental organizations to implement fire for grassland and CRP management in this area. However, landowners in the ecoregion are often reluctant to use fire as a management tool due to fear of losing control of a prescribed fire, lack of training, lack of equipment, and social or cultural constraints (Elmore et al. 2009).

HABITAT REQUIREMENTS

Few scientific studies specifically addressing the ecology and habitat use of Lesser Prairie-Chickens have been conducted in this ecoregion. The knowledge gaps are largely due to the recent recognition of expanded distribution boundaries for the species in the ecoregion (Figure 14.1), as well as the recent realization of the large populations of Lesser Prairie-Chickens in the ecoregion (McDonald et al. 2014). Many research needs exist for this area, with relatively little scientific information to date. Fields (2004) was the first wildlife ecologist to radio-mark individual Lesser Prairie-Chickens (along with Greater Prairie-Chickens) and monitor habitat use and survival in the Short-Grass Prairie/CRP Mosaic Ecoregion (see also Fields et al. 2006). Researchers from Kansas State University began a field study in 2013 with the objectives of collecting new data on patterns of habitat use, space use and movements, and survival (R. Plumb and D. Haukos, unpubl. data).

Diet

No diet studies of Lesser Prairie-Chickens have been conducted in the ecoregion. Grasslands are highly fragmented by cropland throughout the ecoregion (Figure 14.2), and Lesser Prairie-Chicken use has been documented in croplands, especially during late fall and winter months (Fields 2004). Ingested contents in the digestive tract of harvested birds have shown significant use of croplands for food sources during this seasonal period (Dahlgren et al., pers. obs.). The use of grains as a food source, especially during the potential resource bottleneck of winter, may be an important diet consideration, but relative contributions to the diet of Lesser Prairie-Chickens are unknown. It has been suggested that when cropland first became part of the landscape on the plains, prairie chickens may have benefited from waste grain as a novel food source (Chapter 2, this volume). Fields (2004) demonstrated that the use of forb-rich areas across vegetation types by broods suggests that forbs and associated invertebrates were important food items for chicks as reported in the literature from other regions (Taylor et al. 1980, Hagen et al. 2005).

Lekking

Lek sites in the region are predominantly in grasslands, but sometimes in croplands such as winter wheat or fallow fields. Few leks are found within CRP fields due to habitat conditions with taller vegetation height and greater stem density. However, when the vegetation is removed, such as after an emergency haying event, leks have been documented in these areas (R. Plumb, Kansas State University, pers. obs.). The region has Greater Prairie-Chicken, Lesser Prairie-Chicken, and mixed-species leks, with a species dominance gradient ranging from Greater Prairie-Chickens in the north to Lesser Prairie-Chickens in the south (Figure 14.3). Additionally, hybrid prairie chickens have been detected at multiple lek sites (Bain and Farely 2002). Jarnevich and Laubhan (2011) used maximum entropy modeling and known lek locations to produce a probability map for lekking habitat of Lesser Prairie-Chickens that included the Short-Grass Prairie/CRP Mosaic Ecoregion. Lek and prairie chicken densities can be relatively high in some areas, with as many as 12 leks along a 16.1 km route (10-mile) in an area of 51.8 km^2 (20 mile^2), and up to 6.18 birds per km^2 (16 birds per mile^2, see Gove Route, Pitman 2011). The Kansas Department of Wildlife, Parks, and Tourism (KDWPT) conducts lek surveys each spring across Kansas, and currently has four lek routes within the ecoregion (Pitman 2013).

A recent study was conducted in the northeastern portion of the Lesser Prairie-Chicken range in a four-county area of Kansas, including Graham, Rooks, Trego, and Ellis counties. V. Cikanek (unpubl. data) investigated lek sites of both prairie chicken species in relation to surrounding landscape characteristics at three spatial scales: close to the lek, a 1.5 km radius from leks, and a 3 km radius from leks. At the smallest scale, lek sites were further from paved roads and higher in elevation than random locations. At the larger scales of 1.5–3 km, leks were associated with larger and more contiguous patches of grasslands (including CRP), less oil structure development, and more CRP compared to random locations.

Nesting Ecology

Nests of prairie chickens in the region have been found in CRP (70%; n = 42 of 60), grassland (27%; n = 16 of 60), and only rarely in cropland (3%; n = 2 of 60, nest sites of Lesser and Greater Prairie-Chickens combined, Fields 2004). In the vegetation types, nest sites were found in western wheatgrass, little bluestem, big bluestem, and switchgrass (Fields 2004). CRP fields that were interseeded with forbs after grass establishment or not treated were used for nest sites in greater proportion than their relative availability (Fields 2004). Fields et al. (2006) reported apparent nest success of 48.3% for nests where ≥1 egg hatched (n = 29 of 59). In this study, vegetation cover in CRP, grassland, or cropland did not influence daily nest survival (Fields et al. 2006). Age of the nest and seasonal timing had the greatest influence on daily nest survival. There was a progressive decline in nest survival among early-, mid-, and late-season nests (Fields et al. 2006). Fields et al. (2006) also found that increasing temperature led to decreased nest survival. About 50% of complete nest losses were attributed to depredation by mammals (Fields 2004).

Lek searches and monitoring were not initiated until after CRP plantings were established across much of the ecoregion. However, anecdotal evidence suggests that Lesser Prairie-Chickens may have been present in the ecoregion, but abundance and occupancy expanded following the initiation of CRP (Pitman 2013, McDonald et al. 2014). Many biologists have concluded that the population increase was largely due to the increase in available nesting habitat in the form of CRP plantings that were conducted at a large landscape scale within the region (Figure 14.2). Nest survival has been shown to be the most important vital rate impacting population dynamics of Lesser Prairie-Chickens (Hagen et al. 2009). However, in their 2-year study with a relatively small sample of nests, Fields et al. (2006) did not find a difference in nest survival rates between vegetation types. The result suggests that the increase in the quantity of nesting habitat in the region may have been most influential in increasing population abundance and range.

Brood-Rearing Ecology

Habitats used by broods of prairie chickens have mainly been native grasslands, followed by CRP, and then rarely in cropland (Fields et al. 2006). Previous research demonstrated that grassland sites have relatively high forb cover and open areas for the movement of small chicks (Fields 2004). While CRP provides nesting habitat, the vegetative structure can be too dense to provide good brood-rearing habitat in the absence of ecological disturbances from grazing, haying, or fire. CRP that contains significant forb cover has been shown to provide important brooding habitat (Fields 2004). From field observations, it appears that the transitional edge between CRP and grasslands can be an important interface for prairie chicken broods (R. Rodgers, pers. obs.).

In their 2-year study, Fields et al. (2006) reported that only 7 of 25 (28%) monitored broods had ≥1 chick survive more than 60 days after hatch. Brood success from hatch to 60 days was much greater for broods attended by adult females (0.49, SE = 0.19) compared to broods attended by subadult females (0.05, SE = 0.03, Fields et al. 2006). The 2-year study was conducted in years with above-average temperatures and one of the years was extremely dry during the nesting and brood-rearing period.

Water

Lesser Prairie-Chickens have been documented using free water sources in other ecoregions (Boal et al. 2014, Grisham et al. 2014). However, it is unknown whether they benefit from free water, or if they simply use these water sources when available. There are multiple natural and artificial water sources in the ecoregion. Livestock grazing and associated water facilities have increased the availability of surface water sources and distribution across the landscape well beyond historic conditions. Additionally, water facilities for wildlife or "wildlife guzzlers" have been constructed within CRP fields as part of the USDA contracts. Kansas NRCS guidelines specifically state that guzzlers within CRP fields are intended for upland game birds. However, the impact of guzzlers on game birds and other wildlife is poorly understood (Boal et al. 2014).

In a study during the summers of 2011 and 2012, cameras associated with guzzlers were used to obtain data for occupancy modeling of

DLCcovert.com   04–11–2012 16:39:54

Figure 14.5. Female Lesser Prairie-Chickens at a CRP guzzler site in Gove County, Kansas. The guzzler was monitored from late March to early May 2012 with a motion-sensing camera (DLC Covert). All photos during this pilot study indicated attendance by females only. The open ground-level water tank allowed access to prairie-chickens without entry under the guzzler roof. The design was unique compared to other monitored 23 guzzlers that had 50 gal (189.3 L) drums situated under the guzzler roof, where only one visit was recorded for a Lesser Prairie-Chicken. Each 50 gal drum had a small opening cut out with a wildlife ramp leading down into the water source.

upland game birds, mesopredators, and other wildlife (B. Calderon, unpubl. data). The study area had multiple guzzler and paired nonguzzler sites in CRP fields in a five-county area of Kansas in Logan, Gove, Trego, Ellis, and Russell counties. Sites spanned an east-to-west precipitation gradient across the northern extent of the range of Lesser Prairie-Chicken. During the study, no detections of Lesser or Greater Prairie-Chickens were recorded at any guzzler sites. However, preliminary data indicated that raccoons (*Procyon lotor*) and other mesopredators had higher detection and occupancy rates at guzzler compared to nonguzzler sites.

In a pilot study conducted during the spring of 2012, motion-sensing cameras were placed at guzzlers in CRP fields in Gove County in an area of relatively high lek densities of both species of prairie chickens and in the same focal study area used by Fields (2004; D. Dahlgren, unpubl. data). Twenty-four cameras and guzzler locations were used. Monitoring occurred from late March to mid-May during nest initiation and start of the incubation period. Prairie chickens were detected at only 2 of 24 guzzler sites during this period (8%) and only once at 1 of the 2 sites. The remaining site had multiple and regular detections of female Lesser Prairie-Chickens until water levels in the tank dropped below accessibility

(Figure 14.5). Notably, the guzzler with the most prairie chicken use had a different design than others monitored with a ground-level open water tank compared to all other guzzlers, which had 50 gal (189.3 L) drums with small access openings and wildlife escape ramps (Figure 14.5). The use of specifically designed water sources has been similarly documented in other ecoregion (Boal et al. 2014). Mesopredators were also detected at guzzler sites during the pilot study.

POPULATION DYNAMICS

Little information is available regarding population dynamics of Lesser Prairie-Chickens within the Short-Grass Prairie/CRP Mosaic Ecoregion. The KDWPT currently conducts spring lek surveys for both species of prairie chickens in March–April along four 16.1 km routes (10-mile). Routes in the region occur in Hodgeman, Ness, Gove, and Logan counties (Pitman 2013; Chapter 4, this volume). Each spring, biologists conduct two samples with 11 stops that are 1.6 km apart (1 mile) along the route listening and looking for leks of prairie chicken. All detected leks within 1.6 km (1 mile) on either side of the route are located and flushed at least once, and preferably twice during the survey period. Lek survey data provide both regional and statewide trends for populations of prairie chickens. For the Short-Grass Prairie/CRP Mosaic Ecoregion, Lesser and Greater Prairie-Chickens are combined for trend information. The Logan route was tested in 2012 and established in 2013, but not used here because of a lack of trend information. Based on three of the four current routes, lek densities and numbers of birds per lek have been relatively stable since the inception of these routes in the early 2000s (Figure 14.6). However, the trend has been slightly down in recent years, likely due to implications of region-wide drought conditions on recruitment (Figure 14.6).

CONSERVATION AND MANAGEMENT OF LESSER PRAIRIE-CHICKENS IN GRASSLANDS NORTH OF THE ARKANSAS RIVER

Large blocks of native grasslands provide the most valuable habitat in other three ecoregions in the range of the Lesser Prairie-Chicken. Populations of the species in the Short-Grass Prairie/CRP Mosaic Ecoregion, by necessity, depend more on the interspersion of native grassland and CRP tracts (Figure 14.2). CRP grasslands provide suitable nesting cover while native grasslands can often

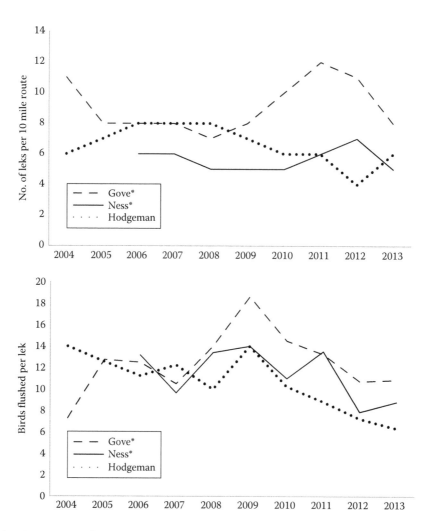

Figure 14.6. Lek survey trends for Lesser and Greater Prairie-Chickens with both species combined for routes in the Short-Grass Prairie/CRP Mosaic Ecoregion. Logan County route was not included due to establishment in 2013 and lack of trend information. Gove and Ness routes have an * to indicate that both Lesser and Greater Prairie-Chickens are observed during annual surveys along each route. Data were provided by the Kansas Department of Wildlife, Parks, and Tourism, and the most recent reports for lek surveys of prairie chickens are available at the agency website. (From Kansas Department of Wildlife, Parks, and Tourism 2014.)

offer better brood-rearing habitat (Fields 2004), allowing females to exploit the edges where these two vegetation types occur adjacent to each other. The addition of CRP grasslands has also been sufficient in many areas to shift the overall grassland–cropland ratio from a landscape of mostly cropland to one that is mostly grassland, likely benefiting Lesser Prairie-Chickens throughout their annual cycle.

Management priorities have been created within the Conservation Reserve Program that provides greater opportunity and incentives for the establishment of CRP grasslands adjacent to or near extant native grasslands. Targeted recruitment has been accomplished through the establishment of carefully targeted Conservation Priority Areas, which provide additional Environmental Benefit Index points for appropriate lands enrolled through the competitive General Signup process of the CRP. The State Acres for Wildlife practice (CP38E) of the Continuous Signup process has also provided opportunity and incentives to target CRP grasslands toward key focal areas where enrollments are most likely to benefit Lesser Prairie-Chickens. The Lesser Prairie-Chicken Range-wide Conservation Plan developed by the Western Association of Fish and Wildlife Agencies (WAFWA) has also delineated critical focal areas in the region and established an avoidance, minimization, and mitigation process whereby greater habitat protections, better habitat management, and new habitat establishment can best be targeted (Van Pelt et al. 2013).

Based on a broad assessment of where CRP grasslands had apparently benefited prairie grouse populations in the western United States, Rodgers and Hoffman (2005) recommended desired CRP stand heights of 30–75 cm (12–30 in.) or roughly shin-to-thigh high. The authors also recommended that CRP stands be established using species that would produce a diverse, clumpy stand structure. Such grasslands will provide three critical habitat structural requirements for prairie chickens to be able to easily (1) hide and be concealed, (2) move without obstruction, and (3) see approaching danger (G. Horak, pers. comm.). Fields (2004) found that vegetation at the nest is generally much taller, and often twice as tall as the surrounding habitat conditions (e.g., 45.7 cm versus 22.9 cm), and that vegetation at successful nests was often taller than unsuccessful nests. The results suggest that habitat patchiness at small scales may

be important for successful nesting. Small-scale patchiness might be more easily achieved in endemic grasslands than in CRP, under current CRP management policy. In summary, it is evident that Lesser Prairie-Chickens need adequate concealment in a grassland stand that is neither too dense nor too tall. In the Short-Grass Prairie/CRP Mosaic Ecoregion, bunchgrasses, such as little bluestem and sideoats grama, can best provide such grassland structure, particularly when complimented with forbs.

Rodgers and Hoffman (2005) strongly discouraged the use of invasive exotic grasses such as smooth brome (Bromus inermis) and Caucasian or yellow bluestem (Bothriochloa ischaemum) in the establishment of new stands of CRP fields. Without a program policy that allows for grazing of CRP stands during the established nesting season of birds (April 15–July 15), cool-season grasses such as western wheatgrass and smooth brome have shown a capacity to outcompete native warm-season species in CRP stands in this semiarid region, resulting in monocultures with little heterogeneous structure. If native warm-season grasses have been heavily invaded by smooth brome or western wheatgrass, the recommended treatment is a November application of glyphosate (plus surfactant) active ingredient at 0.42–0.83 kg per ha (6–12 oz per acre) followed by 1–2 hard freezes at <-4°C when the temperature has warmed at least 12°C. The treatment has been shown to be effective at killing smooth brome and would likely be effective on western wheatgrass while not harming the already-dormant warm-season species (M. Bain, pers. obs.).

Prescribed fire is another useful tool in managing CRP stands for prairie chickens by providing for grassland succession across the landscape, reducing litter accumulation, and controlling undesirable plant species. In this semiarid region, new CRP stands with warm-season grasses generally do not reach maturity until at least the fourth growing season after seeding (R. Rodgers and M. Bain, pers. obs.). Excessive litter accumulation typically will not occur until at least 2 years after maturity is reached. Consequently, initial prescribed burns of new CRP stands typically are not needed until at least six years after the stand was seeded, perhaps longer, depending on stand density. Following the first burn, we suggest that prescribed fire repeated at 4–6 year intervals, depending on precipitation, would likely maintain suitable vegetation vigor and structure to satisfy the habitat needs of Lesser Prairie-Chickens.

The use of prescribed fire during the late summer months of mid-July to August has the potential to benefit Lesser Prairie-Chickens as well. Summer burns can reduce excessive height that occurs in some CRP stands and have been shown to favor forb production in subsequent growing seasons (Howe 1994), particularly if only portions of a landscape are burned resulting in heterogeneity of vegetation structure and composition (Fuhlendorf and Engle 2004). The use of summer burns also extends the potential days available in the year for conducting prescribed fire and can provide an enhanced margin of safety if surrounding vegetation is still green and less flammable.

Prescribed grazing can be an important CRP management tool. In addition to cool-season management, high-intensity grazing for short duration during the dormant season may increase forb production and the vigor of warm-season grasses, but this management tool needs more testing. Typical CRP grazing in July–November generally occurs too late in the growing season to effectively control cool-season species of grasses. Dormant or early season high-intensity, short-term grazing may reduce current year nesting cover, but would at least improve nesting habitat in subsequent years if given adequate rest during the growing season.

The habitat provided by much of the intact native grassland in this ecoregion has been reduced in quality and quantity by high grazing intensity with little or no grazing rest for more than a century. However, little bluestem, sideoats grama, and other grass species selected by Lesser Prairie-Chickens are sometimes still present in such pastures but in a suppressed condition. Reducing grazing pressure by using lighter stocking rates and periodic rest over time can improve the quantity and quality of habitat available to Lesser Prairie-Chickens.

Grassland grazing plans that address the most common limiting factor of quality nesting habitat are especially important in areas lacking CRP. The USDA programs such as Lesser Prairie-Chicken Initiative (LPCI) and the Lesser Prairie-Chicken Range-wide Conservation Plan under the Western Association of Fish and Wildlife Agencies (WAFWA) have provided landowners with incentive payments for prescribed

grazing (Van Pelt et al. 2013). The 3–5-year plans of the LPCI and the 5–10-year plans of WAFWA are relatively short term, with the frequency of long-term adoption by landowners after contract expiration unknown. In the arid region with common, prolonged drought, changes in species composition as a vegetative response to management changes generally require decades rather than years. Only long-term management programs over 15–30 years or in perpetuity can ensure habitat improvements. Moderate stocking rates and rest rotations can be used to achieve at least a minimum amount and appropriate distribution of quality nesting habitat. To efficiently target nesting habitat, management units that include grass species capable of producing structure for nesting must first be identified. The most common species capable of producing suitable structure are sideoats grama, little bluestem, sand dropseed, and western wheatgrass, but some ecological sites simply do not support these plant species. The most effective grazing plans begin with identifying sites to target the development of nesting structure and then provide the stocking rate or rest that best expedite the development of that structure. Grazing plans should complement producers' long-term goals for their operations, ensuring benefits to both the producer and wildlife habitat. Depending upon environmental conditions and existing species composition and structure, deferment, rest, or light to moderate stocking rates with 25%–40% total utilization should be used in these areas. Research is needed to confirm and increase the efficacy of grazing recommendations.

Haying CRP fields may also produce some benefits to habitat similar to fire or grazing. However, the negative impact of long-term loss of cover, especially if haying is followed by drought, may outweigh any benefits. Habitat loss may especially be true if haying occurs at large scales. Generally, we do not recommend haying to benefit Lesser Prairie-Chickens.

More information is needed concerning the impacts of energy development on Lesser Prairie-Chicken, but based on the negative impacts to other species of prairie grouse (Hovick et al. 2014), anthropogenic development should probably be minimized in priority Lesser Prairie-Chicken habitat such as the WAFWA Lesser Prairie-Chicken focal areas (Van Pelt et al. 2013). Approaches for discouraging energy development are highly controversial, but the WAFWA Lesser Prairie-Chicken Range-wide Conservation Plan currently offers a mechanism for accomplishing this goal. Recent evidence suggests that the established design of guzzlers designed to provide open water sources for Lesser Prairie-Chickens is not effective and could be benefiting mesopredators that are detrimental to grassland birds (B. Calderon, unpubl. data). Guzzlers are no longer incentivized by USDA CRP signup programs.

INFORMATION NEEDS AND GAPS

As with the three other ecoregions for Lesser Prairie-Chickens, the most urgent research needs are related to identifying limiting factors of populations and associated habitats at multiple spatial scales and how known threats influence these factors (U.S. Fish and Wildlife Service 2012). To our knowledge, the only published population and habitat related research for Lesser Prairie-Chickens specifically within the Short-Grass Prairie/CRP Mosaic Ecoregion are presented in Fields (2004) and Fields et al. (2006). An example of needed information for our area of interest is given in Hagen et al. (2009), where population modeling was conducted to identify vital rates that may be drivers of population growth. Once population and habitat information becomes readily available, the development of meaningful population, habitat recovery, and long-term conservation objectives will become possible. Economic incentives necessary to obtain these objectives need to be determined, as well as the most appropriate mechanisms to implement these incentives. For a wide-ranging species that occurs primarily on private lands, the effectiveness of the strategy using voluntary conservation is of primary concern.

Perhaps the most urgent information needs in this region are associated with understanding the effects of CRP and adjacent grasslands on population persistence, occupancy, and growth. Anecdotal observations indicate that areas within the region with the highest Lesser Prairie-Chicken densities are comprised of ~60% grassland, 20% CRP, and 20% cropland. Quantification of these land use attributes at landscape scales and their functional influence on demographic rates are urgently needed. To efficiently target and implement conservation, the minimum amount of nesting structure that is required in a given area must be identified for local property management and at a landscape scale. At a landscape scale, information on habitat requirements is critical for efficient and targeted implementation of programs such as CRP. Conversely, with the reduction of the CRP acreage cap and competitive commodity prices, knowing how much grassland (including CRP) is needed in a given area to support populations of Lesser Prairie-Chickens and the optimal size and configuration of available patches will be critically important.

If Lesser Prairie-Chickens truly depend on CRP to persist in the Short-Grass Prairie/CRP Mosaic Ecoregion, the current habitat availability is in a precarious situation where >50% of the range-wide population depends upon the existence of a relatively short-term (10–15-year) program subject to political support and a dynamic financial market driven by commodity prices. If a process to develop long-term solutions for maintaining or increasing CRP and the subsequent mosaic of vegetative structure in this landscape is not found, we could eventually lose the ecoregion as a remaining stronghold for the species. Identifying the process and creating long-term solutions based on market-based or conservation programs will require knowledge of financial drivers that influence land use decisions, such as annual payments that are competitive with current market values. In addition, policy development and subsequent communication must focus on maintaining or increasing landowner participation in CRP. The potential threatened status of Lesser Prairie-Chickens has created landowner fear of participation in CRP. If policy is not developed to alleviate this fear and that policy is not clearly communicated to landowners, a threatened status could have a net negative effect on populations of Lesser Prairie-Chickens in the region. The effort must ultimately identify the level of incentives required to guarantee at least a minimum area of CRP and other grasslands in the landscape for longer periods than current programs, and up to 30 or 50 years to perpetuity. Conservation planning could include identifying an incentive-based path for landowners to transition current CRP into permanent grazing lands. For example, provisions in the "Farm Bill" and other conservation programs, such as Agricultural Land Easements and the WAFWA Lesser Prairie-Chicken Range-wide Conservation Plan (Van Pelt et al. 2013), could be used to secure long-term easements, which then could be modified and targeted to focus conservation measures for Lesser Prairie-Chickens. If such solutions are not found, the owners of CRP and native grassland that currently provide Lesser Prairie-Chicken habitat may ultimately succumb to economic pressures to convert to alternative types of land use.

Further investigation into the trade-off between spatial extent and duration of conservation practices is needed to develop sound strategies for conservation, particularly related to adaptive management. In our ecoregion, long-term trends suggest range expansion and increasing densities remain possible for Lesser Prairie-Chickens. In such areas, conservation priorities that focus on securing long-term maintenance of intact, quality habitat, such as perpetual easements on CRP-grassland complexes, may be more critical than producing additional habitat. Once a certain threshold of habitat is protected over the long term, or participation and demand for long-term conservation has been met, a shift toward restoring potential habitat may be warranted. In areas where little habitat is available to conserve, conservation priorities should focus on restoring as much habitat as possible, as quickly as possible. Once a certain threshold of habitat is restored, it may be more efficient to shift strategies and begin long-term conservation easements for some of the most functional habitat. The greatest information needs associated with these issues are related to the appropriate proportions of short- and long-term habitat at landscape scales and triggers that might be useful in optimizing cost-efficiency and effectiveness over time.

Energy development is increasing in the Short-Grass Prairie/CRP Mosaic Ecoregion. Specifically, oil and natural gas development is creating new roads and associated infrastructure throughout prairie chicken habitat. Additionally, pressure to develop wind energy resources is likely to increase in the region. Similar to other ecoregions within Lesser Prairie-Chicken range, we currently lack a full understanding of the potential impacts that energy developments could have on populations of Lesser Prairie-Chickens.

Management prescriptions currently in place for Lesser Prairie-Chickens within our ecoregion need more experiential and research information to help guide and improve the methods in the future. Policy and resulting management practices must be flexible enough to prescribe management needs for individual fields or habitat patches and provide suitable options for landowners. First, research must be conducted that identifies the most cost-efficient management practices that adequately address needs for nesting habitat, survival, or other limiting factors within the management unit. For example, research and subsequent policy could be developed related to the following list of management information questions: Is flash grazing with high-intensity stocking for a short duration during the dormant season an effective means of reducing grass biomass and improving vegetative structure and plant density? What are the optimal stocking rates to maximize the quality of nesting and brood-rearing habitat? Is summer burning the most effective management tool for mature stands of warm-season grasses with few forbs that do not provide adequate brood rearing? In more arid portions of the region, when is the initial burn appropriate for restored CRP grasslands in terms of stand age or plant structure? For the range of existing stand types, what is the optimal time to burn? Would short-term, intensive grazing or disking have a greater disturbance response and subsequently be preferable to burning? Under what environmental conditions does burning reduce nesting habitat and for how long? If cool-season grass dominance is the primary threat to the stand, are spring burning, glyphosate application in autumn, or other management practices effective at control? At what level of dominance by smooth brome, western wheatgrass, or other cool-season grasses is a prescribed burn during nesting period worth the benefit of habitat restoration versus potential loss of production by Lesser Prairie-Chickens? Is potential nest loss by spring burning offset by renesting or long-term benefits in increased production? For the range of existing stand types, what is the most effective means of seedbed preparation for interseeding with forbs—disking, burning, spraying, or mowing? For established stands, are there any scenarios where seedbed preparation is not necessary? What is the optimum seed mixture and rate or seeding for interseeding of legumes or other priority forb species? We recognize that waiting for consideration of a laundry list of applied research needs is not an option, and managers must move forward based on the best available science and their own knowledge base from personal experience.

However, a need for an urgent response does not preclude the need for research-based information concerning management practices, especially if certain practices become popularized in management prescriptions.

Currently, livestock grazing is considered a compatible use with habitat requirements (USFWS 2012). However, different grazing systems and stocking rates will have variable impacts. Identifying the most appropriate stocking rates and grazing systems, including species of grazer, timing, and duration, that are most compatible with requirements of Lesser Prairie-Chickens will be important to future conservation in this ecoregion. A focus on grazing would be especially important if the management of CRP fields could include livestock grazing as a future management practice. Although not specific to Lesser Prairie-Chickens, much guidance can be garnered from the existing literature on grazing impacts to vegetation.

In addition to grazing, fire is another ecological driver that can be managed. Prescribed fire has rarely been applied to arid grasslands in the region due to ongoing fire suppression and concerns of landowners, and consequently, effects of fire on habitats of Lesser Prairie-Chickens are largely unknown. Prescribed fire on CRP lands is more common, and where undisturbed vegetation likely limits brood rearing, it is likely the most appropriate management tool. Additionally, where tree encroachment is an issue, prescribed fire is often the most cost-effective method to maintain open grasslands. Information is needed to determine the optimum timing and environmental conditions, and appropriate role of prescribed fire in existing grassland fragments and CRP for the purpose of Lesser Prairie-Chicken conservation.

Habitat requirements of prairie chickens within their sympatric range are likely to be similar to allopatric populations, and habitat selection by Lesser Prairie-Chickens and Greater Prairie-Chickens is similar in the contact zone. However, if differences were detected, those findings would likely shed light on important, limiting factors related to habitat for each species. Research that detects similar habitat use for both species would be just as insightful. Currently, little is known about the potential role of hybridization between Lesser and Greater Prairie-Chickens or implications for future conservation measures for either species. Research should be designed to assist in the determination of whether genetic distinctness between Lesser Prairie-Chickens and Greater Prairie-Chickens should be a recovery goal. Therefore, understanding hybridization rates between these two species is an important research need. Vocalizations that are intermediate to the parental species suggest that hybrids produce viable offspring; however, these observations have not been confirmed genetically. Reproductive isolation between Lesser and Greater Prairie-Chickens appears to be weak in this region, but behavioral isolating mechanisms could include female choice, male competition, lek segregation, and other lek attendance attributes and display behaviors. In fact, within the 96 leks described by Bain and Farley (2002), segregation by species was greater than would be predicted by chance alone. The effects of hybridization on production by Lesser Prairie-Chickens, extent, and direction of genetic introgression, and other negative or positive effects are unknown and require more research attention.

Shifts in distribution and occupied range for both Lesser and Greater Prairie-Chickens and a subsequent increase in area of the sympatric zone will likely continue until new range limits are reached or changes in land use, especially conversion of native grasslands or CRP to cropland, occur at larger scales in this region. The dynamic nature of the distribution of Lesser Prairie-Chickens in relation to environmental conditions, particularly in this region, clearly has been influenced by humans and suggests that historic distribution distinctions may offer little toward recovery planning. Rather, research needs to be developed that identify current and projected functional populations and the limiting effects of land use, climate change, or other factors in those areas. Therefore, understanding climate change and its potential impact on this species is warranted (Channell 2010; Chapter 12, this volume). Specifically, how will predicted increases in temperature and evapotranspiration influence life history traits or demographic parameters? To help managers, climate models could be incorporated into recovery efforts for the Lesser Prairie-Chicken to help direct resources to areas with a higher probability of persistence. Information on climate change and distribution shifts that includes potential hybridization with Greater Prairie-Chickens in northern reaches might help identify important recovery areas, perhaps beyond the recognized current distribution.

CONCLUSION

The Short-Grass Prairie/CRP Mosaic Ecoregion north of the Arkansas River in Kansas represents a unique portion of the current distribution for Lesser Prairie-Chickens in their five-state occupied range. Based on current evidence, Lesser Prairie-Chickens in this area use CRP and associated grasslands to meet their seasonal life cycle needs (Fields 2004, Fields et al. 2006; Figure 14.2). The introduction of CRP grasslands at landscape scales is believed to be the cause of population increases in this area over the past three decades, but particularly in the late 1990s and early 2000s. However, it has been documented that Lesser Prairie-Chickens occurred in this region, but presumably at much lower population levels before the Conservation Reserve Program was started (Jensen et al. 2000, Hagen 2003). The CRP provides a short-term contract for grasslands on a field-by-field basis and does not currently address long-term landscape-scale certainty for persistence of the species in this area. Long-term security is a significant need for the species, especially considering increasing commodity prices and other competing issues for CRP contract renewal and expansion in the region. The recognized distribution of Lesser Prairie-Chickens has been moving northward in recent years with new data from spring lek searches conducted by biologists since 1999. Much of this portion of the range of Lesser Prairie-Chickens is sympatric with Greater Prairie-Chickens (Figure 14.3). Hybridization between Lesser and Greater Prairie-Chickens has been documented, but the potential effects on the Lesser Prairie-Chicken are poorly understood (Bain and Farley 2002). More research and information are needed on population genetics, habitat use at various scales, management prescriptions, human dimensions, energy development, climate change, and population dynamics for Lesser Prairie-Chicken in this region, where ~65% of the remaining range-wide population is found (McDonald et al. 2014). If conservation goals can be met, the Short-Grass Prairie/CRP Mosaic Ecoregion may remain one of the last strongholds for Lesser Prairie-Chicken as a species.

ACKNOWLEDGMENTS

We thank the two reviewers who helped to improve this chapter. We also appreciate our two volume editors for their efforts herein. We thank M. Smith, KDWPT Farm Bill Biologist, for his expertise and contributions. We would like to thank all KDWPT personnel who participated in lek surveys for prairie chickens and the reporting process over the years. We also thank the many landowners in this region who have participated in management and allowed access to private lands for the purpose of gathering new data on prairie chicken ecology.

LITERATURE CITED

Aldrich, J. W., and A. J. Duvall. 1955. Distribution of American gallinaceous game birds. Circular 34. U.S. Fish and Wildlife Service, Washington, DC.

American Ornithologists' Union. 1998. Check-list of North American birds, 7th edn. American Ornithologists' Union, Washington, DC.

Andrews, R., and R. Righter. 1992. Colorado birds: a reference to their distribution and abundance. Denver Museum of Natural History, Denver, CO.

Archer, S., and F. E. Smeins. 1991. Ecosystem-level processes. Pp. 109–139 in R. K. Heitschmidt, and J. W. Stuth (editors), Grazing management: an ecological perspective, Timbers Press, Portland, OR.

Bailey, A. M., and R. J. Niedrach. 1965. Birds of Colorado. Denver Museum of Natural History, Denver, CO.

Bain, M. R. 2002. Male-male competition and mating success on leks attended by hybrid prairie chickens. M.S. thesis, Fort Hays State University, Hays, KS.

Bain, M. R., and G. H. Farley. 2002. Display by apparent hybrid prairie chickens in a zone of geographic overlap. Condor 104:683–687.

Boal, C. W., P. K. Borsdorf, and T. Gicklhorn. 2014. Assessment of Lesser Prairie-Chicken use of wildlife water guzzlers. Bulletin of the Texas Ornithological Society 46:10–18.

Bragg, T. B., and A. A. Steuter. 1996. Prairie ecology-the mixed prairie. Pp. 53–65 in F. B Samson, and F. L. Knopf (editors), Prairie conservation: preserving North Americas most endangered ecosystem, Island Press, Washington, DC.

Brockway, D. G., R. G. Gatewood, and R. B. Paris. 2002. Restoring fire as an ecological process in shortgrass ecosystems: initial effects of prescribed burning during the dormant and growing seasons. Journal of Environmental Management 65:135–152.

Channell, R. 2010. Analyses of the distribution and population of Lesser Prairie-Chicken (*Tympanuchus pallidicinctus*): with special reference to Kansas populations. Unpublished Report, Fort Hays State University, Hays, KS.

Coppock, D. L., J. K. Detling, J. E. Ellis, and M. I. Dyer. 1983. Plant-herbivore interactions in North American mixed-grass prairie. II. Responses of bison to modification of vegetation by prairie dogs. Oecologia 56:10–15.

Crawford, J. A. 1978. Morphology and behavior of Greater and Lesser Prairie Chicken hybrids. Southwestern Naturalist 23:591–596.

Daubenmire, R. 1968. Ecology of fire in grasslands. Pp. 209–266 in J. B. Cragg (editor), Advanced ecological restoration, vol. 5, Academic Press, New York, NY.

Ellsworth, D. L., R. L. Honeycutt, and N. J. Silvy. 1995. Phylogenetic relationships among North American grouse inferred from restriction endonuclease analysis of mitochondrial DNA. Condor 97:492–502.

Ellsworth, D. L., R. L. Honeycutt, N. J. Silvy, K. D. Rittenhouse, and M. H. Smith. 1994. Mitochondrial-DNA and nuclear-gene differentiation in North American prairie grouse (genus *Tympanuchus*). Auk 111:661–671.

Elmore, R. D., T. G. Bidwell, and J. R. Weir. 2009. Perceptions of Oklahoma residents to prescribed fire. Pp. 50–61 in K. M. Robertson, K. E. M. Galley, and R. E. Masters (editors), The future of prescribed fire: public awareness, health, and safety. Proceedings of the 24th Tall Timbers Fire Ecology Conference: Tall Timbers Research Station, Tallahassee, FL.

Fields, T. L. 2004. Breeding season habitat use of Conservation Reserve Program (CRP) land by Lesser Prairie-Chickens in west central Kansas. M.S. thesis, Colorado State University, Fort Collins, CO.

Fields, T. L., G. C. White, W. C. Gilgert, and R. D. Rodgers. 2006. Nest and brood survival of Lesser Prairie-Chickens in west central Kansas. Journal of Wildlife Management 70:931–938.

Frost, C. C. 1998. Presettlement fire frequency regimes of the United States: a first approximation. Pp. 70–81 in T. L. Pruden, and L. A. Brennan (editors), Fire in ecosystem management: shifting paradigm from suppression to prescription.

Proceedings of the Twentieth Tall Timbers Fire Ecology Conference, Tall Timbers Research Station, Tallahassee, FL.

Fuhlendorf, S. D., and D. M. Engle. 2004. Application of the fire-grazing interaction to restore a shifting mosaic on tallgrass prairie. Journal of Applied Ecology 41:604–614.

Gibson, R. M., J. W. Bradbury, and S. L. Vehrencamp. 1991. Mate choice in lekking sage grouse revisited: the roles of vocal display, female site fidelity, and copying. Behavioral Ecology 2:165–180.

Grant, P. R., and B. R. Grant. 1996. Speciation and hybridization in island birds. Philosophical Transactions of the Royal Society of London 351:765–772.

Grisham, B. A., P. K. Borsdorf, C. W. Boal, and K. K. Boydston. 2014. Nesting ecology and nest survival of Lesser Prairie-Chickens on the Southern High Plains of Texas. Journal of Wildlife Management 78:857–866.

Hagen, C. A. 2003. A demographic analysis of Lesser Prairie-Chicken populations in southwestern Kansas: survival, population viability, and habitat use. Ph.D. dissertation, Kansas State University, Manhattan, KS.

Hagen, C. A., B. E. Jamison, K. M. Giesen, and T. Z. Riley. 2004. Guidelines for managing Lesser Prairie-Chicken populations and their habitats. Wildlife Society Bulletin 32:69–82.

Hagen, C. A., G. C. Salter, J. C. Pitman, R. J. Robel, and R. D. Applegate. 2005. Lesser Prairie-Chicken brood habitat in sand sagebrush: invertebrate biomass and vegetation. Wildlife Society Bulletin 33:1080–1091.

Hagen, C. A., B. K. Sandercock, J. C. Pitman, R. J. Robel, and R. D. Applegate. 2009. Spatial variation in Lesser Prairie-Chicken demography: a sensitivity analysis of population dynamics and management alternatives. Journal of Wildlife Management 73:1325–1332.

Haukos, D. A., and L. M. Smith. 1994. The importance of playa wetlands to biodiversity of the Southern High Plains. Landscape and Urban Planning 28:83–98.

Holechek, J. L., R. D. Pieper, and C. H. Herbel. 2000. Range management: principles and practices, 4th edn. Prentice Hall College Division, Upper Saddle River, NJ.

Hovick, T. J., R. D. Elmore, D. K. Dahlgren, S. D. Fuhlendorf, and D. M. Engle. 2014. Evidence of negative effects of anthropogenic structures on wildlife: a review of grouse survival and behavior. Journal of Applied Ecology 51:1680–1689.

Howe, H. F. 1994. Response of early- and late-flowering plants to fire season in experimental prairies. Ecological Applications 4:121–133.

Jarnevich, C. S., and M. K. Laubhan. 2011. Balancing energy development and conservation: a method utilizing species distribution models. Environmental Management 47:926–936.

Jensen, W. E., D. A. Robinson, and R. D. Applegate. 2000. Distribution and population trend of Lesser Prairie-Chicken in Kansas. Prairie Naturalist 32:169–176.

Johnsgard, P. A. 1983. The Grouse of the World. Croom Helm, London, U.K.

Jones, R. E. 1964. The specific distinctness of the Greater and Lesser Prairie Chickens. Auk 81:65–73.

Kansas Department of Wildlife, Parks, and Tourism. [online]. 2014. Prairie-Chicken Lek Survey - 2014. <http://ksoutdoors.com/content/download/43606/428313/version/2/file/PcLeks14+%281%29.pdf> (11 August 2014).

Kansas Native Plant Society. [online]. Vegetation cover map. 2014. kansasnativeplantsociety.org/ecoregions.php. (August 11, 2014).

Krueger, K. 1986. Feeding relationships among bison, pronghorn, and prairie dogs: an experimental analysis. Ecology 67:760–770.

Küchler, A. W. 1974. A new vegetation map of Kansas. Ecology 55:586–604.

Lucchini, V., J. Hoglund, S. Klaus, J. Swenson, and E. Randi. 2001. Historical biogeography and a mitochondrial DNA phylogeny of grouse and ptarmigan. Molecular Phylogenetics and Evolution 20:149–162.

McDonald, L., G. Beauprez, G. Gardner, J. Griswold, C. Hagen, D. Klute, S. Kyle, J. Pitman, T. Rintz, and B. Van Pelt. 2014. Range-wide population size of the Lesser Prairie-Chicken: 2012 and 2013. Wildlife Society Bulletin 38:536–546.

McPherson, G. R. 1995. The role of fire in the desert grassland. Pp. 131–151 in M. P. McClaran, and T. R. Van Devender (editors), The desert grassland, University of Arizona Press, Tucson, AZ.

Moore, C. T. 1972. Man and fire in the central North American grassland, 1535–1890: a documentary historical geography. Ph.D. dissertation, University of California, Los Angeles, CA.

Oyler-McCance, S. J., J. St. John, and T. W. Quinn. 2010. Rapid evolution in lekking grouse: implications for taxonomic definitions. Ornithological Monographs 67:114–122.

Pitman, J. C. 2011. Prairie chicken lek survey—2011. Performance Report, Statewide Wildlife Research and Surveys, Grant W-39-R-17. Kansas Department of Wildlife and Parks, Emporia, KS.

Pitman, J. C. 2012. Prairie chicken lek survey—2012. Performance Report, Statewide Wildlife Research and Surveys, Grant W-39-R-17. Kansas Department of Wildlife, Parks, and Tourism, Emporia, KS.

Pitman, J. C. 2013. Prairie chicken lek survey—2013. Performance Report, Statewide Wildlife Research and Surveys, Grant W-39-R-17. Kansas Department of Wildlife, Parks, and Tourism, Emporia, KS.

Ridgway, R. 1885. Some amended names of North American birds. Proceedings of the US National Museum 8:354.

Rodgers, R. D. 1999. Why haven't pheasant populations in western Kansas increased with CRP? Wildlife Society Bulletin 27:654–665.

Rodgers, R. D., and R. W. Hoffman. 2005. Prairie grouse population response to Conservation Reserve Grasslands: an overview. Pp. 120–128 in A. W. Allen, and M. W. Vandever (editors), The conservation reserve program–planting for the future. Proceedings of the National Conference, Fort Collins, CO, 2004, U.S. Geological Survey, Biological Resources Division, Scientific Investigation Report 5145, Reston, VA.

Samson, F. B., and F. L. Knopf (editors). 1996. Prairie conservation: preserving north America's most endangered ecosystem. Island Press, Washington, DC.

Sexson, M. L. 1980. Destruction of sandsage prairie in southwest Kansas. Pp. 113–115 in C. L. Kucera (editor), Proceedings of the Seventh North American Prairie Conference, Southwestern Missouri State University, Springfield, MO.

Shaughnessy, M. 2014. Biological opinion: implementation of the Conservation Reserve Program (CRP) within the occupied range of Lesser Prairie-Chicken. FWS/R2/ER/057136, U.S. Department of the Interior, U.S. Fish and Wildlife Service, Albuquerque, NM.

Shiflet, T. N. 1994. Rangeland cover types of the United States. Society for Range Management, Denver, CO.

Sparling, Jr., D. W. 1983. Quantitative analysis of prairie grouse vocalizations. Condor 85:30–42.

Taylor, M. A., and F. S. Guthery. 1980. Status, ecology, and management of the Lesser Prairie Chicken. USDA Forest Service General Technical Report RM-77. USDA Forest Service, Rocky Mountain Forest and Range Experiment Station, Fort Collins, CO.

U.S. Department of Agriculture, Natural Resource Conservation Service. (USDA NRCS). [online]. 2013. Lesser Prairie-Chicken initiative. <http://www.nrcs.usda.gov/wps/portal/nrcs/detail/national/programs/farmbill/initiatives> (11 August 2014).

U.S. Fish and Wildlife Service. (USFWS). 2012. Endangered and threatened wildlife and plants; listing the Lesser Prairie-Chicken as a threatened species, proposed rule. Federal Register 77:73828–73888.

Van Pelt, W. E., S. Kyle, J. Pitman, D. Klute, G. Beauprez, D. Schoeling, A. Janus, and J. Haufler. 2013. The Lesser Prairie-Chicken range-wide conservation plan. Western Association of Fish and Wildlife Agencies, Cheyenne, WY.

Waddell, B., and B. Hanzlick. 1978. The vanishing sandsage prairie. Kansas Fish and Game 35:17–23.

Weaver, T., E. M. Payson, and D. L. Gustafson. 1996. Prairie ecology—The shortgrass prairie. Pp. 67–75 in F. B. Samson, and F. L. Knopf (editors), Prairie conservation: preserving North Americas most endangered ecosystem, Island Press, Washington, DC.

Whicker, A. D., and J. K. Detling. 1988. Ecological consequences of prairie dog disturbances. BioScience 38:778–785.

Wright, H. A., and A. W. Bailey. 1982. Fire ecology: United States and Southern Canada. Wiley-Interscience, New York, NY.

CHAPTER FIFTEEN

Lesser Prairie-Chickens of the Sand Sagebrush Prairie*

David A. Haukos, Aron A. Flanders, Christian A. Hagen, and James C. Pitman

Abstract. The Sand Sagebrush (*Artemisia filifolia*) Prairie Ecoregion once supported the highest densities of Lesser Prairie-Chickens (*Tympanuchus pallidicinctus*), but the estimated population numbers in 2014 were <500 birds in ~15,975 km² of potential available habitat. Contributing to ongoing declines are long-term conversion of sand sagebrush (*Artemisia filifolia*) prairie to row crop agriculture and reductions in the quality of remaining habitat, whereas short-term variation in climatic conditions with droughts and blizzards are the main causes of population fluctuations. Conversion of sand sagebrush prairie occurred later than the conversion of prairie in other ecoregions following the advent of center-pivot irrigation systems for the irrigation of sandy soils in the 1960s and 1970s. Furthermore, the avoidance of anthropogenic structures in the region has greatly reduced the amount of available quality habitat for Lesser Prairie-Chickens. Current populations in the ecoregion are becoming increasingly isolated, requiring consideration of potential corridors or other mechanisms to increase the connectivity to limit localized extinction events. Information on the ecology of Lesser Prairie-Chickens in the ecoregion is limited. However, sand sagebrush has consistently been demonstrated as important throughout the life history of the species in the ecoregion. Provision of quality nesting and brood-rearing habitat is considered the primary management focus for the ecoregion. Restoration methods for sand sagebrush prairie are uncertain and presumably require a lengthy process in a semiarid environment. Applied practices to increase habitat quality include managed grazing, prescribed fire, and judicious use of herbicides to reduce the cover of sand sagebrush and enhance the composition of grasses and forbs. The presence of public lands in the National Grasslands of the U.S. Forest Service in the ecoregion provides additional conservation opportunities not found in other ecoregions. Established habitat and population goals for Lesser Prairie-Chickens in the Sand Sagebrush Prairie Ecoregion will require intensive and innovative approaches to conservation.

Key Words: Artemisia filifolia, Colorado, ecology, Kansas, Tympanuchus pallidicinctus.

* Haukos, D. A., A. A. Flanders, C. A. Hagen, and J. C. Pitman. 2016. Lesser Prairie-Chickens of the Sand Sagebrush Prairie. Pp. 281–298 in D. A. Haukos and C. W. Boal (editors), Ecology and conservation of Lesser Prairie-Chickens. Studies in Avian Biology (no. 48), CRC Press, Boca Raton, FL.

n the Central and Southern Great Plains, the sand sagebrush (*Artemisia filifolia*)-mixed prairie cover type covers ~48,000 km^2 (Berg 1994). Küchler (1964) labeled this vegetation association as sand sagebush/bluestem (*Andropogon* spp.), which is primarily associated with sandy soils in bands 4.8–18.2 km wide that run parallel to drainages of relatively large streams and rivers, but narrower bands parallel smaller drainages (Berg 1994). Sand sagebrush grows to a height of 1 m and can attain 7%–50% canopy cover under natural conditions, but increases in density as grazing pressure increases (Parker and Savage 1944, Costello 1964, Collins et al. 1987). Typically associated with deep sandy soils on landscapes of dunes, sand sagebrush thrives in semiarid environments with average annual precipitation between 31 and 61 cm (Berg 1994). A variety of mid- and tall grasses are associated with sand sagebrush including sand bluestem (*Andropogon hallii*), little bluestem (*Schizachyrium scoparium*), switchgrass (*Panicum virgatum*), prairie sandreed (*Calamovilfa longifolia*), and sand dropseed (*Sporobolus cryptandrus*). Livestock grazing is the primary land use for extant tracts of sand sagebrush prairie.

Sand sagebrush can be found throughout the species range and provides nesting and loafing cover for Lesser Prairie-Chickens (*Tympanuchus pallidicinctus*). Four ecoregions have been identified as relevant for the conservation and management of Lesser Prairie-Chickens (McDonald et al. 2014b). The Sand Sagebrush Prairie Ecoregion portion of the range of Lesser Prairie-Chickens has been defined with three segments associated with drainages in southeastern Colorado, southwestern Kansas, and the western Oklahoma Panhandle (Van Pelt et al. 2013; McDonald et al. 2014a,b; Figure 1.2). In Colorado, the northwest extent of the species range is an expansive sand sagebrush-dominated watershed starting in Lincoln County, but predominantly in Cheyenne County located between the Rush Creek and Big Sandy drainages that run parallel southeast until converging in Kiowa County. Another segment of the range is associated with the Arkansas River starting in Prowers County, Colorado, and extending eastward into portions of Hamilton, Stanton, Kearney, and Finney counties, Kansas (Figure 7.4b). Historically, there were reports of Lesser Prairie-Chickens in the sand sagebrush prairie on the south side of the Arkansas River further west in Bent County, Colorado (J. Reitz,

Colorado Parks and Wildlife, pers. comm.). One exception to the use of sand sagebrush prairie in the ecoregion is Lesser Prairie-Chickens in Prowers County, Colorado, that also use lands enrolled in the Conservation Reserve Program (CRP) of the U.S. Department of Agriculture (USDA). The use of CRP fields in Colorado is a recent occurrence that was first documented in 2003–2004, and Lesser Prairie-Chickens were not recorded using these habitats before 1999 (Giesen 2000). The third segment of Lesser Prairie-Chicken occupancy in the Sand Sagebrush Prairie Ecoregion is primarily associated with the Cimarron River drainage in Morton, Stevens, Grant, and Seward counties, Kansas, and Baca County, Colorado (Figure 7.4b). Habitats of Lesser Prairie-Chickens within the ecoregion are often separated by >60 km, effectively creating isolated populations in an increasingly fragmenting landscape. The principal exception would be that the eastern edge of the Sand Sagebrush Prairie Ecoregion along the Cimarron River grades into the Mixed-Grass Prairie Ecoregion (Figure 1.2). Sand sagebrush associated with the Arkansas and Cimarron river drainages are found in sandy soils usually classified as Tivoli fine sand or Tivoli-Dune land complex within a Choppy Sands range site (Sexson 1980).

Throughout much of the latter half of the 20th century, the Sand Sagebrush Prairie Ecoregion was considered the core range of Lesser Prairie-Chickens despite being centered within lands impacted by the 1930s Dust Bowl era (Chapter 2, this volume). Indeed, Bent (1932:281) stated that "Their range in its entirety [of Lesser Prairie-Chickens] would probably cover no greater area than a fourth of the State of Kansas, and the most abundant nucleus is in Stevens and Morton counties." He provided an account of an estimated flock of ~15,000 to 20,000 Lesser Prairie-Chickens sighted in a single grain field in Seward County, Kansas in the fall of 1904. The landscape recovered from the ecological calamity of the Dust Bowl and has endured several more periods of prolonged drought since, but Lesser Prairie-Chickens had always persisted in the ecoregion despite the natural variation in habitat quality (Schwilling 1955). Unfortunately, the recent drought of 2011–2014 followed widespread landscape changes that greatly reduced population numbers of Lesser Prairie-Chickens in the ecoregion to the point of conservation concern and proposed action.

Compared to big sagebrush (*Artemisia triden-tata*) of the western United States, the ecology of sand sagebrush is relatively unknown, which results in considerable uncertainty for conservation planning. In addition, the Sand Sagebrush Prairie Ecoregion differs from other ecoregions in that considerable public lands are present in two National Grasslands managed by the U.S. Forest Service (USFS)—the Cimarron National Grassland of Kansas and the Comanche National Grassland of Colorado (>2,230 km^2, ~14% of ecoregion; Chapter 10, this volume), which provides opportunities for conservation in addition to the private lands approach that is required in the other three ecoregions. An ownership pattern with a mix of public and private lands should increase conservation options for Lesser Prairie-Chickens. Our objectives are to review the status and ecology of Lesser Prairie-Chickens in the Sand Sagebrush Prairie Ecoregion and provide a potential framework for conservation efforts in the ecoregion.

POPULATION STATUS

In the early to mid-1970s, the estimated numbers of Lesser Prairie-Chickens in the Sand Sagebrush Prairie Ecoregion were greater than for any other ecoregion (Jensen et al. 2000; Chapter 4, this volume). The peak in population numbers occurred in 1975 when a minimum of 86,700 birds were estimated attending leks in the ecoregion (Chapter 4, this volume). Unfortunately, the densities of Lesser Prairie-Chickens in the ecoregion have declined significantly since the mid-1970s (Figures 4.3, 15.1, and 15.2). Drought conditions in recent years of 2011–2014 have expedited the decline. Robel et al. (2004) suggested that 25%–50% of the population of Lesser Prairie-Chickens in Kansas was in the sand sagebrush prairie of Hamilton, Finney, and Kearney counties in the early 2000s. The status has changed because the current estimated number of birds in the ecoregion is the lowest of the four ecoregions comprising the range of the species (Table 1.2; McDonald et al. 2014a). McDonald et al. (2014b) estimated the densities of Lesser Prairie-Chickens in the ecoregion based on 15,975 km^2 of potential available habitat. Van Pelt et al. (2013) established Lesser Prairie-Chicken focal areas within the Sand Sagebrush Prairie Ecoregion totaling 6,725 km^2. Recently, population density and abundance have declined (Table 15.1). An 18% decline in the population estimate between 2012

and 2013 and a 76% decline between 2013 and 2014 were both statistically significant at α = 0.10 (McDonald et al. 2014a). In conjunction with the decline in population estimates, the estimated number of leks in the Sand Sagebrush Prairie Ecoregion has declined 83% from 357 leks in 2013 to 61 leks in 2014.

Seven lek survey routes in a survey area of 334.4 km^2 have been conducted in the Sand Sagebrush Prairie Ecoregion in Kansas since 1988, with individual routes started as early as 1964. The estimated number of birds per km^2 has declined by 99% since 1988 (Figure 15.1), and only 5 leks and 31 Lesser Prairie-Chickens were detected during the 2013 survey, both all-time low counts in 26 years of lek surveys. In Colorado, counts of birds attending leks have been conducted since 1977. During 2014, 101 of the 120 lek sites previously recorded in Colorado were surveyed (n = 224 lek visits). Of a sample of 101 historical lek locations, only 7 leks were active (J. Reitz, Colorado Parks and Wildlife, unpubl. data). Active leks were located in three counties of Colorado: Cheyenne (n = 2), Prowers (n = 3), and Baca counties (n = 2). The total number of Lesser Prairie-Chickens counted on leks in Colorado in 2014 was 40 birds (31 males and 9 females), which is down 91% compared to a high lek count of 448 birds (350 males) in 1989 (Figure 15.2) and is a 59% decline from a count of 97 birds in 2013 (Figure 15.2). Ken Giesen (unpubl. data) estimated the 1989 population of Lesser Prairie-Chickens in Colorado at 1,200–1,500 birds. Van Pelt et al. (2013) set the long-term population goal as 10,000 birds or 0.91 birds per km^2 in suitable habitat, which was exceeded as recently as 2010 (Chapter 4, this volume), but still 95% below the goal in 2014. Population goals based on birds counted at leks are 149 birds for the Comanche National Grassland and 131 birds for the Cimarron National Grassland (U.S. Forest Service 2014).

CAUSES OF POPULATION DECLINES

The main cause of long-term population declines of Lesser Prairie-Chickens in the Sand Sagebrush Prairie Ecoregion has been conversion of sand sagebrush prairie to row crop agriculture with the advent of center-pivot irrigation starting in the 1960s (Sexson 1980). Dryland farming of sand sagebrush prairie failed after settlement in the early 1900s, cumulating in near universal abandonment of farmland in the ecoregion during the extensive

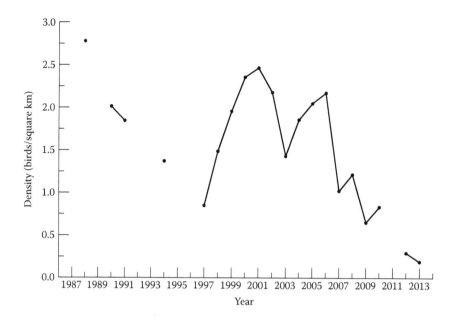

Figure 15.1. Population trends for Lesser Prairie-Chickens in the Southern High Plains Small Game Region of the Kansas Department of Wildlife, Parks, and Tourism, 1988–2013. The region corresponds with the Lesser Prairie-Chicken Sand Sagebrush Prairie Ecoregion in Kansas. The annual regional index (birds/km²) was weighted by the survey area along each route and only calculated when all of the routes were surveyed.

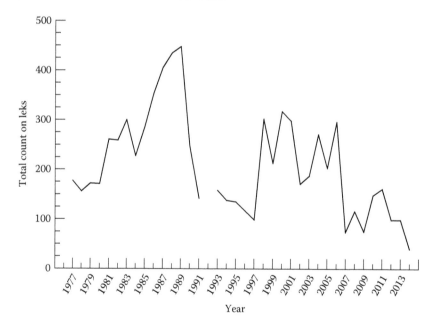

Figure 15.2. Population trends based on lek counts for Lesser Prairie-Chickens in the Sand Sagebrush Prairie Ecoregion in Colorado, 1977–2014.

drought of the 1930s, after which many cultivated areas naturally reverted back to sand sagebrush by the 1960s. Technological innovations in deep well engineering and center-pivot irrigations allowed pumping of ground water from the underlying Ogallala Aquifer to support row crop agriculture in the ecoregion and started in the 1960s (Sexson 1980). Küchler (1974) estimated 5,478 km² of sand sagebrush prairie historically was present in Kansas, but had been reduced by 62% to 2,075 km²

TABLE 15.1

Estimates of density and abundance for Lesser Prairie-Chickens
in the Sand Sagebrush Prairie Ecoregion, 2012–2014.

Year	Density of birds per km² (90% CI)	Abundance (90% CI)
2012	0.1540 (0.0625, 0.3161)	2,460 (998, 5,059)
2013	0.1262 (0.0608, 0.2177)	2,017 (971, 3,477)
2014	0.0299 (0.0151, 0.525)	477 (241, 839)

SOURCES: McDonald et al. (2014a,b).

by 1978. In a core area for Lesser Prairie-Chickens along the Arkansas River in Kearny, Finney, and Gray counties, Kansas, Sexson (1980) reported that sand sagebrush prairie had been reduced 58% from the historical area of 2,093 to 890 km² by 1975. Robel et al. (2004) estimated that of 1,376 km² of sand sagebrush prairie originally in Finney, Kearny, and Hamilton counties, 1,210 km² remained in 1973, which was further reduced when 332 km² were lost to center pivots between 1973 and 2001. The authors estimated that only 867 km² of sand sagebrush prairie remained in these core counties as available habitat for Lesser Prairie-Chickens. Much of the remaining sand sagebrush prairie has been further degraded and fragmented by intensive land use. In most areas of the Sand Sagebrush Prairie Ecoregion, the remaining native prairie is below the ~60% coverage of the landscape necessary to sustain Lesser Prairie-Chicken populations (Crawford and Bolen 1976).

Drought is thought to be a principal determinant of demographic performance and short-term fluctuations in population numbers of Lesser Prairie-Chickens in the Sand Sagebrush Prairie Ecoregion. The region is on the western edge of the species range where environmental conditions can be extreme, with some of the lowest rates of average annual precipitation within the species range (e.g., 40 cm at Lamar, Colorado, 47.5 cm at Elkhart, Kansas), but with extended drought conditions occurring frequently (Grisham et al. 2013). Giesen (2000) found a positive relationship between the number of males on leks as a function of a 2-year lag effect of precipitation where above-average precipitation affected counts of males 2 years later. In addition to low precipitation, other extreme weather events can impact populations of Lesser Prairie-Chickens in the ecoregion. For example, it was estimated that ~80% of the Lesser Prairie-Chickens died overwinter in Prowers County, Colorado, and a ~75% reduction

occurred throughout the Colorado portion of the ecoregion during a major blizzard in the winter of 2006–2007 (J. Reitz, Colorado Division of Parks and Wildlife, pers. comm.). Prowers County was a stronghold for Lesser Prairie-Chickens in Colorado prior to the blizzard (Giesen 2000).

Anthropogenic structures such as oil/gas wells, power lines, roads, center pivots, and buildings have influenced habitat selection by Lesser Prairie-Chickens in sand sagebrush prairie (Robel et al. 2004, Pitman et al. 2005, Hagen et al. 2011). Indeed, the only anthropogenic feature that did not elicit a negative response by Lesser Prairie-Chickens was unimproved roads (Robel et al. 2004). Based on estimated avoidance buffers around anthropogenic structures (Table 15.2), Robel et al. (2004) estimated that 510 km² (58%) of the remaining 867 km² of sand sagebrush prairie in Hamilton, Kearney, and Finney counties of Kansas was not available for nesting Lesser Prairie-Chickens by 2001. The authors concluded that due to conversion to center-pivot irrigation and avoidance of anthropogenic structures, only 26% of historical native sand sagebrush prairie was available as potential nesting habitat for Lesser Prairie-Chickens.

LESSER PRAIRIE-CHICKEN ECOLOGY IN SAND SAGEBRUSH PRAIRIE

Trends in population numbers of Lesser Prairie-Chickens are monitored through annual lek surveys in Kansas and Colorado (Chapter 4, this volume), but information regarding the ecology of the species in sand sagebrush prairie is limited. The primary sources of information for Lesser Prairie-Chickens in the Sand Sagebrush Prairie Ecoregion are results from an intensive 6-year field study from 1997 to 2002 at two sites in Finney County, Kansas (Jamison 2000, Hagen 2003, Pitman 2003). Giesen (1994, 2000) present much

TABLE 15.2

Avoidance distances (m) between anthropogenic structures and the home-range centroids or nests of Lesser Prairie-Chickens in the Sand Sagebrush Prairie Ecoregion.

Structure	Type	Observed mean[c]	SE	Random mean	SE	P	Study
Powerline	Centroids[a,b]	1494	41	594	3		Hagen et al. (2011)
Powerline	Centroids[a,b,d]	709		172		0.001	Hagen et al. (2011)
Powerline	Centroids[a,e]	751	90	582	3		Hagen et al. (2011)
Powerline	Centroids[a,d,e]	123		273		0.43	Hagen et al. (2011)
Powerline	Centroids[b,f]	1388	50	552	3		Hagen et al. (2011)
Powerline	Centroids[b,d,f]	662		176		0.001	Hagen et al. (2011)
Powerline	Centroids[e,f]	1015	137	567	6		Hagen et al. (2011)
Powerline	Centroids[d,f,e]	272		269		0.03	Hagen et al. (2011)
Powerline	Nest[a]	1385	60				Pitman et al. (2005)
Powerline	Nest[f]	1254	69				Pitman et al. (2005)
Wells	Centroids[a]	559	14	490	1		Hagen et al. (2011)
Wells	Centroids[a,d]	242		240		0.011	Hagen et al. (2011)
Wells	Centroids[f]	559	17	551	1		Hagen et al. (2011)
Wells	Centroids[d,f]	320		230		0.007	Hagen et al. (2011)
Wells	Nest[a]	588	18				Pitman et al. (2005)
Wells	Nest[f]	539	27				Pitman et al. (2005)
Buildings	Centroids[a]	1929	38	1987	4		Hagen et al. (2011)
Buildings	Centroids[a,f]	1132		1092		0.005	Hagen et al. (2011)
Buildings	Centroids[f]	2374	46	2179	4		Hagen et al. (2011)
Buildings	Centroids[d,f]	1666		1458		0.001	Hagen et al. (2011)
Buildings	Nest[a]	1951	64				Pitman et al. (2005)
Buildings	Nest[d]	2306	53				Pitman et al. (2005)
Roads	Centroids[a]	1712 m	48 m	1547 m	6 m		Hagen et al. (2011)
Roads	Centroids[f]	2695	141	2019	6		Hagen et al. (2011)
Roads	Centroids[d,f]	990		946		0.016	Hagen et al. (2011)
Roads	Nest[a,g]	224	13				Pitman et al. (2005)
Roads	Nest[a,h]	1526	63				Pitman et al. (2005)
Roads	Nest[d,g]	208	16				Pitman et al. (2005)
Roads	Nest[d,h]	3149	202				Pitman et al. (2005)

[a] Area 1.
[b] Preconstruction.
[c] Average.
[d] 10th percentile.
[e] Post construction.
[f] Area 2.
[g] Unimproved roads.
[h] Improved roads.

of the published information on Lesser Prairie-Chickens in Colorado. Hagen et al. (2009) estimated the finite rate of population growth (λ) for Lesser Prairie-Chickens as 0.54 and 0.74 for two study sites in the Sand Sagebrush Prairie Ecoregion of Kansas. The estimates are well below the rate needed for a stable population ($\lambda = 1$), and factors that had the greatest effect on λ were nest success and chick survival, with additional importance of female survival during the breeding season.

Survival

The annual survival of female prairie chickens differed between study sites in Kansas with higher estimates of survival for yearlings (SY or second-year birds, Site I: S = 0.43 ± 0.12SE, Site II: S = 0.59 ± 0.10) than adults (ASY or after-second year birds, Site I: S = 0.30 ± 0.08, Site II: S = 0.44 ± 0.08). The seasonal estimates of female survival for the 6-month period of April to September were 0.76 ± 0.06SE for yearlings and 0.66 ± 0.06 for adults (Hagen et al. 2007). The mortality rates of female Lesser Prairie-Chickens were greatest during incubation in May to June, and cause-specific mortality was identified as 59% mammal predation, 11% raptor predation with seasonal peaks during spring and fall migration, 18% unknown, 5% accidents including four collisions with powerlines, 3% snake predation, 2% disease, and 1% hunting (Hagen et al. 2007). The survival of incubating females was positively associated with nest sites with greater shrub cover but less vertical vegetation structure (Hagen et al. 2007).

Hagen et al. (2005a) estimated the apparent male annual survival rate as 0.60 ± 0.12SE for yearlings (SY) and 0.44 ± 0.10 for adults (ASY). Thus for both sexes, the annual survival was higher among yearlings than older birds. Hagen et al. (2005a, 2007) hypothesized that costs associated with increased reproductive activity at lek sites may have contributed to lower survival for adult males, and overall greater reproductive effort by older females may have increased their risk of predation.

The estimated daily survival of chicks from hatch to 14 days was 0.95 (95% CI = 0.93–0.97) and from 15 to 60 days was 0.98 (95% CI = 0.97–0.99) for an overall survival estimate from hatch to 60 days of 0.18 (95% CI = 0.03–0.37, Pitman et al. 2006a). The authors also estimated juvenile survival from August 1 to March 31 as 0.70 (95%

CI = 0.47–0.86), where birds were more likely to survive if they had greater than average mass at 50–60 days after hatching. Overwinter survival for the 8-month period of August 1–March 31 was similar between juvenile (0.64) and adult (0.63) prairie chickens. Overall, the annual survival of juveniles from hatch to March 31 of the following year was 0.12 (95% CI = 0.01–0.32).

Nest success for Lesser Prairie-Chickens in the sand sagebrush prairie appears to be lower than for other regions. The average apparent nest success for 227 females (209 nests) during 1997–2002 was 26% ± 3% (Pitman et al. 2006b). Interestingly, Pitman et al. (2005, 2006b) reported little variation in annual nesting effort or success, which differs from the boom–bust reproductive effort reported for other ecoregions (Chapter 17, this volume). Nesting propensity was 94% for juveniles and 92% for adults (Pitman et al. 2006b). Renesting occurred for 31% of females that lost their initial nest (Pitman et al. 2006b). Nest predation was primarily caused by coyotes (45%; *Canis latrans*) and snakes (19%, Pitman et al. 2006b).

Nest Habitat

Sand sagebrush is critical to nest-site selection and success throughout the ecoregion. In Kansas, sand sagebrush plants close to nests of Lesser Prairie-Chickens were 11% greater in number than plants at paired random sites (Pitman et al. 2005). In addition, the cover of sand sagebrush was nearly twice as great at nest sites (15%) compared to paired at random sites (8%). Sand sagebrush density was greater at nest sites (5,064 plants per ha ± 240SE) than random sites (4,129 ± 202; Pitman et al. 2005). Visual obstruction, forb height, and grass height were also greater at nest sites than at paired random sites (Pitman et al. 2005). Vegetative structure appears to lead to higher reproductive success; successful nests had greater sand sagebrush density, mean diameter, and mean height than unsuccessful nests (Pitman et al. 2005). Sand sagebrush was a dominant plant species at nest sites of Lesser Prairie-Chickens in Colorado, where 41% of 29 nests were beneath sand sagebrush and 28% were beneath other shrubs (Giesen 1994). Shrub, forb, and grass height at nests were greater than sites measured at paired random transects: the average maximum vegetation height was 50.7 cm ± 14.7SD with 69% of nests under vegetation ≥40 cm (Giesen 1994).

Brood Habitat

Habitats used by broods have been linked to the availability of invertebrate food resources and sufficient space for movement in sand sagebrush prairie (Jamison et al. 2002, Hagen et al. 2005b). Short-horned grasshoppers (Acrididae: *Melanoplus* spp.) were the main invertebrate prey item, and the total invertebrate biomass were greater at use areas for adults and broods of Lesser Prairie-Chickens than at nonuse areas (Jamison et al. 2002, Hagen et al. 2005b). Jamison et al. (2002) also reported that the biomass of invertebrates was not related to the cover of sand sagebrush, but linked to forb abundance. However, Hagen et al. (2005b) found that low (0.2–0.4 plants per m²) to moderate densities of sand sagebrush (0.5–0.7 plants per m²) supported a greater biomass of acridid grasshoppers. The brood use of an area is related more to vegetation height with greater visual obstruction readings than dominant cover type (Hagen et al. 2005b). Increasing forb cover (>15%) and moderate plant height (2.5–3.0 dm) have been recommended as techniques to increase invertebrate biomass available to Lesser Prairie-Chickens (Jamison et al. 2002, Hagen et al. 2005b).

Movements

The average date of brood breakup was September 13 (range = August 21 to October 6) at ~101 days posthatch for family groups of Lesser Prairie-Chickens in sand sagebrush prairie (Pitman et al. 2006c). Previous to breakup, average minimum daily brood movements increased from 273 m per day for the first 0–14 days after hatching to 312 m per day for 15–60 days after hatching (Pitman et al. 2006c). Dispersal movements were recorded for males and female Lesser Prairie-Chickens between late October and early November, and dispersal movements were greater for females than for males. However, spring dispersal peaked in late February and late March for males and females, respectively (Pitman et al. 2006c). The two periods correspond to a reduction in the number of males at traditional leks with the formation of satellite leks and were just prior to peak attendance by females at leks in other ecoregions (Haukos and Smith 1999). Daily movements of males tend to increase in fall and winter and decrease with onset of the spring lekking period (Jamison 2000). Individual birds and especially females are capable of long-distance movements of up to ~50 km (Hagen et al. 2004), but the demographic and adaptive significance of these movements remains unknown. Pitman et al. (2006c) hypothesized that females are the primary contributors to gene flow and connectivity among populations in the fragmented Sand Sagebrush Prairie Ecoregion.

APPROACHES TO CONSERVATION IN THE SAND SAGEBRUSH PRAIRIE

Preservation, management, and restoration of sand sagebrush prairie are often emphasized for the conservation of Lesser Prairie-Chickens (Giesen 1994, Pitman et al. 2005, Vodehnal and Haufler 2008, Hagen et al. 2013, Van Pelt et al. 2013). Woodward et al. (2001) reported that loss of shrubland habitat correlated to population declines when examining land use changes between 1959 and 1996 in areas of Oklahoma that included sand sagebrush prairie. Due to increasing isolation by distance, Lesser Prairie-Chicken dispersal among populations in different habitat segments of the Sand Sagebrush Prairie Ecoregion is likely to be decreasing. Identification and protection of extant quality habitats in the ecoregion would be appropriate if conservation goals included securing core areas from which populations can expand if habitat is improved adjacent to protected areas and connections are established among populations. Unfortunately, techniques for the restoration of sand sagebrush prairie are uncertain, with little experimental data to guide restoration practices. Anecdotal information indicates that natural restoration of abandoned areas that were used for agriculture is measured in terms of decades rather than years given the dynamic, semiarid environment. Should population numbers of Lesser Prairie-Chickens continue to decline, especially in Colorado, consideration of translocations could be included in short-term conservation planning. Despite several efforts, previous translocations of Lesser Prairie-Chickens have not resulted in the establishment of new populations or recovery of a declining population (Snyder et al. 1999, Giesen 2000). Therefore, an intensive conservation effort may be necessary to curb current population trends of Lesser Prairie-Chickens in the Sand Sagebrush Prairie Ecoregion.

Desirable Ecological Conditions for Lesser Prairie-Chickens

The optimal ecological conditions for the conservation of Lesser Prairie-Chicken populations depend upon the degree to which ecological processes and drivers of site dynamics maintain attributes characteristic of quality habitat. Landscape scale contiguous habitat is required for population persistence of Lesser Prairie-Chickens. The average home range of individual birds has been estimated to range from 120 to 1,035 ha depending upon season, habitat quality, population density, and other parameters (Giesen 1994, Jamison 2000, Walker 2000, Bidwell et al. 2003). Home ranges can overlap and all the birds attending a lek may inhabit an area of 24–49 km² (Giesen 1998). Connected habitat areas in the Sand Sagebrush Prairie Ecoregion are recommended to be >5,250 km² (Van Pelt et al. 2013), if the long-term goal is to maintain connected, resilient populations with an effective population size that can withstand demographic and environmental stochasticity, adapt to changing conditions, endure catastrophes, and prevent buildup of deleterious traits (Franklin and Frankham 1998, Palstra and Ruzzante 2008, Pruett et al. 2011, Jamieson and Allendorf 2012, Manier et al. 2013, Frankham et al. 2014). All other factors being equal, relatively larger grassland habitat areas should be targeted for conservation (Hagen et al. 2010). Large-scale management may be achieved by working on multiple properties within close proximity or forming landowner associations.

Previous studies have reported on nest-site selection, characteristics of successful nest sites, and performed meta-analysis of brood-rearing and nesting habitat (Giesen 1994, Pitman et al. 2005, Hagen et al. 2013), and available data have been used to construct management goals for the nesting ecology of Lesser Prairie-Chickens in the Sand Sagebrush Prairie Ecoregion (Van Pelt et al. 2013). Pitman et al. (2005) recommended a conservation goal of >6,500 mature sand sagebrush plants per ha (18%–20% cover) and 100% visual obstruction reading at >2.7 dm for quality nesting habitat in sand sagebrush prairie. Hagen et al. (2007) concluded that the conservation of Lesser Prairie-Chickens should focus on habitat manipulations that decrease female mortality during incubation and early brood rearing. Recommendations include managing units surrounding leks for 2/3 nesting cover and 1/3 brood cover, where nesting cover is characterized by greater shrub and grass cover (Table 15.3). Quality brood-rearing habitats have greater forb cover than nesting habitats but more open vegetation to facilitate the movement of chicks (Table 15.3). Under the Management Plan for the Cimarron and Comanche National Grasslands, habitat management will be centered on lek sites where quality nesting and brood-rearing habitat will be provided within a 3.2 km radius of existing and historical leks dating back to 2003 (USFS 2014). Vegetation composition and structure standards within a lek-centered area are 15%–20% cover of sand sagebrush, 40%–50% cover of native grasses, 15%–25% cover of native forbs, 100% visual obstruction between 25.4 and 45.7 cm, and average grass height between 30.5 and 45.7 cm.

Unfortunately, metrics for desired ecological conditions for other aspects of the life cycle of Lesser Prairie-Chickens have not yet been developed. The recommendations for nesting and brooding cover are similar to the state-and-transition model community phase 2 for ecological sites such as Sandy 16–22″ R077A and Sand Hills 12–17″ R077B, which are influenced and maintained by soils, climate, fire, and grazing as the primary processes (Figure 15.3; Moseley 2012). Preferred native grasses for Lesser Prairie-Chicken habitat in sand sagebrush prairie include big bluestem (*Andropogon gerardii*), Indian grass (*Sorghastrum nutans*), little bluestem (*Schyzachrium scoparium*), sand bluestem (*A. hallii*), and switchgrass (*Panicum virgatum*, Van Pelt et al. 2013).

Ecological Conditions Impacting Populations of Lesser Prairie-Chicken

Nesting and brooding habitat are widely cited as being limiting factors for Lesser Prairie-Chicken populations and should be considered the focus of management planning (Pitman et al. 2006b, Hagen et al. 2009, Van Pelt et al. 2013). Lek sites are not considered to be a limiting habitat type for population recovery (Van Pelt et al. 2013), but known lek locations are vital for generating management plans because females usually nest and raise young within 3.2 km of an active lek (Copelin 1963, Giesen 1994, Kukal 2010). Also, females have site fidelity to traditional leks compared to new lek sites (Campbell 1972, Haukos and Smith 1999). Leks serve as a convenient central point for

TABLE 15.3

Recommendations for habitat components of sand sagebrush prairie needed to support populations of
Lesser Prairie-Chickens in southwestern Kansas and eastern Colorado.

Habitat component	Nesting	Brood
Sagebrush cover	15% canopy cover[a] 18%–20%[b] 15%–30%[c]	8% canopy cover[a] 10%–25%[c]
Sagebrush density	>6,500 per ha (2631 per acre)[b]	ND[d]
Sagebrush height	47.6 cm (18 in.), ≥40 cm (15 in.)[e] 51.4 cm (20 in.)[b] 42.8 cm (16.8 in.)[a]	ND
Sagebrush diameter	86 cm (34 in.)[b]	ND
Grass cover	37%[a,b] >30%[c]	13%[a] >20%[c]
Grass height	>27.7 cm (10.9 in.)[a] >38 cm (15 in.) average[c]	>38 cm (15 in.) average[c]
Forb cover	7%[a,b] >10%[c]	14%[a] >20%[c]
Forb height	22 cm (8.7 in.)[a]	ND
100% visual obstruction reading	>2.7 dm (10.6 in.)[a,b]	2.7 dm (10.6 in.)[a]
Bare ground	38%[b]	Open enough for chick mobility[c]
Absolute cover shrubs, grasses, and forbs	>60%[a]	36%[a]
Interspersion surrounding lek	65%–75%[a]	35%–25%[a]

[a] Hagen et al. (2013).
[b] Pitman et al. (2005).
[c] Van Pelt et al. (2013).
[d] ND means that no data were available.
[e] Giesen (1994).

a local management unit. For example, of the 810 ha (2,000 acre) within a 1.6 km (1-mile) radius of a lek, 65%–75% should be comprised of nesting habitat with the remainder being interspersed brood habitat (Hagen et al. 2013). Using this example, ~540 ha (1,333 acres) of nesting habitat and 270 ha (667 acres) of brood-rearing habitat should be available for the 4.5-month period of March 1–July 15. Interspersion of brood habitat within 243–304 m (800–1,000 ft) of nesting habitat has been recommended based on movements of broods less than 14-days old (Jamison 2000).

Permanent Protection

Permanent protection of core quality habitat is critical to conservation planning for Lesser Prairie-Chickens (USFWS 2014). In general, support and funding for perpetual protection and management are limited due to the large areas needed for population persistence, cost of managing such large areas, and limited interest by private landowners in selling sufficiently large sections of land. However, attempts at long-term protection and management of Lesser Prairie-Chicken habitats have been applied in the Sand Sagebrush Prairie Ecoregion. For example, The Nature Conservancy and Colorado Parks and Wildlife have secured ~119 km² of sand sagebrush prairie under permanent conservation easements (J. Rietz, Colorado Parks and Wildlife, pers. comm.). Established in 1916, the 1,522 ha (3,760 acres) Sandsage Bison Range located just south of Garden City, Kansas, and operated by Kansas Department of Wildlife, Parks, and Tourism once supported an estimated breeding density of 19.3 Lesser Prairie-Chickens per km². More recently, a Lesser Prairie-Chicken Management Plan was approved in 2014 for the Cimarron and Comanche National Grasslands under the administration of

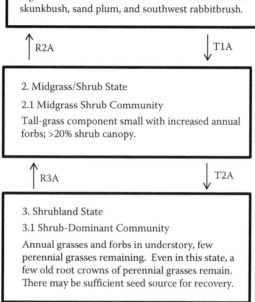

1. Tall/Midgrass State

1.1. Tall/Midgrass Community

Dominated by tall grasses and midgrasses with few perennial forbs and scattered shrubs. Dominant tall grasses include sand bluestem and little bluestem. Dominant shrubs include sand sagebrush and sand shinoak with lesser amounts of skunkbush, sand plum, and southwest rabbitbrush.

↑ R2A   ↓ T1A

2. Midgrass/Shrub State

2.1 Midgrass Shrub Community

Tall-grass component small with increased annual forbs; >20% shrub canopy.

↑ R3A   ↓ T2A

3. Shrubland State

3.1 Shrub-Dominant Community

Annual grasses and forbs in understory, few perennial grasses remaining. Even in this state, a few old root crowns of perennial grasses remain. There may be sufficient seed source for recovery.

Figure 15.3. Ecological Site Description state-and-transition model for sandy soils in 16–22″ precipitation sites. *Notes:* T1A, Heavy Continuous Grazing, No Fire, Brush Invasion; R2A, Prescribed Grazing, Brush Management, Pest Management; T2A, Heavy Continuous Grazing, No Brush Management, No Fire; R3A, Prescribed Grazing, Brush Management, Pest Management, Range Planing. (From Moseley 2012.)

the Pike and San Isabel National Forest (USFS 2014). While these efforts contribute toward long-term persistence of the species, conservation efforts limited to only public lands throughout the range of the Lesser Prairie-Chicken will be insufficient to conserve the species at goal levels (Van Pelt et al. 2013).

Applied Practices to Improve Habitat

Grazing

Grazing is a predominant land use of sand sagebrush prairie that has not been converted to other uses. Thus, integrating nesting and brood-rearing habitats of Lesser Prairie-Chickens with grazing management will be vital for population

recovery. A 20-year study of stocking rates and cow-calf production in the sand sagebrush region found that sand sagebrush canopy cover was not affected by the stocking rates of 1 Animal Unit Month (AUM) per 4.5 ha (11 acres), 6.7 ha (16.4 acres), or 9 ha (22 acres, Gillen and Sims 2006). During this study, sand sagebrush cover averaged 37% (ranging 25%–55%) and was above recommended goals for nest and brood cover for 90% of the 20 years (Table 15.3, Gillen and Sims 2006). However, the range did not drop below minimum cover thresholds necessary for breeding by Lesser Prairie-Chickens (Table 15.3). Climatic variability was more of a primary factor influencing sand sagebrush than stocking rates (Gillen and Sims 2006). Gillen and Sims (2006) also reported stocking rate impacts on grass species that are important for Lesser Prairie-Chicken nesting and brooding habitat. However, the authors measured basal cover, indicating that the standing biomass of grasses in pastures with greater stocking rates was conspicuously less than those subjected to lower stocking rates, and wildlife cover could be negatively affected even though they reported basal cover of grasses unaffected for many species. Hagen et al. (2004) recommended light to moderate stocking rates that yielded residual cover of 60%–70% of key herbaceous species for nesting and incorporated rest periods once every 3–5 years. The Natural Resources Conservation Service (NRCS) in Kansas and Colorado has proposed numerous grazing prescriptions to create habitat for Lesser Prairie-Chickens during the nesting period of April 15 to mid-June, and the brood-rearing period of mid-June through mid-August (Table 15.4; Kansas Range Technical Note KS-9 2014).

Drought planning is critical for livestock producers in the Great Plains to have healthy, productive grasslands that create suitable habitats for Lesser Prairie-Chickens. Significant drought occurs in western Kansas once every five years and 25% of drought years are followed by another drought year (Mosely 2012). Previous studies have reported the influence of weather on grazing and herbaceous production in sand sagebrush prairie (Gillen and Sims 2002, 2006). There is only an 11% chance of being within 2.5 cm (1 in.) of the long-term average rainfall in any one year. Thus, a grazing plan based on average or above-average precipitation will be overstocked half the time. In anticipation of drought, grazing management plans with established trigger dates and actions could be completed and implemented to reduce impacts to the habitats of Lesser Prairie-Chickens (Schwab 2013).

An analysis conducted by Rondeau and Decker (2010) found that vegetation conditions at the Comanche National Grasslands were generally within the range of suitable habitats for Lesser Prairie-Chickens, but lacking in preferred species of bluestem grasses. The Management Plan for Comanche and Cimarron National Grasslands has instituted changes in grazing rotations in designated grazing allotments to enhance habitat conditions for Lesser Prairie-Chickens. The National Grasslands have also closed grazing around active leks of Lesser Prairie-Chickens to improve adjacent nesting habitat.

Herbicide

The herbicide 2,4-D has been commonly used as a control agent to reduce structure, density, and cover of sand sagebrush since the late 1940s (McIlvain and Savage 1949, Thacker et al. 2012). The application of the herbicide did not increase the density of perennial forbs or forb species richness in the short term, but the density of annual forbs did increase in pastures that had been treated >25 years earlier, and the intervening time allowed annual forbs to recover posttreatment. The use of the herbicide

TABLE 15.4

Recommendations for grazing prescriptions to create nesting and brood-rearing habitats for Lesser Prairie-Chickens in Kansas.

Stocking levels	Duration	Habitat type provided
Light[a] to moderate[b]	Until August 15th or no later than October 31st	Nesting
Light to moderate	Dormant season only	Nesting
Light to moderate during late season	Rotational, skipping paddocks	Nesting
Intensive early stocking	Ending July 15th	Nesting and current brooding
Patch burn graze[c]		Nesting and brooding
Light to moderate	Two pasture switch back with first rotation mid- to late June	Nesting and brooding
Light to moderate	Full season	Brooding

SOURCE: Kansas Natural Resources Conservation Service (2014).

[a] Light stocking rate is defined as 16.5% harvest efficiency.

[b] Moderate stocking rate is defined as 20%–25% harvest efficiency.

[c] Not recommended in precipitation zones <71 cm (<28 in).

typically suppressed densities of sand sage-brush for >20 years, but did not increase the abundance of grasshoppers or improve habitat conditions for Lesser Prairie-Chickens (Thacker et al. 2012). Consequently, Thacker et al. (2012) recommended against the use of the herbicide 2,4-D to manage sand sagebrush for improving habitats for Lesser Prairie-Chickens.

Fire

Fire is considered an ecological driver in sand sage-brush prairie. However, little is known about fire frequency, vegetation response, return intervals, or the type and duration of management necessary following the application of fire. Historically, the frequency of lightning-induced fires was highest from June to August (Knapp et al. 2009). Sand sagebrush density and structure typically return to preburn conditions within 3–4 years, but may take as long as seven years following fire (Hagen et al. 2004, Thacker et al. 2013). Reports of mortality in sand sagebrush after fire are conflicted, but evidence suggests that the plant will resprout after fire (McWilliams 2003). Postfire seedling establishment and fire regimes have not been documented. Herbaceous response to fire within the sand sagebrush communities will be largely dependent upon weather conditions preceding, during, and after a fire.

Sand sagebrush will have ~93% resprouting rate after spring (April) and fall (November) prescribed burns without negative impacts to carbohydrate reserves (Vermeire et al. 2001). Research has shown that fire-return intervals of no less than five years on 25%–35% of a management area can promote quality brood-rearing habitat. To avoid disturbance or mortality of nests during incubation, prescribed burns should be avoided during the nesting season of April 15–July 15. To maintain adequate sand sagebrush cover for nesting, fire-return intervals may need to be longer, approaching 10 years. Prescribed burn regimes will need to be adapted to weather patterns and be based upon habitat requirements.

In a patch-burn grazing system study in Oklahoma, temporal habitat measures in sand sagebrush habitat were similar to habitat requirements for nesting and brooding habitat (Doxon et al. 2011). Management included burning ~1/3 of a pasture in March, and stocking of cattle at 6.85 ha (17 acres) per AU for the 5.5-month period of April 1–September 15. Current year burned-grazed areas resembled brood habitat requirements with 17% grass cover, 12% forb cover, 50% bare ground, 4% shrub cover, and 2.5 visual obstruction reading (VOR). Total grass, forb, and shrub cover in current year burned-grazed areas was 31%. Patches that had not been burned in ≥36 months and were available to grazing were estimated to have 39% grass cover, 9% forb cover, 20% bare ground, 14% shrub cover, and 14.4 VOR (cm). Total grass, forb, and shrub cover were 62%. Shrub height and grass height were within nesting and brood-rearing habitat goals primarily in transitional (12–24 months post burn) and older (≥36 months post burn) burned areas. The study highlights similarities to brood and nesting habitat measures within pastures utilizing patch-burn grazing where 1/3 of the area is brood habitat and 2/3 nesting, which is a recommended pattern of interspersion (Hagen et al. 2013).

Salt Cedar

Salt cedar (*Tamarix ramosissima*) is an invasive woody tree within riparian, subirrigated, and other wetland type habitats that has spread to >202 km² (50,000 acres) within Kansas, primarily along southwest Kansas rivers and tributaries in the sand sagebrush region (Kansas Water Office 2005). Impacts of salt cedar on Lesser Prairie-Chickens have not been documented, but the shrub has begun to invade grasslands in the Sand Sagebrush Prairie Ecoregion (D. Sullins, unpubl. data). Research in southwest Kansas reported 100% control with cut-stump herbicide treatment (Fick and Geyer 2010). Cut-stump treatment would be appropriate for mature, rough-barked salt cedars. Low-volume basal bark treatments of an oil-based spray mixture of 61.6% Triclopyr Butoxyethyl ester herbicide, following herbicide label directions, have been shown to control smooth bark plants (Tamarisk Coalition 2009). Tamarisk leaf beetle (*Diorhabda* spp.) is a biological control agent for tamarisk and has been documented within the sand sagebrush range of Lesser Prairie-Chickens in Kansas and Colorado.

Conservation Planning

Knowledge of the relationships among land cover types, amounts, configuration, and occupancy of available habitats by Lesser Prairie-Chickens would enhance conservation planning for the

Sand Sagebrush Prairie Ecoregion. Native extant prairies are key focal areas for enhancement practices (Van Pelt et al. 2013). However, little information is available to assess Lesser Prairie-Chicken population responses to managed grazing, prescribed fire, and other enhancement practices. The Lesser Prairie-Chicken Range-wide Conservation Plan restoration area goal within focal areas of the Sand Sagebrush Prairie Ecoregion is >1,760 km², which is greater than all other ecoregions (Van Pelt et al. 2013). Croplands within the region represent potential opportunities for restoration. However, little is known about the most effective restoration techniques for sand sagebrush prairie in the Southern Great Plains. The assessment of restoration strategies, including desired ecological states, self-design versus intensive site preparation and planting, time to reach desired ecological state, and management necessary to maintain the desired ecological state, would remove considerable uncertainty in regard to increasing sand sagebrush prairie for Lesser Prairie-Chickens.

Somewhat unique to Lesser Prairie-Chickens of the Sand Sagebrush Prairie Ecoregion is the relationship between the land cover and levels of the underlying Ogallala Aquifer. Many crop areas rely on irrigation in this semiarid ecoregion and the mean annual declines in water levels have exhibited significant downward trends, >38 m in some areas, within southwestern Kansas (McGuire 2013). In fact, ~25% of the irrigated cropland area in the sand sagebrush prairie of Kansas are below the minimum saturated thickness required to support well yields for irrigation, and an additional 32% will reach that point in the next 40 years (Smith et al. 2014). In addition to declining water availability, the region of southwest Kansas and eastern Colorado also has a large proportion of cropland acreage with an erodibility index of 20 or greater, making it some of the nation's most environmentally sensitive lands. The CRP Highly Erodible Land Initiative targets to enroll >3,000 km² of land in these areas. The Lesser Prairie-Chicken range in the Sand Sagebrush Prairie Ecoregion is estimated to have >890 km² eligible for this CRP initiative. Coupling the declining aquifer resources for irrigation with the highly erodible land initiative, opportunities exist for policy-makers, natural resource practitioners, and landowners to restore grasslands in the sand sagebrush prairie to benefit air quality, carbon sequestration, groundwater recharge, and wildlife habitat.

The U.S. Department of Agriculture Farm Service Agency has a goal to maintain >5,380 km² of CRP enrollment in native cover acres within the Sand Sagebrush Prairie Ecoregion of the Lesser Prairie-Chicken, which is the highest goal for acreage among the four ecoregions in the range of Lesser Prairie-Chickens (FSA 2014, USFWS 2014). The target goals illustrate the opportunity and importance of CRP as a conservation tool for Lesser Prairie-Chickens across the ecoregions.

CONCLUSIONS

Established habitat and population goals for Lesser Prairie-Chickens in the Sand Sagebrush Prairie Ecoregion will require intensive and innovative approaches to conservation.

Lesser Prairie-Chickens in the Sand Sagebrush Prairie Ecoregion depend on shrubs throughout their annual life cycle. The semiarid environmental conditions frequently limit annual production of herbaceous cover, and sand sagebrush provides reliable vegetation structure. Shrubs provide the necessary habitat structure and cover in a semiarid environment when herbaceous cover is unavailable due to drought, extensive grazing pressure, or a slow response to vegetation removal by fire, mowing, or haying. Restoration of shrubs and associated prairie in semiarid environment is an uncertain proposition, with little guidance available to ensure success. Therefore, identification and conservation of extant sand sagebrush prairie may be considered an initial step for increasing populations of Lesser Prairie-Chickens in the ecoregion. Management for quality nesting and brood-rearing habitats would likely be the most cost-effective means to increase population abundance and habitat occupancy (Hagen et al. 2013). Connectivity is diminishing faster within the Sand Sagebrush Prairie Ecoregion compared to other ecoregions within the range of the Lesser Prairie-Chicken. Lack of connectivity is resulting in increased isolation by distance to the point that movements by Lesser Prairie-Chickens within the metapopulation are likely decreasing as well. The net result is an increase in the conservation issues related to population isolation, including reduced genetic diversity, susceptibility to stochastic events, and declining population viability (Chapters 4 and 5, this volume). The establishment of corridors connecting extant populations would likely reduce issues of low population size and isolation

(Van Pelt et al. 2013). Further, the occurrence of USFS National Grasslands where management of Lesser Prairie-Chickens is a priority will provide a foundation for the establishment of a conservation plan for the ecoregion (Chapter 10, this volume).

ACKNOWLEDGMENTS

J. Reitz provided helpful input regarding Lesser Prairie-Chickens in sand sagebrush prairie of Colorado. A. Godar compiled information on avoidance of anthropogenic structures by Lesser Prairie-Chickens. The chapter benefited from editorial comments by C. Boal.

LITERATURE CITED

Bent, A. C. 1932. Life histories of North American gallinaceous birds. U.S. National Museum Bulletin 162. Dover Publications, New York, NY.

Berg, W. A. 1994. Sand sagebrush-mixed prairie: SRM 722. P. 99 in T. N. Shiflet (editor), Rangeland cover types of the United States, Society for Range Management, Denver, CO.

Bidwell, T., S. Fuhlendorf, B. Gillen, S. Harmon, R. Horton, R. Manes, R. Rodgers, S. Sherrod, and D. Wolfe. 2003. Ecology and management of the Lesser Prairie-Chicken in Oklahoma. Oklahoma State University Extension Circular E-970, Oklahoma Cooperative Extension Unit, Stillwater, OK.

Campbell, H. 1972. A population study of Lesser Prairie Chickens in New Mexico. Journal of Wildlife Management 36:689–699.

Collins, S. L., J. A. Bradford, and P. L. Sims. 1987. Succession and fluctuation in *Artemisia* dominated grassland. Vegetation 73:89–99.

Copelin, F. F. 1963. The Lesser Prairie Chicken in Oklahoma. Oklahoma Department of Wildlife Technical Bulletin 6, Oklahoma City, OK.

Costello, D. F. 1964. Vegetation zones in Colorado. Pp. iii–x in H. D. Harrington (editor), Manual of the plants of Colorado, Sage Books, Denver, CO.

Crawford, J. A., and E. G. Bolen. 1976. Effects of land use on Lesser Prairie Chickens in Texas. Journal of Wildlife Management 40:96–104.

Doxon, E. D., C. A. Davis, S. D. Fuhlendorf, and S. L. Winter. 2011. Aboveground macroinvertebrate diversity and abundance in sand sagebrush prairie managed with the use of pyric herbivory. Rangeland Ecology and Management 64:394–403.

Farm Service Agency. [online]. 2014. Conservation Reserve Program Lesser Prairie-Chicken biological opinion. <http://www.fsa.usda.gov/Internet/FSA_File/lesser_prairie_chicken.pdf> (28 July 2014).

Fick, W. H., and W. A. Geyer. 2010. Cut-stump treatment of saltcedar (*Tamarix ramosissima*) on the Cimarron National Grasslands. Transactions of the Kansas Academy of Science 113:223–226.

Frankham, R., C. J. A. Bradshaw, and B. W. Brook. 2014. Genetics in conservation management: revised recommendations for the 50/500 rules, Red List criteria and population viability analyses. Biological Conservation 170:56–63.

Franklin, I. R., and R. Frankham. 1998. How large must populations be to retain evolutionary potential? Animal Conservation 1:69–73.

Giesen, K. M. 1994. Movements and nesting habitat of Lesser Prairie-Chicken hens in Colorado. Southwestern Naturalist 39:96–98.

Giesen, K. M. 2000. Population status and management of Lesser Prairie-Chicken in Colorado. Prairie Naturalist 32:137–148.

Gillen, R. L., and P. L. Sims. 2002. Stocking rate and cow-calf production on sand sagebrush rangeland. Journal of Range Management 55:542–550.

Gillen, R. L., and P. L. Sims. 2006. Stock rate and weather impacts on sand sagebrush and grasses: a 20-year record. Rangeland Ecology and Management 59:145–152.

Grisham, B. A., C. W. Boal, D. A. Haukos, D. M. Davis, K. K. Boydston, C. Dixon, and W. R. Heck. 2013. The predicted influence of climate change on Lesser Prairie-Chicken reproductive parameters. PLoS One 8:e68225.

Hagen, C. A. 2003. A demographic analysis of Lesser Prairie-Chicken populations in southwestern Kansas: survival, population viability, and habitat use. Ph.D. dissertation, Kansas State University, Manhattan, KS.

Hagen, C. A., B. A. Grisham, C. W. Boal, and D. A. Haukos. 2013. A meta-analysis of Lesser Prairie-Chicken nesting and brood rearing habitats: implications for habitat management. Wildlife Society Bulletin 37:750–758.

Hagen, C. A., B. Jamison, K. Giesen, and T. Riley. 2004. Guidelines for managing Lesser Prairie-Chicken populations and their habitats. Wildlife Society Bulletin 32:69–82.

Hagen, C. A., J. C. Pitman, T. M. Loughin, B. K. Sandercock, R. J. Robel, and R. D. Applegate. 2011. Impacts of anthropogenic features on habitat use by Lesser Prairie-Chickens. Pp. 63–76 in B. K. Sandercock, K. Martin, and G. Segelbacher (editors),

Ecology, conservation, and management of grouse. Studies in Avian Biology (no. 39), University of California Press, Berkeley, CA.

Hagen, C. A., J. C. Pitman, B. K. Sandercock, R. J. Robel, and R. D. Applegate. 2005a. Age-specific variation in apparent survival rates of male Lesser Prairie-Chickens. Condor 107:78–86.

Hagen, C. A., J. C. Pitman, B. K. Sandercock, R. J. Robel, and R. D. Applegate. 2007. Age-specific survival and probable causes of mortality in female Lesser Prairie-Chickens. Journal of Wildlife Management 71:518–525.

Hagen, C. A., G. C. Salter, J. C. Pitman, R. J. Robel, and R. D. Applegate. 2005b. Lesser Prairie-Chicken brood habitat in sand sagebrush: invertebrate biomass and vegetation. Wildlife Society Bulletin 33:1080–1091.

Hagen, C. A., B. K. Sandercock, J. C. Pitman, R. J. Robel, and R. D. Applegate. 2009. Spatial variation in Lesser Prairie-Chicken demography: a sensitivity analysis of population dynamics and management alternatives. Journal of Wildlife Management 73:1325–1332.

Haukos, D. A., and L. M. Smith. 1999. Effects of lek age on age structure and attendance of Lesser Prairie-Chickens (*Tympanuchus pallidicinctus*). American Midland Naturalist 142:415–420.

Jamieson I. G., and F. W. Allendorf. 2012. How does the 50/500 rule apply to MVPs? Trends in Ecology and Evolution 27:578–584.

Jamison, B. E. 2000. Lesser Prairie-Chicken chick survival, adult survival, and habitat selection and movements of males in fragmented rangelands of southwestern Kansas. M.S. thesis, Kansas State University, Manhattan, KS.

Jamison, B. E., R. J. Robel, J. S. Pontius, and R. D. Applegate. 2002. Invertebrate biomass: associations with Lesser Prairie-Chicken habitat use and sand sagebrush density in southwestern Kansas. Wildlife Society Bulletin 30:517–526.

Jensen, W. E., D. A. Robinson Jr., and R. D. Applegate. 2000. Distribution and population trend of Lesser Prairie-Chicken in Kansas. Prairie Naturalist 32:169–175.

Kansas Natural Resources Conservation Service (Kansas NRCS). [online]. 2014. Identifying and creating Lesser and Greater Prairie-Chicken habitat. Kansas Range Technical Note, KS-9, Natural Resources Conservation Service, Salina, KS. <http://www.nrcs.usda.gov/Internet/FSE_DOCUMENTS/nrcs142p2_032476.pdf> (13 August 2014).

Kansas Water Office. 2005. 10-year strategic plan for the comprehensive control of tamarisk and other non-native phreatophytes. Topeka, KS.

Knapp, E. E., B. L. Estes, and C. N. Skinner. 2009. Ecological effects of prescribed fire season: a literature review and synthesis for managers. General Technical Report PSW-GTR-224, U.S. Department of Agriculture, Forest Service, Pacific Southwest Research Station, Albany, CA.

Küchler, A. W. 1964. Potential natural vegetation of the conterminous United States. Special Publication Number 36, American Geographical Society, New York, NY.

Küchler, A. W. 1974. A new vegetation map of Kansas. Ecology 55:586–604.

Kukal, C. A. 2010. The over-winter ecology of Lesser Prairie-Chickens (*Tympanuchus pallidicinctus*) in the northeast Texas Panhandle. M.S. thesis, Texas Tech University, Lubbock, TX.

Manier, D. J., D. J. A. Wood, Z. H. Bowen, R. M. Donovan, M. J. Holloran, L. M. Juliusson, K. S. Mayne et al. 2013. Summary of science, activities, programs, and policies that influence the rangewide conservation of Greater Sage Grouse (*Centrocercus urophasianus*). U.S. Geological Survey Open-File Report 2013-1098, Reston, VA.

McDonald, L., K. Adachi, T. Rintz, G. Gardner, and F. Hornsby. [online]. 2014a. Range-wide population size of the Lesser Prairie-Chicken: 2012, 2013, and 2014. Western EcoSystems Technology, Inc., Laramie, WY. <http://www.wafwa.org/documents/LPC-aerial-survey-results-2014.pdf> (12 August 2014).

McDonald, L., G. Beauprez, G. Gardner, J. Griswold, C. Hagen, D. Klute, S. Kyle, J. Pitman, T. Rintz, and B. Van Pelt. 2014b. Range-wide population size of the Lesser Prairie-Chicken: 2012 and 2013. Wildlife Society Bulletin 38:536–546.

McGuire, V. L. 2013. Water-level and storage changes in the High Plains aquifer, predevelopment to 2011 and 2009–11. U.S. Geological Survey Scientific Investigations Report 2012–5291, Reston, VA.

McIlvain, E. H., and D. A. Savage. 1949. Spraying 2,4-D by airplane on sand sagebrush and other plants of the Southern Great Plains. Journal of Range Management 2:43–52.

McWilliams, J. [online]. 2003. *Artemisia filifolia*. Fire Effects Information System, Fire Sciences Laboratory, Rocky Mountain Research Station, Forest Service, U.S. Department of Agriculture, Washington, DC. <http://www.fs.fed.us/database/feis/> (17 August 2014).

Moseley, M. 2012. Ecological site description sandy 16–22″, rangeland, R077AY666TX. Natural Resource Conservation Service, United States Department of Agriculture, Washington, DC.

Palstra, F. P., and D. E. Ruzzante. 2008. Genetic estimates of contemporary effective population size: what can they tell us about the importance of genetic stochasticity for wild population persistence? Molecular Ecology 17:3428–3447.

Parker, K. W., and D. A. Savage. 1944. Reliability of the line interception method in measuring vegetation on the Southern Great Plains. Journal of American Society of Agronomy 36:97–110.

Pitman, J. C. 2003. Lesser Prairie-Chicken nest site selection and nest success, juvenile gender determination and growth and juvenile survival and dispersal in southwestern Kansas. M.S. thesis, Kansas State University, Manhattan, KS.

Pitman, J. C., B. E. Jamison, C. A. Hagen, R. J. Robel, and R. D. Applegate. 2006c. Brood break-up and juvenile dispersal of Lesser Prairie-Chicken in Kansas. Prairie Naturalist 38:85–99.

Pitman, J. C., C. A. Hagen, B. E. Jamison, R. J. Robel, T. M. Loughin, and R. D. Applegate. 2006a. Survival of juvenile Lesser Prairie-Chickens in Kansas. Wildlife Society Bulletin 34:675–681.

Pitman, J. C., C. A. Hagen, E. Jamison, R. J. Robel, T. M. Loughin, and R. D. Applegate. 2006b. Nesting ecology of Lesser Prairie-Chickens in sand sagebrush prairie of southwestern Kansas. Wilson Journal of Ornithology 118:23–35.

Pitman, J. C, C. A. Hagen, R. J. Robel, T. M. Loughin, and R. D. Applegate. 2005. Location and success of Lesser Prairie-Chicken nests in relation to vegetation and human disturbance. Journal of Wildlife Management 69:1259–1269.

Pruett, C. L., J. A. Johnson, L. C. Larsson, D. H. Wolfe, and M. A. Patten. 2011. Low effective population size and survivorship in a grassland grouse. Conservation Genetics 12:1205–1214.

Robel, R. J., J. A. Harrington, Jr., C. A. Hagen, J. C. Pitman, and R. R. Reker. 2004. Effect of energy development and human activity on the use of sand sagebrush habitat by prairie-chickens in southwestern Kansas. Transactions of the North American Wildlife and Natural Resource Conference 69:251–266.

Rondeau, R., and K. Decker. [online]. 2010. Lesser Prairie-Chicken habitat assessment, Comanche National Grassland. Colorado Natural Heritage Program, Colorado State University, Fort Collins, CO. <http://www.cnhp.colostate.edu/download/reports.aspx> (9 August 2014).

Schwab, J. C. 2013. Planning and drought. Planning Advisory Service Report No. 574, American Planning Association, Chicago, IL. <drought.gov/drought/content/resources/planning-home> (14 August 2014).

Sexson, M. L. 1980. Destruction of sandsage prairie in southwest Kansas. Pp. 113–115 in C. L. Kucera (editor), Proceedings of the Seventh North American Prairie Conference, Southwestern Missouri State University, Springfield, MO.

Smith, J., G. A. Ludvigson, H. Harlow, and B. Platt. 2014. Ogallala-High Plains Aquifer Special Study Phase III: lithologic calibration of practical saturated thickness in the Ogallala-High Plains Aquifer. Open-File Report 2014-2, Kansas Geological Survey, Kansas Water Office, Topeka, KS.

Snyder, J. W., E. C. Pelren, and J. A. Crawford. 1999. Translocation histories of prairie grouse in the United States. Wildlife Society Bulletin 27:428–432.

Schwilling, M. D. 1955. A study of the Lesser Prairie-Chicken in Kansas: August 1953–July 1955. Kansas Forestry, Fish, and Game Commission, Pratt, KS.

Tamarisk Coalition. [online]. 2009. Colorado River Basin Tamarisk and Russian Olive Assessment. <http://www.tamariskcoalition.org/sites/default/files/files/TRO_Assessment_FINAL%2012–09.pdf> (10 August 2014).

Thacker, E., D. Elmore, and B. Reavis. [online]. 2013. Management of sand sagebrush rangelands. Publication NREM-2892, Oklahoma Cooperative Extension Service, Oklahoma State University, Stillwater, OK. <http://pods.dasnr.okstate.edu/docushare/dsweb/Get/Document-8783/NREM-2892web.pdf> (10 August 2014).

Thacker, E. T., S. A. Gunter, R. L. Gillen, and T. L. Springer. 2012. Sand sagebrush control: implications for Lesser Prairie-Chicken habitat. Range Ecology and Management 28:204–212.

U.S. Fish and Wildlife Service. 2014. Endangered and threatened wildlife and plants; special rule for the Lesser Prairie-Chicken. Federal Register 79:19973–20071.

U.S. Forest Service. [online]. 2014. Lesser Prairie-Chicken management plan: cimarron and Comanche National Grasslands. <http://www.fs.usda.gov/Internet/FSE_DOCUMENTS/stelprd3804315.pdf> (31 August 2014).

Van Pelt, W. E., S. Kyle, J. Pitman, D. Klute, G. Beauprez, D. Schoeling, A. Janus, and J. Haufler. 2013. The Lesser Prairie-Chicken range-wide conservation plan. Western Association of Fish and Wildlife Agencies, Cheyenne, WY.

Vermeire, L. T., R. B. Mitchell, and S. D. Fuhlendorf. 2001. Sand sagebrush response to fall and spring prescribed burns. Pp. 233–235 in E. D. McArthur, and D. J. Fairbanks (Compilers). Shrubland ecosystem genetics and biodiversity. Proc. RMRS-P-21, Rocky Mountain Research Station, Forest Service, U.S. Department of Agriculture, Ogden, UT.

Vodehnal, W. L., and J. B. Haufler (editors). 2008. A grassland conservation plan for Prairie Grouse. North American Grouse Partnership, Fruita, CO.

Walker, T. L., Jr. 2000. Movements and productivity of Lesser Prairie-Chickens in southwestern Kansas. Final Report Federal Aid Project W-47-R, Kansas Department of Wildlife and Parks, Pratt, KS.

Woodward, A. J. W., S. D. Fuhlendorf, D. M. Leslie, and J. S. Shackford. 2001. Influence of landscape composition and change on Lesser Prairie-Chicken populations. American Midland Naturalist 145:261–274.

CHAPTER SIXTEEN

The Lesser Prairie-Chicken in the Mixed-Grass Prairie Ecoregion of Oklahoma, Kansas, and Texas*

Donald H. Wolfe, Lena C. Larsson, and Michael A. Patten

Abstract. The Mixed-Grass Prairie Ecoregion of northeastern Texas, panhandle of northwestern Oklahoma, and south-central Kansas is the geographic center of the extant range of Lesser Prairie-Chickens (*Tympanuchus pallidicinctus*), and historically contained the highest density of the species. Currently, only the short-grass and sand sagebrush prairies of western Kansas support larger populations of Lesser Prairie-Chickens. Much of the mixed-grass prairie was severely fragmented by homesteading over a century ago, and fragmentation is ongoing due to oil and gas development, wind power development, transmission lines, highways, and expansion of invasive plants such as eastern red-cedar (*Juniperus virginiana*). Road and fence densities in the mixed-grass prairie are also high compared to other ecoregions in the range of Lesser Prairie-Chickens; fencelines pose a high risk for collision mortalities in the Mixed-Grass Prairie Ecoregion. Densities of Lesser Prairie-Chickens in the mixed-grass prairie have declined considerably over the past century. The combination of a lek breeding system and isolation of subpopulations has resulted in vulnerability to stochastic events such as drought, heat waves, and epizootics, as well as a reduced effective population size and increased risk of inbreeding depression. Prescribed fire and moderated grazing regimes have been shown to promote vegetation floristic and structural diversity that is needed by Lesser Prairie-Chickens. We suggest the use of management guides to evaluate habitat quality and determine best practices for the conservation of Lesser Prairie-Chickens. We review the known issues in the mixed-grass prairie, focusing on conservation measures that will allow persistence of prairie chickens as an "icon of the prairie."

Key Words: anthropogenic impacts, genetic structure, habitat use, life history, threats, *Tympanuchus pallidicinctus.*

The Mixed-Grass Prairie Ecoregion of southwestern Kansas, northwestern Oklahoma, and northeastern Texas Panhandle provides crucial habitat for Lesser Prairie-Chickens (*Tympanuchus pallidicinctus*). The southern mixed-grass prairies as a whole formerly covered ~120,000 km², of which just a little over half remains (Samson et al. 2004). With a current population recently estimated at <4,000 birds, the Mixed-Grass Prairie Ecoregion supports the second largest extant

* Wolfe, D. H., L. C. Larsson, and M. A. Patten. 2016. The Lesser Prairie-Chicken in the Mixed-Grass Prairie Ecoregion of Oklahoma, Kansas, and Texas. Pp. 299–314 in D. A. Haukos and C. W. Boal (editors), Ecology and conservation of Lesser Prairie-Chickens. Studies in Avian Biology (no. 48), CRC Press, Boca Raton, FL.

TABLE 16.1

TABLE 16.1

Nesting habitats of Lesser Prairie-Chickens with habitat at nest sites compared to availability.

Habitat type	No. of nests	Successful nests	Successful nests (%)	In habitat type (%)	Habitat available (%)[a]
Native grasslands	30	13	43.3	46.2	51.8
Nonnative habitats	35	15	42.9	53.8	43.9
All CRP fields	25	12	48.0	38.5	19.6
Old world bluestem	11	5	45.5	16.9	12.7
Native seed mixture	14	7	50.0	21.5	6.9
Agricultural fields	10	3	30.0	15.4	23.8
Fallow	9	3	33.3	13.8	10.4
Crops present	1	0	0.0	1.5	13.4
Total	65	28	43.1		

[a] Additional 4.3% was not included as vegetation habitat, such as ponds and roads. Data were collected in the Mixed-Grass Prairie Ecoregion of northwestern Oklahoma. We compared vegetation sampling at nest sites to vegetation at >6,700 randomly generated sampling points from 1999 through 2003 (Sutton Avian Research Center, unpubl. data).

population of Lesser Prairie-Chickens, and only the population of the Short-Grass Prairie/CRP Ecoregion of northwestern Kansas is estimated to be at least twice as large (McDonald et al. 2014).

Upland soils in the mixed-grass prairie tend to be deep and loamy sands, which support vegetation characterized by a mix of sand sagebrush (*Artemisia filifolia*) and perennial grasses (Dhillion et al. 1994). Woody vegetation includes some sand plum (*Prunus* sp.) as well as sand shinnery oak (*Quercus havardii*) and yucca (*Yucca* spp.) in the southern and western portions of the region. The terrain is relatively flat with a prevailing climate of cold and dry winters, hot and humid summers, and seemingly perpetual wind, typically from the south (Fonstad et al. 2003). The annual precipitation averages ~60 cm and, although the total varies considerably from year to year, precipitation is more reliable in the mixed-grass prairie than in other portions of the range of Lesser Prairie-Chicken with a 70% probability of 36–61 cm precipitation in any given year (U.S. Department of Agriculture, National Resources Conservation Services, esis.sc.egov.usda.gov/). The Natural Resources Conservation Service (NRCS) designated the majority of associated Major Land Resource Area as "Southern High Plains, Breaks" (MLRA 77E) and "Central Rolling Red Plains, Eastern Part" (MLRA 78C; Bliss 1989). About half of the 27,000 km² in MLRA 77E and a fifth of the 51,000 km² in MLRA 78C are part of the geographic range of Lesser Prairie-Chickens.

Roughly half of the region supports habitat that can be classified as native mixed-grass prairie, with the remainder converted for agricultural production or other uses (Table 16.1; Drummond 2007). Conservation Reserve Program (CRP) fields occupy between 10% and 20% of the Mixed-Grass Prairie Region (Peterson et al. 2001, Rao et al. 2007). CRP lands in Oklahoma and the northeastern panhandle of Texas are dominated by exotic grasses including Old World bluestem (*Bothriochloa* spp.) and lovegrass (*Eragrostis* spp.), but native species of warm-season grasses were used for CRP plantings in Kansas (Ripper et al. 2008). In addition to current CRP fields, 17%–25% of the region is now devoted to wheat and other crops, and a slight amount of the region has been converted to "improved" pastures with nonnative forage plants. Improved pastures are mainly expired CRP contracts or agricultural fields seeded with grasses for livestock forage.

HABITAT REQUIREMENTS

Lesser Prairie-Chicken populations require vast tracts of treeless prairie with a full range of plant succession stages and >65%–70% noncultivated vegetation (Elmore et al. 2009). Home ranges are large and average 18.0 km² ± 9.9SE for radio-marked females in western Oklahoma (V. Winder et al. 2015). The landscape must support a complex of leks or "gobbling grounds" as communal areas where males congregate to display and mate with visiting females. Leks can be traditional sites used for consecutive years or ephemeral satellite leks, and a lek complex can be a network of sites that are linked by dispersal. Females select nest sites <2 km from the nearest lek, although not necessarily the lek where she mated. Applegate and Riley (1998) recommended complexes of at least six leks with interlek distances of 1.9 km or less. Relative to availability and across the species' range, female Lesser Prairie-Chickens select nest sites with greater vertical structure of shrubs or bunch grasses and a lower cover of bare ground (Hagen et al. 2013). The removal or reduction of shrub cover in otherwise suitable prairie greatly reduces the probability that a female will place a nest in the habitat: 0–4 nests per 65 ha study block, over 5-year period when shrub cover <20% and 4–10 nests when cover >20% (Patten and Kelly 2010). On average, the total horizontal cover exceeds 35% in brood-rearing habitat and 58% in nesting habitat, whereas canopy cover for these habitats was estimated as >35% and 39%, respectively (Hagen et al. 2013).

Floristics and physiognomy are both important components of habitat for Lesser Prairie-Chickens. Shrubs in the mixed-grass prairie likely provide protection from predators as well as shade needed in the summer for thermoregulation (Larsson et al. 2013); adult survival is positively correlated with 10%–20% shrub canopy cover (Patten et al. 2005a). One exception to a requirement for shrub cover is periods of the year when birds attend leks. Males need to be seen to attract a potential mate and must be vigilant for approaching predators. Thus, lek sites are usually in relatively open sites with patches of bare ground and less woody vegetation (Copelin 1963, Jones 1963, Cannon and Knopf 1979, Taylor and Guthery 1980). Most leks are located on broad ridge tops or relatively high elevation sites with good visibility (Hagen and Giesen 2005). Black-tailed prairie-dog (*Cynomys ludovicianus*) towns, with hard pan soils and shorter vegetative component, are sometimes used where the species co-occur, although the occurrence of prairie chickens is not related to presence of prairie dogs (Lomolino and Smith 2003). It has been argued that grass cover alone can provide sufficient habitat for Lesser Prairie-Chickens (McCleery et al. 2007). Studies of brood habitat and adult survival in sand sage prairie (Hagen et al. 2005b, 2009) and nest placement and brood habitat in sand shinnery oak prairie (Johnson et al. 2004, Davis 2009, Bell et al. 2010, Patten and

Kelly 2010) have shown that away from leks, the species responds positively to shrub cover in the range of 10%–20% (Patten et al. 2005a).

Many of the plants most closely associated with the species occurrence are potential food sources for Lesser Prairie-Chickens and likely support arthropods during the warmer summer months (Jones 1964). For example, beetles and weevils (Coleoptera) and grasshoppers and crickets (Orthoptera) are seasonally primary food sources for Lesser Prairie-Chickens, especially for young birds (Jamison et al. 2002, Hagen et al. 2005b). Broom snakeweed (*Gutierrezia sarothrae*) is a subshrub or large perennial forb that has been identified as a winter source of food for Lesser Prairie-Chicken (Jones 1963). Snakeweed is an indicator of disturbed short-grass prairie and was positively associated with Lesser Prairie-Chicken leks in New Mexico, but negatively associated with habitat occupancy of prairie chickens in Oklahoma (Hunt and Best 2010, Larsson et al. 2013). Indian blanket (*Gaillardia pulchella*) and sunflower (*Helianthus* spp.) are both negatively associated with the occurrence of Lesser Prairie-Chickens (Larsson et al. 2013). The two plant species thrive in disturbed habitats along roadsides and fences, which are avoided by Lesser Prairie-Chickens (Patten et al. 2005b; Pitman et al. 2005; Pruett et al. 2009a,b; Hagen et al. 2011). The presence of such structures rather than specific plant species may explain apparent avoidance of a given site.

Lesser Prairie-Chickens occasionally feed in cultivated fields on waste grains, including grain sorghum (*Sorghum* spp.), wheat (*Triticum* spp.), and corn (*Zea mays*; Salter et al. 2005). Wheat is negatively associated with occupancy throughout the year, with a slight positive association during fall when the availability of arthropods declines, and unharvested grains or fresh sprouts may provide an alternative food source (Larsson et al. 2013). By contrast, the species' occurrence correlates positively with alfalfa (*Medicago* spp.), although it is unclear if the association was related to forage, invertebrates, or moisture associated with frequently irrigated alfalfa (Lindenmayer et al. 2011, Larsson et al. 2013).

ANTHROPOGENIC IMPACTS

The bulk of the mixed-grass prairie was homesteaded over a century ago (Chapter 2, this volume). Settlement proceeded rapidly, and the prairie was subdivided into tracts of 65 ha (160 acres or a quarter section) and occasionally 130 ha (320 acres or a half section). Many fences were erected to contain or exclude livestock, protect crops, and designate property boundaries (Patten et al. 2005b). Settlement led to the rapid removal of shrubs to plant row crops and in a perceived effort to improve forage for livestock. In addition to land use changes, prairie chickens were hunted extensively for food or killed as a "pest" species thought to destroy crops (Fleharty 1995).

Road density is fairly high following settlement in quarter section parcels of land (Patten et al. 2005b). The footprint of oil and gas development remains high, with 1.5–4.6 wells per km^2 (4–12 per $mile^2$) in parts of the mixed-grass prairie. The spread of wind energy development has claimed additional prairie in the region, and the expansion of new extra-high voltage transmission lines to keep pace with energy development has further fragmented the prairie (Pruett et al. 2009a,b). Wind energy development is relatively new in the mixed-grass prairie, and we do not yet know the full effect that development may have on Lesser Prairie-Chickens. Detrimental effects of wind production facilities have been observed for other lekking grouse where leks were abandoned or shifted away from wind farms (Zeiler and Grünschachner-Berger 2009, Hagen et al. 2011, Winder et al. 2014). Despite a lack of specific data for Lesser Prairie-Chickens, habitat suitability models for the species suggest negative impacts from the footprint of energy development (Jarnevich and Laubhan 2011).

Lesser Prairie-Chickens avoid tall structures, oil and gas wells, roads, and outbuildings (Robel et al. 2004, Pitman et al. 2005, Pruett et al. 2009b, Hagen et al. 2011). An increase in any of these features on the landscape may result in the perception by the species that associated habitat may be unsuitable even if it contains an adequate mix of grass and shrub cover. The habitat use for nesting and brood rearing is estimated to be affected up to 1.6 km (1 mile) from man-made structures (Robel et al. 2004). The direct effects of noise from oil and gas pumps have not been investigated fully; however, Blickley et al. (2012) reported a reduction in lek attendance by Greater Sage-Grouse (Centrocercus urophanianus) when noise from oil and gas drilling and roads was experimentally introduced. Hunt (2004) noted that

drilling noise was greater at abandoned leks than at active leks in New Mexico, but concluded that the noise levels alone did not cause lek abandonment.

A major factor driving the loss of mixed-grass prairie is the westward spread of the eastern red-cedar (Juniperus virginiana) during the past several decades (Briggs et al. 2002). Oklahoma Forestry Services estimated the average rate of expansion of eastern red-cedar in 2002 to be 308 ha per day (762 acres per day) or 1,214 km^2 per year (469 $mile^2$ per year). As much as 8,000–16,000 km^2 (2–4 million acres) would need to be treated "aggressively" to restore lands affected by cedar encroachment in Oklahoma (Drake and Todd 2002). Substantial efforts have been made to slow expansion and remove eastern red-cedars from the landscape within the range of Lesser Prairie-Chicken (Bidwell et al. 2002). Unfortunately, the effort has had limited success in many areas, partly because some agencies have promoted the use of eastern red-cedar for erosion control and wind protection, and some landowners remain unaware or unconcerned about the effect of invasive species on native flora and fauna of mixed-grass prairie. Recent outreach and education efforts aim to increase eastern red-cedar management, including the creation of local burn associations to conduct prescribed burns.

Cattle grazing does not appear to affect nest success, but the presence of cattle or effects of grazing may influence nest-site selection (Wolfe et al. 2003). Low-density continuous grazing systems may help to create and maintain a mosaic that includes areas of increased grass and brush cover necessary for suitable nesting habitat, and applications of patch burning may even further enhance the necessary mosaic (Fuhlendorf et al. 2006, 2009; Winter et al. 2011). Permanent cross-fencing to manipulate grazing pressure increases the risk of collision for low-flying Lesser Prairie-Chickens in Oklahoma (Patten et al. 2005b).

Some populations of Lesser Prairie-Chickens have been negatively affected by Ring-necked Pheasants (Phasianus colchicus), an Asian bird species introduced as a game bird in the mixed-grass prairie. Pheasants occasionally parasitize Lesser Prairie-Chicken nests and disrupt leks (Hagen et al. 2002, Holt et al. 2010). Beyond affecting mate access, lek disruption presents the opportunity for hybridization, which would affect fitness.

There are no known hybrids of Lesser Prairie-Chicken and Ring-necked Pheasant, but pheasants have hybridized with Greater Prairie-Chickens (*T. cupido*, Lincoln 1950) and other species of grouse (Peterle 1951).

Population Trends

Human settlement of the Southern Great Plains during the past century dramatically altered the landscape and habitat and greatly affected populations of many grassland species. Unregulated harvest, landscape conversion, fire suppression, deliberate or accidental import of nonnative invasive species, energy development, and other impacts have all contributed to changes in population abundance and occurrence of Lesser Prairie-Chickens. Recent population trends of the Lesser Prairie-Chicken in Oklahoma and in the panhandle of Texas are sharply negative (Horton 2000, Sullivan et al. 2000). Population sizes in Kansas have been more stable overall (Rodgers 2009), but bird numbers in the Sand Sagebrush Prairie Ecoregion in the southwestern portion of the state have also declined (Jensen et al. 2000, Hagen et al. 2009; Chapter 4, this volume).

Data from western Oklahoma illustrate the population trends in mixed-grass prairie. Lek counts on a 4,144 ha (10,240-acre) study area in southern Ellis County from 1932 to 1939 recorded between 266 and 606 males in a given year (Davison 1940). Assuming an equal sex ratio, these counts yielded an estimate of 13–29 Lesser Prairie-Chickens per km^2 (33–76 birds per mile^2). The density of birds plunged 20 years later with the number of males ranging from 131 to 291 in a given year from 1958 to 1961 (Copelin 1963) and an estimated density of 6–14 prairie chickens per km^2 (16–36 birds per mile^2). Results from the Christmas Bird Count at Arnett, Ellis County for the 45-year period from 1966 to 2010 indicate an even greater decline from the 1960s through the 1990s, with an average of 205.5 Lesser Prairie-Chickens counted per year in the 1960s, 27.1 per year in the 1970s, 5.1 per year in the 1980s, and 1.8 per year in the 1990s. Counts since the 1990s have remained low but held steady.

Population declines of Lesser Prairie-Chickens are chiefly associated with loss of shrub and native grass cover, typically as a function of the rate at which land has been converted from one vegetation type or land use to another (Woodward et al. 2001).

Historical landscape changes affected the population trends, where private ownership of smaller tracts of land in Oklahoma resulted in greater changes that impacted the Lesser Prairie-Chickens more than adjacent populations in Texas (Woodward et al. 2001). Analyses at a broad spatial extent suggest that the amount of cropland, tree encroachment, and general landscape changes have the greatest effects on populations of Lesser Prairie-Chickens in mixed-grass prairie, while change in edge density affected the population trend at a finer spatial scale (Fuhlendorf et al. 2002). Furthermore, lek persistence is correlated inversely to the level of habitat fragmentation (Fuhlendorf et al. 2002).

Life History

Annual estimates of adult survival vary between 0.23 and 0.62, depending on the ecoregion and quantitative methods used to estimate rates (Jamison 2000; Hagen et al. 2005a, 2007; Patten et al. 2005b). Adult survival does not appear to vary with gender in some populations, but female survival is considerably lower than male survival in the southern mixed-grass prairie. For example, female Lesser Prairie-Chickens in Oklahoma had less than half the probability of surviving two years compared to males (Patten et al. 2005b), a pattern unusual in birds. Birds tend to have constant rates of annual survival after they reach maturity, but survival rates decrease with age in Lesser Prairie-Chickens, a typical pattern in grouse (Hagen et al. 2005a, 2007). If one were to predict a linear rate after maturity, survivorship the second year should equal the square of the first year's estimate. However, second-year survival rates of Lesser Prairie-Chickens can be up to 50% lower, implying that survival estimates on the basis of data collected from only one year could be misleading.

With 11–14 eggs per nest, the mean clutch size is relatively high for Lesser Prairie-Chickens nesting in the mixed-grass prairie. The average clutch size of Lesser Prairie-Chickens is two eggs larger on average for birds nesting in mixed-grass prairie versus the sand shinnery oak prairie (Patten et al. 2005b, Pruett et al. 2011). Geographic variation in clutch size has been attributed to a life history trade-off in which adult females in Oklahoma select for an increased reproductive effort and produce larger clutch sizes at the cost of experiencing relatively higher mortality rates (Patten et al. 2005b, Pruett et al. 2011). Clutch size is not associated with rainfall (Patten et al. 2005b) or nutrition, as measured from blood screening (Sutton Avian Research Center, unpubl. data). Likewise, nest success is not associated with rainfall (Figure 16.1).

Male mortality peaks during spring months of March to early May, when males defend small territories at leks and are exposed to predation risk by their conspicuous displays and low vegetation cover at gobbling grounds (Patten et al. 2005b). Lekking activity is also energetically expensive: males often lose >10% of their body mass during the 2–3 months of active displays (Sutton Avian Research Center, unpubl. data). In contrast, female mortality peaks slightly later during late spring or early summer in the months of May and June. The timing coincides with two periods of extensive movements when females both visit leks to choose a "mate" and select a suitable nest site. The typical female visits two or more leks, which are seldom <1 km of each other. During both periods, females are more vulnerable to collision or other mortality risks, such that peak mortality coincides with peak motility (Patten et al. 2005b, Hagen et al. 2007).

A principal cause of mortality responsible for >40% of deaths in the southern mixed-grass prairie of Oklahoma is not only collisions with anthropogenic objects, mainly with fence wires, but also vehicles and powerlines (Patten et al. 2005b; Wolfe et al. 2007). We posit that collisions with fences and other man-made structures are additive to mortality from predation, especially in dry years when recruitment is inadequate to offset low adult survivorship. Scavenging by mammals can occur within days at >50% of carcasses (Bumann and Stauffer 2002), and collisions with fences or powerlines may occur at a higher rate than is evident from necropsy. Bias in cause-specific mortality results can be due to lack of direct observation or misinterpretation of sign during necropsy. In Oklahoma, >60% of carcasses were recovered <50 m from a fence but if mortalities were randomly distributed across the landscape, only 12% of carcasses would be expected <50 m from a fence (Sutton Avian Research Center, unpubl. data). Nevertheless, in southwestern Kansas and other regions where

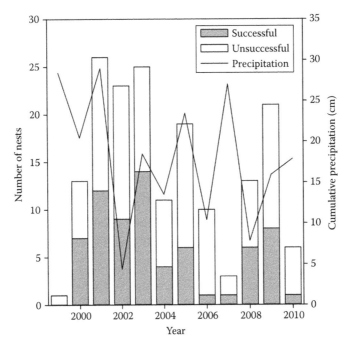

Figure 16.1. Nest success of Lesser Prairie-Chickens in the Mixed-Grass Prairie Ecosystem relative to cumulative precipitation in the 6-month period before the nesting season (October–March). Nest data were from northwestern Oklahoma (Sutton Avian Research Center, unpubl. data) and the northeastern portion of the Texas Panhandle (Jones 2009, Holt 2012). Climatic data were from the panhandle of Oklahoma (Southern Regional Climate Center, www.srcc.lsu.edu, January 20, 2014). The proportion and number of successful nests were not related to cumulative precipitation.

fence densities are lower, collision mortalities are encountered less frequently (Pitman et al. 2006). The concept of collision mortalities is not new: over 60 years ago, Ligon (1951) opined that Lesser Prairie-Chickens would be affected negatively by collisions with powerlines in New Mexico and noted anecdotal reports of railroad workers subsisting on prairie chickens collected from under telegraph wires in Kansas and Nebraska. Copelin (1963) also named powerline collisions as a potential threat to the species in Oklahoma. Even so, there are no data on the extent of serious, debilitating injuries to Lesser Prairie-Chickens from fence or powerline collisions that may cause infections or make a bird more susceptible to predation or thermal stress, to say nothing of how fences may alter predator density or patterns of movement.

Mortality from raptor predation in the mixed-grass prairie is nearly as great as that attributable to collisions, but mortality from mammals is markedly lower, accounting for about 20% of deaths at some sites (Wolfe et al. 2007). Regional differences in cause-specific mortality rates may be important given that predation by mammals exceeds that by raptors in the sand sagebrush prairie of Kansas (Pitman et al. 2006). Lesser Prairie-Chickens are prone to depredation by migratory raptors that move through the Mixed-Grass Prairie Ecoregion (Wolfe et al. 2007). When average daily temperature drops below 10°C–15°C, Lesser Prairie-Chickens occur more in the open, presumably to absorb solar radiation and avoid hypothermia (Larsson et al. 2013), a behavior that may render the species more vulnerable to raptor predation.

CURRENT THREATS

Threat Rankings

Over 50% of the historical extent of southern mixed-grass prairie has already been lost (Samson et al. 2004), but habitat loss remains a key threat to Lesser Prairie-Chickens. Continued conversion of prairie for cultivation or energy development will exacerbate steep population declines of the species in the region. The amount of suitable habitat is further reduced by continued spread of invasive plants, both woody plants like eastern redcedar and herbaceous plants such as nonnative grasses. Smaller populations are more vulnerable

to stochastic events that may lead to a spiral of decline, exacerbated by inbreeding depression and a reduced ability to adapt to environmental changes (Gilpin and Soulé 1986). Two additional threats, habitat fragmentation and climate change, deserve special consideration in light of human activities in the Mixed-Grass Prairie Ecoregion of the Southern Great Plains.

Habitat Fragmentation and Degradation

Habitat fragmentation from roads, fences, and powerlines are not necessarily considered in conventional studies of landscape fragmentation, but may be a greater factor affecting survival and population persistence than previously thought (Woodward et al. 2001, Fuhlendorf et al. 2002). Permanent cross-fencing, especially for cell-type grazing systems, is common in the mixed-grass prairie. In a single section (259 ha or 640 acres), a pasture can include 6.5 km (4.0 mile) of boundary or exterior fencing, meaning a prairie chicken would encounter a fence less than once every km on average. In an eight-pasture cell system of the same size, the amount of fences would more than double to 14.2 km (8.8 mile) and increase the potential encounter rate. Habitat fragmentation can affect population genetics via reduced gene flow and an increased risk of inbreeding depression and population viability via reduced recruitment or immigration (Fahrig 2003, Pruett et al. 2011).

We posit that the negative association with some native grasses (Larsson et al. 2013), such as gramas (*Bouteloua* spp.), is an effect of physiognomy rather than floristics, especially in areas where fire suppression has allowed ground-level vegetation to become dense (Fuhlendorf and Engle 2001), limiting sight and terrestrial movements of Lesser Prairie-Chickens. It is increasingly accepted that the Mixed-Grass Prairie Ecoregion is being converted to woodlands and forest by fire suppression and alterations of a natural fire regime (Engle et al. 2008, Van Auken 2009). Prescribed fire has the potential to restore many prairie ecosystems, and fire and grazing can be managed to interact in a way that simulates random historical cycles that created habitat heterogeneity and provide the foundation for biological diversity and ecosystem function of North American grasslands (Fuhlendorf et al. 2009). The concept of "pyric herbivory" promotes floristic and structural diversity across the landscape and across the entire disturbance gradient from undisturbed to heavily disturbed, creating a gradient of habitats that are needed by Lesser Prairie-Chickens in different seasons (Fuhlendorf et al. 2009). Prescriptions of patch burning have a larger effect on heterogeneity of vegetation structure in the mixed-grass region than on sites dominated by short grasses (Winter et al. 2011).

Population Viability

Habitat loss and fragmentation create a patchwork of suitable habitat in which a spatially structured population or metapopulation is divided up into smaller subpopulations. Connectivity among subpopulations is crucial for population viability. As populations and occupied habitats shrink in numbers or area, population persistence is vulnerable to increases in stochastic variation in births and deaths and impacts of catastrophic events, such as severe weather or disease outbreak. Fewer individuals increase the probability that close relatives will interbreed. Some researchers have questioned the relative importance of genetic effects of inbreeding depression compared to potential vulnerability to environmental and demographic effects (Keller and Waller 2002), but pervasive evidence to the contrary has accumulated. Inbreeding depression negatively affects individual performance and reduces survival, reproduction, and resistance to environmental stress (Crnokrak and Roff 1999, Keller and Waller 2002, Armbruster and Reed 2005). A relict population of Greater Prairie-Chickens in Illinois suffered concurrent declines in population size, fitness, and genetic diversity (Westemeier et al. 1998). Poor productive performance manifested through lower fertility and hatching rates prompted a translocation program to genetically rescue the population and increase reproduction (Westemeier et al. 1998). Periodic translocations continue because isolated populations of small size and inadequate habitat are otherwise not viable.

Lesser Prairie-Chickens in the Mixed-Grass Prairie Ecoregion retain high levels of genetic diversity (Van Den Bussche et al. 2003, Hagen et al. 2010, Corman 2011, Pruett et al. 2011; Chapter 5, this volume), although the available data cannot distinguish between retention of historical diversity and recent gene flow. A

significant pattern of isolation by distance occurs where genetic differences increase with geographic distance from Oklahoma into Kansas and Colorado (Hagen et al. 2010). Estimates of genetic effective population size (N_e) were low for both northeastern Texas and Oklahoma (Corman 2011, Pruett et al. 2011). The N_e estimates suggest that the maintenance of evolutionary potential may be compromised (Corman 2011, Pruett et al. 2011). On the basis of demographic parameters, only a third (34.1%) of Lesser Prairie-Chickens in Oklahoma contributed genes to the next generation (Pruett et al. 2011). If N_e remains low, genetic diversity loss will lead to decreased ability to adapt and increased risk of inbreeding depression in the center of the geographic distribution of Lesser Prairie-Chickens.

Climate Change

Increased variability in temperature and rainfall are predicted consequences of global climate change in the central United States (Knapp et al. 2002, Schär et al. 2004). Shrub cover may decrease with increased aridity of the Southern Great Plains given that shrubs require more moisture than grasses for growth and reproduction (Knoop and Walker 1985, Harrington 1991, Seager et al. 2007). For example, seasonal precipitation is projected to either decrease (Seager et al. 2007) or become more variable in the Great Plains (Knapp et al. 2002). How the Lesser Prairie-Chicken will respond to changes in precipitation is unclear—loss of shrub cover affects both nest site selection and adult survival (Johnson et al. 2004, Patten et al. 2005b, Patten and Kelly 2010), whereas chick survival decreases with increased precipitation (Fields et al. 2006). In general, the loss of vegetative cover is likely to have a negative effect on the reproduction of prairie grouse (Flanders-Wanner et al. 2004). We predict that the survival of adults and juveniles will decrease both as heat stress increases from rising temperatures and precipitation becomes more sporadic or scant. Field data, physiological experiments, and mathematical modeling are needed to explore the effects of climate change in the mixed-grass prairie, but models of climate change in Sand Shinnery Oak Ecoregion predict that nest survival will fall below the threshold for population persistence (Grisham et al. 2013; Chapter 12, this volume).

MANAGEMENT AND CONSERVATION IN THE MIXED-GRASS PRAIRIE

Habitat Needs

The life cycle of Lesser Prairie-Chickens requires vast expanses of relatively intact prairie with a cluster of leks and core habitat complex requiring ~10,000 ha (25,000 acres) of high-quality habit (Van Pelt et al. 2013). Accordingly, the effect of each additional form of habitat fragmentation is magnified, whether from energy development, habitat loss, fencing, or the invasion of woody plants. Increased fragmentation likely inhibits crucial connectivity among subpopulations of Lesser Prairie-Chickens. Patch occupancy is associated with interpatch distance among other grouse species, suggesting that increased distances can limit dispersal (Fritz 1979). The management of the mixed-grass prairie through grazing and prescribed fire such that the vegetation community is maintained in middle to late seral stages of plant succession (native tall grasses, forbs, and legumes) interspersed with early stages of plant succession (native annual forbs) is optimal for Lesser Prairie-Chickens. A prescribed fire regime of burning 20%–30% of a management unit in a given year results in a 3–5-year fire-return interval, which provides quality nesting and brood-rearing habitat, while preventing invasion of undesirable woody plants like eastern red-cedar (Elmore et al. 2009). Rotational grazing systems for cattle, such as patch-burn grazing or use of salt licks, have been developed to simulate historical grazing patterns by large herbivores such as the American bison (*Bison bison*) and elk (*Cervus elaphus*). Similarities in grazing behavior are incomplete because wild ungulates could move freely across the landscape to graze sites with the highest-quality forage and additional fencing can affect mortality rates of prairie chickens (Wolfe et al. 2007, 2009). Historical accounts and contemporary research indicate that given a choice, grazing animals are attracted to new growth in a recently burned or grazed area and remain in suitable patches until higher quality forage is available (Fuhlendorf and Engle 2001).

The idea of providing quality forage for livestock grazing is consistent with improving habitat quality for Lesser Prairie-Chickens. Nesting females typically favor areas with more sand sagebrush canopy, a vertical structure of perennial bunch grass, and a low extent of bare ground relative to availability in surrounding grassland (Patten et al. 2005b, Pitman et al. 2005, Hagen et al. 2013), and successful nests usually have even greater shrub and grass cover (Lyons et al. 2011). In regions where sagebrush or other shrubs are sparse or nonexistent, bunch grasses alone may provide adequate nesting cover if grazing is excluded or conducted at a light to moderate intensity. CRP fields usually lack a shrub component, but can be used for nesting by Lesser Prairie-Chickens because ungrazed bunch grasses can provide adequate nesting structure (Sutton Avian Research Center, unpubl. data). Grazing affects prairies by changing the amount, kind, and pattern of residual grass in the system (Milchunas et al. 1988). If stocking rates are light to moderate (Klipple and Bement 1961), seasonal or year-round grazing can create an interspersion of short grass, forbs, bare ground, and tall, lightly grazed bunches of grass with a structural diversity that provides easily traversed lanes for broods and abundant insects (Hagen et al. 2005b). Patch burning and resulting patch grazing also produce such variation (Fuhlendorf et al. 2006), as well as interspersed forbs, which leads to more invertebrates (Jamison et al. 2002). Hagen et al. (2005b) hypothesized that site differences in body mass and survival of Lesser Prairie-Chicken chicks at two study areas in Kansas were likely related to local differences in invertebrate abundance. Grasslands subjected to light to moderate stocking rates and pyric herbivory may produce more insects and other food items with greater habitat diversity than either ungrazed or heavily grazed areas (Doxon et al. 2011).

In areas of the mixed-grass prairie in Oklahoma and Texas, exotic plants have been introduced to provide forage or erosion control, including CRP fields that were established at the start of the program. Common examples include Old World bluestem (*Bothriochloa ischaemum*), weeping lovegrass (*Eragrostis curvula*), Bermuda grass (*Cynodon dactylon*), and Johnsongrass (*Sorghum halepense*). None of the exotic species provide more benefit than native grasses for Lesser Prairie-Chickens, and invasive grasses have often outcompeted native vegetation to create exotic monocultures with little ecological value. The control of exotic species may require the use of herbicides, but we suggest that herbicide use be restricted to the target plants or fields, and never for native mixed-grass prairie where desirable grasses and forbs, and arthropods that depend on them, may be affected negatively. Lesser Prairie-Chickens in sand shinnery oak prairie treated with herbicides had lower body mass and reduced fat reserves compared to birds collected at untreated sites. Birds from treated areas also differed in the morphology of the small intestine, cecum, and gizzard because the diet includes more foliage and fewer insects or acorns (Olawsky 1987).

When making management decisions for a given patch of habitat, we advocate greater effort to use "check lists" or general management guides for the evaluation of habitat suitability for Lesser Prairie-Chicken habitats (Elmore et al. 2009, Van Pelt et al. 2013). Elmore et al. (2009) provided grass height and shrub density recommendations specifically for Oklahoma, and Van Pelt et al. (2013) adapted the recommendations for the species' entire range. Additionally, vegetation requirements developed for the Greater Prairie-Chicken in Kansas could provide a preliminary set of guidelines that could be adapted for Lesser Prairie-Chickens in the mixed-grass prairie (Horak 1985).

Land Use Policies

It will be imperative that wind turbines and associated high-voltage transmission lines are located in sites that will avoid direct and indirect impacts to mixed-grass prairie occupied by Lesser Prairie-Chickens. Siting guidelines may include a setback distance of at least 1 km as suggested by Pruett et al. (2009b), but the greater distances of 8 km suggested by the U.S. Fish and Wildlife Service for siting of wind power facilities might be more effective (Manville 2004). Siting guidelines will likely vary with terrain and habitat configuration, so an interactive scheme should be developed for case-by-case assessment of impacts.

Lesser Prairie-Chickens have apparently responded positively to CRP plantings in Kansas and Oklahoma, especially native mix plantings. Where the forb or shrub component is sparse or nonexistent, overseeding with alfalfa has provided structural diversity that facilitates the better use of CRP fields by prairie chickens (R. Rodgers, pers. comm.). We nonetheless suggest that alfalfa seeding should be viewed as a short-term solution and that more emphasis ought to be placed on establishing sand sagebrush as well as other native shrubs and forbs. As CRP

contracts expire, fields demonstrated to be suitable habitat for prairie chickens and sites located adjacent to native rangeland should be renewed, and future Farm Bills need to account for those designations. Alternatively, financial support from mitigation efforts could be used to provide financial incentives for landowners to allow expiring CRP fields to remain in native vegetation.

The mixed-grass prairie is a fire-adapted ecosystem. Fire removes rank vegetation, slows or inhibits invasion of invasive woody plants, limits size and structure of shrubs, and provides a rapid response of grass and forb growth that often increases arthropod abundance (Swengel et al. 2001). The location and size of a burned patch in relationship to adjacent unburned patches are critical to ensure that adequate habitats for nests, cover, and forage are available. Fire plans should follow a 5-year return interval, with 20%–30% of a management unit burned in a given year, a type of patch burning that increases plant and animal diversity without negatively affecting livestock production (Fuhlendorf et al. 2006). Importantly, burns can be reconfigured and relocated depending on needs and conditions. In areas where fire alone is insufficient to control the spread of invasive woody plants or density of native shrubs, the use of mechanical thinning by roller chopping or herbicides may be needed (Haukos 2011). However, we caution that widespread application of recommended herbicide treatment rates may reduce the shrub component below a threshold that the species perceives as suitable habitat (~20% shrub cover; Patten and Kelly 2010). Shrubs also provide the benefit of being a more persistent cover type, especially during periods of drought, than herbaceous cover.

Female Lesser Prairie-Chickens often move considerable distances to suitable nesting habitat (Hagen and Giesen 2005). For example, the mean distance between lek of capture and nest sites in Oklahoma was 3–4 km, with a maximum distance of 22 km (Larsson et al. 2013). In areas with high levels of fragmentation, female movement may be even more pronounced and risk of fence collision may increase with more obstructions (Patten et al. 2011). Over much of northwestern Oklahoma, county roads are laid out along section lines, and most pastures are fenced in quarter sections of 65 ha (160 acres) in a pattern that results in 3.7 linear km of fences per km². A bird moving through the landscape would encounter a fence line every 800 m (0.5 mile) or twice as often than if the pasture was fenced as a full section. Unnecessary fences in Oklahoma, such as those surrounding CRP fields, should be removed, and the probability of fence collisions should be taken under consideration before advocating any further cross-fencing of pastures. One alternative is the use of electric fences to control grazing pressure, as lower fence heights of 50–60 cm can be used to restrict livestock movements. If permanent barbed wire fences must remain, a prudent management action would be to mark them so that they are more visible to flying Lesser Prairie-Chickens (Wolfe et al. 2009). No data are available regarding the effectiveness of fence marking to reduce mortality in the Lesser Prairie-Chicken, but similar methods have reduced collision rates by 50% in Red Grouse (*Lagopus lagopus scoticus*), by 64% in Western Capercaillie (*Tetrao urogallus*), by 91% in Black Grouse (*Tetrao tetrix*) in Scotland (Baines and Andrew 2003), and by 90% in Greater Sage-Grouse in the western United States (Stevens et al. 2012). However, marking is costly and would likely exceed beyond what most landowners could accomplish without subsidies. Additionally, the vast linear distances of fences in the occupied or potential range of Lesser Prairie-Chickens makes marking all fences impractical, but efforts could be focused on areas where collision risks are highest, such as near leks or known foraging areas. State and federal agencies should consider subsidizing landowners who wish to mark their fences.

In 2010, the Oklahoma Field Office of the Bureau of Land Management, which controls the minerals in parts of the Lesser Prairie-Chicken range in Oklahoma and Kansas, modified their "conditions of approval" for oil and gas producers that drill on BLM-controlled areas, to either mark 6.4 km (4 miles) of fence or pay a mandatory mitigation fee to cover the expense of removing or marking fences off-site. The policy has generated mitigation fees that have allowed for the marking of over 300 km of fences in suitable range of Lesser Prairie-Chickens. Voluntary mitigation for wind power facility development, transmission lines, and oil/gas development, as outlined in Horton et al. (2010) and Van Pelt et al. (2013) have led to the creation of new funds for acquisition, maintenance, and enhancement of critical habitats. Mitigation funds are a promising development for the future and could allow for tens of thousands of additional hectares to be acquired or enhanced for future conservation of Lesser Prairie-Chickens.

ACKNOWLEDGMENTS

Much of the data used in this chapter resulted from the research funded by the Oklahoma Department of Wildlife Conservation, the U.S. Fish and Wildlife Service, and Sutton Avian Research Center supporters. D. Haukos, G. Kramos, and A. Flanders reviewed an earlier version of this chapter and provided valuable recommendations to improve the final version.

LITERATURE CITED

Applegate, R. D., and T. Z. Riley. 1998. Lesser Prairie-Chicken management. Rangelands 20:13–15.

Armbruster, P., and D. H. Reed. 2005. Inbreeding depression in benign and stressful environments. Heredity 95:235–242.

Baines, D., and M. Andrew. 2003. Marking of deer fences to reduce frequency of collisions by woodland grouse. Biological Conservation 110:169–176.

Bell, L. A., S. D. Fuhlendorf, M. A. Patten, D. H. Wolfe, and S. K. Sherrod. 2010. Lesser Prairie-Chicken hen and brood habitat use on sand shinnery oak. Rangeland Ecology and Management 63:478–486.

Bidwell, T. G., J. R. Weir, and D. M. Engle. 2002. Eastern redcedar control and management—Best Management Practices to Restore Oklahoma's Ecosystems. Oklahoma Cooperative Extension Service NREM-2876, Oklahoma State University, Stillwater, OK.

Blickley, J. L., D. Blackwood, and G. L. Patricelli. 2012. Experimental evidence for the effects of chronic anthropogenic noise on abundance of Greater Sage-Grouse at leks. Conservation Biology 26:461–471.

Bliss, N. B. 1989. A National Natural Resource Data Base: techniques for Linking the Major Land Resource Area map, the 1982 National Resources Inventory and the Soils Interpretations Record Data Bases in a Geographic Information System. Eros Data Center, Sioux Falls, SD.

Briggs, J. M., G. A. Hoch, and L. C. Johnson. 2002. Assessing the rate, mechanisms, and consequences of the conversion of tallgrass prairie to *Juniperus virginiana* forest. Ecosystems 5:578–586.

Bumann, G. B., and D. F. Stauffer. 2002. Scavenging of Ruffed Grouse in the Appalachians: influences and implications. Wildlife Society Bulletin 30:853–860.

Cannon, R. W., and F. L. Knopf. 1979. Lesser Prairie Chicken responses to range fires at the booming ground. Wildlife Society Bulletin 7:44–46.

Copelin, F. F. 1963. The Lesser Prairie Chicken in Oklahoma. Oklahoma Department of Wildlife Conservation Technical Bulletin 6:1–58.

Corman, K. S. 2011. Conservation and landscape genetics of Texas Lesser Prairie-Chicken: population structure and differentiation, genetic variability, and effective size. M.S. thesis, Texas A&M University–Kingsville, Kingsville, TX.

Crnokrak, P., and D. A. Roff. 1999. Inbreeding depression in the wild. Heredity 83:260–270.

Davis, D. M. 2009. Nesting ecology and reproductive success of Lesser Prairie-Chickens in shinnery oak-dominated rangelands. Wilson Journal of Ornithology 121:322–327.

Davison, V. E. 1940. An 8-year census of Lesser Prairie Chickens. Journal of Wildlife Management 4:55–62.

Dhillion, S. S., M. A. Mcginley, C. F. Friese, and J. C. Zak. 1994. Construction of sand shinnery oak communities of the Llano Estacado: animal disturbances, plant community structure, and restoration. Restoration Ecology 2:51–60.

Doxon, E. D., C. A. Davis, S. D. Fuhlendorf, and S. L. Winter. 2011. Aboveground macroinvertebrate diversity and abundance in sand sagebrush prairie managed with the use of pyric herbivory. Rangeland Ecology and Management 64:394–403.

Drake, B., and P. Todd. 2002. [online]. A strategy for control and utilization of invasive juniper species in Oklahoma—Final Report of the "Redcedar Task Force". Oklahoma Department of Agriculture, Food, and Forestry, Oklahoma City, OK. <http://www.forestry.ok.gov/Websites/forestry/Images/rcstf.pdf> (8 April 2014).

Drummond, M. A. 2007. Regional dynamics of grassland change in the western Great Plains. Great Plains Research 17:133–144.

Elmore, D., T. Bidwell, R. Ranft, and D. H. Wolfe. 2009. Habitat evaluation guide for the Lesser Prairie-Chicken. Oklahoma Cooperative Extension Service E-1014, Oklahoma State University, Stillwater, OK.

Engle, D. M., B. R. Coppedge, and S. D. Fuhlendorf. 2008. From the Dust Bowl to the green glacier: human activity and environmental change in Great Plains grasslands. Pp. 253–271 in O. W. Van Auken (editor), Western North American Juniperus communities: a dynamic vegetation type, Springer, New York, NY.

Fahrig, L. 2003. Effects of habitat fragmentation on biodiversity. Annual Review of Ecology, Evolution, and Systematics 34:487–515.

Fields, T. L., G. C. White, W. C. Gilgert, and R. D. Rodgers. 2006. Nest and brood survival of Lesser Prairie-Chickens in west central Kansas. Journal of Wildlife Management 70:931–938.

Flanders-Wanner, B. L., G. C. White, and L. L. McDaniel. 2004. Weather and prairie grouse: dealing with effects beyond our control. Wildlife Society Bulletin 32:22–34.

Fleharty, E. D. 1995. Wild animals and settlers on the Great Plains. University of Oklahoma Press, Norman, OK.

Fonstad, M., W. Pugatch, and B. Vogt. 2003. Kansas is flatter than a pancake. Annals of Improbable Research 9:16–17.

Fritz, R. S. 1979. Consequences of insular population structure: distribution and extinction of spruce grouse populations. Oecologia 42:57–65.

Fuhlendorf, S. D., and D. M. Engle. 2001. Restoring heterogeneity on rangelands: ecosystem management based on evolutionary grazing patterns. BioScience 51:625–632.

Fuhlendorf, S. D., D. M. Engle, J. Kerby, and R. Hamilton. 2009. Pyric herbivory: rewilding landscapes through the recoupling of fire and grazing. Conservation Biology 23:588–598.

Fuhlendorf, S. D., W. C. Harrell, D. M. Engle, R. G. Hamilton, C. A. Davis, and D. M. Leslie Jr. 2006. Should heterogeneity be the basis for conservation? Grassland bird response to fire and grazing. Ecological Applications 16:1706–1716.

Fuhlendorf, S. D., A. J. W. Woodward, D. M. Leslie, Jr., and J. S. Shackford. 2002. Multi-scale effects of habitat loss and fragmentation on Lesser Prairie-Chicken populations. Landscape Ecology 17:617–628.

Gilpin, M. E., and M. E. Soulé. 1986. Minimum viable populations: processes of extinction. Pp. 19–34 in M. E. Soulé (editor), Conservation biology: the science of scarcity and diversity, Sinauer Associates, Sunderland, MA.

Grisham, B. A., C. W. Boal, D. A. Haukos, D. M. Davis, K. K. Boydston, C. Dixon, and W. R. Heck. 2013. The predicted influence of climate change on Lesser Prairie-Chicken reproductive parameters. PLoS One 8:e68225.

Hagen, C. A., and K. M. Giesen. [online]. 2005. Lesser Prairie-Chicken (Tympanuchus pallidicinctus). No. 364 in A. Poole (editor), The Birds of North America Online, Cornell Lab of Ornithology, Ithaca, NY. <http://bna.birds.cornell.edu/bna/species/364> (9 November 2013).

Hagen, C. A., B. A. Grisham, C. W. Boal, and D. A. Haukos. 2013. A meta-analysis of Lesser Prairie-Chicken nesting and brood-rearing habitats: implications for habitat management. Wildlife Society Bulletin 37:750–758.

Hagen, C. A., B. E. Jamison, R. J. Robel, and R. D. Applegate. 2002. Ring-necked Pheasant parasitism of Lesser Prairie-Chicken nests in Kansas. Wilson Bulletin 114:522–524.

Hagen, C. A., J. C. Pitman, T. M. Loughin, B. K. Sandercock, R. J. Robel, and R. D. Applegate. 2011. Impacts of anthropogenic features on habitat use by Lesser Prairie-Chickens. Pp. 63–75 in B. K. Sandercock, K. Martin, and G. Segelbacher (editors), Ecology, conservation, and management of grouse. Studies in Avian Biology (no. 39), University of California Press, Berkeley, CA.

Hagen, C. A., J. C. Pitman, B. K. Sandercock, R. J. Robel, and R. D. Applegate. 2005a. Age-specific variation in apparent survival rates of male Lesser Prairie-Chickens. Condor 107:78–86.

Hagen, C. A., J. C. Pitman, B. K. Sandercock, R. J. Robel, and R. D. Applegate. 2007. Age-specific survival and probable causes of mortality in female Lesser Prairie-Chickens. Journal of Wildlife Management 71:518–525.

Hagen, C. A., J. C. Pitman, B. K. Sandercock, D. H. Wolfe, R. D. Applegate, and S. J. Oyler-McCance. 2010. Regional variation in mtDNA of the Lesser Prairie-Chicken. Condor 112:29–37.

Hagen, C. A., G. C. Salter, J. C. Pitman, R. J. Robel, and R. D. Applegate. 2005b. Lesser Prairie-Chicken brood habitat in sand sagebrush: invertebrate biomass and vegetation. Wildlife Society Bulletin 33:1080–1091.

Hagen, C. A., B. K. Sandercock, J. C. Pitman, R. J. Robel, and R. D. Applegate. 2009. Spatial variation in Lesser Prairie-Chicken demography: a sensitivity analysis of population dynamics and management alternatives. Journal of Wildlife Management 73:1325–1332.

Harrington, G. N. 1991. Effects of soil moisture on shrub seedling survival in a semi-arid grassland. Ecology 72:1138–1149.

Haukos, D. A. 2011. Use of tebuthiuron to restore sand shinnery oak grasslands of the Southern High Plains. Pp. 103–124 in M. Naguib and A. E. Hasaneen (editors), Herbicide: mechanisms and mode of action, Intech, Rijeka, Croatia.

Holt, R. D. 2012. Breeding season demographics of a Lesser Prairie-Chicken (Tympanuchus pallidicinctus) population in the northeast Texas Panhandle. Ph.D. dissertation, Texas Tech University, Lubbock, TX.

Holt, R. D., M. J. Butler, W. B. Ballard, C. A. Kukal, and H. Whitlaw. 2010. Disturbance of lekking Lesser Prairie-Chickens (Tympanuchus pallidicinctus) by Ring-necked Pheasants (Phasianus colchicus). Western North American Naturalist 70:241–244.

Horak, G. J. 1985. Kansas prairie chickens. Wildlife Bulletin No. 3, Kansas Forest, Fish, and Game Commission, Pratt, KS.

Horton, R. E. 2000. Distribution and abundance of Lesser Prairie-Chicken in Oklahoma. Prairie Naturalist 32:189–195.

Horton, R., L. Bell, C. M. O'Meilia, M. McLachlan, C. Hise, D. Wolfe, D. Elmore, and J. D. Strong. 2010. A spatially-based planning tool designed to reduce negative effects of development on the Lesser Prairie-Chicken (Tympanuchus pallidicinctus) in Oklahoma: a multi-entity collaboration to promote Lesser Prairie-Chicken voluntary habitat conservation and prioritized management actions. Oklahoma Department of Wildlife Conservation, Oklahoma City, OK.

Hunt, J. L. 2004. Investigation into the decline of populations of the Lesser Prairie-Chicken (Tympanuchus pallidicinctus Ridgeway) in southeastern New Mexico. Ph.D. dissertation, Auburn University, Auburn, AL.

Hunt, J. L., and T. L. Best. 2010. Vegetative characteristics of active and abandoned leks of Lesser Prairie-Chickens (Tympanuchus pallidicinctus) in southeastern New Mexico. Southwestern Naturalist 55:477–487.

Jamison, B. E. 2000. Lesser Prairie-Chicken chick survival, adult survival, and habitat selection and movements of males in fragmented rangelands of southwestern Kansas. M.S. thesis, Kansas State University, Manhattan, KS.

Jamison, B. E., R. J. Robel, J. S. Pontius, and R. D. Applegate. 2002. Invertebrate biomass: associations with Lesser Prairie-Chicken habitat use and sand sagebrush density in southwest Kansas. Wildlife Society Bulletin 30:517–526.

Jarnevich, C. S., and M. K. Laubhan. 2011. Balancing energy development and conservation: a method utilizing species distribution models. Environmental Management 47:926–936.

Jensen, W. E., D. A. Robinson Jr., and R. D. Applegate. 2000. Distribution and population trend of Lesser Prairie-Chicken in Kansas. Prairie Naturalist 32:169–175.

Johnson, K., B. H. Smith, G. Sadoti, T. B. Neville, and P. Neville. 2004. Habitat use and nest site selection by nesting Lesser Prairie-Chickens in southeastern New Mexico. Southwestern Naturalist 49: 334–343.

Jones, R. E. 1963. Identification and analysis of Lesser and Greater Prairie Chicken habitat. Journal of Wildlife Management 27:757–778.

Jones, R. E. 1964. Habitat used by Lesser Prairie Chickens for feeding related to seasonal behavior of plants in Beaver County, Oklahoma. Southwestern Naturalist 9:111–117.

Jones, R. S. 2009. Seasonal survival, reproduction, and use of wildfire areas by Lesser Prairie-Chickens in the northeastern Texas panhandle. M.S. thesis, Texas A&M University, College Station, TX.

Keller, L. F., and D. M. Waller. 2002. Inbreeding effects in wild populations. Trends in Ecology and Evolution 17:230–241.

Klipple, G. E., and R. E. Bement. 1961. Light grazing—Is it economically feasible as a range improvement practice? Journal of Range Management 14:57–62.

Knapp, A. K., P. A. Fay, J. M. Blair, S. L. Collins, M. D. Smith, J. D. Carlisle, C. W. Harper, B. T. Danner, M. S. Lett, and J. K. McCarron. 2002. Rainfall variability, carbon cycling, and plant species diversity in a mesic grassland. Science 298:2202–2205.

Knoop, W. T., and B. H. Walker. 1985. Interactions of woody and herbaceous vegetation in a southern African savanna. Journal of Ecology 73:235–253.

Larsson, L. C., C. L. Pruett, D. H. Wolfe, and M. A. Patten. 2013. Fine-scale selection of habitat by the Lesser Prairie-Chicken. Southwestern Naturalist 58:135–149.

Ligon, J. S. 1951. Prairie chickens, highways, and powerlines. New Mexico Magazine 9:29.

Lincoln, F. C. 1950. A Ring-necked Pheasant × prairie chicken hybrid. Wilson Bulletin 62:210–212.

Lindenmayer, R. B., N. C. Hansen, J. Brummer, and J. G. Pritchett. 2011. Deficit irrigation of alfalfa for water-savings in the Great Plains and Intermountain West: a review and analysis of the literature. Agronomy Journal 103:45–50.

Lomolino, M. V., and G. A. Smith. 2003. Terrestrial vertebrate communities at black-tailed prairie dog (Cynomys ludovicianus) towns. Biological Conservation 115:89–100.

Lyons, E. K., R. S. Jones, J. P. Leonard, B. E. Toole, R. A. McCleery, R. R. Lopez, M. J. Peterson, S. J. DeMaso, and N. J. Silvy. 2011. Regional variation in nest success of Lesser Prairie-Chickens in Texas. Pp. 223–231 in B. K. Sandercock, K. Martin, and G. Segelbacher (editors), Ecology, conservation, and management of grouse, Studies in Avian Biology (no. 39), University of California Press, Berkeley, CA.

Manville, A. M., II. 2004. Prairie grouse leks and wind turbines: U.S. Fish and Wildlife Service justification for a 5-mile buffer from leks; additional grassland songbird recommendations. Division of Migratory Bird Management, U.S. Fish and Wildlife Service, Arlington, VA.

McCleery, R. A., R. R. Lopez, and N. J. Silvy. 2007. Transferring research to endangered species management. Journal of Wildlife Management 71:2134–2141.

McDonald, L., G. Beauprez, G. Gardner, J. Griswold, C. Hagen, D. Klute, S. Kyle, J. Pitman, T. Rintz, and B. Van Pelt. 2014. Range-wide population size of the Lesser Prairie-Chicken: 2012 and 2013. Wildlife Society Bulletin 38:536–546.

Milchunas, D. G., O. E. Sala, and W. K. Lauenroth. 1988. A generalized model of the effects of grazing by large herbivores on grassland community structure. American Naturalist 132:87–106.

Olawsky, C. D. 1987. Effects of shinnery oak control with tebuthiuron on Lesser Prairie-Chicken populations. M.S. thesis, Texas Tech University, Lubbock, TX.

Patten, M. A., and J. F. Kelly. 2010. Habitat selection and the perceptual trap. Ecological Applications 20:2148–2156.

Patten, M. A., C. L. Pruett, and D. H. Wolfe. 2011. Home range size and movements of Greater Prairie-Chickens. Studies in Avian Biology 39:51–62.

Patten, M. A., D. H. Wolfe, E. Shochat, and S. K. Sherrod. 2005a. Effects of microhabitat and microclimate on adult survivorship of the Lesser Prairie-Chicken. Journal of Wildlife Management 36:1270–1278.

Patten, M. A., D. H. Wolfe, E. Shochat, and S. K. Sherrod. 2005b. Habitat fragmentation, rapid evolution, and population persistence. Evolutionary Ecology Research 7:235–249.

Peterle, T. J. 1951. Intergeneric galliform hybrids: a review. Wilson Bulletin 63:219–224.

Peterson, D. L., A. M. Stewart, S. L. Egbert, C. M. Lauver, E. A. Artinko, K. P. Price, and S. Park. 2001. Identifying and understanding land use/land cover changes in Kansas. Proceedings of ASPRS 2001: Gateway to the New Millennium, St. Louis, American Society for Photogrammetry & Remote Sensing, Bethesda, MD.

Pitman, J. C., C. A. Hagen, B. E. Jamison, R. J. Robel, T. M. Loughin, and R. D. Applegate. 2006. Survival of juvenile Lesser Prairie-Chickens in Kansas. Wildlife Society Bulletin 34:675–681.

Pitman, J. C., C. A. Hagen, R. J. Robel, T. M. Loughin, and R. D. Applegate. 2005. Location and success of Lesser Prairie-Chicken nests in relation to vegetation and human disturbance. Journal of Wildlife Management 69:1259–1269.

Pruett, C. L., J. A. Johnson, L. C. Larsson, D. H. Wolfe, and M. A. Patten. 2011. Low effective population size constraints rapid demographic evolution in a grassland grouse. Conservation Genetics 12:1205–1214.

Pruett, C. L., M. A. Patten, and D. H. Wolfe. 2009a. It's not easy being green: wind energy and a declining grassland bird. BioScience 59:257–262.

Pruett, C. L., M. A. Patten, and D. H. Wolfe. 2009b. Avoidance behavior of prairie grouse: implications for wind energy development. Conservation Biology 23:1253–1259.

Rao, M., G. Fan, J. Thomas, G. Cherian, V. Chudiwale, and M. Awawdeh. 2007. A web-based GIS Decision Support System for managing and planning USDA's Conservation Reserve Program (CRP). Environmental Modelling & Software 22:1270–1280.

Ripper, D., M. McLachlan, T. Toombs, and T. VerCauteren. 2008. Assessment of Conservation Reserve Program fields within the current distribution of Lesser Prairie-Chicken. Great Plains Research 18:205–218.

Robel, R. J., J. A. Harrington Jr., C. A. Hagen, J. C. Pitman, and R. R. Reker. 2004. Effect of energy development and human activity on the use of sand sagebrush habitat by Lesser Prairie-Chickens in southwestern Kansas. Transactions of the North American Wildlife and Natural Resources Conference 69:251–266.

Rodgers, R. 2009. A summary of population and habitat changes, and conservation of the Lesser Prairie-Chicken in Kansas. Kansas Department of Wildlife and Parks, Hays, KS.

Salter, G. C., R. J. Robel, and K. E. Kemp. 2005. Lesser Prairie-Chicken use of harvested corn fields during fall and winter in southwestern Kansas. Prairie Naturalist 37:1–9.

Samson, F. B., F. L. Knopf, and W. R. Ostlie. 2004. Great Plains ecosystems: past, present, and future. Wildlife Society Bulletin 32:6–15.

Schär, C., R. L. Vidale, D. Lüthi, C. Frei, C. Häberli, M. A. Liniger, and C. Appenzeller. 2004. The role of increasing temperature variability in European summer heatwaves. Nature 427:332–336.

Seager, R., M. Ting, I. Held, Y. Kushnir, J. Lu, G. Vecchi, H.-P. Huang, N. Harnik, A. Leetmaa, N. C. Lau, C. H. Li, J. Velez, and N. Naik. 2007. Model projections of an imminent transition to a more arid climate in southwestern North America. Science 316:1181–1184.

Stevens, B. S., K. P. Reese, J. W. Connelly, and D. M. Musil. 2012. Greater Sage-Grouse and fences: does marking reduce collisions? Wildlife Society Bulletin 36:297–303.

Sullivan, R. M., J. P. Hughes, and J. E. Lionberger. 2000. Review of the historical and present status of the Lesser Prairie-Chicken (*Tympanuchus pallidicinctus*) in Texas. Prairie Naturalist 32:178–188.

Swengel, A. B. 2001. A literature review of insect responses to fire compared to other conservation managements of open habitat. Biodiversity and Conservation 10:1141–1169.

Taylor, M. A., and F. S. Guthery. 1980. Status, ecology, and management of the Lesser Prairie Chicken. USDA Forest Service General Technical Report RM-77, Rocky Mountain Forest and Range Experiment Station, Fort Collins, CO.

Van Auken, O. W. 2009. Causes and consequences of woody plant encroachment into western North American grasslands. Journal of Environmental Management 90:2931–2942.

Van Den Bussche, R. A., S. F. Hoofer, D. A. Wiedenfeld, D. H. Wolfe, and S. K. Sherrod. 2003. Genetic variation within and among fragmented populations of Lesser Prairie-Chickens (*Tympanuchus pallidicinctus*). Molecular Ecology 12:675–683.

Van Pelt, W. E., S. Kyle, J. Pitman, D. Klute, G. Beauprez, D. Schoeling, A. Janus, and J. Haufler. 2013. The Lesser Prairie-Chicken range-wide conservation plan. Western Association of Fish and Wildlife Agencies, Cheyenne, WY.

Westemeier, R. L., J. D. Brawn, S. A. Simpson, T. L. Esker, R. W. Jansen, J. W. Walk, E. L. Kershner, J. L. Bouzat, and K. N. Paige. 1998. Tracking and long-term decline and recovery of an isolated population. Science 282:1695–1698.

Winder, V. L., L. B. McNew, A. J. Gregory, L. M. Hunt, S. M. Wisely, and B. K. Sandercock. 2014. Effects of wind energy development on survival of female Greater Prairie-Chickens. Journal of Applied Ecology 51:395–405.

Winder, V. L., K. M. Carrlson, A. J. Gregory, C. A. Hagen, D. A. Haukos, D. C. Kesler, L. C. Larsson, T. W. Matthews, L. B. McNew, M. A. Patten, J. C. Pitman, L. A. Powell, J. A. Smith, T. Thompson, D. H. Wolfe, and B. K. Sandercock. 2015. Factors affecting female space use in ten populations of prairie chickens. Ecosphere 6:art166.

Winter, S. L., S. D. Fuhlendorf, C. L. Goad, C. A. Davis, and K. R. Hickman. 2011. Topoedaphic variability and patch burning in sand sagebrush shrubland. Rangeland Ecology and Management 64:633–640.

Wolfe, D. H., M. A. Patten, and S. K. Sherrod. 2003. Factors affecting nesting success and mortality of Lesser Prairie-Chickens in Oklahoma. Federal Aid in Wildlife Restoration Project W-146-R Final Report. Oklahoma Department of Wildlife Conservation, Oklahoma City, OK.

Wolfe, D. H., M. A. Patten, and S. K. Sherrod. 2009. Reducing grouse collision mortality by marking fences. Ecological Restoration 27:141–143.

Wolfe, D. H., M. A. Patten, E. Shochat, C. L. Pruett, and S. K. Sherrod. 2007. Causes and patterns of mortality in Lesser Prairie-Chickens and implications for management. Wildlife Biology 13(Suppl. 1):95–104.

Woodward, A. J., S. D. Fuhlendorf, D. M. Leslie Jr., and J. Shackford. 2001. Influence of landscape composition and change on Lesser Prairie-Chicken (*Tympanuchus pallidicinctus*) populations. American Midland Naturalist 145:261–274.

Zeiler, H. P. and V. Grünschachner-Berger. 2009. Impact of wind power plants on Black Grouse, *Lyrurus tetrix* in Alpine regions. Folia Zoologica 58:173–182.

CHAPTER SEVENTEEN

Ecology and Conservation of Lesser Prairie-Chickens in Sand Shinnery Oak Prairie*

Blake A. Grisham, Jennifer C. Zavaleta, Adam C. Behney, Philip K. Borsdorf, Duane R. Lucia, Clint W. Boal, and David A. Haukos

Abstract. Sand shinnery oak (*Quercus havardii*) prairie is a unique ecosystem endemic to sandy soils of eastern New Mexico, northwestern Texas, and western Oklahoma. The prairie system provides important habitat and overlaps with the historic and current distribution of Lesser Prairie-Chickens (*Tympanuchus pallidicinctus*). Populations of Lesser Prairie-Chickens in sand shinnery oak prairie of the Southern Great Plains have declined substantially since the late 1980s, following conversion of nesting and brood-rearing habitat to row crop agriculture and extended periods of drought. In addition to the universal threats throughout the species distribution, populations in sand shinnery oak are susceptible to a changing climate in an ecoregion that is already an extreme environment for ground-nesting birds. Recent studies of Lesser Prairie-Chicken ecology in sand shinnery oak prairie have expanded our knowledge on the ecology and management of the species, but a thorough review of the historic and current literature is lacking. Current management guidelines focus on Lesser Prairie-Chickens in mixed-grass and sand sagebrush prairie, but recommendations are lacking for the species in sand shinnery oak prairie. Different approaches might be required for management given the unique aspects of the vegetation community, relative ecosystem drivers, and environmental variation in sand shinnery oak prairie and recent listing of Lesser Prairie-Chickens as a threatened species under the U.S. Endangered Species Act. We provide a new synthesis of available information on the life history, habitat requirements and management, and population management for Lesser Prairie-Chickens in sand shinnery oak prairie. We also provide specific management suggestions for the species in the ecoregion and highlight current and future research needs. We focus on two recent long-term investigations into the ecology of Lesser Prairie-Chickens in the Sand Shinnery Oak Prairie Ecoregion including a 6-year population study of Lesser Prairie-Chickens in Roosevelt County, New Mexico, and Cochran, Hockley, Terry, and Yoakum counties, Texas, 2006–2012, and a 10-year vegetation dataset collected in Roosevelt County, New Mexico, 2001–2011.

Key Words: New Mexico, Southern High Plains, Texas, *Tympanuchus pallidicinctus*.

* Grisham, B. A., J. C. Zavaleta, A. C. Behney, P. K. Borsdorf, D. R. Lucia, C. W. Boal, and D. A. Haukos. 2016. Ecology and conservation of Lesser Prairie-Chickens in Sand Shinnery Oak Prairie. Pp. 315–344 in D. A. Haukos and C. W. Boal (editors), Ecology and conservation of Lesser Prairie-Chickens. Studies in Avian Biology (no. 48), CRC Press, Boca Raton, FL.

S and shinnery oak (*Quercus havardii*; hereafter shinnery oak) prairie is a unique ecosystem associated with sandy soils of the Southern Great Plains in eastern New Mexico, western Oklahoma, and northwestern Texas (Peterson and Boyd 1998, Haukos 2011; Figure 1.2). Shinnery oak prairie is associated with areas characterized by almost 100% sandy soils (Alfisols and Entisols in the Brownfield series; Peterson and Boyd 1998, Haukos 2011). The systems are believed to have evolved with environmental conditions associated with sandy soil deposits that were formed between 4,000 and 7,000 years ago (see Haukos 2011 for specific geomorphological formation). Shinnery oak can hybridize with post oak (*Q. stellata*) to form "mottes" or groves of taller trees (~6 m) embedded in a prairie community dominated by short stands of shinnery oak (~1 m), sand sagebrush (*Artemisia filifolia*), and mid- and tall grasses. Mottes are more common in the eastern portion of the distribution of shinnery oak (Peterson and Boyd 1998) and are likely the result of a precipitation gradient that runs west to east between xeric conditions in eastern New Mexico to arid conditions in western Oklahoma (Figure 12.9).

Shinnery oak–grassland communities provide an ecological gradient that follows spatial gradients in precipitation (Peterson and Boyd 1998). Plant communities in the western distribution are dominated by shinnery oak with an interspersion of grasses and sand sagebrush. Conversely, the eastern distribution of the plant community is a matrix of sand sagebrush–dominated areas interspersed with grasses and stands of shinnery oak and shinnery oak mottes (Peterson and Boyd 1998). The focus of our chapter is the shinnery oak prairie of the Southern High Plains of New Mexico and Texas. With the exception of state-owned Prairie Chicken Areas and federally owned Bureau of Land Management lands in New Mexico, the majority of shinnery oak prairie on the Southern High Plains is privately owned. The remaining patches of shinnery oak prairie have become isolated, relict communities because the surrounding short-grass prairie of the Southern High Plains having been converted to row crop agriculture or fragmented by oil and gas exploration and urban development (Peterson and Boyd 1998, Wester 2007, Haukos 2011). Peterson and Boyd (1998) estimated that shinnery oak prairie historically covered 6,070 km^2 in New Mexico and 14,000 km^2 in Texas. Estimates from Texas suggest that 5,000 km^2

were converted by 1972 and 1,300 km^2 were treated with the herbicide tebuthiuron (N-[5-(1,-dimethylethyl)-1,3,4-thiadiazol-2-yl]-N,N′-dimethylurea) by 1995 (Wester 2007), causing outright loss and degradation of shinnery oak prairie. Technological advances in irrigated row crop agriculture have led to recent losses of shinnery oak prairie habitat in eastern New Mexico and west Texas (D.A. Haukos, pers. obs.).

On the Southern High Plains, presettlement composition of shinnery oak in the extant vegetation community was estimated at 5%–25% (Haukos 2011), but decades of unmanaged grazing, fire suppression, and periods of intense drought have converted much of the remaining shinnery oak prairie to stands dominated by shinnery oak (~50%–100%, Peterson and Boyd 1998, Haukos 2011). Scientific opinion has debated the relative importance of shinnery oak prairie for Lesser Prairie-Chickens (*Tympanuchus pallidicinctus*; McCleery et al. 2007). Haukos (2011) suggested that habitats supported by sandy soils comprise the core of the historic and current population range on the Southern High Plains (Figure 1.2). The core habitat perspective is supported by population indices indicating that numbers of Lesser Prairie-Chickens in shinnery oak prairie have remained stable since the mid-1990s, compared to ongoing declines among birds in Mixed- and Short-Grass Prairie Ecoregions (Davis et al. 2008). Despite habitat conversion and degradation, shinnery oak prairie remains the primary vegetation community and dominant habitat for Lesser Prairie-Chickens on the Southern High Plains. Thus, conservation and management of Lesser Prairie-Chickens in shinnery oak prairie are a high priority, regardless of debate regarding historical and current distributions.

The Sand Shinnery Oak Prairie Ecoregion of the Southern High Plains represents the extreme southwest portion of the Lesser Prairie-Chicken distribution and is geographically disconnected from populations elsewhere in the species distribution (Hagen and Giesen 2005). The region is characterized by a hot, dry climate with relatively less fragmentation than the rest of the range of Lesser Prairie-Chickens (Patten et al. 2005a, Grisham et al. 2013). The Southern High Plains population faces unique challenges associated with an area that is representative of an extreme environment for ground-nesting birds (Saalfeld et al. 2012; Grisham et al. 2013, 2014). For example, the average annual precipitation in the ecoregion (~38–43 cm) is half of that in the northern extent of the birds' range (~76–86 cm). The majority of precipitation events occur as intense, localized convective thunderstorms between June and October (Newman 1964, Grisham 2012). During the breeding season, mean temperatures are 7°C higher and relative humidity is 7% lower at the onset of incubation compared to more northerly populations (30-year means for Lubbock, Texas versus Dodge City, Kansas; Chapter 12, this volume). Populations of Lesser Prairie-Chickens in shinnery oak prairie face threats that are common throughout the distribution of the species, including habitat destruction, habitat degradation, anthropogenic development, and invasive woody species. However, a unique feature of the ecoregion is that populations may be more susceptible to a changing climate in an ecosystem that already has extreme environment conditions (Grisham et al. 2013).

Populations of Lesser Prairie-Chickens in shinnery oak prairie in the Southern High Plains are one distinct genetic population (Corman 2011). The objectives for our chapter are threefold: (1) to synthesize available information on the life history, habitat requirements and management, and population management for Lesser Prairie-Chickens in the Sand Shinnery Oak Prairie Ecoregion; (2) to provide management suggestions and guidelines for Lesser Prairie-Chickens in sand shinnery oak; and (3) to highlight current and future research needs for prairie chickens for the southernmost ecoregion of the species range. We focus on two recent long-term investigations into the ecology of Lesser Prairie-Chickens in the Sand Shinnery Oak Prairie Ecoregion including a 6-year population study of Lesser Prairie-Chickens in Roosevelt County, New Mexico, and Cochran, Hockley, Terry, and Yoakum counties, Texas, 2006–2012, and a 10-year vegetation dataset collected in Roosevelt County, New Mexico, 2001–2011.

CHARACTERISTICS OF SHINNERY OAK PRAIRIE

Plant Associations

The community composition of shinnery oak prairie includes a matrix of shrubs (shinnery oak and sand sagebrush), various mid- to tall grasses, and forbs. Common and important grasses include sand bluestem (*Andropogon gerardii*

hallii), big bluestem (*A. geradii*), little bluestem (*Schizachyrium scoparium*), sand dropseed (*Sporobolus cryptandrus*), purple three-awn (*Aristida purpurea*), and sand paspalum (*Paspalum setaceum*, Pettit 1979, Woodward et al. 2001, Zavaleta 2012). Common forbs include silverleaf nightshade (*Solanum elaeagnifolium*), spectacled pod (*Dimorphocarpa wislizeni*), Indian blanket (*Gaillardia pulchella*), wooly locoweed (*Astragalus mollissimus*), common sunflower (*Helianthus annuus*), scarlet gaura (*Gaura coccinea*), and halfshrub sundrop (*Oenothera serrulata*). Vegetation taxonomy follows the guidelines, names, and descriptions in *The Flora of the Great Plains* (Barkley 1986).

Ecological Drivers

Major ecosystem drivers in shinnery oak prairie include drought, grazing, and fire (Peterson and Boyd 1998, Haukos 2011). Of the three, drought is the only driver that cannot be controlled by human activities. The magnitude and frequency of drought are highly variable in space, time, and magnitude (Chapters 12 and 13, this volume). Interactions between drought and other ecological processes can be important. For example, the success of herbicide application, prescribed burns, and other management actions in shinnery oak prairie can depend upon environmental conditions (Peterson and Boyd 1998, Haukos 2011). Drought ultimately controls community structure through multiple natural and human-related mechanisms.

Drought

Community recovery after a drought event depends on timing and duration of precipitation. Precipitation drives community structure and the ecology of Lesser Prairie-Chickens in shinnery oak prairie (Zavaleta 2012, Grisham et al. 2014). Climatic conditions in the winter months of November–March are important predictors of soil moisture and plant community structure. Above-average temperatures and below-average precipitation versus 30-year means had the largest and most negative influence on plant community structure and composition (Zavaleta 2012). La Niña events bring drought conditions to the Southern High Plains at ~5–10-year intervals but can have variable effects on community structure (Trenberth et al. 1988, Atlas et al. 1993, Chen and Newman 1998, Cook et al. 2004). The 2011 drought provides insight to how an extreme La Niña event affects shinnery oak prairie. For the 10.5-month period from October 15, 2010 to August 31, 2011, the total precipitation was 2.5 cm in Yoakum and Cochran counties, Texas, which was the worst drought and warmest La Niña event on record. The drought of 2011 was so severe that sand shinnery oak and grasses on the study site did not leaf out and failed to provide any vegetation cover.

Grazing

Shinnery oak prairie of the Southern High Plains coevolved with seasonal grazing by large mammals, principally American bison (Bison bison), and also pronghorn (Antilocapra americana), elk (Cervus canadensis), and mule deer (Odocoileus hemionus; Peterson and Boyd 1998). Today, the primary grazers are domestic cattle, which often graze for longer periods and at higher densities compared to historic patterns of grazing by nomadic species. High stocking densities and unmanaged grazing have transformed the composition and structure of much of the shinnery community (Lauenroth et al. 1994). Long-term, high-intensity grazing, especially during drought, causes landscape degradation as pressure on grasses leads to increases in shrub cover (Fuhlendorf 1999, Beck and Mitchell 2000, Derner et al. 2009, Haukos 2011). However, domestic livestock can function as ecosystem engineers to positively influence vegetation coverage and structure and improve habitat conditions for Lesser Prairie-Chickens (Derner et al. 2009). Cattle production is an important socioeconomic and cultural driver on the Southern High Plains, and it is imperative to understand the role of grazing as a tool for restoration and management (Ethridge et al. 1987, Fuhlendorf et al. 2012).

Fire

Fire was a major ecological process that influenced the structure, composition, and productivity in the prairie ecosystems of the Great Plains (Fuhlendorf and Engle 2004, Wester 2007, Haukos 2011). With implementation of high-intensity grazing by domestic livestock, the natural regime has been altered to favor drought-resistant, disturbance-resistant plants, including shinnery oak (Brockway et al. 2002). Fire is an essential disturbance process vital to sustaining long-term ecosystem health of many prairie systems, and reintroduction of fire into historical fire-adapted ecosystems can enhance species richness, diversity, and productivity (Brockway et al. 2002). Fire played a role within shinnery oak prairie, but return intervals were likely longer than in short-grass prairie because of the longer periods needed to accumulate a fuel load sufficient to carry a fire across patchy vegetation on sandy soils (Haukos 2011).

Historic Management

Ecological Site Description

A majority of remaining stands of shinnery oak prairie has been degraded to monotypic stands of shinnery oak (Hagen et al. 2004, Haukos 2011, Grisham 2012, Zavaleta 2012). Ecological Site Descriptions (ESD), as described by the Natural Resources Conservation Service of the U.S. Department of Agriculture (hereafter, NRCS), provide an estimate of the historic climax community composition. The estimate likely represents long-term averages for percent grass, shrubs, and forbs for shinnery oak prairie (Figure 17.1; NRCS 2011) and lacks a measure of temporal variation in relative composition because of the annual variation in precipitation (Zavaleta 2012). Under historic, natural conditions, grasses dominated shinnery oak communities in density and percent composition compared to shinnery oak (Moldenhauer et al. 1958). The majority of remaining stands of shinnery oak prairie have been degraded to extensive stands of shinnery oak (Hagen et al. 2004, Haukos 2011, Grisham 2012, Zavaleta 2012). Contemporary estimates of shinnery oak composition range between 50% and 90% in most stands (Biondini et al. 1986, Dhillion et al. 1994).

Use of Herbicides

In the past, the application of herbicides was the management tool that was commonly used to eradicate or reduce shinnery oak (Pettit 1979, Doerr and Guthery 1980, Peterson and Boyd 1998). Early research to control shinnery oak involved applications of phenoxy, propionic, benzoic, and picolinic acids (Greer et al. 1968, Peterson and

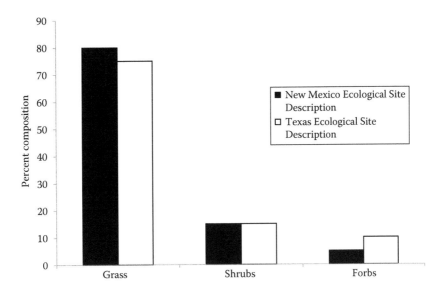

Figure 17.1. Standards for percent composition in Ecological Site Descriptions for sand shinnery oak prairie in New Mexico and Texas. (Data from Natural Resources Conservation Service 2013.)

TABLE 17.1

Herbicides used to reduce or eliminate sand shinnery oak in the Central United States before the development of tebuthiuron.

Herbicide type	Market names	Chemical composition	Result (%)	Reapplication (Years)[a]	Sources
Acetic acid	Numerous	2,4-Dichlorophenoxy-acetic acid	Top kill (85–95)	2–3[b]	Greer et al. (1968)
		2,4,5-Trichlorophenoxy-acetic acid[c]	Top kill (85–95)	2–3	Greer et al. (1968)
Propionic acid	Silvex	2–2,4,5-Trichlorophenoxy-propionic acid	Top kill (85–95)	2–3	Greer et al. (1968)
Benzoic acid	Oracle, Diablo, or Vanquish	3,6-Dichloro-2-methoxybenzoic acid	Top kill (85–95)	2–3	Greer et al. (1968)
Picolinic acid	Picloram	4-Amino-3,5,6-trichloro-2-pyridinecarboxylic acid	Top kill (85–95)	2–3	Greer et al. (1968), Peterson and Boyd (1998), Haukos (2011)

[a] Application recommendation for all herbicides was 0.5 kg/ha of 2,4,5-T or Silvex with 3.8 L of diesel oil with water to make a 17.6 L/0.405 ha solution.
[b] Greer et al. (1968); Scifres (1972); Jacoby and Meadors (1982); Peterson and Boyd (1998).
[c] Banned by the Environmental Protection Agency in 1985.

Boyd 1998, Haukos 2011; Table 17.1). Temporary control of shinnery oak was relatively expensive compared to the short-term benefits of improved production of forage grasses. The need for long-term control and reduction of shinnery oak subsequently led to the widespread use of tebuthiuron as a herbicide in shinnery oak prairie (trade names: Spike or Graslan).

Current Management

Herbicides

Tebuthiuron was introduced in 1974 and has at least three advantages: it is relatively nontoxic to nontarget species (Emmerich 1985), it requires only one application to be effective (Scifres et al. 1981), and it is a dry pelleted herbicide that does not result in overspray to nontarget areas (Peterson and Boyd 1998, Haukos 2011). Tebuthiuron attacks the root system of shinnery oak and causes repeated defoliation, which ultimately limits rhizome production and subsequently suppresses shinnery oak for several years (Pettit 1979, Jones and Pettit 1984, Haukos 2011). The long-term effects of tebuthiuron make it the most effective and economically feasible way to initially control shinnery oak, but the economic advantages depend on herbicide cost, rainfall, depth of fine sand, and beef prices (Ethridge et al. 1987, Peterson and Boyd 1998). The most conspicuous aspect of tebuthiuron application is the resulting vegetation community and related structure. For example, the reduction of shinnery oak via tebuthiuron application followed by immediate grazing pressure will result in stands dominated by sand sagebrush with interspersions of forbs due to continuous grazing pressure on grasses (Peterson and Boyd 1998). In contrast, the reduction of shinnery oak followed by deferred grazing results in an interspersion of grasses, forbs, and shrubs because grazing pressure on grasses is reduced (Zavaleta 2012).

Grazing

Grazing management is a complex topic with many descriptive terms (Hart et al. 1988; Table 17.2). Terms such as "overgrazing" and "heavy grazing"

can refer to stocking rates and grazing intensity but are often used to describe negative grazing management. The terms are often confused with "continuous grazing," which is a grazing system. The different terms describe different practices for rangeland management that have different impacts on vegetation composition and structure. Long-term, high-intensity grazing is detrimental to desirable plants and range condition through reductions in leaf area that damage root systems and result in limited plant growth on managed rangelands (Briske et al. 2008). Plant species richness is reduced because more palatable species are consumed first, which increases annual grasses and forbs at the expense of perennial grasses (Peterson and Boyd 1998). Deferment, on the other hand, indicates that periodic exclusion of grazing can promote the recovery and maintenance of greater leaf area, especially during periods of rapid growth (Holechek et al. 2006, Briske et al. 2008).

Precipitation is a driving factor influencing herbaceous production in native grasslands (Biondini et al. 1998, Holechek et al. 2006, Briske et al. 2008), but magnitude of grazing systems, stocking rates, and environmental factors on vegetation production, composition, and structure in shinnery oak prairie remain contested (Sullivan 1980, Biondini et al. 1998, Holechek et al. 2006). Stocking rates may have a greater influence on herbaceous production

TABLE 17.2

Common terminology and definitions of grazing systems and stocking rates.

Treatment	Definition	Sources
Grazing system		
Continuous	Cattle are confined to one pasture without subdivisions or temporal variation	Hart et al. (1988)
Rotationally deferred	Cattle are moved among pastures, but some pastures are deferred or rested for a predetermined time	Hart et al. (1988)
Short duration	Subdivided pasture or paddocks are grazed for a short period in each with frequent rotation within the same season	Hart et al. (1988)
Stocking rates[a]		
Light	20%–40% consumed[b,c]	Van Poollen and Lacey (1979)
Moderate	40%–60% consumed	Van Poollen and Lacey (1979)
Heavy	>60% consumed	Van Poollen and Lacey (1979)

[a] Light, moderate, and heavy are arbitrary distinctions that often depend on range condition and relative amounts of annual precipitation.
[b] Available vegetation.
[c] Consumption represents utilization but does not include vegetation lost to trampling or excretion.

than the type of grazing system (Van Poollen and Lacey 1979, Briske et al. 2008). Responses of herbaceous plants to grazing are more variable in the southwest due to less precipitation (Van Poollen and Lacey 1979), and predicting herbaceous production is difficult in shinnery oak prairie. The effects of herbaceous production and grazing regime are not well understood, but grazing system and stocking rate can both affect the vegetative structure, a critical factor that determines the habitat use by Lesser Prairie-Chickens (Knopf et al. 1988, Peterson and Boyd 1998, Beck and Mitchell 2000, Hagen et al. 2013).

Fire

Given the short-term composition changes and potential soil erosion following a burn, especially if little precipitation occurs to facilitate subsequent plant growth, prescribed fire is not currently considered or used as a viable strategy for long-term habitat restoration of degraded shinnery oak prairie in the western part of the Southern High Plains. However, fire may be used to create lek sites or after a tebuthiuron treatment to maintain a mosaic of grasses, forbs, and shrubs in shinnery oak prairie.

HABITAT REQUIREMENTS

Diet

Lesser Prairie-Chickens forage nearly exclusively on shinnery oak leaves and catkins in spring (Suminski 1977, Davis et al. 1980, Doerr and Guthery 1983b; Table 17.3), but transition to insects when they become more available in the summer (Davis et al. 1980). Grasshoppers (Acrididae) and treehoppers (Membracidae) were the main summer foods for adults, subadults, and chicks (Davis et al. 1980, Riley et al. 1993). Lesser Prairie-Chicken chicks forage almost exclusively on insects during the first 10 weeks of life (Davis et al. 1980, Savory 1989). In one study, insect availability increased with forb availability (Davis et al. 1980), suggesting an importance of forbs, which are typically lacking in shinnery oak monocultures (Grisham 2012, Zavaleta 2012).

Information on the diet of Lesser Prairie-Chickens in shinnery oak prairie comes from the mid- to late 1970s, and mainly focuses on autumn and winter foods from hunter-harvested birds.

In addition, the surrounding landscape once consisted of grain sorghum and winter wheat, but has subsequently been converted to mainly cotton.

Breeding Habitats

Lekking

Leks are generally located in areas interspersed with native vegetation and bare ground for display (Crawford and Bolen 1976a). Lek locations can include areas with bare ground created by anthropogenic disturbance, including abandoned oil pads, dirt roads, cultivated field adjacent to native grassland, recently burned areas, and areas surrounding stock tanks (Taylor 1979, Locke 1992, Hunt and Best 2010, C.W. Boal, unpubl. data). Hunt and Best (2010) found that active leks had more bluestem and less dropseed compared to abandoned leks. Abandoned leks were closer to areas dominated by honey mesquite (*Prosopis glandulosa*). Of 39 abandoned leks, 30 sites (76.9%) supported large mesquite trees that were >60 cm in height, while only 9 of 33 active leks (27.3%) had large trees at the lek.

Nesting Habitats

Female Lesser Prairie-Chickens usually select nest sites within 360 m of previous nests from the same or previous year and sites <1.9 km from lek of capture (Grisham 2012). Nest sites typically consisted of similar structural characteristics regardless of specific plant type (Haukos and Smith 1989, Grisham 2012, Hagen et al. 2013, Grisham et al. 2014), but interspersion of grasses and shrubs may have a high ecological value for nesting females. Grisham (2012) found 34 of 74 nests (46%) in little bluestem from 2001 to 2010. Similarly, Haukos and Smith (1989) and Riley et al. (1992) reported that nest sites of Lesser Prairie-Chicken were mainly associated with native grasses such as purple three-awn, little bluestem, and sand bluestem. In New Mexico, Riley et al. (1992) found that 24 of 37 nests (65%) of Lesser Prairie-Chickens were with native grass species—*Andropogon* spp. being the most common. More recent studies have found that females use a wider variety of plants as nesting cover, including sand sagebrush and shinnery oak (Leonard 2008, Davis 2009, Grisham et al. 2014). For example, 25 of 36 nests (69%) were located in sand sagebrush in Texas (Grisham et al. 2014) and the average percent shrubs at nest was a good

TABLE 17.3
Diet of Lesser Prairie-Chickens in sand shinnery oak prairie of the Central United States.

Date	Season	Method	Age	Measurement	Mast and seed	Vegetation	Insect	Source
						Percentage (%)		
1971–1973	Autumn	30 crops	Adult	% Frequency	N/A	57	42.9	Crawford and Bolen (1976b)
				% Mass	N/A	89.9	10	Crawford and Bolen (1976b)
				% Volume	N/A	80.9	19	Crawford and Bolen (1976b)
1976	Spring	9 crops	Adult		27.5	60.4	12.1	Suminski (1977)
1976	Summer and autumn	12 crops	Chicks 20 weeks		70.7	11.9	17.4	Suminski (1977)
1976	Autumn	9 crops	Adult		66	27.4	6.6	Smith (1979)
1977	Autumn	17 crops			20.5	49.9	29.6	Smith (1979)
1976–1977	Winter	6 crops			69.3	26	4.7	Smith (1979)
1976–1978	Spring	21 crops	Adult	% Comp.	15.5 ± 5.8	78.7 ± 7.6	5.9 ± 3.8	Davis et al. (1980)
	Summer	18 crops	Adult		21.4 ± 8.2	23.3 ± 7.2	55.3 ± 9.3	Davis et al. (1980)
		10 crops	Chicks 1–4 weeks		0	0	100	Davis et al. (1980)
		17 crops	Chicks 5–8 weeks		0.6 ± 0.6	0.1	99.3 ± 6.3	Davis et al. (1980)
1978–1979	Summer	30 droppings	Adult	% Frequency	0	40	59.6	Doerr and Guthery (1983b)
	Autumn				4.4	30.6	64.9	Doerr and Guthery (1983b)
	Winter				63.1	29	7.6	Doerr and Guthery (1983b)
	Spring				15.9	54.4	26.7	Doerr and Guthery (1983b)
1984	Summer (teb)[a]	6 crops		% Volume	N/A	98.6	1.38	Olawsky (1987)
	Summer (no teb)	9 crops			N/A	81.3	18.6	Olawsky (1987)
1985	Summer (teb)	15 crops			N/A	65.8	34.1	Olawsky (1987)
	Summer (no teb)	12 crops			N/A	68.2	31.7	Olawsky (1987)
1976	Autumn	9 crops	Adult	% Comp.	75	28	7	Riley et al. (1993)
1977	Autumn	17 crops			21	49	30	Riley et al. (1993)
1976–1977	Winter	6 crops			69	26	5	Riley et al. (1993)

[a] Areas treated with tebuthiuron.

predictor of nest-site selection. On average, percent cover of shrubs at nests (38.6%) was almost nine times greater than random points (4.4%) in Texas (Grisham et al. 2014). Similarly, percent composition of shrubs at nests in New Mexico averaged 53% (Patten and Kelly 2010). Vegetation heights at nests ranged between 30 and 65 cm tall, and visual obstruction readings were between 25 and 40 cm for all nests, regardless of whether the nest site was located in grasses or shrubs (Haukos and Smith 1989, Riley et al. 1992, Leonard 2008, Davis 2009, Grisham et al. 2014).

In Texas, nest-site selection was influenced by the proximity to human activity. Nests were located farther from unimproved roads and utility poles than what would be expected at random based on simulated datasets (Grisham et al. 2014). Nests were located farthest from buildings (range = 922 m to 10 km), improved roads (range = 671 m to 9 km), and pumpjacks at oil wells (432 m to 7 km). In contrast, 50% of all Lesser Prairie-Chicken females nested within 1.5 km of stock tanks, which was closer than what was expected at random, despite high levels of human activity in these areas (Grisham et al. 2014).

Brood-Rearing Habitats

The vegetation composition of brood-rearing habitats varies among studies, but brood sites tend to be a mixture of dense shrubs or grasses interspersed with forbs and abundant bare ground. Brood locations are typically characterized by tall grasses and shrubs (~29–31 cm), moderate horizontal cover (~2–3 dm), that are dominated by bare ground (~35%–40%) and litter (~40%–45%), with moderate cover of grasses and shrubs (~20%–30%) or forbs (~7%–10%, Davis et al. 1979, Ahlborn 1980, Riley and Davis 1993, Grisham 2012). Shinnery oak is an important habitat for brood rearing, most likely due to microclimate effects that alleviate thermal stress for chicks (Bell et al. 2010). Bell et al. (2010) found that broods selected habitats that were warmer than random points during cool periods and sites that were cooler than random points when temperatures were ≥26.4°C.

Areas that support high invertebrate biomass are important components of brood habitat (Davis et al. 1979, Ahlborn 1980, Riley and Davis 1993, Grisham 2012). Percent canopy cover for forbs in studies from shinnery oak prairie is well below the recommendation of 16%–23% forb cover suggested by Hagen et al. (2013). Grisham (2012) suggested that a lack of forbs (<3% cover) in shrub monocultures was a limiting factor for brood survival in Texas. Broods were more likely to survive to beyond 30 days posthatch at a restored ranch in New Mexico, compared to ranches dominated by monocultures of shinnery oak in Texas. The canopy cover of forbs at brood locations was ~7% in New Mexico versus ~3% in Texas (2006–2011; Grisham 2012).

Autumn and Winter Habitats

Lesser Prairie-Chickens of the Southern High Plains historically occurred in both mixed-grass and shinnery oak prairie, but at higher densities in the latter ecosystem (Henika 1940). Populations of Lesser Prairie-Chicken in Texas are considered to have the highest numbers in the early 1900s. Concomitantly, dry-land grain sorghum was becoming a common crop planted across the Southern High Plains. The availability of supplemental waste grain is thought to have been partly responsible for increased population numbers and range expansion by Lesser Prairie-Chickens during that period (Jackson and DeArment 1963, Crawford and Bolen 1976a, Ahlborn 1980, Taylor and Guthery 1980).

Shinnery oak is an important winter food resource, but in terms of cover, native bluestem grasses (*Andropogon* spp. and *Schizachyrium scoparium*) are critical components of the plant community due to the deciduous nature of shinnery oak (Taylor 1978, Doerr and Guthery 1983a, Riley et al. 1993, Pirius 2011). In New Mexico, Riley et al. (1993) reported that habitats with grasses were used for general cover or winter roost sites, and shinnery oak was used as a foraging area. Taylor (1978) found that Lesser Prairie-Chickens used the shinnery oak–sand sagebrush types over the shinnery oak in all autumn or winter months. Mesquite–shinnery oak, short-grass prairie, and reverted cropland were mostly avoided. Pirius et al. (2013) found that Lesser Prairie-Chickens selected areas of shinnery oak dominated by prairie grasses (≥70%) and avoided areas dominated by sand sagebrush during winter. The authors suggested conservation of shinnery oak prairie within 4.8 km of leks to provide adequate cover and foraging habitat for Lesser Prairie-Chickens in winter.

Water Resources

The use of free surface water by galliform birds is common, but the importance of water resources has been debated because the physiological requirements for water by galliform birds in North America can be met via preformed water in ingested foods or metabolic water created by the oxidation of fat or carbohydrates (Guthery 1999). It has been assumed that Lesser Prairie-Chickens do not require surface water and obtain their moisture requirements from preformed or metabolic sources (Henika 1940, Snyder 1967). However, Lesser Prairie-Chickens frequently use surface water sources when available (Schwilling 1955; Copelin 1963; Jones 1963, 1964; Crawford and Bolen 1973; Davis et al. 1979; Sell 1979; Locke 1992). The use of surface water by Lesser Prairie-Chickens can be frequent enough that researchers have targeted trapping efforts at both water sources and leks (Schwilling 1955, Copelin 1963, Davis et al. 1980, Holt 2012). The species will use surface water when available, but the ecological importance of free surface water on recruitment of Lesser Prairie-Chickens has not been assessed, particularly during periods of drought.

Coats (1955) discussed the preference of drinking droplets of water, compared to trough water, by pen-reared chicks of Lesser Prairie-Chickens. Schwilling (1955) was the first to note an increase in the use of surface water sources by females leading up to the nesting period in southwest Kansas. Copelin (1963) noted that small groups of Lesser Prairie-Chickens in Oklahoma began watering at surface water sources in September. Jones (1964) also noted the use of surface water by Lesser Prairie-Chickens during the late summer and autumn. In contrast, Crawford and Bolen (1973), Davis et al. (1979), and Sell (1979) reported the increased use of surface water by Lesser Prairie-Chickens primarily during spring on the Southern High Plains.

Boal et al. (unpubl. data) used motion-activated cameras to record the use of livestock tanks and overflows from March 2009 through August 2011. Depending on year, they detected 50%–78% of female visits between March and July, whereas 69%–75% of male visits occurred between November and February. The potential value of surface water was revealed during a drought in 2011 with only 2.23 cm of annual precipitation (Grisham 2012). The average number of visits by Lesser Prairie-Chicken to water sources was 156.2 ± 85.3SE visits per 100 trap days during the summer months, compared to only 5.7 ± 7.0 and 23.8 ± 31.1 visits per 100 trap days during the previous two years. The annual differences in water use suggest that green food and abundant insects can meet the moisture requirements of Lesser Prairie-Chickens in shinnery oak prairie habitats in normal years, but not during drought conditions.

POPULATION DYNAMICS

Population Trends

Estimates from New Mexico suggest Lesser Prairie-Chickens have been extirpated from 56% of their historical range. The remaining core distribution of Lesser Prairie-Chickens on the Southern High Plains of east-central New Mexico is the shinnery oak prairie ecosystem. Historical estimates of population numbers of Lesser Prairie-Chickens in New Mexico are not accurate (Bailey and Williams 2000, Davis et al. 2008), but various estimates range from up to 50,000 birds between 1949 and 1961 (Sands 1968) to density estimates as high as 0.53 birds/ha in 1985–1986 (Olawsky and Smith 1991). Recent estimates suggest the New Mexico population has the greatest number of known leks in Roosevelt County (Davis et al. 2008), which is dominated by shinnery oak prairie.

In Texas, evidence from 2011 suggests that <2,600 Lesser Prairie-Chickens persist in two distinct geographic regions: the Southern High Plains and the northeastern panhandle (Timmer et al. 2013). The current range is a significant reduction from an estimate of 12,000 Lesser Prairie-Chickens in the 1940s (Henika 1940, Texas Game, Fish, and Oyster Commission 1945), or recent density estimates of >0.50 birds/ha during peak population counts of the mid-1980s (Olawsky and Smith 1991). In the northeastern panhandle of Texas, the average number of males per lek decreased from 11 to 5 birds from 1998 to 2007 (Davis et al. 2008). In contrast, Davis et al. (2008) estimated a lek density of 1.9 leks/km² (0.74 leks/mile²) with 7.9 males per lek on the Southern High Plains of Texas. Population numbers of Lesser Prairie-Chickens in the Sand Shinnery Oak Prairie Ecoregion appear to have been stable since the mid-1990s, up until the recent drought of 2011 (Chapter 4, this volume). Results from aerial surveys have estimated the population numbers of Lesser Prairie-Chickens to be 2,946 birds in 2012 (90% CI = 1,325–7,973) and 1,967 birds in 2013 (90% CI = 844–3,754). The estimated number of leks for Lesser Prairie-Chickens in the Sand Shinnery Oak Prairie Ecoregion was 366 leks in 2012 (90% CI = 117 = 987) and 118 leks in 2013 (90% CI = 2–355; McDonald et al. 2014).

Population Drivers

Drought is known to have a negative effect on most populations of Lesser Prairie-Chickens, but the negative effects are exacerbated on the Southern High Plains because the ecoregion is characterized by a hotter, drier climate compared to the rest of the species distribution (Grisham et al. 2013). The drought conditions of the 1930s and 1950s have been cited as having significant, negative effects on populations of Lesser Prairie-Chickens (Henika 1940, Hamerstrom and Hamerstrom 1961, Crawford and Bolen 1975, Crawford 1980, Applegate and Riley 1998, Bailey and Williams 2000). Drought conditions impact Lesser Prairie-Chickens indirectly through the degradation of suitable nesting cover and food (Henika 1940, Davis et al. 1979, Olawsky 1987, Giesen 2000). Home range size increases by 20%–50% in years following drought (Merchant 1982, Borsdorf 2013). Recruitment can be significantly reduced due to low breeding propensity, where up to 80% of radio-marked females may fail to initiate a nest (Grisham et al. 2014), or to poor nest survival among females that do produce a clutch (0%–20%; Merchant 1982; Hagen

and Giesen 2005; Grisham 2012; Grisham et al. 2013, 2014).

Recent evidence suggests that extreme drought conditions can exceed the thermal stress threshold for an incubating female and her eggs, resulting in nest failure and annual variation in demographic parameters (Table 17.4; Patten et al. 2005a, Grisham 2012, Grisham et al. 2014). Females breeding in the sand sage prairie of Kansas typically have a high probability of nest initiation (0.93) and renesting after clutch loss (0.31, Pitman et al. 2006). In contrast, females breeding in shinnery oak prairie in Texas during a drought year (2011) had dismal productivity with a much lower probability of nest initiation (0.20) and no renesting (Grisham et al. 2014). In the same study in Texas, females had good nest success (70%) in years with greater precipitation and cooler spring temperatures (Grisham et al. 2014). Nest abandonment may be more common in the Southern High Plains than other regions. Elsewhere, Holt (2012) reported that 1 of 24 nests (4%) were abandoned in the mixed-grass prairie of the northeastern Texas Panhandle, and Pitman et al. (2005b) reported that 4 of 200 nests (2%) were abandoned in sand sagebrush prairie of Kansas.

Population Vital Rates

Lesser Prairie-Chickens in the Southern High Plains invest more in survival and less in reproduction compared to populations at the northern extent of the species range (Patten et al. 2005a, Grisham et al. 2014). The regional variation in life

TABLE 17.4

Components of reproductive effort for female Lesser Prairie-Chickens breeding in sand shinnery oak prairie on the Southern High Plains of the Central United States.

State	Clutch size	Apparent nest success	Nest initiation rate	Renesting rate	Nest initiation date	Sources
Texas	7.4	46	0.2–1.0	0.07	May 15	Grisham et al. (2014)
Texas		38	0.62		May 25	Lyons et al. (2009)
Texas	7.8	15				Haukos (1988)
New Mexico	6–12	76	1		April 25–May 7	Davis (2009)
New Mexico	9.5–10.1	0–50	0.73–0.92	0–1.0	April 29–May 10	Merchant (1982)
New Mexico				0.5		Grisham (2012)
New Mexico		28				Riley et al. (1992)
New Mexico		42			May 12	Patten et al. (2005b)

history traits is illustrated by some of the highest reported annual and seasonal survival rates occurring on the Southern High Plains (Table 17.5, Pirius 2011, Grisham 2012), but with lower or more variable rates of nest initiation, clutch size, nest success, and renesting than other areas of the species range (Table 17.4, Patten et al. 2005a, Pitman et al. 2006, Davis 2009, Grisham 2012, Grisham et al. 2014). Reproductive vital rates vary from year to year with annual variation in climatic conditions on the Southern High Plains. In other regions, however, vital rates are relatively constant among years (Patten et al. 2005a; Pitman et al. 2005b, 2006). High survival likely allows Lesser Prairie-Chickens on the Southern High Plains to take advantage of good breeding years whenever they occur, as opposed to investing heavily in reproduction every year regardless of environmental conditions (Patten et al. 2005a).

Limiting Factors for Populations

Brood survival from hatch to 14 days posthatch is the main limiting factor for the population viability of Lesser Prairie-Chickens in shinnery oak prairie habitats (Davis 2009, Grisham 2012). Grisham (2012) reported 0% survival of broods monitored during a 4-year study in Texas, and most brood losses occurred 0–14 days after hatching. In New Mexico, Grisham (2012) found that only 4 of 18 broods (22%) survived until 30 days of age, with 93% of losses occurring 0–14 days after hatching. Davis (2009) found that annual rates of brood success were 0% and 80% during a 2-year study during 2004–2005 in New Mexico. Again, most brood mortality occurred within 14 days of hatching. Merchant (1982) found that 27% and 0% of females in New Mexico produced a brood that survived until independence in 1979 and 1980, respectively. The estimates are consistent with studies outside the Southern High Plains where brood survival was lowest during the 2-week period after hatching (Fields et al. 2006, Pitman et al. 2006, Holt 2012). In New Mexico, chicks grew to 90% of asymptotic body mass by 50 days and asymptotic wing chord length by 35 days (Bell et al. 2007). Chicks in sand sage prairie of Kansas took longer to grow to 90%

TABLE 17.5
Survival estimates for Lesser Prairie-Chickens in New Mexico, the Southern High Plains of Texas (Texas SHP), and the panhandle of northeast Texas (Texas NE).

| Location | Breeding | | | Nonbreeding | Annual | | | Type of estimate | Sources |
	Female	Male	Pooled	Pooled	Female	Male	Pooled		
Texas SHP	0.71–0.89[a]	0.57[b]		0.72	0.51–0.64	0.41		Seasonal	Pirius (2011) (nonbreeding), Grisham (2012) (breeding)
Texas SHP			0.87	0.93			0.31	Monthly, Annual	Lyons et al. (2009)
New Mexico	0.44							Seasonal	Merchant (1982)
New Mexico	0.79							Seasonal	Grisham (2012)
New Mexico					0.64	0.60		Annual	Pruett et al. (2011)
Texas SHP	0.58[c]							Seasonal	Haukos et al. (1989)
Texas NE			0.93	0.96			0.52	Monthly, Annual	Lyons et al. (2009)
Texas NE					0.34	0.26–0.55		Annual	Kukal (2010) (nonbreeding), Holt (2012) (breeding)

[a] Breeding season for females defined as March 15–August 31.
[b] Breeding season for males defined as March 1–August 31.
[c] Breeding season defined as March 15–May 15.

of asymptotic body mass (54 days, Pitman et al. 2005a), but growth rates were similar in both populations. The small difference in the duration of growth could be related to the variation in asymptotic body mass between Lesser Prairie-Chickens in New Mexico and Kansas.

Winter food availability is also a limiting factor for Lesser Prairie-Chicken populations in shinnery oak prairie. Shinnery oak acorns are the primary food source for the species during winter (Doerr and Guthery 1983a, Peterson and Boyd 1998). If acorns are not available through herbicide application, drought, or natural fluctuations in abundance, food resources may limit populations through increased movements, decreased body condition, and increased mortality (Merchant 1982, Olawsky 1987, Borsdorf 2013). The outright loss of shinnery oak eliminates a primary natural food source during winter. Peak population numbers during the 1980s could have been due to the availability of waste grain from dry-land sorghum, which supplemented natural food availability during the winter months (Olawsky and Smith 1991).

Causes of Mortality

Mortality Patterns

The seasonal survival of female Lesser Prairie-Chickens is generally greater during nonbreeding than breeding seasons (Patten et al. 2005a, Hagen et al. 2007). The same pattern of seasonal mortality appears to hold for both sexes in the Southern High Plains (Wolfe et al. 2007, Lyons et al. 2009, Grisham 2012, Pirius et al. 2013). Lyons et al. (2009) reported that the monthly survival of females was highest during the nonbreeding season (0.93), followed by late (0.89) and early periods (0.85) of the breeding season. Wolfe et al. (2007) found that peak mortality occurred during the breeding season in eastern New Mexico. Male mortality peaked in March and April, whereas female mortality was highest in May. Within the breeding season, Grisham (2012) found that most mortality occurred June and July for males and May and June for females. Haukos et al. (1989) also reported female mortality during breeding season peaked in early May.

Depredation of Lesser Prairie-Chickens at leks has been suggested to be the cause of low male survival during the breeding season, due to male conspicuousness and predators attracted by male lekking activity (Hagen et al. 2005a; Wolfe et al. 2007). However, evidence suggests that the predation at leks on the Southern High Plains is an uncommon event (Haukos and Broda 1989, Behney et al. 2011; Chapter 8, this volume). Infrequent mortality at leks is consistent with Grisham's (2012) finding that peak male mortality occurred at the end or after the male lekking season. Lekking entails an energetic cost with both increased activity and reduced foraging time, and males lose body mass during lekking (Wolfe et al. 2007). The cumulative effects of energy expenditure during the lekking season may interact with other environmental stressors such as drought and contribute to reduced male survival following the end of the lekking period in mid-May (Grisham 2012).

Depredation events for both sexes of Lesser Prairie-Chickens during the breeding season were mainly due to mammalian predators (males: 50%, females: 57%), followed by avian (males: 39%, females 29%), and unknown predators (11%, 14%, Grisham 2012). During the nonbreeding season, Pirius et al. (2013) attributed 77% of mortalities to avian predators, 15% to mammalian predators, and 8% to unknown predators. Avian predators accounted for a greater percentage of mortality during the nonbreeding season at the Southern High Plains than the northeastern Texas Panhandle (47%, Kukal 2010), presumably due to a greater diversity and abundance of overwintering raptors on the Southern High Plains (Behney et al. 2012). Similarly, avian predators were also responsible for a greater percentage of mortality on the Southern High Plains than in Oklahoma (Patten et al. 2005a; Wolfe et al. 2007). In New Mexico, raptors accounted for most Lesser Prairie-Chicken mortality (47%) followed by mammals (36.4%), collisions with fences or powerlines (16.1%), and accidental drowning (0.5%; Patten et al. 2005a, Wolfe et al. 2007). Haukos et al. (1989) reported that 62% of female breeding season mortalities were due to raptor predation and 38% due to predation by coyotes (Canis latrans).

Collisions with fences and powerlines have been implicated as a substantial source of Lesser Prairie-Chicken mortality in Oklahoma and to a lesser extent, New Mexico (Patten et al. 2005b, Wolfe et al. 2007). However, none of the other population studies on the Southern High Plains have found evidence of collision mortality (Merchant 1982, Haukos et al. 1989, Davis 2009, Grisham 2012, Pirius et al. 2013). Collision risk is likely a function of fence density and most areas on the Southern High Plains consist of large parcels of land with relatively low fence densities compared to areas studied in Oklahoma (Patten et al. 2005a, Wolfe et al. 2007). Therefore, collision mortality may be less of a concern on the Southern High Plains than other regions where habitat fragmentation is more pronounced.

Predation at leks appears to be uncommon on the Southern High Plains (Haukos and Broda 1989, Behney et al. 2011). Behney et al. (2011) did not observe any successful raptor or mammalian predation during 650 h of lek monitoring in spring 2007 and 2008 on the Southern High Plains. The authors observed 61 raptor encounters at leks but only 15 encounters resulted in an attack with a raptor attempting to capture a Lesser Prairie-Chicken. Northern Harriers (*Circus cyaneus*) were the most common raptor (50%) encountered at leks, followed by Swainson's Hawks (*Buteo swainsoni*, 18%), other *Buteo* hawks (15%), falcons (*Falco* spp., 8%), accipiters (*Accipiter* spp., 3%), and unknown raptors (6%). During the nesting period of Lesser Prairie-Chickens, Swainson's Hawk were the most abundant raptor present at the Southern High Plains (Behney et al. 2012), but Behney et al. (2010) did not observe any prairie-chickens in the diet of nestling Swainson's Hawks in areas where the two species coexist. During the nonbreeding season, prairie-chickens are at risk from a more diverse and abundant suite of raptors (Behney et al. 2012).

Ranking Limiting Factors

Brood survival is the primary factor influencing Lesser Prairie-Chicken population demography in shinnery oak prairie. Management should focus on improving brood survival, because it is paramount to population persistence in these vegetation communities. Improving annual nest survival beyond 40% and maintaining current survival rates have the potential to alleviate pressure on brood survival, but restoration of habitats out of postclimax, shinnery oak monocultures is necessary to conserve the species in shinnery oak prairie.

CONSERVATION AND MANAGEMENT OF LESSER PRAIRIE-CHICKENS IN SHINNERY OAK PRAIRIE

Desired Ecological State for Shinnery Oak Prairie

Shinnery oak is a critical component of the habitat of Lesser Prairie-Chickens (Davis et al. 1980, Doerr and Guthery 1983a, Olawsky 1987, Haukos and Smith 1989, Riley et al. 1993, Bell et al. 2010). Given the widespread loss of shinnery oak prairie in the past century and increasing recognition of the ecological role of the shrub in these ecosystems, desired ecological conditions include shinnery oak as a key species but embedded within a matrix of native grasses, forbs, and shrubs consistent with published Ecological Site Descriptions (ESD, Olawsky and Smith 1991, Riley et al. 1993, Hagen et al. 2004, Smythe and Haukos 2009, Zavaleta 2012, Grisham et al. 2014; Figure 17.1).

Habitat Management

Here, we discuss habitat management strategies that can be used to obtain community structure consistent with ESDs and specific information relevant to the management of Lesser Prairie-Chicken. Land management at localized scales of <500 ha that focus on providing adequate cover for nesting and brood rearing will be beneficial (Jamison 2000; Hagen 2003; Hagen et al. 2004, 2013; Grisham et al. 2014). Conservation planning at the landscape level of the Southern High Plains within the time frame of >20 years would be prudent to maintain populations of Lesser Prairie-Chickens on the Southern High Plains. Thus, improvement of brood rearing and nesting habitat by reduction of shinnery oak monocultures and buffering against the negative influence of drought has high ecological value. Therefore, we consider options specific to areas with known leks and adjacent areas (600–1,500 ha) in shinnery oak prairie, with the ultimate goal of scaling up to a landscape-level conservation plan for the Southern High Plains. At a local spatial scale, improving brood-rearing and nesting habitats while maintaining no negative impact to annual survival, food resources, nonbreeding season habitat, and avoiding anthropogenic development near leks is the initial step of the process for the creation of landscape-level conservation plans.

Tebuthiuron

Several studies have examined the effects of tebuthiuron applications and have concluded that timing and application rate have different effects on vegetation composition and structure (Peterson and Boyd 1998, Haukos 2011). Application rates >1.0 kg/ha completely eradicate shinnery oak and are not recommended for the management of Lesser Prairie-Chicken habitat (Peterson and Boyd 1998, Haukos 2011). Lesser Prairie-Chickens require different plant structure during their annual life cycle; therefore, we suggest creating a mosaic of monotypic stands of shinnery oak and tebuthiuron-treated areas within 1.6 km of known leks would improve nesting and brood-rearing habitat (Hagen et al. 2013). Suggested rates for tebuthiuron application range from 0.2 to 0.6 kg/ha because moderate rates reduce shinnery oak and increase grass production without reducing forb and seed production and overall vegetation cover (Doerr and Guthery 1983a, Jones and Pettit 1984, Biondini et al. 1986, Zavaleta 2012). The application rates of 0.2–0.6 kg/ha can provide adequate nesting cover and improve forb production for brood-rearing cover (Grisham 2012, Zavaleta 2012). Additionally, incorporating the interspersion of untreated areas will provide summer thermal refugia (Bell et al. 2010) and ensure continued availability of shinnery oak catkins, galls, and acorns as food resources (Doerr and Guthery 1983a).

Nesting habitats are maximized when they are interspersed with brood habitats to facilitate easier movements of broods among locations and to reduce mortality from predation, exposure, or starvation (Pitman 2003, Hagen et al. 2013). The application of tebuthiuron had no apparent effect on the amount of bare ground in treated plots (Zavaleta 2012). More ecologically significant, a reduced application rate of tebuthiuron at 0.6 kg/ha yielded an increased forb component by three years after application. Forbs are an important characteristic of brood-rearing habitat because of their association with invertebrate food resources, particularly grasshoppers (Jones 1963; Riley et al. 1992; Jamison et al. 2002; Hagen et al. 2004, 2005b). Chicks eat almost exclusively insects for the first four weeks of their life, and their diet consists of up to 80% grasshoppers (Davis et al. 1980). Grasshopper biomass increased by 200% cumulatively over eight years in treated plots, suggesting an increased capacity to provide food sources to broods (Zavaleta 2012). Thus, brood-rearing habitats are maximized when adjacent to nesting habitat and consisting of not <30% of targeted areas (Hagen et al. 2013).

Adult Lesser Prairie-Chickens eat a combination of insects, leaf material, seeds, and shinnery oak (Jones 1963, Crawford and Bolen 1976b, Smith 1979, Davis et al. 1980, Doerr and Guthery 1983b, Olawsky 1987, Riley et al. 1993). Tebuthiuron treatment reduces the amount of shinnery oak, which has been reported to comprise up to 84% of the diet in winter and spring (Jackson and DeArment 1963, Davis et al. 1980, Bell 2005). Acorns, catkins, and galls are dominant winter foods that exhibit interannual variability and are a primary limiting factor for overwinter survival and spring recruitment. However, loss of these food items would not be detrimental to prairie chicken populations if insects and seeds increased after tebuthiuron treatment, or if treatments are applied adjacent to nontreated areas (Doerr and Guthery 1983b). For example, Zavaleta (2012) found that seed production increased by 1,698% in sand paspalum and 273% in sand dropseed in plots treated with tebuthiuron versus untreated plots. However, there was no difference in seed production of bluestems or grama species between untreated areas and sites treated with tebuthiuron. The trade-offs of food availability in treated and untreated areas are not well understood, but Olawsky (1987) found that Lesser Prairie-Chickens in treated areas had lower lipid levels, possibly due to lack of acorns in these areas. Maintaining availability of untreated areas interspersed with treated areas should provide adequate food resources for all life stages.

Grazing Management

An adaptive grazing strategy that creates a mosaic of vegetation in shinnery oak prairie is crucial, because Lesser Prairie-Chickens require structural diversity for critical habitat features—residual grasses for nesting cover, bare ground as travel lanes for broods, abundant grasses and forbs for access to seeds and insects, and shrubs for close escape and thermal cover (Peterson and Boyd 1998, Beck and Mitchell 2000, Derner et al. 2009, Smythe and Haukos 2009, Bell et al. 2010). A strategy that adjusts stocking rates and grazing systems to maintain middle to late stages of plant succession interspersed with early stages of plant succession is supported in the literature for Lesser Prairie-Chickens (Bidwell et al. 2003, Hagen et al. 2004). Grazing systems that incorporate periods of rest to allow for residual nesting cover as well as maintain vegetation heterogeneity are also important to conserving habitat for Lesser Prairie-Chickens (Applegate and Riley 1998, Derner et al. 2009). Grazing systems with rest are especially important in Sand Shinnery Oak Prairie because the ecoregion experiences more frequent and greater magnitude of swings in temperature and precipitation compared to other ecoregions (Grisham et al. 2014).

Grazing alters the horizontal patterning of the lower vegetation layers and has an effect on vertical and horizontal cover (Knopf et al. 1988). High-intensity grazing reduces concealment cover for nests (Bidwell et al. 2003). Light or moderate grazing that maintains >50% of herbaceous production for residual nesting cover can be an important management tool to create and maintain nesting habitat (Holechek et al. 2006).

Grazing also influences the percent composition of bare ground, which influences nesting and brood-rearing habitats for Lesser Prairie-Chickens (Hagen et al. 2013, Grisham et al. 2014). Broods would benefit from bare ground patches to use as travel lanes interspersed among forbs and escape cover (Jones 1963, Riley et al. 1992, Bidwell et al. 2003, Hagen et al. 2013). However, bare ground composition above ~50% does not benefit brood or nesting habitat (Hagen et al. 2005b, 2013; Patten et al. 2005b). Zavaleta (2012) found 28% more bare ground in grazed than in nongrazed pastures in a system that incorporated rotation and moderate-intensity grazing. Hart et al. (1988) found that bare ground was significantly less on continuous lightly grazed pastures as compared to continuous, high-intensity, grazed pastures.

Another important effect of grazing on habitat structure for Lesser Prairie-Chickens is forb cover, though the amount of forbs is influenced to a greater extent by precipitation compared to grazing management (Biondini et al. 1998, Zavaleta 2012). Forbs are an important food source for adult Lesser Prairie-Chickens (Davis et al. 1980, Doerr and Guthery 1983b), with a strong relationship between forb availability and grasshopper abundance (Joern 2004, Hagen et al. 2005b). However, relationships among grazing and invertebrate availability are unclear. Joern (2004) reported that grasshopper densities were 2.5 times greater in grazed than in ungrazed watersheds, and seven of nine invertebrate species increased their densities in grazed areas. Zavaleta (2012), however, found no influence of moderate grazing within shinnery oak prairie on grasshopper biomass.

Fuhlendorf and Engle (2004) and Bidwell et al. (2003) recommended that ~20%–30% of management pastures should be rested annually on a ~3–5-year rotation. Rotational grazing is a sound approach in arid ecosystems because alternation of grazing and rest creates vegetation structure needed for various stages of the life cycle of Lesser Prairie-Chickens. In addition, rotational grazing will facilitate landscape planning in context of a dynamic environment, providing an investment for landowners to maintain habitat for Lesser Prairie-Chickens during years of poor grass production. Rotational grazing includes provision of horizontal cover to facilitate nest concealment and provide thermal refugia, open spaces to lek, and some bare ground for brood travel lanes (Peterson and Boyd 1998, Beck and Mitchell 2000, Derner et al. 2009, Smythe and Haukos 2009, Bell et al. 2010).

Tebuthiuron and Grazing

Vegetation biomass, composition, and structure were similar between tebuthiuron-treated– grazed and tebuthiuron-treated–ungrazed areas under a conservative herbicide rate and moderate grazing intensity (Zavaleta 2012). However, the influence of these practices on the ecology of Lesser Prairie-Chickens is not straightforward (Patten et al. 2006, Leonard 2008, Grisham 2012, Borsdorf 2013). Leonard (2008) found that Lesser Prairie-Chickens used tebuthiuron-treated– grazed areas slightly more than untreated–grazed areas, but females nested almost exclusively in ungrazed grassland. Similarly, Haukos and Smith (1989) and Patten et al. (2006) found that females did not choose to nest in tebuthiuron-treated areas, although brood survival rates were not affected by tebuthiuron treatment or grazing. Conversely, Grisham (2012) found that Lesser Prairie-Chickens nested in similar densities among combinations of grazing, tebuthiuron, tebuthiuron-grazed, and control because vegetation structure had sufficiently developed for nesting during a 6-year period after herbicide treatment, and the grazing plan resulted in little difference in residual vegetation among treatments (Table 17.6).

Grisham (2012) found a reduced application of tebuthiuron at <0.60 kg/ha, and moderate-intensity and short-duration grazing did not negatively affect nest-site selection or nest survival. Habitat structure did not differ between nest sites and random points (n = 78) in habitats that were 6–10 years after herbicide treatment and 5–9 years following the establishment of a grazing treatment in eastern New Mexico. The results differed from other studies where grazing was conducted at a much greater intensities (Haukos and Smith 1989, Riley et al. 1992, Davis 2009, Grisham et al. 2014). In addition, a greater diversity of plant species was used for nesting in tebuthiuron-grazed areas compared to other treatment types. Lesser Prairie-Chickens have a similar response to applied tebuthiuron and grazing treatments because resulting vegetation structure and composition are suitable for nesting.

Herbicide and grazing practices can improve the suitability of landscapes for Lesser Prairie-Chickens, but the species requires a mosaic of habitat structure and composition at a large spatial scale (Davis et al. 1979; Sell 1979; Taylor and Guthery1980; Hagen et al. 2004, 2013; Bell et al. 2010). It is important to create a landscape with a mixture of shinnery oak–dominated tracts (>100 ha; Hagen et al. 2013) that are interspersed with herbicide-treated–grazed areas (Grisham 2012, Zavaleta 2012). Hagen et al. (2004, 2013) emphasized the importance of conserving shrub-dominated or grassland communities that were <1.6 km of lek sites because most nesting occurs within this area (Giesen 1994, Woodward et al. 2001, Pitman 2003). Currently, tebuthiuron treatment is considered the best management option for the restoration of shinnery oak prairie (Peterson and Boyd 1998, Haukos 2011). However, because Lesser Prairie-Chickens require a matrix of vegetation structure and composition of shrubs, grasses, and forbs, tebuthiuron

TABLE 17.6

The observed proportion of nests observed for four herbicide and grazing treatments (95% CI), expected proportions for each treatment, and selection assessment for four leks in Roosevelt County, New Mexico, 2003–2010.

Lek	Treatment[a]	Observed	95% Confidence Interval	Expected	Outcome
1	NT–G	0.14	0.08–0.20	0.53	Used less
	NT–NG	0.50	0.41–0.59	0.31	No difference
	T–G	0.14	0.08–0.20	0.08	No difference
	T–NG	0.14	0.08–0.20	0.08	No difference
4	NT–G	0.56	0.48–0.64	1.00	Used less
	NT–NG	0.06	0.02–0.10	0.00	No difference
	T–G	0.38	0.30–0.46	0.00	Used more
	T–NG	0.00	0.00–0.00	0.00	No difference
7	NT–G	0.32	0.24–0.40	0.00	Used more
	NT–NG	0.04	0.01–0.07	0.00	Used more
	T–G	0.64	0.56–0.72	0.93	Used less
	T–NG	0.04	0.01–0.07	0.07	No difference
17	NT–G	0.08	0.04–0.12	0.63	Used less
	NT–NG	0.67	0.59–0.75	0.29	Used more
	T–G	0.25	0.18–0.32	0.08	Used more
	T–NG	0.00	0.00–0.00	0.00	No difference

SOURCE: Data from Grisham (2012).
[a] NT–G, Not Treated and Grazed; NT–NG, Not Treated and Not Grazed; T–G, Treated and Grazed; T–NG, Treated and Not Grazed.

and grazing treatments are best suited to provide the heterogeneity of habitat at the intermediate spatial scale of <500 ha for a typical ranch rather than as an absolute management strategy for landscape-level conservation in shinnery oak prairie (<500 ha; Hagen et al. 2013).

Prescribed Fire

The effects of prescribed fire on Lesser Prairie-Chickens in shinnery oak prairie are poorly known. Cannon and Knopf (1979) observed two adjacent leks and reported that one lek in a burned area had greater male attendance the year following a fire, whereas a lek in an unburned area was abandoned. The anecdotal observation suggests that fire can create a vegetative structure that is preferred for lekking. However, nesting habitat and thermal cover are minimized because of a reduction in overhead and horizontal cover caused by fire, especially in the spring (Boyd and Bidwell 2001). The specific responses of shinnery oak to fire are not well understood, but tend to vary depending upon environmental conditions prior to and after the burn (Peterson and Boyd 1998). Spring burns in years of abundant rainfall will induce resprouting of top-killed shinnery oak, whereas a spring burn during a dry year decreased shinnery oak density (McIlvain and Armstrong 1966). Controlled burns are not recommended to eradicate or reduce shinnery oak because fire does not kill the root system and allows individual plants to vigorously resprout, resulting in only a temporary reduction of shinnery oak.

Fire has different effects on the important food sources for Lesser Prairie-Chickens. After a burn, acorn density decreased by 94% the first year after the burn, but there was no difference in acorn densities two years after a fire (Boyd and Bidwell 2001, Leonard 2008). The most positive influence of fire for Lesser Prairie-Chickens in sand shinnery oak prairie appears to be by an increase in forbs. Fire can increase forb composition by up to 190% for 2 years after a burn, with concurrent increases in grasshopper density (Boyd and Bidwell 2001).

Anthropogenic Development

Disturbance from anthropogenic activity, even at low densities of infrastructure, can directly and indirectly reduce or degrade the available habitat for Lesser Prairie-Chickens in shinnery oak prairie (Hunt 2004; Robel 2004; Pitman et al. 2005b; Pruett et al. 2009a,b; Hagen 2010; Hagen et al. 2011; Grisham et al. 2014). A meta-analysis of 22 studies revealed anthropogenic features displace the species in all biological seasons, with the largest effects on nest placement. (Hagen 2010; Table 17.7). Pitman et al. (2005b) found that female Lesser Prairie-Chickens avoided anthropogenic features, but distance from structures was not necessarily a predictor of nest success. Similarly, Grisham et al. (2014) used Monte Carlo simulations to show that nest locations were farther than expected from buildings, pumpjacks, improved roads, and utility poles in shinnery oak prairie. Grisham et al. (2014) were not able to identify specific distance thresholds for all features in the assessment, but 85% of nests were located >0.5 km from all anthropogenic features, except unimproved roads and stock tanks. The distribution of nests suggests that the species would benefit if further anthropogenic development were avoided within occupied habitats of Lesser Prairie-Chickens. An overview of specific setback distances for various anthropogenic features can be found in Hagen et al. (2011).

The conversion of native prairie to row crop agriculture has had both positive and negative effects for Lesser Prairie-Chickens. Some agriculture surrounding native areas originally was considered positive because Lesser Prairie-Chickens consume waste grain as a food resource (Crawford 1974; Chapter 2, this volume). However, current row crop agriculture is dominated by unsuitable habitats such as cotton (Wester 2007); therefore, contemporary agricultural is no longer considered beneficial to Lesser Prairie-Chickens. The conversion of native shinnery oak prairie to row crop agriculture is the main factor responsible for declines in population numbers of Lesser Prairie-Chickens because habitat loss directly limits available nesting habitat (Hagen 2003, Hagen et al. 2004).

Landscape Management for Sustainable Populations

Evidence suggests that sand shinnery oak grasslands can be restored to presettlement standards through a combination of tebuthiuron treatment at 0.2–0.6 kg/ha and moderate grazing (30%–40%),

TABLE 17.7

Mean and closest distance between anthropogenic features and nests of Lesser Prairie-Chickens for birds breeding in sand sagebrush and sand shinnery oak prairie in the Central United States.

Feature	Ecoregion	Mean distance (±SE) from nest to feature (m)	Closest distance from nest to feature (m)	Comments
Powerlines				
	Sand Sagebrush	1,385 ± 60	263	Study area 1
	Sand Sagebrush	1,254 ± 69	144	Study area 2
	Sand Shinnery Oak	1,621 ± 234	102	
Oil wells				
	Sand Sagebrush	588 ± 18	140	Study area 1
	Sand Sagebrush	539 ± 27	54	Study area 2
	Sand Shinnery Oak	2,409 ± 237	432	
Buildings				
	Sand Sagebrush	1,951 ± 64	503	Study area 1
	Sand Sagebrush	2,306 ± 53	1,019	Study area 2
	Sand Shinnery Oak	4,293 ± 479	922	Nest
Roads				
	Sand Sagebrush	224 ± 13	9	Study area 1; Unimproved roads
	Sand Sagebrush	1,526 ± 63	252	Study area 1; Improved roads
	Sand Sagebrush	208 ± 16	11	Study area 2; Unimproved roads
	Sand Sagebrush	3,149 ± 202	465	Study area 2, Improved roads
	Sand Shinnery Oak	554 ± 76	10	Unimproved roads
	Sand Shinnery Oak	4,822 ± 427	671	Improved roads

SOURCES: Data from Pitman et al. (2005b), Grisham et al. (2014).

followed by ≥2 years of grazing deferment following herbicide treatment. A mosaic of treated and untreated areas ~65 ha in size within 1.6 km of leks would maximize food availability and thermal refugia for nests and broods of Lesser Prairie-Chickens (Davis et al. 1979, Taylor and Guthery 1980, Bell et al. 2010, Grisham et al. 2014). Managers should assess stocking rates of cattle, and other livestock, based on annual precipitation and current vegetation conditions (Biondini et al. 1998). Moreover, grazing can be used as a tool to maintain heterogeneity in vegetative structure (Van Poollen and Lacey 1979). Prescribed fire is not an effective management tool for the restoration of shinnery oak prairie from monocultures of shinnery oak (Haukos 2011).

Several incentive programs are available for the management of private lands (Table 17.8). We provide a suggested prioritization of management recommendations based on the available literature for long-term, landscape-level conservation within the framework of adaptive resource management (Table 17.9, Figure 17.2). Maintaining, restoring, and connecting large tracts (>200 km²) of shinnery oak prairie are necessary to facilitate movement among patches and buffer against drought effects on the survival of nests, broods, and adult birds (Grisham 2012; Grisham et al. 2013, 2014). Conservation easements and habitat management plans with a time frame of less than a decade are unlikely to be sufficient for the conservation of Lesser

TABLE 17.8

Incentive programs for private lands that are specific to habitat management for Lesser Prairie-Chickens.

Program	Organization	Goal	Length	Incentives and practices supported
Environmental Quality Incentives Program	National Resource Conservation Service (USDA)	Plan and implement conservation practices that address natural resources concerns and opportunities to improve soil, water, plant, animal, air, and released resources on agricultural lands	10 years	Financial (50% cost share) and technical assistance to agricultural producers. Early-successional habitat development, restoration and management of declining habitat, upland wildlife habitat management.
Lesser Prairie-Chicken Initiative	National Resource Conservation Service (USDA)	Designed to decrease Lesser Prairie-Chicken habitat loss, increase habitat connectivity, and ensure the continued viability of western ranching	N/A	Cost-share funding on practices and scheduling, cost share is usually offered at 75% with the private landowner supplying the other 25% of the funding. Brush management, early-successional habitat development, firebreaks, fence, obstruction removal, prescribed grazing, prescribed burning, range planting, restoration of rare and declining habitats, watering facility.
Partners for Fish and Wildlife	U.S. Fish and Wildlife Service	Restore or enhance fish and wildlife habitats for Federal trust species	10 years	Cost-share opportunities with the private landowners for range management practices, cost-share opportunities with the private landowners for range management practices. Habitat establishment, habitat restoration, habitat enhancement.
Landowner Initiative Program	Texas Parks and Wildlife	The Texas Landowner Incentive Program is a collaborative effort by Texas Parks and Wildlife Department to meet the needs of private, nonfederal landowners wishing to enact good conservation practices on their lands for the benefit of healthy terrestrial and aquatic ecosystems.	2 years with the possibility of extension	Texas Parks and Wildlife will contribute between 50% and 75% of total project costs, while the applicant is expected to contribute the balance. Early-successional habitat development, restoration and management of declining habitat, upland wildlife habitat management

(Continued)

TABLE 17.8 (*Continued*)

Incentive programs for private lands that are specific to habitat management for Lesser Prairie-Chickens.

Program	Organization	Goal	Length	Incentives and practices supported
Working Lands for Wildlife	National Resource Conservation Service (USDA)	Restore populations of declining wildlife species, provide farmers, ranchers, and forest managers with regulatory certainty that conservation investments they make today help sustain their operations over the long term, and strengthen and sustain rural economies by restoring and protecting the productive capacity of working lands	≥30 years	Provides technical and financial assistance to rural landowners, farmers, ranchers, and forest owners. Core practices: Prescribed grazing, upland wildlife habitat management; numerous other supporting practices
Candidate Conservation Agreement and Candidate Conservation Agreement with Assurances	Center for Excellence for Hazard Material Management	Development, coordination, and implementation of conservation actions that reduce or eliminate known threats to Lesser Prairie-Chickens in New Mexico on federal, state and private surface and minerals, support ongoing efforts to reestablish and maintain viable populations of Lesser Prairie-Chickens in currently occupied and suitable habitats, and encourage the development and protection of suitable Lesser Prairie-Chicken habitat by giving participating cooperators incentives to implement specific conservation measures	Permanent	CEHMM provides fund management and administration and is responsible for implementing, monitoring, and reporting on projects completed with candidate conservation agreement funds. Brush management, water facility, fence, habitat reclamation, reseeding, upland wildlife habitat management, powerline removal

TABLE 17.9

An adaptive resource management framework for conservation of Lesser Prairie-Chickens in sand shinnery oak prairie.

Phase	Problem	Design	Implement	Monitor	Evaluate	Adjust
1	Loss and degradation of sand shinnery oak prairie	Restoration of habitats at localized scale (<500 ha), identification, protection, and connectivity of localized habitats at the landscape level	Create heterogeneous habitats around existing leks (<1.6 km radius) through a combination of grazing and herbicide application, consider the population-level influence of anthropogenic structures	Vegetation response (composition and structure) and associated Lesser Prairie-Chicken response (vital rates, movement patterns, lek counts)	Habitats evaluated April (nesting cover) and June (brood-rearing cover), lek counts surveys annually in spring	Fifth- and tenth-year posthabitat manipulation
2	Localized, short-term management (≤10 years) does not meet landscape level or temporal suitability needed to protect populations	Habitat management should be at temporal periods that incorporate population booms and busts (≥10 years)	Long-term conservation easements on private lands (Table 17.8), identify and protect focal areas and connectivity zones, identify and avoid development in existing and adjacent habitats	Lek persistence, #males on lek, lek density to assess population trends, monitor development	Lek surveys conducted annually in spring	Implemented immediately, adjusted on 3–7-year basis because of potential changes in energy demands, agriculture production and market value, and population booms and busts
3	Short-term management (≤10 years) does not account for changes in vegetation composition and structure and subsequently population-level response due to climate change	Connectivity of habitats at scales (≥20,000 ha) needed for large-scale movements to buffer against drought and disturbance	Extend habitat management beyond current distribution on the Southern High Plains (Similar implementation as Phase 2)	Dispersal and movement, lek formation, and persistence	Landover suitability (native grasslands, Conservation Reserve Program, focal areas for development)	See Phase 2
4	Long-term population and habitat monitoring lacking	Standardize research and monitoring protocols in Phases 1–3	Implement research and monitoring protocols range-wide	Phases 1–3	Phases 1–3	Implement immediately and adjust annually based on new information from previous year's efforts
5	Potential for genetic bottlenecks and local extirpations due to climate change	Phases 1–4 at range-wide scale	Phases 1–4 at range-wide scale	Phases 1–4	Phases 1–4	

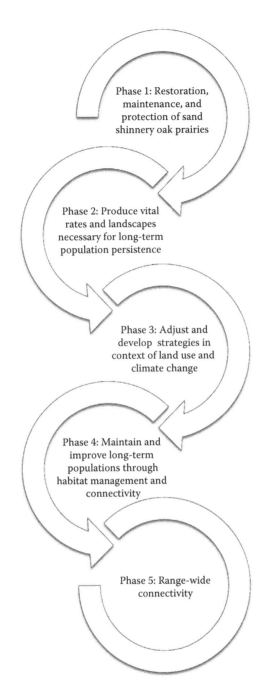

Figure 17.2. Phases 1–5 of an adaptive resource management framework necessary for the conservation of Lesser Prairie-Chickens in sand shinnery oak prairie.

Prairie-Chickens on the Southern High Plains because of natural population fluctuations and the life history strategy of the species. Anthropogenic development and siting of vertical structures in shinnery oak prairie should be evaluated to protect remaining lek and nesting habitat. Future landscape-level planning that includes an understanding of land use and habitat priorities is needed to conserve Lesser Prairie-Chickens. Development within the current existing tracts of shinnery oak prairie is likely counterproductive to conservation.

INFORMATION NEEDS AND GAPS

Lesser Prairie-Chickens in shinnery oak habitats are among the best-studied populations within the species range, but knowledge gaps remain in critical areas. Stock tanks are common features across ranch lands of the Southern High Plains, and Lesser Prairie-Chickens will use tanks as a source of surface water. However, it remains unclear whether surface water holds any positive or negative influences for the species at a population level. Research assessing the influences of surface water on behavior, reproduction, and survival of Lesser Prairie-Chickens is warranted. Informed management decisions require information on whether stock tanks have potential for an increased risk of predation, play a role in disease transmission, or have a buffering effect against drought conditions. The conservation of prairie chickens could be enhanced by increasing the availability and access to surface water.

A comprehensive investigation of the prevalence and intensity of infectious agents of Lesser Prairie-Chickens in shinnery oak prairie is needed (Hagen et al. 2004, Peterson 2004; Chapter 9, this volume). Current information on pathogens and parasites is mainly descriptive (Stabler 1978, Wiedenfeld et al. 2002, Smith et al. 2003). An assessment of the population-level effects of the viruses that cause Newcastle disease, reticuloendotheliosis, and infectious bronchitis is considered a research priority (Peterson 2004). In addition, the prevalence and population-level effects of West Nile virus (WNV) is a concern for Lesser Prairie-Chickens in shinnery oak prairie because birds have tested positive for WNV antibodies in Texas (Chapter 9, this volume). Microparasites (bacteria, protozoans, and fungi) and macroparasites (cestode tapeworms, nematodes, and parasitic arthropods) have been reported from Lesser Prairie-Chickens in populations at the northern extent of their range (Hagen et al. 2002, Peterson et al. 2002, Robel et al. 2003, Peterson 2004; Chapter 9, this volume). Most of these organisms have not been detected in prairie chickens in shinnery oak prairie, with the exception of avian malaria (*Plasmodium [Giovannolaia] pedioeceti*; Pence and Sell 1979, Pence et al. 1983).

An emerging and potentially critical issue is the discovery of numerous leks of Lesser Prairie-Chickens located on lands enrolled in the Conservation Reserve Program (CRP) in the Southern High Plains. Leks occurred primarily on CRP lands in the CP-1 conservation practice, which were dominated by introduced grass species. Initial enrollment of CRP often used CP-1 plantings because the use of CP-2 with >51% native plants was not required until the 1996 Farm Bill (or Federal Agriculture Improvement and Reform Act). The discovery of leks on CP-1 tracts was unexpected given the wide-spread view that Lesser Prairie-Chickens would avoid CP-1 lands without suitable nesting and brood-rearing habitat for the species (Hagen et al. 2004, Hagen and Giesen 2005, Grisham 2012). Expiring CRP contracts is a concern for Lesser Prairie-Chicken population persistence because of additional changes in land cover (Woodward et al. 2001). Based on previous work in the Texas High Plains, many contract holders will return expired CRP to crop production rather than maintain some or all previous CRP acreage as some form of vegetation cover for haying or grazing (Johnson et al. 1997). Future landowner decisions and incentive programs regarding CRP could be guided by further research on Lesser Prairie-Chickens on CRP lands to determine the optimal habitat configuration for the species. Given the federal listing of Lesser Prairie-Chickens as a threatened species under the U.S. Endangered Species Act, and a need to develop conservation strategies for CRP-dominated landscapes in the Southern High Plains, an assessment of the breeding ecology, survival, and habitat use in the CRP areas is needed.

Updated information on diet and available food resources for Lesser Prairie-Chickens in shinnery oak prairie would improve conservation efforts because available information on diet was collected more than 30 years ago in the 1960–1980s. The availability of food resources including invertebrates, acorns and leaves of shinnery oak, and seed abundance and availability in different seasons would improve conservation planning. In addition, the effects of CRP plantings, herbicide applications, grazing, prescribed fire, and other management practices on the availability of food resources should be a focus of future investigations for the conservation of Lesser Prairie-Chickens in the Sand Shinnery Oak Prairie Ecoregion.

ACKNOWLEDGMENTS

We thank a multitude of private landowners in Texas for private land access. We thank the Grasslans Charitable Foundation and Weaver Ranch in New Mexico for access to study sites and logistical and financial support. We thank a plethora of field technicians who assisted with field data collection and M. Patten and P. McDaniel for reviewing earlier drafts of this chapter. We thank A. J. Godar for compiling data on anthropogenic avoidance by Lesser Prairie-Chickens. Financial and logistical support was provided by Texas Tech Department of Natural Resources Management, U.S. Geological Survey, Texas Parks and Wildlife, Great Plains Landscape Conservation Cooperative, and The Nature Conservancy. We report research results from a 6-year dataset on Lesser Prairie-Chickens that was collected in Roosevelt County, New Mexico and Cochran, Hockley, Terry, and Yoakum counties, Texas, 2006–2012. Field protocols for the use of wild birds in research were approved by the Institutional Animal Care and Use Committee at Texas Tech University (Protocol no. 10052-08).

LITERATURE CITED

Ahlborn, C. G. 1980. Brood-rearing habitat and fall–winter movements of Lesser Prairie Chickens in eastern New Mexico. M.S. thesis, New Mexico State University, Las Cruces, NM.

Applegate, R. D., and T. Z. Riley. 1998. Lesser Prairie-Chicken management. Rangelands 20:13–15.

Atlas, R., N. Wolfson, and J. Terry. 1993. The effect of SST and soil moisture anomalies on GLA model simulation of the 1988 United States summer droughts. Journal of Climate 6:2034–2048.

Bailey, J. A., and S. O. Williams III. 2000. Status of the Lesser Prairie-Chicken in New Mexico in 1999. Prairie Naturalist 32:157–168.

Barkley, T. M. 1986. Flora of the Great Plains. University Press of Kansas, Lawrence, KS.

Beck, J. L., and D. L. Mitchell. 2000. Influences of livestock grazing on sage grouse habitat. Wildlife Society Bulletin 28:993–1002.

Behney, A. C., C. W. Boal, H. A. Whitlaw, and D. R. Lucia. 2010. Prey use by Swainson's Hawks in the Lesser Prairie-Chicken range of the Southern High Plains of Texas. Journal of Raptor Research 44:317–322.

Behney, A. C., C. W. Boal, H. A. Whitlaw, and D. R. Lucia. 2011. Interactions of raptors and Lesser Prairie-Chickens at leks in the Texas Southern High Plains. Wilson Journal of Ornithology 123:332–338.

Behney, A. C., C. W. Boal, H. A. Whitlaw, and D. R. Lucia. 2012. Raptor community composition in the Texas Southern High Plains Lesser Prairie-Chicken range. Wildlife Society Bulletin 36:291–296.

Bell, L. A. 2005. Habitat use and growth and development of juvenile Lesser Prairie-Chickens in southeast New Mexico. M.S. thesis, Southeastern Oklahoma State University, Durant, OK.

Bell, L. A., S. D. Fuhlendorf, M. A. Patten, D. H. Wolfe, and S. K. Sherrod. 2010. Lesser Prairie-Chicken hen and brood habitat use on sand shinnery oak. Rangeland Ecology and Management 63:478–486.

Bell, L. A., J. C. Pitman, M. A. Patten, D. H. Wolfe, S. K. Sherrod, and S. D. Fuhlendorf. 2007. Juvenile Lesser Prairie-Chicken growth and development in southeastern New Mexico. Wilson Journal of Ornithology 119:386–391.

Bidwell, T., B. Fuhlendorf, S. Harmon, R. Manes, R. Rodgers, S. Sherrod, and D. Wolfe. 2003. Ecology and management of the Lesser Prairie-Chicken in Oklahoma. Oklahoma Cooperative Extension Service, E-970. Oklahoma State University, Stillwater, OK.

Biondini, M. E., B. D. Patton, and P. E. Nyren. 1998. Grazing intensity and ecosystem processes in Northern Mixed-Grass Prairie. US. Ecological Applications 8:469–479.

Biondini, M., R. D. Pettit, and V. Jones. 1986. Nutrition value of forages on sandy soils as affected by tebuthiuron. Journal of Range Management 39:396–399.

Borsdorf, P. K. 2013. Lesser Prairie-Chicken habitat selection across varying land use practices in eastern New Mexico and west Texas. M.S. thesis, Texas Tech University, Lubbock, TX.

Boyd, C. S., and T. G. Bidwell. 2001. Influence of prescribed fire on Lesser Prairie-Chicken habitat in shinnery oak communities in western Oklahoma. Wildlife Society Bulletin 29:938–947.

Boyd, C. S., L. T. Vermeire, T. G. Bidwell, and R. L. Lochmiller. 2001. Nutritional quality of shinnery oak buds and catkins in response to burning or herbivory. Southwestern Naturalist 46:295–301.

Briske, D. D., J. D. Derner, J. R. Brown, S. D. Fuhlendorf, W. R. Teague, K. M. Havstad, R. L. Gillen, A. J. Ash, and W. D. Willms. 2008. Rotational grazing on rangelands: reconciliation of perception and experimental evidence. Rangeland Ecology and Management 61:3–17.

Brockway, D. G., R. G. Gatewood, and R. Paris. 2002. Restoring fire as an ecological process in shortgrass prairie ecosystems: initial effects of prescribed burning during the dormant and growing seasons. Journal of Environmental Management 65:135–152.

Cannon, R. W., and F. L. Knopf. 1979. Lesser Prairie Chicken response to range fires at the booming ground. Wildlife Society Bulletin 7:44–46.

Chen, P., and M. Newman. 1998. Rossby wave propagation and the rapid development of upper-level anomalous anticyclones during the 1988 US drought. Journal of Climate 11:2491–2504.

Coats, J. 1955. Raising Lesser Prairie Chickens in captivity. Kansas Fish and Game 13:16–20.

Cook, E. R., C. A. Woodhouse, C. M. Eakin, D. M. Meko, and D. W. Stahle. 2004. Long-term aridity changes in the Western United States. Science 306:1015–1018.

Copelin, F. F. 1963. The Lesser Prairie Chicken in Oklahoma. Oklahoma Department of Wildlife Technical Bulletin 6. Oklahoma Department of Wildlife, Oklahoma City, OK.

Corman, K. S. 2011. Conservation and landscape genetics of Texas Lesser Prairie-Chicken: population structure and differentiation, genetic variability, and effective size. M.S. thesis, Texas A&M University–Kingsville, Kingsville, TX.

Crawford, J. A. 1974. The effects of land use on Lesser Prairie Chicken populations in west Texas. Ph.D. dissertation, Texas Tech University, Lubbock, TX.

Crawford, J. A. 1980. Status, problems, and research needs of the Lesser Prairie-Chicken. Pp. 1–7 in P. A. Vohs and F. L. Knopf (editors), Proceedings of the Prairie Grouse Symposium, Oklahoma State University, Stillwater, OK.

Crawford, J. A., and E. G. Bolen. 1973. Spring use of stock ponds by Lesser Prairie-Chickens. Wilson Bulletin 85:471–472.

Crawford, J. A., and E. G. Bolen. 1975. Spring lek activity of the Lesser Prairie Chicken in west Texas. Auk 92:808–810.

Crawford, J. A., and E. G. Bolen. 1976a. Effects of land use on Lesser Prairie Chickens in Texas. Journal of Wildlife Management 40:96–104.

Crawford, J. A., and E. G. Bolen. 1976b. Fall diet of the Lesser Prairie Chickens in west Texas. Condor 78:142–144.

Davis, C. A., T. Z. Riley, J. F. Schwarz, H. R. Suminski, and M. J. Wisdom. 1980. Live trapping female prairie chickens on spring leks. Pp. 64–67 in P. A. Vohs and F. L. Knopf (editors), Proceedings of the Prairie Grouse Symposium, Oklahoma State University, Stillwater, OK.

Davis, C. A., T. Z. Riley, H. R. Suminski, and M. J. Wisdom. 1979. Habitat evaluation of Lesser Prairie Chickens in eastern Chaves County, New Mexico. Final Report to Bureau of Land Management, Roswell, Contract YA-512-CT6-61. Department of Fishery and Wildlife Sciences, New Mexico State University, Las Cruces, NM.

Davis, D. M. 2009. Nesting ecology and reproductive success of Lesser Prairie-Chickens in shinnery oak–dominated rangelands. Wilson Journal of Ornithology 121:322–327.

Davis, D. M., R. E. Horton, E. A. Odell, R. D. Rodgers, and H. A. Whitlaw. 2008. Lesser Prairie-Chicken conservation initiative. Unpublished Report. Lesser Prairie-Chicken Interstate Working Group, Colorado Division of Wildlife, Fort Collins, CO.

Derner, J. D., W. K. Lauenroth, P. Stapp, and D. J. Augustine. 2009. Livestock as ecosystem engineers for grassland bird habitat in western Great Plains of North America. Rangeland Ecology and Management 62:111–118.

Dhillion, S. S., M. A. McGinley, C. F. Friese, and J. C. Zak. 1994. Construction of sand shinnery oak communities of the Llano Estacado: animal disturbances, plant community structure, and restoration. Restoration Ecology 2:51–60.

Doerr, T. B., and F. S. Guthery. 1980. Effects of shinnery oak control on Lesser Prairie Chicken habitat. Pp. 59–63 in P. A. Vohs and F. L. Knopf (editors), Proceedings of the Prairie Grouse Symposium, Oklahoma State University, Stillwater, OK.

Doerr, T. B., and F. S. Guthery. 1983a. Effect of tebuthiuron on Lesser Prairie-Chicken habitat and foods. Journal of Wildlife Management 47:1138–1142.

Doerr, T. B., and F. S. Guthery. 1983b. Food selection by Lesser Prairie-Chickens in northwest Texas. Southwestern Naturalist 28:381–383.

Emmerich, W. 1985. Tebuthiuron—Environmental concerns. Rangelands 7:14–16.

Ethridge, D., R. Pettit, R. Sudderth, and A. Stoecker. 1987. Optimal economic timing of range improvement alternatives: Southern High Plains. Journal of Range Management 40:555–559.

Fields, T. L., G. C. White, W. C. Gilgert, and R. D. Rodgers. 2006. Nest and brood survival of Lesser Prairie-Chickens in west central Kansas. Journal of Wildlife Management 70:931–938.

Fuhlendorf, S. D. 1999. Ecological considerations for woody plant management. Rangelands 21:12–15.

Fuhlendorf, S. D., and D. M. Engle. 2004. Application of the fire-grazing interaction to restore a shifting mosaic on tallgrass prairie. Journal of Applied Ecology 41:604–614.

Fuhlendorf, S. D., D. M. Engle, R. D. Elmore, R. F. Limb, and T. G. Bidwell. 2012. Conservation of pattern and process: developing an alternative paradigm of rangeland management. Rangeland Ecology and Management 65:579–589.

Giesen, K. M. 1994. Movements and nesting habitat of Lesser Prairie-Chicken hens in Colorado. Southwestern Naturalist 39:96–98.

Giesen, K. M. 2000. Population status and management of Lesser Prairie-Chicken in Colorado. Prairie Naturalist 32:137–148.

Greer, H. A. L., E. H. McIlvain, and C. G. Armstrong. 1968. Controlling shinnery oak in western Oklahoma. Oklahoma State University Extension Facts No. 2765. Oklahoma State University, Stillwater, OK.

Grisham, B. A. 2012. The ecology of Lesser Prairie-Chickens in shinnery oak–grassland communities in New Mexico and Texas with implications toward habitat management and future climate change. Ph.D. dissertation, Texas Tech University, Lubbock, TX.

Grisham, B. A., P. K. Borsdorf, C. W. Boal, and K. K. Boydston. 2014. Nesting ecology and nest survival of Lesser Prairie-Chickens on the Southern High Plains of Texas. Journal of Wildlife Management 78:857–866.

Grisham, B. A., C. Boal, D. Haukos, D. Davis, K. Boydston, C. Dixon, and W. Heck. 2013. The potential influence of climate change on Lesser Prairie-Chicken reproductive parameters. PLoS One 8:e68225.

Guthery, F. S. 1999. The role of free water in bobwhite management. Wildlife Society Bulletin 27:538–542.

Hagen, C. A. 2003. A demographic analysis of Lesser Prairie-Chicken populations in southwestern Kansas: survival, population viability, and habitat use. Ph.D. dissertation, Kansas State University, Manhattan, KS.

Hagen, C. A. 2010. Impacts of energy development on prairie grouse ecology: a research synthesis. Transactions of the North American Wildlife and Natural Resources Conference 75:96–103.

Hagen, C. A., S. S. Crupper, R. D. Applegate, and R. J. Robel. 2002. Prevalence of mycoplasma antibodies in Lesser Prairie-Chicken sera. Avian Diseases 46:708–712.

Hagen, C. A., and K. M. Giesen. 2005. Lesser Prairie-Chicken (Tympanuchus pallidicinctus). in A. Poole (editor), The birds of North America (No. 364). Cornell Lab of Ornithology, Ithaca, NY.

Hagen, C. A., B. Grisham, C. Boal, and D. Haukos. 2013. A meta-analysis of Lesser Prairie-Chicken nesting and brood rearing habitats: recommendations for habitat management. Wildlife Society Bulletin 37:750–758.

Hagen, C. A., B. E. Jamison, K. M. Giesen, and T. Z. Riley. 2004. Guidelines for managing Lesser Prairie-Chicken populations and their habitats. Wildlife Society Bulletin 32:69–82.

Hagen, C. A., J. C. Pitman, T. M. Loughin, B. K. Sandercock, R. J. Robel, and R. D. Applegate. 2011. Impacts of anthropogenic features on habitat use by Lesser Prairie-Chickens. Pp. 63–75 in B. K. Sandercock, K. Martin, and G. Segelbacher (editors), Ecology, conservation, and management of grouse. Studies in Avian Biology (no. 39), University of California Press, Berkeley, CA.

Hagen, C. A., J. C. Pitman, B. K. Sandercock, R. J. Robel, and R. D. Applegate. 2005a. Age specific variation in apparent survival rates of male Lesser Prairie-Chickens. Condor 107:78–86.

Hagen, C. A., G. C. Salter, J. C. Pitman, R. J. Robel, and R. D. Applegate. 2005b. Lesser Prairie-Chicken brood habitat in sand sagebrush: invertebrate biomass and vegetation. Wildlife Society Bulletin 33:1080–1091.

Hagen, C. A., J. C. Pitman, B. K. Sandercock, R. J. Robel, and R. D. Applegate. 2007. Age-specific survival and probable causes of mortality in female Lesser Prairie-Chickens. Journal of Wildlife Management 71:518–525.

Hamerstrom, F. N., Jr., and F. Hamerstrom. 1961. Status and problems of North American grouse. Wilson Bulletin 73:284–294.

Harrell, W. C., S. D. Fuhlendorf, and T. G. Bidwell. 2001. Effects of prescribed fire on sand shinnery oak communities. Journal of Range Management 54:685–690.

Hart, R. H., M. J. Samuel, P. S. Test, and M. A. Smith. 1988. Cattle, vegetation, and economic responses to grazing systems and grazing pressure. Journal of Range Management 41:282–286.

Haukos, D. A. 1988. Reproductive ecology of Lesser Prairie-Chickens in west Texas. M.S. thesis, Texas Tech University, Lubbock, TX.

Haukos, D. A. 2011. Use of tebuthiuron to restore sand shinnery oak grasslands of the Southern High Plains. Pp. 103–124 in M. Naguib and A. E. Hasaneen (editors), Herbicide: mechanisms and mode of action. Intech, Rijeka, Croatia.

Haukos, D. A., and G. S. Broda. 1989. Northern Harrier (Circus cyaneus) predation of Lesser Prairie-Chicken (Tympanuchus pallidicinctus). Journal of Raptor Research 23:182–183.

Haukos, D. A., and L. M. Smith. 1989. Lesser Prairie-Chicken nest site selection and vegetation characteristics in tebuthiuron-treated and untreated sand shinnery oak in Texas. Great Basin Naturalist 49:624–629.

Haukos, D. A., L. M. Smith, and G. S. Broda. 1989. The use of radio-telemetry to estimate Lesser Prairie-Chicken nest success and hen survival. Pp. 238–243 in C. J. Amlaner (editor), Proceedings of the 10th International Symposium on Biotelemetry, University of Arkansas Press, Fayetteville, AR.

Henika, F. S. 1940. Present status and future management of the prairie chicken in region 5. Texas Game, Fish, and Oyster Commission, Division of Wildlife Restoration Special Report, Project 1-R, Region 5, Lubbock, TX.

Holechek, J., R. Pieper, and C. Herbel. 2001. Range management: principles and practices. Prentice Hall, Upper Saddle River, NJ.

Holt, R. D. 2012. Breeding season demographics of a Lesser Prairie-Chicken (*Tympanuchus pallidicinctus*) population in the northeast Texas panhandle. Ph.D. dissertation, Texas Tech University, Lubbock, TX.

Hunt, J. L. 2004. Investigation into the decline of populations of the Lesser Prairie-Chicken (*Tympanuchus pallidicinctus*) in southeastern New Mexico. Ph.D. dissertation, University of Auburn, Auburn, AL.

Hunt, J. L., and T. Best. 2010. Vegetative characteristics of active and abandon leks of Lesser Prairie-Chicken (*Tympanuchus pallidicinctus*) in southeastern New Mexico. Southwestern Naturalist 55:477–487.

Jackson, A. S., and R. DeArment. 1963. The Lesser Prairie Chicken in the Texas panhandle. Journal of Wildlife Management 27:733–737.

Jacoby, P. W., and C. H. Meadors. 1982. Control of sand shinnery oak (*Quercus havardii*) with pelleted picloram and tebuthiuron. Weed Science 30:594–597.

Jamison, B. E. 2000. Lesser Prairie-Chicken chick survival, adult survival, and habitat selection and movements of males in fragmented rangelands of southwestern Kansas. M.S thesis, Kansas State University, Manhattan, KS.

Jamison, B. E., R. J. Robel, J. S. Pontius, and R. D. Applegate. 2002. Invertebrate biomass: associations with Lesser Prairie-Chicken habitat use and sand sagebrush density in southwestern Kansas. Wildlife Society Bulletin 30:517–526.

Joern, A. 2004. Variation in grasshopper (Acrididae) densities in response to fire frequency and bison grazing in tallgrass prairie. Environmental Entomology 33:1625.

Johnson, P. N., S. K. Misra, and R. T. Ervin. 1997. A qualitative choice analysis of factors influencing post-CRP land use decisions. Journal of Agricultural and Applied Economics 29:163–173.

Jones, R. E. 1963. Identification and analysis of Lesser and Greater Prairie Chicken habitat. Journal of Wildlife Management 27:757–778.

Jones, R. E. 1964. Habitat used by Lesser Prairie Chickens for feeding related to seasonal behavior of plants in Beaver County, Oklahoma. Southwestern Naturalist 9:111–117.

Jones, V. E., and R. D. Pettit. 1984. Low rates of tebuthiuron for control of sand shinnery oak. Journal of Range Management 37:488–490.

Knopf, F. L., J. A. Sedgwick, and R. W. Cannon. 1988. Guild structure of riparian avifauna relative to seasonal cattle grazing. Journal of Wildlife Management 52:280–290.

Kukal, C. A. 2010. The over-winter ecology of Lesser Prairie-Chicken (*Tympanuchus pallidicinctus*) in the northeast Texas panhandle. M.S. thesis, Texas Tech University, Lubbock, TX.

Lauenroth, W. K., D. G. Milchunas, J. L. Dodd, R. H. Hart, R. K. Heitschmidt, and L. R. Rittenhouse. 1994. Effects of grazing on ecosystems of the Great Plains. Pp. 69–100 in M. Vavra, W. A. Laycock, and R. D. Piper (editors), Ecological implications of livestock herbivory and the West. Society for Range Management, Denver, CO.

Leonard, J. P. 2008. The effects of shinnery oak removal on Lesser Prairie-Chicken survival, movement, and reproduction. M.S thesis, Texas A&M University, College Station, TX.

Locke, B. A. 1992. Lek hypotheses and the location, dispersion, and size of Lesser Prairie-Chicken leks. Ph.D. dissertation, New Mexico State University, Las Cruces, NM.

Lyons, E. K., B. A. Collier, N. J. Silvy, R. R. Lopez, B. E. Toole, R. S. Jones, and S. J. Demaso. 2009. Breeding and non-breeding survival of Lesser Prairie-Chickens *Tympanuchus pallidicinctus* in Texas, USA. Wildlife Biology 15:89–96.

McCleery, R. A., R. R. Lopez, and N. J. Silvy. 2007. Transferring research to endangered species management. Journal of Wildlife Management 71:2134–2141.

McDonald, L., G. Beauprez, G. Gardner, J. Griswold, C. Hagen, D. Klute, S. Kyle, J. Pitman, T. Rintz, and B. Van Pelt. 2014. Range-wide population size of the Lesser Prairie-Chicken: 2012 and 2013. Wildlife Society Bulletin 38:536–546.

McIlvain, E. H., and C. G. Armstrong. 1966. A summary of fire and forage research on shinnery oak rangelands. Proceedings of the Tall Timbers Fire Ecology Conference 5:127–129.

Merchant, S. S. 1982. Habitat-use, reproductive success, and survival of female Lesser Prairie Chickens in two years of contrasting weather. M.S. thesis, New Mexico State University, Las Cruces, NM.

Moldenhauer, W. C., J. R. Coover, and M. E. Everhart. 1958. Control of wind erosion in the sandy lands of the southern High Plains of Texas and New Mexico. ARS 41-20. USDA Agricultural Research Service, Washington, DC.

Natural Resources Conservation Service. [online]. 2011. Ecological Site Description for R077DY045TX. <esis.sc.egov.usda.gov/ESDReport/fsReport.aspx?id=R077DY045TX> (15 November 2013).

Newman, A. L. 1964. Soil survey of Cochran County, Texas. USDA Soil Conservation Service, Washington, DC.

Olawsky, C. D. 1987. Effects of shinnery oak control with tebuthiuron on Lesser Prairie-Chicken populations. M.S. thesis, Texas Tech University, Lubbock, TX.

Olawsky, C. D., and L. M. Smith. 1991. Lesser Prairie-Chicken densities on tebuthiuron-treated and untreated sand shinnery oak rangelands. Journal of Range Management 44:364–368.

Patten, M. A., and J. F. Kelly. 2010. Habitat selection and the perceptual trap. Ecological Applications 20:2148–2156.

Patten, M. A., D. H. Wolfe, and S. K. Sherrod. 2006. The effects of shrub control and grazing on habitat quality and reproductive success of Lesser Prairie-Chickens. Final Report to New Mexico Department Game and Fish. Sutton Avian Research Center, Bartlesville, OK.

Patten, M. A., D. H. Wolfe, E. Shochat, and S. K. Sherrod. 2005a. Habitat fragmentation, rapid evolution and population persistence. Evolutionary Ecology Research 7:235–249.

Patten, M. A., D. H. Wolfe, E. Shochat, and S. K. Sherrod. 2005b. Effects of microclimate and microclimate selection on adult survivorship of the Lesser Prairie-Chicken. Journal of Wildlife Management 69:1270–1278.

Pence, D. B., J. T. Murphy, F. S. Guthery, and T. B. Doerr. 1983. Indications of seasonal variation in the helminth fauna of the Lesser Prairie-Chicken, *Tympanuchus pallidicinctus* (Ridgway) (Tetraonidae), from northwestern Texas. Proceedings of the Helminthological Society of Washington 50:345–347.

Pence, D. B., and D. L. Sell. 1979. Helminths of the Lesser Prairie Chicken, *Tympanuchus pallidicintus* [sic] (Ridgway) (Tetraonidae), from the Texas panhandle. Proceedings of the Helminthological Society of Washington 46:146–149.

Peterson, M. J. 2004. Parasites and infectious diseases of prairie grouse: should managers be concerned? Wildlife Society Bulletin 32:35–55.

Peterson, M. J., P. J. Ferro, M. N. Peterson, R. M. Sullivan, B. E. Toole, and N. J. Silvy. 2002. Infectious disease survey of Lesser Prairie-Chickens in North Texas. Journal of Wildlife Diseases 38:834–839.

Peterson, R. S., and C. S. Boyd. 1998. Ecology and management of sand shinnery communities: a literature review. General Technical Report RMRS-GTR-16. USDA Forest Service, Rocky Mountain Research Station, Fort Collins, CO.

Pettit, R. D. 1979. Effects of picloram and tebuthiuron pellets on sand shinnery oak communities. Journal of Range Management 32:196–200.

Pirius, N. E. 2011. The non-breeding season ecology of Lesser Prairie-Chickens (*Tympanuchus pallidicinctus*) in the Southern High Plains of Texas. M.S. thesis, Texas Tech University, Lubbock, TX.

Pirius, N. E., C. W. Boal, D. A. Haukos, and M. C. Wallace. 2013. Winter habitat use and survival of Lesser Prairie-Chickens in West Texas. Wildlife Society Bulletin 37:759–765.

Pitman, J. C. 2003. Lesser Prairie-Chicken nest site selection and nest success, juvenile gender determination and growth, and juvenile survival and dispersal in southwestern Kansas. M.S. thesis, Kansas State University, Manhattan, KS.

Pitman, J. C., C. A. Hagen, B. E. Jamison, R. J. Robel, T. M. Loughin, and R. D. Applegate. 2006. Nesting ecology of Lesser Prairie-Chickens in sand sagebrush prairie of southwestern Kansas. Wilson Journal of Ornithology 118:23–35.

Pitman, J. C., C. A. Hagen, R. J. Robel, T. M. Loughin, and R. D. Applegate. 2005a. Gender identification and growth of juvenile Lesser Prairie-Chickens. Condor 107:87–96.

Pitman, J. C., C. A. Hagen, R. J. Robel, T. M. Loughin, and R. D. Applegate. 2005b. Location and success of Lesser Prairie-Chicken nests in relation to vegetation and human disturbance. Journal of Wildlife Management 69:1259–1269.

Pruett, C. L., M. A. Patten, and D. H. Wolfe. 2009a. Avoidance behavior by prairie grouse: implications for development of wind energy. Conservation Biology 23:1253–1259.

Pruett, C. L., M. A. Patten, and D. H. Wolfe. 2009b. It's not easy being green: wind energy and a declining grassland bird. BioScience 59:557–262.

Pruett, C. L., J. A. Johnson, L. C. Larson, D. H. Wolfe, and M. A. Patten. 2011. Low effective population size and survivorship in a grassland grouse. Conservation Genetics 12:1205–1214.

Riley, T. Z., and C. A. Davis. 1993. Vegetative characteristics of Lesser Prairie-Chicken brood foraging sites. Prairie Naturalist 25:243–238.

Riley, T. Z., C. A. Davis, and R. A. Smith. 1993. Autumn–winter habitat use of the Lesser Prairie-Chicken (*Tympanuchus pallidicinctus*) (Galliformes: Tetraonidae). Great Basin Naturalist 53:409–411.

Riley, T. Z., C. A. Davis, M. Ortiz, and M. J. Wisdom. 1992. Vegetative characteristics of successful and unsuccessful nests of Lesser Prairie-Chickens. Journal of Wildlife Management 56:383–387.

Robel, R. J. 2004. Summary remarks and personal observations of the situation by an old hunter and researcher. Wildlife Society Bulletin 32:119–122.

Robel, R. J., T. J. Walker, Jr., C. A. Hagen, R. K. Ridley, K. E. Kemp, and R. D. Applegate. 2003. Internal helminth parasites of Lesser Prairie-Chicken *Tympanuchus pallidicinctus* in southwestern Kansas: incidence, burdens, and effects. Wildlife Biology 9:341–349.

Saalfeld, S. T., W. C. Conway, D. A. Haukos, and W. P. Johnson. 2012. Snowy Plover nest site selection, spatial patterning, and temperatures in the Southern High Plains of Texas. Journal of Wildlife Management 76:1703–1711.

Sands, J. L. 1968. Status of the Lesser Prairie Chicken. Audubon Field Notes 22:454–456.

Savory, C. J. 1989. The importance of invertebrate foods to chicks of gallinaceous species. Proceedings of the Nutrition Society 48:113–133.

Schwilling, M. D. 1955. Study of the Lesser Prairie Chicken in southwest Kansas. Kansas Fish and Game 10–12.

Scifres, C. J. 1972. Herbicide interactions in control of sand shinnery oak. Journal of Range Management 25:386–389.

Scifres, C. J., J. W. Smith, and R. W. Bovey. 1981. Control of oaks (*Quercus* spp.) and associated woody species on rangeland with tebuthiuron. Weed Science 29:270–275.

Sell, D. L. 1979. Spring and summer movements and habitat use by Lesser Prairie Chicken females in Yoakum County, Texas. M.S. thesis, Texas Tech University, Lubbock, TX.

Smith, B. H., D. W. Buszynski, and K. Johnson. 2003. Survey for coccidian and haemosporidia in the Lesser Prairie-Chicken (*Tympanuchus pallidicinctus*) from New Mexico with description of a new *Eimeria* species. Journal of Wildlife Diseases 39:347–353.

Smith, R. A. 1979. Fall and winter habitat of Lesser Prairie Chickens in southeastern New Mexico. M.S. thesis, New Mexico State University, Las Cruces, NM.

Smythe, L. A., and D. A. Haukos. 2009. Nesting success of grassland birds in shinnery oak communities treated with tebuthiuron and grazing in eastern New Mexico. Southwestern Naturalist 54:136–145.

Stabler, R. M. 1978. *Plasmodium* (*Giovannolaia*) *pedioecetii* from gallinaceous birds of Colorado. Journal of Parasitology 62:539–544.

Sullivan, J. C. 1980. Differentiation of sand shinnery oak communities in West Texas. M.S. thesis, Texas Tech University, Lubbock, TX.

Suminski, H. R. 1977. Habitat evaluation for Lesser Prairie Chickens in eastern Chaves, New Mexico. M.S. thesis, New Mexico State University, Las Cruces, NM.

Taylor, M. A. 1978. Fall and winter movements and habitat use of Lesser Prairie Chickens. M.S. thesis, Texas Tech University, Lubbock, TX.

Taylor, M. A. 1979. Lesser Prairie Chicken use of man-made leks. Southwestern Naturalist 24:706–707.

Taylor, M. A., and F. S. Guthery. 1980. Fall–winter movements, ranges, and habitat use of Lesser Prairie Chicken. Journal of Wildlife Management 44:521–524.

Texas Game, Fish, and Oyster Commission. 1945. Principal game birds and mammals of Texas: their distribution and management. Von Boeckmann-Jones Co. Press, Austin, TX.

Timmer, J. M., M. J. Butler, W. B. Ballard, C. W. Boal, and H. A. Whitlaw. 2013. Abundance and density of Lesser Prairie-Chickens and leks in Texas. Wildlife Society Bulletin 37:741–749.

Trenberth, K. E., G. W. Branstator, and P. A. Arkin. 1988. Origins of the 1988 North American drought. Science 242:1640–1645.

Van Poollen, H. W., and J. R. Lacey. 1979. Herbage response to grazing systems and stock intensities. Journal of Range Management 32:250–253.

Wester, D. B. 2007. The Southern High Plains: a history of vegetation from 1540 to present. Pp. 24–47 in R. E. Sosebee, D. B. Wester, C. M. Britton, E. D. McArthur, and S. G. Kitchen (editors), Proceedings RMRS-P-47: shrubland dynamics—fire and water, USDA Forest Service, Rocky Mountain Research Station, Fort Collins, CO.

Wiedenfeld, D. A., D. H. Wolfe, J. E. Toepfer, L. M. Mechlin, R. D. Applegate, and S. K. Sherrod. 2002. Survey of reticuloendotheliosis viruses in wild populations of Greater and Lesser Prairie-Chickens. Wilson Bulletin 114:142–144.

Wolfe, D. H., M. A. Patten, E. Shochat, C. L. Pruett, and S. K. Sherrod. 2007. Causes and patterns of mortality in Lesser Prairie-Chickens *Tympanuchus pallidicinctus* and implications for management. Wildlife Biology 13:95–104.

Woodward, A. J. W., S. D. Fuhlendorf, D. M. Leslie, Jr., and J. Shackford. 2001. Influence of landscape composition and change on Lesser Prairie-Chicken (*Tympanuchus pallidicinctus*) populations. American Midland Naturalist 145:261–274.

Zavaleta, J. C. 2012. Community response to use of prescribed grazing and herbicide for restoration of sand shinnery oak grasslands. M.S. thesis, Texas Tech University, Lubbock, TX.

CHAPTER EIGHTEEN

Synthesis, Conclusions, and a Path Forward*

Christian A. Hagen and R. Dwayne Elmore

Abstract. The Lesser Prairie-Chicken (*Tympanuchus pallidicinctus*) is an iconic species of the Southern Great Plains that has experienced significant declines in population numbers and distribution since European settlement. Large-scale land use changes such as conversion of prairies to crop lands, suppression of fire, invasive species, and industrialization have led to ongoing declines. Interest in conservation has recently increased due to the federal listing process of Lesser Prairie-Chickens as a threatened species under the Endangered Species Act. The chapters contained in this volume detail the state of knowledge and recommendations for the recovery of this sensitive prairie grouse. The scope of recovery is significant in that conservation plans encompass five states, multiple agencies, thousands of private landowners, and industry partners. While the information necessary for recovery is largely known, the collective will to implement recovery is uncertain.

Key Words: conservation planning, Endangered Species Act, Lesser Prairie-Chicken, *Tympanuchus pallidicinctus*.

The wildlife of today is not ours to do with as we please. The original stock was given to us in trust for the benefit both of the present and the future. We must render an accounting of this trust to those who come after us.

THEODORE ROOSEVELT

From the Dust Bowl to energy booms, the Lesser Prairie-Chicken (*Tympanuchus pallidicinctus*) continues to capture the imagination of the public. The Lesser Prairie-Chicken is an iconic species of the prairies and shrublands of the Southern Great Plains and tied to native plant communities where ecological patterns and processes are still intact. Like most species of grouse, prairie chickens require large, mostly unfragmented landscapes, a rare commodity in the Anthropocene. The current range of Lesser Prairie-Chickens is the one of the most southerly distributions of any species of grouse (Chapter 1, this volume). Only the Attwater's Prairie-Chicken (*T. cupido attwateri*) can claim a distribution that is further south—currently restricted to a small area along the Gulf Coast of Texas. However, this subspecies of Greater Prairie-Chickens is functionally extinct because persistence of wild populations

* Hagen, C. A. and R. D. Elmore. 2016. Synthesis, conclusions, and a path forward. Pp. 345–351 in D. A. Haukos and C. W. Boal (editors), Ecology and conservation of Lesser Prairie-Chickens. Studies in Avian Biology (no. 48), CRC Press, Boca Raton, FL.

depends on continuing releases of captive-reared birds, a story many hope is not a future shared by the Lesser Prairie-Chicken. The history of the Lesser Prairie-Chicken is unique because the distributional limits do not appear to have changed dramatically since the Pleistocene (Chapter 5, this volume), in contrast to other members of the *Tympanuchus* genus. In fact, distributions of Greater Prairie-Chickens (*T. cupido*) and Sharp-tailed Grouse (*T. phasianellus*) have fluctuated greatly since the last glaciation of the Northern Great Plains and, at least in the case of the Greater Prairie-Chicken, the arrival of Europeans to this continent.

While the overall distribution limits of the Lesser Prairie-Chicken appear to have remained somewhat stable, the occupancy within the landscapes of the Southern Great Plains has been in constant flux before and after settlement by Europeans. Estimates are that >90% of the Lesser Prairie-Chicken historic distribution was lost to conversion of native prairies to cropland (Chapter 2, this volume). Another shift in distribution has occurred more recently, with birds occupying some of the restored grasslands established through the U.S. Department of Agriculture, Conservation Reserve Program (CRP). The CRP in Kansas, in particular, has proven beneficial to the species (Chapter 2, this volume) to the extent that >50% of the current population now resides in this region (Chapter 4, this volume). It is perhaps ironic that one of the largest current populations of Lesser Prairie-Chicken is located outside of the boundaries of the area thought to define the historic distribution (Chapter 14, this volume). In addition to historical observations, recent distributional expansion further illustrate that distribution and population numbers are dynamic and recovery plans must consider a nonequilibrium framework. It is clear that land use has direct and significant implications to Lesser Prairie-Chickens and their future on the landscape. Ultimately, the future of Lesser Prairie-Chickens will depend on patterns of land use and land cover change.

Lesser Prairie-Chickens exhibit boom–bust population dynamics as a result of alternating periods of high and low productivity in response to dynamic environmental conditions with some periodic cycling (Chapter 4, this volume). Precipitation is a known driver of population fluctuations, but negative long-term trends in population numbers are greater than the rates that might be expected based on patterns of precipitation alone. Negative trends have been associated with anthropogenic-driven habitat loss primarily caused by the conversion of prairie to cropland, fire suppression, grazing mismanagement, invasion of exotic plant species, and industrialization that degrades the quality of extant habitats. Accurate estimates of historic population numbers are difficult to ascertain for Lesser Prairie-Chickens, but it is generally agreed that the overall population is much reduced from both historic and contemporary numbers. The historic distribution of the species was fairly extensive, but it cannot be assumed that all of the distribution was occupied at any one time. In fact, at least in contemporary times, much of the estimated distribution has included plant communities that are unlikely to be habitats for the Lesser Prairie-Chicken, such as short-grass prairie and juniper woodland.

Within the overall distribution, extremes in population response are tied to land use and subsequent land cover. A shift in distribution and increase in abundance in the Short-Grass Prairie/CRP Mosaic Ecoregion north of the Arkansas River is encouraging; however, populations of Lesser Prairie-Chickens in the Sand Sagebrush Prairie and Mixed-Grass Prairie Ecoregions have struggled to remain stable. The latter two ecoregions exemplify loss and fragmentation of habitat as the primary threat facing the species (Chapters 6, 11, 15, and 16, this volume). On one hand, the Short-Grass Prairie/CRP Mosaic Ecoregion has provided a roadmap of habitat restoration and population recovery and expansion, and on the other hand, the Mixed-Grass Prairie Ecoregion provides a clear illustration of the factors that lead to the extirpation of populations.

There are minor differences in habitat use within the various ecoregions that support Lesser Prairie-Chickens within their overall distribution. For example, birds occupy both shrublands and grasslands that may be dominated by sand shinnery oak (*Quercus havardii*), sand sagebrush (*Artemisia filfolia*), mixed-grass prairie, short-grass prairie, and converted grasslands (Chapters 14 and 16, this volume). Shrublands increase in importance from east to west in concordance with the longitudinal gradient of decreasing precipitation. Despite a general strategy of potential high reproductive output, different populations of Lesser Prairie-Chickens exhibit regional variation in life history strategies. Females from populations in the Sand Shinnery Oak Prairie Ecoregion have elevated annual survival and produce smaller clutches than populations further north (Chapter 17, this volume). Evidence exists that the pattern may be partly explained by differential mortality associated with land settlement patterns and associated anthropogenic structures (Chapter 16, this volume). Climatic conditions, primary productivity, and predator communities also differ among ecoregions and could contribute to the observed variation in life history traits.

Contemporary declines in population numbers of the Lesser Prairie-Chicken and concerns about the species status are not novel and have a long history (Chapter 2, this volume). However, societal concerns have increased recently with the proposed and then subsequent listing of the Lesser Prairie-Chicken as a threatened species under the Endangered Species Act (Chapter 3, this volume) and then the listing decision being vacated based on procedural grounds through a Federal court decision. Any discussion of recovery of the Lesser Prairie-Chicken should include consideration of genetic consequences and goals. The closely related Greater Prairie-Chicken has experienced substantial distribution shifts since the Pleistocene. Shifts in distribution have created dynamic contact zones between Greater and Lesser Prairie-Chickens in the past and at present (Chapter 5, this volume). The Lesser Prairie-Chicken exhibits morphological and behavioral differentiation from the Greater Prairie-Chicken, but molecular evidence suggests that speciation is not clear from a genetic standpoint (Chapters 5 and 14, this volume). At present, little discussion has taken place within the conservation community over the issues of genetic distinctness between species or among populations (Chapter 5, this volume), hybridization, and patterns of current and future occupancy outside of historic distribution. These issues must be clearly addressed before realistic recovery goals can be identified. Further, recovery goals should state what success looks like and provide milestones toward that success. For example, is it a reasonable goal to conserve genetic variation only, or do additional goals related to population abundance and functional occupancy in the community need to be expressed and monitored? If the latter objectives, should the goals be met across the distribution, or only in certain core areas that coincide with genetic distinctness? Unfortunately, the Endangered Species Act offers little guidance on such matters. The law may be considered elegant in its simplicity and generous in its interpretation, but a lack of specificity allows for arbitrary and sometimes unattainable goals. It is therefore incumbent on managers involved with recovery to provide robust attainable objectives that are transparent to the public.

Prior to the listing decision, the state of knowledge for the conservation of Lesser Prairie-Chickens was summarized in a set of Management Guidelines intended to direct science-based conservation of the species (Hagen et al. 2004). More than a decade later, the legal status of the species is in question, but two major conservation initiatives are in place: the Lesser Prairie-Chicken Initiative and the Lesser Prairie-Chicken Range-wide Conservation Plan (RWP, Van Pelt et al. 2015). Thus, it would be timely to evaluate recent progress for the conservation actions proposed by Hagen et al. (2004). The actions or recommendations were divided into three major categories: Conservation Strategies, Habitat Guidelines, and Population Guidelines.

CONSERVATION STRATEGIES

With the implementation of the RWP, the conservation strategy outlined in the Guidelines is largely being implemented. It is clearly too early to tell whether the strategies will succeed, but at minimum, they are in place. A strategic approach enables managers to target conservation actions that address the most pressing threats in each of the "Management Zones" that are based on broad ecoregions. Not only has this approach assisted in focusing conservation efforts but also in identifying variation in costs basis for addressing those threats. Further, explicit designation of Focal Areas and Connectivity Zones, or target areas in the sense of the Guidelines, has provided an avenue for ensuring that conservation being delivered to the areas is likely to have the greatest biological effect on the species. The Focal Areas and Connectivity Zones should be the subject of continual scrutiny to ensure that they have, and continue to incorporate, the best science available and contribute to species recovery. Target areas should not be driven by political or economic whims, as such planning would not likely result in the recovery of the species. Perhaps most importantly, population goals have been established based on ecoregion designations as the ultimate measure of success of the conservation efforts. It is incumbent upon us to improve our understanding of genetics and ascertain the relevance of developing goals of genetic health at multiple scales. One of the next steps in strategy should identify specific conservation targets such as acres of tree removal for each Focal Area, Connectivity Zone, and Ecoregion to provide clear milestones in achieving the intended conservation benefit.

HABITAT GUIDELINES

Defining a minimum patch size for viable Lesser Prairie-Chicken populations has proven to be challenging in the past and continues to be an issue for future conservation planning. Lesser Prairie-Chickens are clearly an area-sensitive species, but empirical evidence to quantify that minimum area has proven to be elusive (Chapters 6 and 11, this volume). Several lines of evidence indicate that population occurrence can be affected by patch size and configuration at spatial scales ranging from 3,000 to ~30,000 ha (Chapter 11, this volume). Thus, it is reaffirming to know that the RWP

identified Focal Areas that average 20,000 ha in size and are inextricably linked by Connectivity Zones. Therefore, target Focal Area appears to be within the minimum area requirement for Lesser Prairie-Chickens, based on the best information currently available. The original guidelines may have underestimated the minimum patch size, but the importance of connecting patches remains. Better information is needed on patch size and particularly what landscape features constitute barriers or filters to movements of Lesser Prairie-Chickens to ensure that Connectivity Zones are valid and relevant. Regardless of current uncertainties, it is clear that Lesser Prairie-Chickens require large landscapes of relatively unfragmented habitat. Therefore, goals at the landscape level instead of the local level must be a primary consideration (Figure 18.1). Providing adequate nesting cover in a pasture may be inadequate if the matrix of the landscape is unsuitable for Lesser Prairie-Chickens. Further, issues such as disease and loss of genetic potential will be exacerbated by further isolation of populations and lack of focus and action at the landscape level (Chapters 5 and 9, this volume).

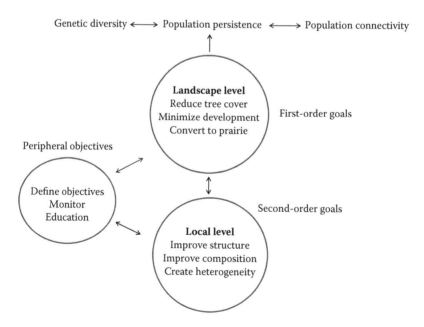

Figure 18.1. Actions to recover the Lesser Prairie-Chicken can be divided into first-order goals, second-order goals, and peripheral objectives. First-order goals are landscape level and long term in nature. These are the most critical for species persistence. Second-order goals are also important at the local population level yet, individually, do less for the probability of persistence across the landscape. Second-order goals would be irrelevant without addressing first-order goals. Peripheral objectives are associated with larger goals but, while critical, are indirectly linked to recovery.

The need for quality habitat within local patches is still important, albeit at a second order than landscape-level considerations (Figure 18.1). The weight of the evidence provides an understanding of the habitat needs of Lesser Prairie-Chickens that are common to all four ecoregions where the species occurs. Heterogeneous structure in the vegetation is vital to fulfilling the life history needs of Lesser Prairie-Chickens—in particular for nesting and brood-rearing stages. Tall denser structure, whether it is comprised entirely of herbaceous plants, or codominant with shrub cover of sand sagebrush or sand shinnery oak, is the vegetation structure usually selected for nesting (Chapter 6, this volume). Vegetative structure provides the dual purpose of concealment cover from predators and thermal refugia from hot temperatures as the optimal conditions for incubation (Chapter 17, this volume). Alternatively, structurally diverse, but generally more open and herbaceous dominant sites are selected by females attending mobile broods of young (Chapter 6, this volume). Our understanding of nest and brood requirements is largely unchanged since the first population studies of the species, but the challenge before us is to identify the management techniques to improve habitat conditions at an appropriate scale for sustainable vital rates and population growth of Lesser Prairie-Chickens. Improved conservation measures are an achievable and measurable objective. However, agencies and managers must not forget the overarching importance of the larger landscape scale, as isolated local conservation actions will not meet the larger goals of recovery (Figure 18.1). Last, the climate of the Southern Great Plains is predicted to undergo some of the most extreme change of any area of North America. Therefore, the effects of climate, the importance of thermal refuge, and the vegetation associated with such refuge need further investigation (Chapter 17, this volume). Emerging evidence suggests that both Lesser and Greater Prairie-Chickens are both highly sensitive to thermal conditions, and especially during incubation, but that variation in microclimate is both predictable and manageable through vegetation manipulation (Hovick et al. 2014; Chapter 17, this volume)

Historically, the synergistic relationship of grazing and periodic fire was the driver of the Southern Great Plains vegetation structure and quality (Chapter 10, this volume). Today, this interaction remains the primary driver, but it has been relegated to infrequent use or use on an insufficient scale to provide the necessary landscape heterogeneity for Lesser Prairie-Chickens and other prairie-dependent species to thrive (Chapter 10, this volume). Changes to patterns of natural disturbance in contemporary prairies are problematic for both public and private lands within the region (Chapter 10, this volume). While the frequency and extent of fire in the more xeric Lesser Prairie-Chicken areas are still being debated, there are regions of Kansas, Oklahoma, and northeast Texas that are being lost at an exponential rate to invasion of mesquite (*Prosopis glandulosa*) or eastern red-cedars (*Juniperus virginiana*) in the absence of fire (Chapter 16, this volume). Thus, we must redouble our efforts to reintroduce disturbance at appropriate temporal and spatial scales to reverse the trends of tree encroachment and improve the quality and quantity of habitat available to Lesser Prairie-Chickens. The removal of woody vegetation is one of the single most direct actions that can increase available habitat and is widely accepted by the ranching community.

Empirical data on the impacts of grazing on Lesser Prairie-Chickens remain elusive. However, the literature is rich with data on vegetation response to various types of grazing management including season-long, rotational, or patch-burn grazing. A synthesis of our knowledge regarding Lesser Prairie-Chicken life history and vegetation response to grazing can provide a framework for grazing management that will serve as a working hypothesis, which can be evaluated accordingly. Ultimately, stocking rates and grazing systems that provide heterogeneity in structure and plant species should optimize habitat quality necessary for Lesser Prairie-Chickens to successfully fulfill the nesting and brood-rearing life history stages (Chapter 6, this volume).

Last, the importance of the larger landscape is illustrated by the observation that the probability of occurrence for a breeding population of Lesser Prairie-Chickens diminishes dramatically when anthropogenic impacts exceed 20% of a landscape (Chapter 11, this volume). Thus, proactive approaches to avoid anthropogenic development in Focal Areas are paramount to ensure population persistence of crucial population centers. The RWP has established a mitigation framework that increases the cost of disturbance proportional to the quality of the habitat, and included in that calculation is a Focal Area multiplier, which provides a minimum of a 2-to-1 unit off-set for every acre impacted within a Focal Area. Early data suggest that this framework has been successful, at least in some instances, in diverting 70% of the development to locations outside Focal Areas (Van Pelt et al. 2015). However, ongoing monitoring and evaluations are necessary to determine the effectiveness of the RWPs mitigation framework for population persistence of Lesser Prairie-Chicken. Ongoing scientific investigations should reveal more refined information relative to the effects of anthropogenic development and how to mitigate and avoid negative effects.

POPULATION GUIDELINES

Since 2004, much has remained the same with regard to population monitoring as the five states with extant populations continue to monitor leks with ground-based surveys, including visits to lek sites or survey routes that were established 40+ years ago. However, new methods for population reconstruction have been developed to examine trend and persistence from count data (Chapter 4, this volume). Perhaps more importantly, a range-wide aerial survey was initiated in 2012 to provide an unbiased estimate of leks and Lesser Prairie-Chicken abundance across the species' distribution (McDonald et al. 2014). Aerial surveys have resulted in annual population estimates and a measure of trend that closely reflects the trend quantified from ongoing ground-based surveys (Chapter 4, this volume). More work is needed to ascertain at what scale population indices reflect changes in management or habitat loss. Additionally, understanding dispersal and dispersal barriers is an area of information that is desperately needed. Monitoring efforts are a peripheral, yet essential objective to ensure that recovery goals are met (Figure 18.1).

Landscape management reintroducing fire across the landscape may be the single most important action that managers can facilitate going forward. Large-scale disturbance will be necessary to convert some areas of woodland to open grasslands or shrublands that can support Lesser Prairie-Chickens. While changes in climate may shift distributions and local persistence, it is paramount that we continue to focus conservation efforts on current populations, as these birds will serve as the source of future generations.

Conservation will require coordinated effort among agencies, conservation organizations, industry, and landowners to minimize any action that fragments existing habitat.

SUMMARY

It is apparent that the current RWP and associated conservation efforts are attempting to address much of what was identified in the Guidelines for Lesser Prairie-Chicken Management. While many knowledge gaps remain, a general framework now exists to start a new phase of recovery for Lesser Prairie-Chickens. The solutions for maintaining or improving habitat conditions for Lesser Prairie-Chickens are relatively straightforward—maintain large open landscapes with minimal fragmentation. Fragmentation is caused by many factors and most notably is associated with encroaching tree cover, conversion to crop fields, and industrial development in native prairies. Generally speaking, we know the where and the how to accomplish recovery. Goals should be focused at a landscape level, but individual properties and landowners are the key to actual implementation. However, the important question is whether or not we, as a society, are willing to implement the key actions that are needed for the Lesser Prairie-Chicken or for any imperiled species. While society in principle may support biodiversity, and the ESA mandates such through the will of the voter, the reality of funding recovery or foregoing other land use is often unpalatable.

Past research has made clear that the public supports the regulation of biodiversity through the Endangered Species Act (Czech and Krausman 1999). However, the bureaucratic administration of endangered species conservation is sometimes not supported at the implementation locale (Elmore et al. 2007). Therefore, transparent and well-defined goals and objectives are paramount to successful recovery of the Lesser Prairie-Chicken, especially given the landownership patterns of the species' distribution.

Greater than 95% of the distribution of the Lesser Prairie-Chicken is in private ownership, and habitat conservation will require voluntary actions by private individuals (Chapter 10, this volume). Even if listed, the absence of a federal nexus for 95% of the distribution provides a significant challenge to those practitioners engaged in habitat conservation for Lesser

Prairie-Chickens. Currently, the bulk of the funding for Lesser Prairie-Chicken conservation is contained in the Food and Securities Act (the Farm Bill), in the context of the Environmental Quality Incentives Program provided by the Natural Resources Conservation Service, and the Conservation Reserve Program of the Farm Services Agency, both agencies of the U.S. Department of Agriculture. The financial resources cannot be realized until a landowner visits a USDA office and requests assistance for technical or financial assistance. Yet, incentives will largely be needed to encourage voluntary conservation on private lands. While the RWP also holds great potential as a funding source for conservation, it is still too early to ascertain what resources will be available. Regardless of the resources, the story is the same—it will require the voluntary efforts of private citizens encouraged by monetary, regulatory, or moral and ethical incentives. Thus, the next step in Lesser Prairie-Chicken conservation is a social experiment in which we will endeavor to educate, encourage, and fund landowners to maintain or improve their operations to benefit habitat conditions, quality of life, and ultimately conditions on the landscape for the species. For without them, the prairies and Lesser Prairie-Chickens that occupy them will likely be lost. The Lesser Prairie-Chicken provides society with a both a challenge and an opportunity. The challenges are well described in this volume of *Studies of Avian Biology*. The opportunity is whether or not society is willing to pay for what, in principle, they have demanded from the ESA. That is, the ideal that biological diversity and the associated function of biotic communities is a fundamental right of the public and necessary.

In retrospect, we can only hear the faint echoes of the last remaining Heath Hen (*T. cupido cupido*) on the eastern shores. After nearly 50 years of protection under ESA, we stand witness to the mere phantom of the Attwater's Prairie-Chicken holding onto a tiny parcel of coastal prairie. Many wonder if a similar fate awaits the Lesser Prairie-Chicken. The opportunity is before us to make a different future, and untested approaches are in motion. Is it enough of the right actions in the right places to maintain and recover this icon of the Southern Great Plains? The final outcome hinges on our collective will to restore the prairie and the dynamic disturbances that maintain it. Only then will our descendants hear those first gobbles at sunrise on a clear and crisp spring morning—knowing that their ancestors learned from the past and held the public trust as something worth the sacrifice.

ACKNOWLEDGMENTS

We thank D. Haukos and C. Boal for providing the impetus and motivation to bring this volume to fruition and for inviting us to attempt to synthesize an outstanding compendium on the ecology of Lesser Prairie-Chickens. We recognize the contributions of all authors in this volume for providing a complete picture of the threats, challenges, and opportunities facing the species and those vested in its conservation. This work was supported in part by a cooperative agreement between Pheasants Forever and Oregon State University (CAH) and by Oklahoma State University Agricultural Experiment Station (RDE).

LITERATURE CITED

Czech, B., and P. R. Krausman. 1999. Public opinion on endangered species conservation and policy. Society and Natural Resources 12:469–479.

Elmore, R. D., T. A. Messmer, and M. W. Brunson. 2007. Perceptions of wildlife damage and species conservation: lessons learned from the Utah prairie dog. Journal of Human Wildlife Conflicts 1:78–88.

Hagen, C. A., B. E. Jamison, K. M. Giesen, and T. Z. Riley. 2004. Guidelines for managing Lesser Prairie-Chicken populations and their habitats. Wildlife Society Bulletin 32:69–82.

Hovick, T. J., R. D. Elmore, B. W. Allred, S. D. Fuhlendorf, and D. K. Dahlgren. 2014. Landscapes as a thermal moderator of thermal extremes: a case study from an imperiled grouse. Ecosphere 5:art35.

McDonald, L., G. Beauprez, G. Gardner, J. Griswold, C. A. Hagen, D. Klute, S. Kyle, J. Pitman, T. Rintz, D. Schoeling, and W. Van Pelt. 2014. Range-wide population size of the Lesser Prairie-Chicken: 2012 and 2013. Wildlife Society Bulletin 38:536–546.

Van Pelt, W. E., S. Kyle, J. Pitman, D. VonDeBur, and M. Houts, 2015. The 2014 Lesser Prairie-Chicken range-wide conservation plan annual progress report. Western Association of Fish and Wildlife Agencies, Boise, ID.

INDEX

A

Abert, James, 16
abundance of prairie chickens
    accounts of early settlers, 19, 282
    current and percent of original estimated
        abundance, 206
    current estimate for Mixed-Grass Prairie
        Ecoregion, 299
    current estimates by state for 2000–2012, 8 (table)
    current in Texas, 324
    decline, 20, 85
    during 2000–2011, 6
    ecoregional from 2012 to 2014, 7 (table), 8
    effect of drought and landscape changes in Sand
        Sagebrush Prairie Ecoregion, 282
    estimates for Sand Sagebrush Prairie Ecoregion, 283,
        285 (table)
    fluctuations and environmental conditions, 236
    historical, 19–20, 108, 283, 324
    impact of CRP, 268
    Kansas as population core, 140, 346
    limiting factors, 103, 268
    population goal by ecoregion, 7 (table), 8
adaptive management, 237
    framework for conservation, 126, 244
Aedes sp. (mosquito), 167
Agricultural Act, 21
Agricultural Land Easements, 273
agriculture. *See also* grazing
    center-pivot irrigation, 21, 22, 24, 228, 260, 284–285,
        294, 316
    climate change, 228, 237
    conversion from grain to cotton production in sand
        shinnery oak prairie, 321
    croplands used as lek sites, 267
    drought, 228
    economics of row-crop agriculture, 25–26
    erodibility index of cropland in Sand Sagebrush Prairie
        Ecoregion, 294
    habitat conversion, 8, 15, 21, 28, 40, 72–73, 85, 100,
        102–103, 124, 188, 195, 206–208, 217, 237, 275,
        285, 305, 316, 346

    industrial-scale feedlots and habitat loss, 22
    livestock grazing, 197–198, 266
    loss of artesian water sources, 124
    loss of historical range, 346
    population decline 19, 283
Akaike's Information Criterion, 54–55
alfalfa (*Medicago sativa*), 262
alfalfa (*Medicago* spp.), 301
analysis
    Akaike's Information Criterion, 54
    annual rate of population change, 53
    assumptions of Ricker and Gompertz population
        growth models, 55
    carrying capacity, 55
    consideration and limitations of lek data, 57–59
    ecoregional population dynamics assessment, 49–76
    Generalized Linear Mixed Model, 212
    index of population size, 53–54
    methods to assess viability of metapopulations, 56–57
    methods to estimate effective population
        size, 59–60
    methods to fit population growth models, 54–55
    minimum population estimate, 52
    population growth models, 54–55
    population reconstruction, 52–54
    ratio estimators to reconstruct abundance
        estimates, 53
    Ricker and Gompertz models, 54–55
    sampling error in lek data, 57
    stochastic population projections, 55–56
Andropogon spp. (bluestem grass)
    nesting cover, 321
    winter cover in shinnery oak communities, 323
Anopheles sp. (mosquito), 167
anthropogenic features, 285
    avoidance buffer distances, 121, 211 (table)
    behavioral avoidance, 121, 171, 208, 302, 332
    distance between anthropogenic features and nest
        site, 333 (table)
    estimates of habitat loss, 208
    nest success, 121
    nesting, 22, 121, 208
    predator response, 146

anthropogenic impacts, 2, 21, 207–209, 332. *See also*
  specific land use
  center-pivot irrigation and habitat loss, 21,
    284–285, 316
  conversion of habitat in Sand Shinnery Oak Prairie
    Ecoregion, 316
  driver of population declines, 2
  energy development, 22, 199–200, 227
  exacerbated by climate change, 216
  fragmentation, 2, 72, 200, 303, 305, 307
  habitat loss due to behavioral avoidance, 121, 211
  human settlement, 16–19, 207
  land use practices and population recovery following
    drought, 206
  landscape threshold, 349
  Mixed-Grass Prairie Ecoregion, 301–303
  noise, 22, 302
  potential lag effects, 217
  proportion of landscape in native prairie to support
    Lesser Prairie-Chickens, 207
  rate of landscape change and lek attendance, 207–208
  Short-Grass Prairie/Conservation Reserve Program
    (CRP) Mosaic Ecoregion, 260
  Southern Great Plains, 303
Arkansas River, 17, 260, 262, 270, 282
Aspergillus flavus (pathogenic fungus), 170
  spp., 169
  parasiticus, 170
Audubon Christmas Bird Count
  Lesser Prairie-Chicken trend data for Oklahoma, 303

B

badger (*Taxidea taxus*), 147
Bankhead–Jones Farm Tenant Act, 20
Beaver River Wildlife Management Area, 189
beetle, tamarisk [saltcedar leaf] (*Diorhabda* spp.)
  biological control agent for salt cedar, 293
behavior, 148, 209
  antipredator strategies, 146, 148
  avoidance of anthropogenic features, 73, 121, 171–172,
    174, 208–209, 216, 238, 302
  avoidance of trees, 104
  brood, 4, 288, 323
  competition, 264
  flight, 2
  flocking, 116, 124, 282
  interspecific nest parasitism, 302
  isolating mechanisms, 264, 275
  lekking and raptor migration, 148
  mechanisms to explain avoidance of vertical structures,
    208–209
  microclimate selection, 323
  nest abandonment, 325
  nest site selection, 120
  plasticity, 237
  predator, 148, 154
  site fidelity, 217, 289
  territorial, 235
  variability in response to avian predators, 148
biofuel, 27–38
bison, American (*Bison bison*), 15, 17–18, 206–207,
    245, 266, 318
  grazing behavior, 207, 307
  restoration of prairie ecosystem functions, 206

Black Kettle National Grassland
  acreage and history, 20
  fragmentation by tree cover, 188
blackhead. *See* histominiasis
blanket, Indian (*Gaillardia pulchella*), 301, 317
bluestem (*Andropogon* spp.), 103, 282
bluestem, big (*Andropogon gerardii*), 125, 230 (table), 262,
    289, 317
  nesting cover, 121, 268
bluestem, little (*Schizachyrium scoparium*), 103, 125, 230
    (table), 262, 272, 282, 289, 317
  important winter cover in shinnery oak
    communities, 323
  nesting cover, 268, 270, 321
bluestem, Old World [Caucasian, yellow] (*Bothriochloa
    ischaemum*), 271, 300, 307
  monotypic stands in CRP, 190
bluestem, sand (*Andropogon gerardii hallii*), 230 (table), 317
bluestem, sand (*Andropogon hallii*), 129, 230 (table),
    282, 289
  nesting cover, 125, 129, 209, 321
bluestem, silver (*Bothriochloa saccharoides*), 230 (table)
bobcat (*Lynx rufus*), 147
Bobwhite, Masked (*Colinus virginianus ridgwayi*), 92
Bobwhite, Northern (*Colinus virginianus*), 163, 170, 173, 176
  female nutritional status and productivity, 169
  Oxyspirura petrowi, 166
  pathogenicity of *Tetrameres* spp., 164
  preincubation losses due to thermal stress, 172
Bonasa sp., 81
Bouteloua spp. (bluestem grass), 305
Brazos River, 245
breeding behavior, 264
  copulation, 4
  courtship display, 2–3 (photo), 4, 78
  female choice, 164, 264
  female lek visitation, 4, 116, 148, 235, 288, 304
  hybrid males, 264
  influence of drought, 235
  lekking, 3–4, 52, 116
  multiple mating by females, 89
  return time to leks following disturbance, 148
bristlegrass, plains (*Setaria vulpiseta*), 125
brome, smooth (*Bromus inermis*), 271
bronchitis, infectious. *See* viruses
brood ecology, 4, 148–150, 288
  Sand Sagebrush Prairie Ecoregion, 122–123
  Sand Shinnery Oak Prairie Ecoregion, 122–123
brood habitat, 267–268, 323, 349
  bare ground composition, 330
  characteristics, 122–123, 289
  forb composition and vegetation structure, 123, 198
  horizontal and canopy cover, 301, 323
  importance of movement ability, 288
  invertebrate food resources, 123, 288, 323
  thermal refugia, 323, 329
  transitional edge between CRP and grasslands, 268
brood survival, 122, 148–150, 287, 323
  environmental conditions, 150
  estimates by state and ecoregion, 122 (table)
  forbs in shrub monocultures as limiting factor, 323
  Greater Sage-Grouse, 150
  herbicide use and grazing, 330
  invertebrate availability and biomass, 236
  precipitation, 231

petroleum extraction practices and environmental contaminants, 171

potential impacts to Lesser Prairie-Chickens, 199

projected future increase, 227–228, 238

voluntary mitigation and funding of conservation efforts, 309

well density in Mixed-Grass Prairie Ecoregion, 302

Wildlife Management Areas, 24

wind energy potential within occupied range, 229 (figure)

Energy Independence and Security Act, 28

Enlarged Homestead Act, 18, 245

environmental contaminants, 170–171

dichlorodiphenyl-trichloroethane (DDT) and Attwater's Prairie-Chickens, 170

estrogens, 175

impacts on bird health, 170

Environmental Quality Incentives Program (EQIP), 34, 191, 248–249, 251, 334 (table), 351

Exotic species

competition with native species, 307

CRP monocultures, 270–271, 307

grass species, 271, 300

habitat degradation, 26

introduced in Oklahoma, 307

F

Falcipennis spp., 81

Falcon, Prairie (*Falco mexicanus*), 147

Farm Service Agency, 190, 262

Federal Agriculture Improvement and Reform Act, 33, 248

Federal Energy Policy Act

biofuel industry, 27

production targets for ethanol, 28

Federal Preemption Act, 17

Federal Soil Bank Program, 21

Federal Timber Culture Act, 18

fences, 18, 150, 208, 302, 327

density in Oklahoma and history of settlement, 18

electric, 309

fence line density, 208, 305, 308–309

localized hazard risk, 172

marking, 309

removal, 309

fescue, six weeks (*Vulpia octoflora*), 230 (table)

fire, 266, 271, 349

ecological driver in prairie ecosystems, 207, 275, 293, 318

effects of suppression, 22, 201, 266

funding allocated to fuels reduction on federal lands, 197

grassland response and moisture, 266, 293

management tool for tree and shrub control, 195–197, 275

patch burning and structural diversity of habitat, 307

prescribed, 2, 195, 302, 332

return interval, 207, 266, 293, 318

role in maintaining Lesser Prairie-Chicken habitat, 266–267

timing and return interval, 293

fitness

energetic demands during breeding season, 108

food availability and quality, 108

potential effect of hybridization, 264

reproductive success and vegetation structure, 287

site variation and invertebrate abundance, 307

food resources

crop fields, 19, 108, 124, 207, 267, 282, 301

effect of temperature and precipitation on invertebrate availability and diversity, 236

forb availability and grasshopper abundance, 330

grain and mycotoxins, 170

grasshopper biomass and sand sagebrush density, 288

insect availability and forbs, 321, 329

invertebrate biomass and forb abundance, 288

limiting factor for populations, 113

movements, 108

sand shinnery oak, 321, 323, 327

spatial and temporal variation, 108, 113

use of grain during severe environmental conditions, 113

Food Security Act, 190

fossil record, 82

fox, gray (*Urocyon cinereoargenteus*), 147

fox, red (*Vulpes vulpes*), 147

fox, swift (*Vulpes velox*), 147

Fremont, John C., 16

fungi. *See* parasites

G

gaura, scarlet [scarlet beeblossom] (*Gaura coccinea*), 317

Gene Howe Wildlife Management Area, 189

genetic diversity, 86, 89

bottlenecks, 88

gene flow and female movements, 288

heterozygosity excess, 88

importance of immigration, 89

inbreeding depression, 306

isolated and peripheral populations, 86

Mixed-Grass Prairie Ecosystem, 306

populations in New Mexico and Texas, 86

rapid range contraction, 89

regional variation, 3

synergistic factors influencing, 91

within and among populations, 86–89

genetics

autosomal genes, 81

central-marginal hypothesis, 86

differentiation within *Tympanuchus* and climate-induced changes, 84

divergence and speciation of *Tympanuchus*, 78–79 (table), 81, 265

effect of historical and recent habitat changes, 82–86

gametic and postzygotic isolating mechanisms, 264

genetic distance and estimated migrants per generation by county and state, 57–58 (table)

intermixing with Greater Prairie-Chicken, 237

introgression, 82, 92, 264–265

lek structure and related males, 89

maximum likelihood phylogeny of grouse mtDNA control region sequences, 79 (figure)

mitochondrial genes, 81

molecular evidence regarding speciation, 347

neutral genetic variation of Lesser Prairie-Chicken, 81, 86, 88

shifts in geographic range, 82

focus on working lands, 34
goals and objectives, 34, 191
Grassland Reserve Program (GRP), 191
importance of partnerships to achieve goals, 255
incidental take and Endangered Species Act, 43–44
influence on grazing management, 266
prescribed grazing incentives, 272
private lands impacted, 34, 192
technical assistance and cost sharing, 34
Lesser Prairie-Chicken Interstate Working Group, 347
conservation efforts among state and federal
agencies, 29–30
Geographical Priority Area in New Mexico, 249
inter-organizational communication, 33–34
New Mexico local working group, 249
objectives, 29–30
participants, 29
public outreach efforts, 29, 246
publications, 29–30
role of energy industries to fund conservation
efforts, 30
role of private landowners in conservation efforts, 29
Lesser Prairie-Chicken Range-wide Conservation Plan
(RWP), 34, 244, 347
core habitat area and connectivity corridors, 34
delineation of critical areas in Short-Grass Prairie/
Conservation Reserve Program (CRP) Mosaic
Ecoregion, 271
energy development, 273
funding prioritization, 193
importance of partnerships to achieve goals, 255
listing decision and incidental take, 44
management purposes, 193
mitigation funding, 34
opposition, 35
population goals range-wide and by ecoregion, 34
prescribed grazing incentives, 272
public input, 34–35
recommendation for size of Focal Area, 348
role of private landowners, 34
voluntary framework, 193
life history strategy
breeding success and age class, 71
clutch size and renesting in Oklahoma, 86, 91
female survival and clutch size, 287, 303, 346–347
geographic variation for chick growth and asymptotic
body mass, 326–327
geographic variation for survival and reproductive
efforts, 325–326 (table)
influence of environment, 5
paternal investment, 4
populations in Sand Shinnery Oak Ecoregion, 71
regional variation, 346
r-selected for Lesser Prairie-Chickens, 89
spatial and temporal variation of reproductive
success, 4
life history traits, 3–5, 26, 303–305. See also clutch size
adult survival rates, 89
annual variation of vital rates on the Southern High
Plains, 326
longevity, 89
mating system, 3–4, 81
sex ratio, 107
spatial variation for average generation time, 5
synchronous hatch, 4

Little Ice Age, 18
lizard, dunes sagebrush (Sceloporus arenicolus), 31, 252
Llano Estacado, 16–17. See also Southern High Plains
description, 245
settlement, 17–18
locoweed, wooly (Astragalus mollissimus), 317
lovegrass (Eragrostis spp.), 300
lovegrass, sand (Eragrostis trichodes), 230 (table)
lovegrass, weeping (Eragrostis curvula), 230 (table), 307
monotypic stands in CRP, 190

M

malaria, avian. See also Plasmodium pediocetii
detection in sand shinnery oak prairie, 338
impacts, 173
potential to regulate bird populations, 173
management practices
anthropogenic structures, 195
efficacy and interactions with drought, 317
efforts north of Arkansas River in Kansas, 270–273
escape ramps for stock tanks, 198
fire suppression, 40, 305
grazing, 40, 195, 291–292 (table)
herbicide use and grazing practices in Sand Shinnery
Oak Ecoregion, 330–332
importance of scale, 121
interseeding with forbs and use of CRP fields for nest
sites, 268
patch burning, 302, 306–307
patch-burn grazing system and nesting and brooding
habitat, 293
Sand Sagebrush Prairie Ecoregion, 291–293
woody plant encroachment, 31, 195–197, 308
management recommendations, 275, 289, 294, 308–309
anthropogenic features, 174, 211 (table), 272
brood habitat, 73, 288, 290, 292 (table), 323
brood survival, 328
connectivity, 237
Conservation Reserve Program lands, 123, 271–272
drought, 228, 237
energy development, 41, 73, 201, 308
exotic grasses, 271–272
fences, 309
grazing prescriptions to create nesting and brood-
rearing habitats in Kansas, 292 (table)
grazing, 126, 198, 272, 292, 307, 329–330
habitat, 73, 120, 125–126, 237, 332–337
harvest, 134
herbicides, 307–308, 329
incorporation of climate models into recovery
efforts, 275
landscape scale, 289, 350
lek complexes and interlek distance, 301
mixed-grass prairie, 307–309
monitoring, 72, 92
nesting ecology in Sand Sagebrush Prairie Ecoregion, 289
nesting habitat, 73, 119–120, 329
predators, 154
prescribed fire, 73, 210, 271–272, 307–308, 321
proportion of nest and brood habitat near lek site, 290
research based information, 274
sand sagebrush prairie, 290 (table)
sand shinnery oak prairie, 321, 329, 331–337
woody plant encroachment, 73

home-range size, 108, 217
long distance, 91, 288
mortality risk, 304, 308
nomadic, 108
range expansions from 1870s to 1900s, 19
rescue effect for Mixed-Grass Prairie and Sand
    Sagebrush Prairie ecoregions, 71–72
seasonal variation for males, 288
Muleshoe National Wildlife Refuge, 189
Mycoplasma spp. (pathogenic bacteria), 167, 173
    gallisepticum, 168 (table)
    meleagridis, 168 (table)
    synoviae, 168 (table)

N

National Fish and Wildlife Foundation, 249
National Grasslands
    guidelines relative to energy development, 199
    Lesser Prairie-Chicken Habitat Requirements, 197
    Lesser Prairie-Chicken Management Plan, 290–291
    locations within Lesser Prairie-Chicken range by
        state, 188–189
    management and Multiple-Use and Sustained Yield
        Act, 194
    management authority, 188
    prescribed fire protocol, 197
    primary land uses within Lesser Prairie-Chicken
        range, 188
National Wildlife Refuges
    locations within historical range, 189
Native Americans, 207, 245, 266
Natural Resources Conservation Service (NRCS), 190
    conditioned conservation practices, 44
    contracts and acreages targeting habitat of Lesser
        Prairie-Chickens in eastern New Mexico, 251
    Lesser Prairie-Chicken Initiative (LPCI), 34
    Major Land Resource Area, 300
needle-and-thread (*Hesperostipa comata*), 230 (table)
nest initiation
    dates and rates for Texas and New Mexico, 325 (table)
    renesting rate in sand shinnery oak prairie, 326 (table)
    renesting, 4, 287
nest site
    characteristics by state, 117–118 (table)
    characteristics in Mixed-Grass Prairie Ecoregion, 121
    cover and density of sand sagebrush, 287
    distance between anthropogenic features and nests
        for birds breeding in Sand Sagebrush and Sand
        Shinnery Oak Prairie, 333 (table)
    distance from anthropogenic features, 121, 323, 332
    distance from lek, 4, 119, 121, 217, 289, 301, 321
    distance from previous nests from the same or different
        year, 321
    distance from stock tanks, 323
    importance of litter in Sand Shinnery Oak Prairie, 120
    importance of structural characteristics, 121
    influence of herbicide use and grazing, 330–331
    litter, 117–118 (table), 120
    microclimate, 119
    percent cover of shrubs in Texas and New Mexico,
        321, 323
    placement and CRP, 121
    placement and grazing, 121
    placement and herbicide treatment, 121

plant species used as cover, 321
selection in Sand Shinnery Oak Prairie Ecoregion, 120
selection of bare ground and litter, 120
use of shrubs, 120
vegetation structure, 120, 271, 287, 323
vegetation types in Short-Grass Prairie/Conservation
    Reserve Program (CRP) Mosaic Ecoregion, 268
vertical structure and bare ground, 301
nest success
    cause-specific nest failure of Lesser Prairie-Chickens,
        149 (table)
    dispersed nesting sites, 154–155
    drought and thermal stress, 325
    factors influencing, 268
    grazing, 302
    importance of small-scale habitat patchiness, 271
    indirect impact of depredation of nesting female, 147
    influence of vegetation structure in Sand Sagebrush
        Prairie Ecoregion, 287
    predation by mammals, 147, 268
    relationship to spring temperature and
        precipitation, 325
    relative to precipitation in 6-month period before
        nesting season, 304 (table)
    Sand Sagebrush Prairie Ecoregion, 287
    sand shinnery oak prairie, 326 (table)
    Short-Grass Prairie/Conservation Reserve Program
        (CRP) Mosaic Ecoregion, 268
    temperature, 231, 268
    timing of incubation, 235–236
    vegetation characteristics, 307
    weather, 235
nesting habitat, 116–122, 125, 289, 301–302, 321–323,
        330, 332
    characteristics of sand sagebrush cover in Sand
        Sagebrush Prairie Ecoregion, 287
    effect of fire in sand shinnery oak prairie, 332
    features in mixed-grass prairie, 300 (table), 307
    grass species that provide suitable structure in Short-
        Grass Prairie/Conservation Reserve Program
        (CRP) Mosaic Ecoregion, 272
    horizontal and canopy cover, 301
    interspersion of grasses and shrubs for nesting, 321
    interspersion with brood habitat, 329
    proportion of nests in CRP and grassland for
        Kansas, 268
    proportion of nests observed for herbicide and grazing
        treatments in New Mexico, 331 (table)
    residual cover, 120, 198
    Sand Sagebrush Prairie Ecoregion, 287
    selected vegetation structure, 349
    use of CRP, 120, 268, 271
    vegetation characteristics of successful nests, 307
New Mexico Department of Game and Fish, 42
New Mexico, 137, 245
    average annual rainfall in eastern portion, 245
    BLM Area of Critical Environmental Concern, 31
    BLM planning efforts to benefit Lesser Prairie-Chickens,
        31–32
    early records of Lesser Prairie-Chicken, 245–247
    early settlement, 245
    estimates for Lesser Prairie-Chickens during
        1960–1995, 138 (figure)
    historical estimates of population numbers, 324
    number and extent of Prairie Chicken Areas, 246

prairie-dog, black-tailed (*Cynomys ludovicianus*), 206, 266, 301
Pratt Sandhills Wildlife Area, 189
precipitation
    amount and pattern for Sand Shinnery Oak Prairie Ecoregion, 317
    annual for current distribution, 103
    annual for eastern New Mexico, 245
    annual for Mixed-Grass Prairie Ecoregion, 300
    annual for Sand Sagebrush Prairie Ecoregion, 285
    annual variation and population trends, 89
    annual within Lesser Prairie-Chicken range, 222
    average for sand sagebrush–mixed prairie, 282
    boundary for semiarid environmental condition, 103
    brood survival, 231
    correlation with incubation timing, 236
    dynamic cycle for Great Plains, 206
    ecological driver of grassland productivity, 222
    extreme events for New Mexico, 245
    gradient across occupied range of Lesser Prairie-Chickens, 103, 227 (figure), 231
    lek counts in Sand Sagebrush Prairie Ecoregion, 285
    level and influence on limiting factors 103
    predicted changes under climate change, 222
    probability for Mixed-Grass Prairie Ecoregion, 300
    soil moisture, 222
    timing in Short-Grass Prairie/Conservation Reserve Program (CRP) Mosaic Ecoregion, 265–266
predation risk, 147
    environmental conditions and nest detection distance, 149
    factors influencing variation, 151, 153
    female lek visitation and presence of raptors, 148, 327
    lekking period, 148, 327
    males, 148, 327
    potential role of disease and parasites, 153
    prey dynamics, 153
    raptor breeding season and migration, 5, 147, 327
    stress-related mortality, 153
    temporal and spatial variation, 150–151, 304
    variation in opportunistic predation risk, 147
predation, 4–5, 126, 150
    adult, 150–151
    avian, 5, 305, 327
    cause-specific mortality of juveniles and adults, 151–152 (table)
    cause-specific nest failure, 149 (table)
    chick, 148–152
    effects and landscape-level impacts of human activities, 146
    factors influencing population-level predation, 146
    functional and numerical responses of predators, 153
    leks in Southern High Plains, 328
    mammal, 5, 150, 305, 327
    nest, 148–150, 268, 287
    number of predation mortalities in Texas during the breeding season, 151 (figure)
    observed raptor encounters, 148, 328
    predator-prey dynamics, 153–154
    rates for radio-marked birds, 171
    relative monthly mammal and raptor predation of Lesser Prairie-Chicken in New Mexico and Oklahoma, 151 (figure)
    seasonal variation, 147, 150, 287, 305, 327
    snake, 5, 147

predators, 5, 147, 287, 328
    cattle, 148
    challenges of control programs, 154–155
    detection at water sources, 269
    difficulty identifying, 146, 153
    mammal species, 5, 147–148, 154, 287, 327
    mammals as key chick predators, 150
    nest, 147
    olfactory detection of nests, 148
    opportunistic, 147, 153
    raptor species, 147–148, 327–328
    seasonal variation, 327
    snakes, 287
    suppression by early settlers, 18
    utility poles as a predictor of raptor density, 154
private lands
    adaptive resource management, 237–238
    Candidate Conservation Agreements (CCA), 192
    conservation banks, 193–194
    conservation programs, 194–195
    disturbance factors, 188
    grasslands landownership, 2, 8
    importance for conservation efforts, 8
    incentive-based conservation, 195, 248, 333–337
    Lesser Prairie-Chicken Initiative enrollments by state, 192
    Lesser Prairie-Chicken Range-wide Conservation Plan, 193
    management focus and landowner objectives, 194
    management practices and land uses, 195–201
    mitigation efforts to fund financial incentives for landowners, 308
    Partners for Fish and Wildlife, 192–193
    potential energy development restrictions, 200
    proportion of the occupied range of Lesser Prairie-Chickens, 188
    recreation, 199
    Safe Harbor agreements, 192
    state programs for conservation efforts, 193
    variability of habitat quality, 188
    woody plant management, 196–197, 201, 349
pronghorn (*Antilocapra americana*), 266, 318
ptarmigan (*Lagopus* spp.), 78
public lands
    hunting, 199
    management practices and land uses, 195–201
    number and acreage of Prairie-Chicken Areas in New Mexico, 24
    optimization of management objectives, 194
    potential to support Lesser Prairie-Chickens, 190
    role of policy in conservation actions, 194
    state and federal, 188–190
    Wildlife Management Areas in Oklahoma, 24
    wind and natural gas development, 24

Q

Quail, Gambel's (*Callipepla gambelii*), 163
Quail, Montezuma (*Cyrtonyx montezumae*), 163
Quail, Scaled (*Callipepla squamata*), 163

R

raccoon (*Procyon lotor*), 269
ragweed, western (*Ambrosia psilostachya*), 230 (table)
raven (*Corvus* spp.), 145, 147–148
Raven, Chihuahuan (*Corvus cryptoleucus*), 147

STUDIES IN AVIAN BIOLOGY
Series Editor: Brett K. Sandercock

34. *Beyond Mayfield: Measurements of Nest-Survival Data.* Jones, S. L., and G. R. Geupel, editors. 2007.

35. *Foraging Dynamics of Seabirds in the Eastern Tropical Pacific Ocean.* Spear, L. B., D. G. Ainley, and W. A. Walker. 2007.

36. *Status of the Red Knot (Calidris canutus rufa) in the Western Hemisphere.* Niles, L. J., H. P. Sitters, A. D. Dey, P. W. Atkinson, A. J. Baker, K. A. Bennett, R. Carmona, K. E. Clark, N. A. Clark, C. Espoz, P. M. González, B. A. Harrington, D. E. Hernández, K. S. Kalasz, R. G. Lathrop, R. N. Matus, C. D. T. Minton, R. I. G. Morrison, M. K. Peck, W. Pitts, R. A. Robinson, and I. L. Serrano. 2008.

37. *Birds of the US–Mexico Borderland: Distribution, Ecology, and Conservation.* Ruth, J. M., T. Brush, and D. J. Krueper, editors. 2008.

38. *Greater Sage-Grouse: Ecology and Conservation of a Landscape Species and Its Habitats.* Knick, S. T., and J. W. Connelly, editors. 2011.

39. *Ecology, Conservation, and Management of Grouse.* Sandercock, B. K., K. Martin, and G. Segelbacher, editors. 2011.

40. *Population Demography of Northern Spotted Owls.* Forsman, E. D., et al. 2011.

41. *Boreal Birds of North America: A Hemispheric View of Their Conservation Links and Significance.* Wells, J. V., editor. 2011.

42. *Emerging Avian Disease.* Paul, E., editor. 2012.

43. *Video Surveillance of Nesting Birds.* Ribic, C. A., F. R. Thompson III, and P. J. Pietz, editors. 2012.

44. *Arctic Shorebirds in North America: A Decade of Monitoring.* Bart, J. R., and V. H. Johnston, editors. 2012.

45. *Urban Bird Ecology and Conservation.* Lepczyk, C. A., and P. S. Warren, editors. 2012.

46. *Ecology and Conservation of North American Sea Ducks.* Savard, J.-P. L., D. V. Derksen, D. Esler, and J. M. Eadie, editors. 2015.

47. *Phenological Synchrony and Bird Migration: Changing Climate and Seasonal Resources in North America.* Wood, E. M., and J. L. Kellermann, editors. 2015.

48. *Ecology and Conservation of Lesser Prairie-Chickens.* Haukos, D. A. and C. W. Boal, editors. 2016.

Milton Keynes UK
Ingram Content Group UK Ltd.
UKHW050454071024
449327UK00015B/377